THE COVID-19 DISRUPTION AND THE GLOBAL HEALTH CHALLENGE

THE COVID-19 DISRUPTION AND THE GLOBAL HEALTH CHALLENGE

VINCENZO ATELLA

PASQUALE LUCIO SCANDIZZO

ELSEVIER

ACADEMIC PRESS
An imprint of Elsevier

Academic Press is an imprint of Elsevier
125 London Wall, London EC2Y 5AS, United Kingdom
525 B Street, Suite 1650, San Diego, CA 92101, United States
50 Hampshire Street, 5th Floor, Cambridge, MA 02139, United States
The Boulevard, Langford Lane, Kidlington, Oxford OX5 1GB, United Kingdom

Notices

Knowledge and best practice in this field are constantly changing. As new research and experience broaden our understanding, changes in research methods, professional practices, or medical treatment may become necessary.

Practitioners and researchers must always rely on their own experience and knowledge in evaluating and using any information, methods, compounds, or experiments described herein. In using such information or methods they should be mindful of their own safety and the safety of others, including parties for whom they have a professional responsibility.

To the fullest extent of the law, neither the Publisher nor the authors, contributors, or editors, assume any liability for any injury and/or damage to persons or property as a matter of products liability, negligence or otherwise, or from any use or operation of any methods, products, instructions, or ideas contained in the material herein.

ISBN: 978-0-443-18576-2

For information on all Academic Press publications
visit our website at https://www.elsevier.com/books-and-journals

Publisher: Stacy Masucci
Acquisitions Editor: Elizabeth A. Brown
Editorial Project Manager: Himani Dwivedi
Production Project Manager: Swapna Srinivasan
Cover Designer: Miles Hitchen

Typeset by VTeX

Working together
to grow libraries in
developing countries

www.elsevier.com • www.bookaid.org

Contents

PART 1 How and why all this had an origin

PART 2 The COVID-19 crisis management: not an easy task!

PART 4 The policy analysis

List of figures

List of tables

List of boxes

Biography

Vincenzo Atella

Vincenzo Atella is a Professor of Economics at the University of Rome 'Tor Vergata', where he teaches Macroeconomics and Applied Health Economics at both graduate and post-graduate levels. In addition, he is an Adjunct Associate at Stanford University's Center for Health Policy and has served as a visiting professor on multiple occasions. He has been Scientific Director of the Farmafactoring Foundation and past President of the Italian Health Economic Association (AIES). Prof. Atella has coordinated several European projects and received funding from the European Science Foundation. His research primarily focuses on applied economics, particularly within the healthcare sector. The findings of his research have been published in numerous international peer-reviewed journals as well as books.

Pasquale Lucio Scandizzo

Pasquale Lucio Scandizzo, Ph.D. from the University of California, Berkeley, is Senior Fellow, member of the advisory board of Tor Vergata Foundation and President of Villa Mondragone Development Association at the University of Rome 'Tor Vergata', President Emeritus of the Italian Association of Development Economists, Scientific Director of OpenEconomics, a university spin-off focusing on cost-benefit analysis and policy impact evaluation. He has widely published on a variety of topics on economic development. Formerly a holder of several academic and government positions, he is presently advising the World Bank and other international institutions on the evaluation of policies for sustainable development.

Introduction

From earliest antiquity, humans have been accustomed to living with the problem of epidemics/pandemics, which have arrived in waves and often lasted for years. Epidemics frequently accompanied famines and wars, alternating with significant periods of cold. The most famous and deadly in Europe were plague, cholera, smallpox, and typhus. The black plague that ravaged the continent from 1347 to 1352 led to the disappearance of between 25% and 50% of the European population, bringing great changes in the economy, geopolitics, and religion.

The last of such deadly events occurred during the final year of World War I, when a virulent form of flu quickly spread across the planet, becoming one of the most lethal events in history. In only 18 months, the flu, known as 'the Spanish flu', infected at least a third of the world's population. Precise estimates of the death toll do not exist, and the existing ones vary enormously, from 20 to 50 or even 100 million victims. To better understand the magnitude of this death toll, consider that the combined total of victims from both World War I and World War II was less than 50 million. Moreover, the Spanish flu was the first known 'global pandemic', primarily driven by the movement of people, particularly troops on the continent during World War I, including American soldiers who came to fight in Europe.

Since then and for over a century, the western world appeared to have become immune to such disastrous pandemics, relegating them to history as unrepeatable phenomena. But appearance often deceives; a lesson that we have recently learned at great cost. As it is now known, at the end of 2019, an unseen enemy lurked in the shadows. Unnoticed and unexpected, the Coronavirus Disease 2019 (COVID-19) outbreak caught the world off guard. It rapidly spread and drastically altered our global society, community organization, daily life, and personal relationships, compelling us to reassess our lives and connections. However, though hidden (probably deep in a forest), the enemy was not entirely unexpected, given the clear messages sent by virologists and infectious disease specialists posited as a sort of Cassandra of the new millennium. If we limit our observation to the last 20 years, the SARS and MERS infections were clear signs that new and ominous pandemics were about to occur. Even HIV, a relatively new disease in human history and for which a vaccine has not yet been found, in just 35 years infected about 70 million people, killing half of them.

Indications of an impending viral threat were evident, yet adequate precautions were not implemented. Given this, it's imperative for us to safeguard humanity's future by meticulously reviewing our responses during the onset of the pandemic. We need to highlight errors and good practices to draw valuable lessons that can help everyone live in a world that perhaps will no longer be the same. Informed by experience, we should remember that unseen challenges can always lurk in the background. History has shown us this.

Today, we are better equipped to face this type of event than in the past. Until 1796, epidemics were considered events for which little could be done, as the lack of knowledge in immunology and microbiology prevented the use of adequate countermeasures. It was in 1796 when Edward Jenner's research on vaccination as a technique for preventing smallpox laid the groundwork for later studies on the nature of infectious diseases that led to the development of immunology in the nineteenth century. Jenner was the initiator of the vaccination campaigns against smallpox, a disease officially declared eradicated by the World Health Organization (WHO) in 1980, almost two centuries after the vaccination program had been formally launched. This result was achieved thanks to large-scale worldwide vaccination campaigns since 1958, with Somalia's last case diagnosed in 1977. Similarly, polio and diphtheria, which mainly affect children under the age of 5 years, are now under control in many parts of the world. Even measles, although highly contagious, no longer circulates in all regions where the vaccination rate is high enough.

Since the 1970s, according to the WHO (WHO, 2018a), successes in vaccination and antibiotics have led many experts to believe that infectious diseases were no longer a problem. In 1969, Jesse Steinfeld, then US Surgeon General, declared that it was 'time to close the book on the problem of infectious diseases', as if humanity had won the war against microorganisms. Ten years later, his successor, Julius B. Richmond, announced that infectious diseases could be regarded as the 'predecessors' of the degenerative diseases that supplanted and replaced them. We then need to wait until the mid-1980s when Everett Koop, the new Surgeon General, amid the HIV crisis, tried to awaken the consciences of health professionals by bringing attention to infectious diseases and sending to 107 million U.S. households a brochure to explain what HIV infection was and how to defend themselves.

What is astonishing about this behavior is the extent of the underestimate of the potential impact of infectious diseases by physicians, despite the fact that, since the 1970s, more than 1500 new pathogens had been discovered (of which 70% turned out to be of animal origin (WHO, 2018a)). A signal that despite advances in medicine, 'epidemics are a fact of life and the world remains vulnerable. We do not know where or when the next global pandemic will occur, but we know that it will take a terrible toll, both in terms of human lives and the global economy' (Ghebreyesus, 2018). Nature never withdrew, and although non-communicable diseases have garnered much attention over time, the battle against viruses and bacteria is far from over.

The first 20 years of the new millennium were the scene of at least two major pandemics (SARS and MERS). Today, humanity finds itself exposed to a new virus, formerly unknown to the scientific community, which in a few weeks was able to block the whole world, causing millions of cases and hundreds of thousands of victims. This virus has shown to be insidious not so much for its deadliness - history has witnessed more lethal epidemics - but in its method of transmission between individuals. Its frequent and unforeseeable changes sometimes result in undetected infections, characterized by a large proportion of symptom-free cases and clinical positives. An enemy in the shadows who can undermine trust between people. This last point is the prominent challenge people worldwide have faced in the last three years: the greater the level of contagion, the greater the distrust! This is something almost all living individuals on earth have never seen before: all other major epidemics have hit only a limited number of countries in the last few decades (SARS in 2009 was less dangerous than expected, while Zika, Ebola, and HIV were mostly restricted to specific groups of people), while the COVID-19 pandemic has directly affected everyone, changing the nature of everyday life.

For these reasons, COVID-19 is an important watershed for modern society, more than the attack on the twin towers in 2001. The unique characteristics of this event are its origin in the environment-animal-human interface and its rapid explosion following unprecedented levels of mobility and interconnection of global trade. The world will no longer be the same: almost all economic and social activities must be reviewed, rethought, and adapted to the new context. To this, we need to add another aspect of uniqueness: the immediate impact of the infection spread on all forms of social relationships and public policy assessment. If a government takes a direction in the fight against climate change, the effects will not be felt for years, and even then, analyzing them will be difficult. At that point, assessing the choices made and identifying those responsible will be challenging. In contrast, if a government announces that everyone can take a test at a designated time to detect an infection, and the very next day that test is unobtainable or out of reach, then determining both the oversight and the party at fault becomes much more straightforward. In a pandemic, the consequences of the failure of public health agencies, the loss of skills, or the malfunction of hospitals are no longer weighed solely on the vague indicators of public approval regarding the decisions made. Rather, they appear as tangible and critical symptoms and manifestations of the disease, the undeniable presence of which is objectively evident. Further, pandemics often exhibit a level of democracy, in the sense that even those who are usually protected from adverse events due to their privileges and power may find themselves, or their loved ones, susceptible to infection or, worse, losing their lives. At the same time, pandemics may dramatize income inequalities, creating disruptive social differences that 'decouple' the rich from the poor far more drastically than differences in consumption or income.

Furthermore, there is another reason why the COVID-19 pandemic deserves to be reported for its peculiarity as an enemy in the shadows: the moment it launched its attack, it found no organized defense. Since the outbreak began, effective epidemic management was lacking for a long time. This generated a widespread feeling of being in a situation with no clear ideas on how to approach the phenomenon, both in the medical-epidemiological and the economic-organizational-logistical domain. Nearly every country in the world was impacted. The global nature of the pandemic also highlighted a broader and more sensitive issue regarding the global governance of the crisis: the failure of the international community to recognize and anticipate the problem.

The past 15 years have witnessed several indicators of a potential pandemic, highlighting the need for enhanced global preparedness and response mechanisms. In November 2008, the National Intelligence Council (NIC), an affiliate of the US Central Intelligence Agency (CIA), released the latest in a series of futuristic publications to guide the then-inaugurating Obama administration. Looking at its analytical crystal ball, in a report entitled Global Trends 2025, the NIC report asserted: 'If a pandemic disease emerges, it probably will first occur in an area marked by high population density and close association between humans and animals, such as many areas of China and Southeast Asia, where human populations live in close proximity to livestock'. The only missing element in the CIA forecasts was a more specific topographic indication: a market in Wuhan. What U.S. intelligence analysts feared most was 'a novel, highly transmissible, and virulent human respiratory illness' (National Intelligence Council, 2018, p.75).

In a 2009 article in the Prehospital and Disaster Medicine journal, a group of researchers warned that 'it is a common belief that a flu pandemic is not only inevitable but imminent. It is further believed by some [...], that such a pandemic will herald an end to life as we know it. Are such claims hyperbolic, or is a pandemic the most significant threat to public health in the new millennium?' (Perrin et al., 2009).

Bill Gates' 2015 TED (Technology, Entertainment and Design) talk represents another striking example in this direction. Moving from his experience fighting the Ebola epidemic, the former CEO of Microsoft remarked that humanity had been fortunate in that instance. The epidemic had broken out in West Africa, a poorly connected region outside of large urban concentrations. Furthermore, the international organizations had been able to act quickly, and, above all, the disease was transmitted by contact with fluids and had a high mortality rate, which is why the sick had not had much time to infect the healthy. In his conclusions, Gates warned that an epidemic could cost millions of lives and have a substantial economic impact if not well prepared internationally. In particular, he added that the epidemic could be much more insidious than Ebola, as it could 'spread into the air, reaching people who would only feel the first symptoms late and would still be able to travel by train and plane'.

In his book, 'Global Catastrophes and Trends', Vaclav Smil wrote that after the 1958–59, 1968, and 2009 pandemics, 'The likelihood of another influenza pandemic during the next 50 years is virtually 100%, but quantifying probabilities of mild, moderate, or severe events remains largely a matter of speculation because we simply do not know how pathogenic a new virus will be and what age categories it will preferentially attack' (Smil et al., 2008).

Among multilateral institutions, the World Health Organization (WHO) has long warned about the risk of a global pandemic caused by a new virus. In a 2005 report, the WHO stated that 'the world is overdue for a major, global epidemic of infectious diseases' (WHO et al., 2005). The report noted that the risk of a pandemic was increasing due to factors such as global travel and trade, urbanization, and climate change, which can increase the probability of viruses spreading from animals to humans. Similarly, in a 2018 report, the World Bank emphasized the risk of a pandemic as one of the several global risks with major impacts on the world economy (Collins, 2018). The report noted that a pandemic could have significant economic impacts, including disruptions in global supply chains and reductions in international trade and tourism.

The latest example is the Annual Meeting for Disease Prioritization under the auspices of WHO's R&D Blueprint held in Geneva on 6 and 7 February 2018 to review the list of priority diseases for the WHO's R&D project. The project focused on emerging severe infectious diseases that can generate a public health crisis and for which there are no satisfactory or effective preventive solutions or treatments. At the 2015 meeting, an agreement was reached on the first list of diseases that meet these requirements and for which further investment in research and development was urgently needed. Subsequently, the list was updated in January 2017. In a February 2018 meeting, a third revision was undertaken, leading to an updated list of relevant illnesses. These conditions could lead to public health crises, and given the lack of effective treatments and/or vaccines, immediate research efforts were deemed essential. Among the various diseases indicated were SARS, MERS, Ebola, and 'Disease X'.

Disease X was envisioned as something that was predicted to occur in some parts of the world and whose characteristics were unknown. When describing the disease, the experts agreed that it would be caused by an animal virus and would emerge somewhere on the planet where economic development brings people closer to wildlife. In the first stage, disease X would probably be confused with other diseases and, therefore, spread rapidly and silently through travel and trade networks. It would spread as quickly as seasonal influenza but would have a higher mortality rate. It would shake financial markets even before it reached pandemic status: in a nutshell, a description of COVID-19.

Governments' reaction around the world was inadequate. Even after experiencing the first wave of the pandemic, the management of the second wave seemed to manifest an equal propensity to failure. This occurred despite the resurgence of the virus had been widely anticipated in the narrative before and after the first lockdown.

All of these events, combined with additional evidence, suggest that despite multiple alerts, adequate defense measures were scarcely developed and available when required. Why has COVID-19 seen such a lack of coordinated international response? It could be argued that the world lacks past experiences of pandemics compared to other events (mainly economic or humanitarian): therefore, with policy makers' attention more focused on other issues, health risks have never been seen as a trigger for significant international mobilization. Unfortunately, this lack of preparedness has generated grave and unprecedented problems and social costs, given that the pandemic has changed many of the rules of the game, opening new scenarios, threats, and opportunities never imagined before. This has been partly caused by the existence of a propitious terrain for spreading fears on our economy and civilization's future. Following this course, COVID-19 has changed our perception of the world and is perhaps changing the world itself in many forms, ranging from expectations and behavior patterns to a new emphasis on science and technology.

One of the reasons COVID-19 has caused sudden changes and, at the same time, a poor organized response, lies perhaps in its unusual economic impact. This impact resulted from a sequence of asymmetric supply and demand shocks, which were selectively determined, first through lockdowns and restrictive measures, and then through more profound and subtle changes in behavioral patterns and business models, a suspension, and in some cases, a transformation of ordinary activities. This unusual series of changes has affected urban regions in particular, with significant repercussions on city centers and core activities, including services of all kinds. Lockdowns and social isolation measures have stimulated forms of consumer substitution based on digitization, but have also been dramatically accelerating some processes of recession from primary social interaction, as well as, on the virtuous side, from the use and abuse of natural resources. These processes include, for example, home delivery logistics and smart working, dramatic reduction in short- and long-haul travel, and other phenomena that have released the pressure on the environment of ordinary human activities. These phenomena have sometimes reversed previous patterns of behavior, while in other cases, they have continued an existing trend, but with a leap of unpredictable proportions.

The trend towards a territorial poly-centrism, already in place before the pandemic, but strongly polarized between centers and suburbs, urban and rural areas, was reinforced by the lockdowns and the health emergency policies. However, its pattern appears to have been transformed by the unprecedented enhancement of residential housing, as places of family self-sufficiency that internalize communication and information services for work and leisure, rendering obsolete many current structures of work and consumption within and outside the city walls. As a result, housing in our old cities is hit by a wave of research, reconstruction, and multi-functional use. This is reminiscent of similar historical events, when the demand for self-sufficiency of residences in rural areas was stimulated by wars or other security emergencies. It is also suggestive of the ever-present tendency to transform public into private services in a society that aims at mass opulence, but at the same time is ready to retrench in response to environmental threats, when these become high and present dangers.

These and other effects appear to follow from the unfolding experience with COVID-19 in ways that are transformative, but still largely unpredictable, as the pandemic ghost is still raging in different forms, in its extension over space and time. At the same time, these effects do not address the changes necessary to reform our institutions and governance systems in ways that more systematically incorporate the lessons that can be derived from the pandemic. These lessons can only be drawn by squarely facing two core questions: *i)* why the pandemic has unfolded along so sudden and menacing lines and *ii)* why we have been so unprepared to handle it.

Concerning the first question, a widespread consensus has formed that the increasing epidemic and pandemic outbreak trends around the world are related to the environmental damage caused by the rising anthropogenic pressure on the bio-ecosystem. The synergistic balance of all existing animal species may have been broken, exceeding critical ecological thresholds. Human population levels have risen rapidly in a few years, occupying more and more spaces that were previously for wild animals, thus fragmenting ecosystems to create housing and productive settlements, intensively raising domestic animals, practicing intensive agriculture, and often sacrificing native biodiversity. Anthropogenic pressure on the environment is especially high and the future is 'crowded', with a population expected to reach 9 billion in 2050: people who will have to be properly fed and, therefore, will increasingly put pressure on the limited resources of the planet. These events are responsible for the higher risk of spillovers of infectious diseases from animals to humans (zoonoses), as evidenced by the increasing number of new outbreaks of infections that have occurred in the last 20 years. As such, climate change receives the greatest attention, but many other planetary transformations are evident, including interruptions in the nitrogen cycle, disruption of ecosystems, loss of biodiversity, and land over-utilization.

Why we have been so unprepared to manage this phenomenon mainly relates to procrastination, which is considered one of the most problematic and unresolved aspects of human behavior. There is a huge literature on this topic, with procrastination ranging from delaying preventive actions at the individual level (that is, refraining from smoking, improving one's diet, and avoiding all forms of health risky habits), to collective choices based on public policy interventions to reduce the exposure of individuals to potential risks (that is, earthquakes, floods, pandemics, etc.). Procrastination is often combined with an important cognitive bias: the tendency to underestimate the dangers of low probability and high impact events. An interesting discussion in this respect is provided by Perrow (2011) who cites the possibility, albeit remote, that planet Earth could be hit by a swarm of asteroids. For these events, even though people expect protection from their governments, policymakers tend to ignore the related risks until they are forced to react, even if the price of prevention is small. Perrow states that this attitude represents 'an abdication of responsibility and a betrayal of the future'.

The unprecedented nature of the COVID-19 pandemic may account for the government's insufficient attention to the issue, but it does not justify it. Protection is one of the main tasks for which the State exists, and governments must intervene to prevent and mitigate low probability-high-stakes dangers because these are a problem of lack of public good provision where markets typically fail to provide acceptable solutions. In the case of the current pandemic, very few countries were able to put together the winning combination of practical plans, the intervention kits required by those plans, and the bureaucratic capacity to implement them. The few countries that have been able to react in accordance with the plans prepared as, for example, South Korea, Taiwan, or New Zealand, have benefited considerably from this capacity, which has allowed them to proceed efficiently and with no improvisation. The simple lesson to be learned in this respect is that crisis management works if we are prepared and have time to think about, try, and eventually change the initial plans. South Korea's success can be attributed to the 'track-test-treat' system, which wasn't established overnight. Instead, it was developed over a span of at least 15 years, dating back to when the country first encountered SARS. Moreover, this system has been consistently refined and checked with meticulous care and commitment.

A major reason for the puzzling unpreparedness that included both developed and developing countries was also the widespread decline in public investment, which collapsed after the Great Recession (2008), and caused a dramatic increase in global uncertainty and in systemic risks. On the other hand, the pandemic also suggests that the organization of production and investment has gradually become unsustainable for two main reasons. First, the search for short-term allocative efficiency has dramatically lengthened the production chains, resulting in an agile and streamlined linear structure, at minimum unit costs, but at the same time highly vulnerable to unforeseen events and especially extreme ones. The so-called 'black swans', although individually extremely unlikely, seem to have become more and more frequent as a whole, so we are continually surprised by the appearance of new catastrophic threats. As in the case of the Great Recession, COVID-19 is thus an example of the growing systemic risk in a globalized and hyper-connected society, where even minor disturbances tend to propagate with unpredictable speed and consequences.

Second, the pandemic has revealed a dramatic lack of social infrastructures, for example, but not only, in the health sector, civil protection, assistance to the poor and the elderly, food security, and, in general, the supply of essential goods and strategic materials. There are many reasons for these shortcomings, but the main one appears to be the ignored interdependence between private goods and public goods, the cause of a simultaneous failure of the state and the market. Despite the dangers presented by extensive state involvement, it has been evident during the various phases of the pandemic how the response to the emergency depended on the ability of public administrations to offer not only rules and regulations, but also and above all strategic direction and coordination activities.

More generally, the pandemic has revealed an unknown aspect of the relationship between public and private goods. In fact, over the years, the role of private entrepreneurship has progressively extended to the provision of public goods in appropriate circumstances due to the presence of private characteristics (rivalry and excludability) in the so-called impure public goods. But the libertarian-inspired search for a minimalist condition of the State has meant that the symmetrical aspect, that is, the presence of public characteristics in private goods, has been progressively neglected. The pandemic has made it clear that an efficient State and an adequate transnational organization are crucial components of the ability to respond to the explosion of systemic risks in the globalized society. Health security, in particular, has emerged as the most critically endangered global public good in a hyper-connected world, where health facilities are unequally distributed in endowment and efficiency, and there is no coordination mechanism based on effective international governance.

Against this disheartening scenario, vaccines have represented a great achievement. The race for a vaccine to protect against the SARS-CoV-2 virus started in January 2020, after the virus genetic code was cracked in China and made available to the scientific community. In just 9 months, researchers were able to develop a vaccine that looked very effective, an unprecedented feat, given that in the past the average time for vaccine development had been about 10 years. All this was made possible both by recent technological advances and by a massive financial commitment. The role of technology was crucial, marked by significant progress in bio-technological methods. In addition to deciphering the virus genome in record time, several vaccines were soon developed, with diverse manufacturing techniques, using different technology platforms that involve the insertion of genetic material from the virus into a proven delivery vector. Using these technologies not only results in faster development but also ensures that more is known about its vaccine safety profile from the start.

The development of vaccines in a record time was also achieved with a great commitment at the logistical level. In fact, while advances in technology have simplified product development and data recording, the massive use of social networks has made it easier to recruit trial participants. Thanks to public financial aid, the production of vaccines was also carried out in parallel with clinical trials in the hope that the trials would be successful.

On the other hand, the distribution of vaccines has also exposed many failures and limitations on the State and the market, both at country and international level. The uneven distribution of vaccines during critical phases of the pandemic, with some countries having much greater access than others, has exposed the inequities deriving from the present fragmented bio-infrastructure that follows the national lines of the richest countries. It has highlighted the absence a worldwide network of R&D, production, transportation and storage systems that can ensure that vaccines are timely developed, produced and reach those who need them.

* * * * * * * *

To find a proper answer to the momentous questions raised by the pandemic, we need to change the way we approach and handle health problems. This is the main message of this book. We need a new paradigm that embodies the concept that in a globalized world, phenomena like health, economics, finance, climate change, demographic pressure, and pollution, among others, cannot be considered independent factors and are addressed jointly. In particular, the causes of the COVID-19 pandemic, and their possible solutions, should be addressed within a global approach, which finds its theoretical foundations in bio-economics and in the new notion of 'One Health - One World'. According to Manour Mohammadian, 'bio-economics is the science determining the socioeconomic activity threshold for which a biological system can be utilized effectively and efficiently without destroying the conditions for its regeneration and therefore its sustainability. In other words, bio-economics in addition to being the science of supply, demand, and prices, is also the science of accounting for the biological, economic, social, environmental, and ethical realities of resource depletion, wealth inequality, social inequity, environmental contamination, and ethical misconduct'.[1] Bio-economics approaches the investigation of environmental, socioeconomic and biological issues in an integrative and holistic manner and leads to the concept of 'One Health' as a paradigm of harmonious well being for all living things.

In this book, we argue on the need to invest and develop the bio-economy as the basis of a new global and integrated health system, pulling together conservation, resource economics, and health (as originated through both preventive and curative medicine). In this perspective, the experience of the COVID-19 pandemic offers an important opportunity, since it is a somewhat spectacular event that accompanies profound environmental changes. These are associated with environmental degradation and climate change, as well as with the exponentially growing anthropogenic pressure on the world bio-ecosystem. As it was repeatedly noted in several conferences and policy statements around the world, while the bio-economy has been the oldest major economic activity in which men have been engaged, need and opportunity are transforming it into the newest one. At the frontier of the bio-economy, and both spurred and constrained by the search for a new sustainable relationship with nature, biotechnology holds a high promise of transformative progress. The developments in diagnostics, vaccines, and cures during COVID-19 have indeed shown remarkable success, but most of all they have demonstrated the potential of the capabilities and commitment of the world scientific community to face a global challenge with unity of purposes and resolve.

In this book, we undertake the challenging task of finding a middle ground between simplistic explanations and specialized knowledge, aiming to provide a general audience with an economic perspective on the pandemic and its aftermath. With the aim of reorganizing the reflection around the pandemic, the text seeks to shed light in the shadows of a dense, crowded, often badly framed debate in which we feel that a new

---

[1] More information can be found at the following web address: https://www.scienceofbioeconomics.com/bioeconomics.

awareness needs to be brought about. Complexity is part of nature and the world and cannot be trivialized. Images, examples, and anecdotes are available to return articulated concepts, but these must be fully disclosed. This is the only way to show awareness and bring home lessons for tomorrow. Otherwise, the debate, instead of spreading knowledge, will be limited to the crowding of confused narratives and suggestions, as it has appeared to be during most of the time of the pandemic's development.

Unlike previous epidemic experiences, COVID-19 came at a time when the ability to spread and be heard by large audiences was much wider than in the past. Since the first outbreaks in Wuhan, there was no media coverage of the outbreak, until a point was reached just a few weeks after the WHO declaration of the 'pandemic', when communication was so intense and widespread that it began to be referred to as '*infodemics*' or a form of information pathology. 'Infodemics' represents the overwhelming flood of information, which can often include inaccuracies and can lead to confusion among the public (Kluge, 2020).

The disorderly spread of information was only partly due to excessive media activity. Perhaps surprisingly, it was also the effect of unprecedented scholarly attention. According to Google Scholar, in March and April 2020 alone, the scientific community produced more than 36,000 papers on the subject. At the end of May 2020, just three months after the outbreak of the pandemic, the COVID-19 Scientific Gateway, which automatically collects all COVID-19 and Sars-CoV-2 research products, reported more than 39,000 scientific publications, more than 3000 databases accessible to all for research, 240 software and more than 3200 other research products. An unprecedented effort. In addition, there were thousands of contributions from communicators and experts' opinions. At the end of November 2020, the number of scientific publications had exceeded 150,000, more than 9000 databases, more than 600 software programs, and more than 6000 other research products were available.

Within a flurry of unprecedented attention of the scientific community to a single health emergency, the book seeks to provide a perspective that encompasses two interconnected worlds: life sciences and economics. In the opinion of the authors, a comprehensive understanding of a phenomenon such as a pandemic can only be developed by exploring these two worlds, their interconnections, and the rules that govern them. The global perspective was profoundly altered by the COVID-19 pandemic, as it touched the lives of over 3 billion people worldwide. The pandemic also highlighted the interconnectedness and vulnerability of the global production system, as shortages of both low-tech items such as masks and high-tech equipment such as ventilators were evident. This realization showed that while global supply chains can be efficient in normal circumstances, they can also be a liability in times of crisis. This was the first time that the global nature and the hidden fragilities of production were brought to the forefront in ways similar to the global financial system in its recurring crises.

The COVID-19 experience underscored the global nature of production systems in several ways. The most significant was the impact of production and supply chain

disruptions on the availability of personal protective equipment (PPE) and other medical supplies. The demand for masks, gloves and gowns surged as the pandemic spread,
leading to shortages in many parts of the world. This was partly because much of the
production of these items was concentrated in a few countries, such as China. Disruptions in these production centers caused ripple effects on the global supply chain.
Additionally, the pandemic exposed the dependence of many countries on a global network of production and trade, as national lockdowns and travel restrictions disrupted
the flow of goods and materials. This led to shortages of essential items in some areas
and excesses in others, again highlighting the interconnectedness of global production
systems. Finally, the pandemic brought attention to the vulnerabilities of global supply
chains, as they were disrupted by both the pandemic and the measures taken to mitigate
its spread. This prompted demands for more robust global distribution networks and an
increased emphasis on local manufacturing of crucial goods in numerous nations.

Drawing from a range of topics from biology to economics and informed by the
authors' firsthand experience as economists during the pandemic, the book aims to
offer a comprehensive view of COVID-19 and its deep societal impacts, addressing
some pivotal questions:

- Why the epidemic originated: including the factors that contributed to the emergence and spread of the virus, such as animal reservoirs, human behavior, and global
interconnectedness.
- How the virus proliferated, by examining the transmission dynamics of the virus,
including how it spreads from person to person and the role of various factors such
as travel and social gatherings.
- How the pandemic was managed, by looking at the response of governments,
healthcare systems, and other organizations to the outbreak, including implementing measures such as lockdowns, testing, and contact tracing.
- What effects the pandemic has had on the economy, analyzing the economic impacts of the pandemic, including job losses, business closures, and declines in GDP.
- Who has been the most affected by the pandemic, examining the disproportionate
impacts of the pandemic on certain groups, such as older adults, essential workers,
and marginalized communities?
- How the new reality emerging from the pandemic will change people and transform
their way of seeing the world, by exploring the long-term societal and cultural
impacts of the pandemic, such as changes in work, education, and social interactions.

In addressing these questions, we seek to present content objectively and impartially,
with a focus on the facts and evidence. This may be an over-ambitious goal, which is
difficult both to pursue and to achieve because of the nature of much of the information available and our own prejudices. However, throughout the writing of the book,
we have remained committed to providing a balanced account of events, presenting different viewpoints where possible, and allowing the readers to form their own informed
opinion on what happened.

In addition to presenting an objective account of events, we also seek to explore possible solutions to the crisis being discussed to find ways to resolve the many controversial issues raised by the pandemic. In this task, our aim is to provide a comprehensive approach to the topic, covering both the chronicle and the commentary on the events that have occurred and the potential ways to move forward.

Before diving into the book, it's pertinent to underscore two key considerations. Firstly, discussing the general unpreparedness of governments and organizations in the face of the pandemic is not aimed at leveling blame or criticism against any party. In contrast, after having observed, read and analyzed many contributions, interviews, and interventions, we find it valuable to clearly and systematically outline the issues encountered and the approaches taken to address and navigate them. The book is based on evidence that should hopefully answer many of the various doubts and questions that have emerged in the last months on such a complex issue. The second point is that although scientific knowledge about COVID-19 is growing, it remains critically incomplete. Fundamental issues such as the long-term implications of the virus, the strength and duration of post-infection antibodies, and the precise mechanisms of transmission of infection were still unknown at the time this book was completed. As a result, many of the issues discussed here following the public debate are still open, as are the measures that should be put in place. More importantly, social science research on human behavior during a pandemic is in its infancy. It is usually a difficult task to draw causal inferences from social sciences and even more so to do so in the current context. Based on a small set of studies, some of which provide only correlations, it is almost impossible.

Finally, at the time of publication of this book, it seems that the SARS-CoV-2 infection is here to stay long, even though it may perhaps gradually evolve into an endemic form of influenza. Inevitably, as the pandemic continues, more problems will emerge in the coming months, and many dilemmas will be raised that are not covered in this book, or that may render obsolete some of the evidence reported here. This is quite normal, given the attempt to describe a still-evolving phenomenon. However, far from giving definitive answers to evolving problems, relying on currently available scientific evidence, it is our hope that the book will help readers navigate the complex and still often confusing world of the COVID-19 pandemic. This entails offering background and rationale for the events and their causes, along with scrutinizing the measures executed (or overlooked) in reaction to the pandemic and the motivations behind such decisions. Consequently, the book aspires to furnish a thorough yet nuanced grasp of the pandemic, rather than assert final conclusions.

The portrayal of the pandemic as a hidden enemy that caught humanity off guard complicates efforts to present a neutral, analytical narrative. This metaphor demands a dramatic representation of the pandemic's dynamics, consequences, and the responsibility and lessons it entails. Simultaneously, it suggests that our behavior may be the true adversary rather than a specific disease or environmental disturbance. The pandemic is not a battle between humans and microbes; it serves as a potent reminder that humanity's future hinges on finding a sustainable balance between nature and nurture.

* * * * * * * * *

The writing of this book would not have been possible without the help of many people who, at different times and different stages of writing, have contributed directly and indirectly to help us continue to complete the project. A first thank you goes to a large group of general practitioners belonging to the Italian Society of General Practice (SIMG). During the first weeks of the pandemic, they allowed us to participate in interesting webinars, during which we learned much useful information about the nature of the virus, how it is transmitted, and the main issues that have been faced among the patients. Furthermore, we gained some understanding of the drama that community and hospital workforce were experiencing day after day. One of the most critical aspects of this participation was the realization that managing a pandemic of this type is a highly complex phenomenon whose analysis cannot be fully carried out within the boundaries of any single discipline, but requires a comprehensive multi- and trans-disciplinary approach. This challenge has been one of the main reasons behind our decision to write a book that delves into the complexity of pandemic management, while also remaining accessible to a general audience.

During the book's writing, we were helped by various colleagues who, at multiple times, read preliminary parts, sometimes incomplete, and provided valid indications on how to improve the contents and exposition of the text. Therefore, in strict alphabetical order, we would like to thank Carlo Caltagirone, Stefano De Nicolai, Giuseppe Novelli, Federico Perali, Beniamino Quintieri, and Giovanni Tria. Special thanks go to Joanna Kopinska and Andrea Piano Mortari, who shared with us existing material, with some chapters based on previous joint work with them. We have also benefited from the technical capacity of Alessandro Rosi, who had the patience to generate and edit several versions of the graphics presented in the book. Of course, we are solely responsible for any possible errors, inaccuracies, or other problems that may occur in the current version of the text.

Last but not least, we thank our families who, at such a particular moment, had the patience to take care and support us. Despite the challenges and uncertainties, with the assistance and support of our loved ones, we have been able to find the inner peace and motivation required to successfully finish a demanding task that has taken up a significant amount of our time.

VA and PLS
Rome
September 2023

PART 1

How and why all this had an origin

CHAPTER 1

The origins of infections
Some basic concepts

Contents

The etymology of the word infection can be traced to the Latin *infectus*, from the verb *inficere*, which literally means 'put in', but was mostly used to indicate an action that would stain or corrupt the recipient. Although the concept of some corrupting agent was always present in the attempt to explain the emergence of diseases, the history of medicine is characterized by the persistent belief that the root of the pathologies was rooted in the organism of the affected person rather than in the injection of an outside agent.

The modern concept of infection is related to the emergence of germ theory and especially to the work of Louis Pasteur and Robert Koch in the middle of the 19th century. Pasteur proved that spoiled food was caused by contamination from microscopic organisms, which also caused infectious diseases, and not from spontaneous generation. Koch famously identified the microorganisms that cause anthrax, tuberculosis, and cholera.

Although the notion of infection is very general, it is often used as a metaphor with many ambiguities and multiple meanings. However, three critical components may be selected as identifiers. First, infection generally refers to parasitic organisms, not necessarily microscopic, which may affect human and animal health. Second, infection is also characterized by a greater or a smaller degree of transmissibility, including migration of parasites between individuals and species. Third, infection is associated with the outbreak of diseases. The combination of parasitic and pathogenic characteristics excludes from infection the many symbiotic associations of organisms and, generally, the contribution of biodiversity to the ecological equilibria. It also characterizes the relationship between the infecting and the infected organisms as a predator-prey relationship, which is highly entangled across different species, with predators and prey often swapping positions as a consequence of complex evolutionary processes at different scales of infection and transmission.

Infection occurs when microorganisms, such as bacteria, viruses, fungi, or protozoa, enter the body and multiply. The type of microorganism and how it enters the body can determine the type of infection that develops. The infection is called exogenous if the microorganism comes from an external source. The infection is called endogenous if the microorganism is already present in the body. Exogenous infections can enter the body through various pathways, including the skin, respiratory, and digestive systems. An infectious disease is the manifestation of an infection in the body and the severity of the disease can vary depending on the type and virulence of the microorganism and the health of the individual.

The life of micro-organisms within the human body depends on several factors; not always the effects of their presence are harmful to man. Many micro-organisms are beneficial to human health and compose the so-called human microbiota, which consists of the 10 to 100 trillion symbiotic microbial cells harbored by each person, primarily bacteria in the gut (Ursell et al., 2012). Some microorganisms can also survive without causing damage in a so-called latency state (sub-clinical infection) or, after an incubation period in which they settle and multiply, they can cause an infection that develops rapidly (acute) or, if the organism infected does not completely heal, that lasts in time (chronic).

In the course of evolution, a number of barriers and mechanisms have developed to defend the human organism from the attack of harmful micro-organisms. Barriers are mainly physical in nature and are represented by the skin and mucous membranes, which, together with sebum and mucus, carry out important antimicrobial activities. If these barriers are overcome, a number of chemical and biological mechanisms activate internal defenses. For example, temperature rises (fever), which is useful in relation to viruses; phagocytosis (the biological process by which certain types of cells can destroy external material) increases the production of phagocytes that engulf infectious agents and neutralize them. An especially effective but potentially damaging mechanism is inflammation, through which the body produces cytokines, proteins that trigger the immune response to act against microorganisms locally or generally.

All these functions are part of the so-called innate immunity, which is implemented regardless of the nature of the infectious agent. If the micro-organism survives this general reaction, a second-level response follows, specific immunity, based on antibody production. This second type of response, which unfortunately takes time to develop, plays an important role in the case of infections by microorganisms that successfully overcome the first barrier of general immunity. As we shall see, it corresponds to the concept of immunization after infection (including, for example, via vaccine).

1.1. Bacteria, fungi, protozoa, and viruses: what are they?

Before we describe what a virus is, it is useful to underline the fact that there is a radical difference between a virus and other microorganisms along multiple dimensions.

Bacteria are single-celled microorganisms that have a cell wall and contain all the necessary components for survival and reproduction. They can be found almost everywhere on Earth and play an important role in various ecosystems. Most bacteria are harmless and can even be beneficial to human health, as they help to maintain a healthy balance in the body. However, a small percentage of bacteria can cause diseases such as pneumonia, meningitis, and food poisoning. Antibiotics are used to treat bacterial infections, but are not effective against viruses. Some bacterial infections can be prevented through vaccination. The human body contains a diverse community of bacteria, including the gut microbiome, which is made up of tens of trillions of bacteria, and the skin microbiome, which contains trillions of bacteria.

The microbiome is the collective name of the microorganisms that live in a particular environment, such as the human body. These microorganisms include bacteria, viruses, fungi, and protozoa. The human microbiome refers to the microorganisms that live in or on the human body, including those found in the skin, mouth, respiratory system, gastrointestinal tract, and genitourinary system. The microbiome plays a crucial role in maintaining the health and function of the body by performing various functions, such as helping digestion, helping protect against infections, and influencing the immune system. An imbalance in the microbiome, known as dysbiosis, can contribute to the development of diseases. Scientists are still learning about the complexity and diversity of the microbiome and its role in human health.

Some chronic diseases, such as cancer and heart disease, have sometimes been associated with poor oral health, often due to an imbalance of bacteria within our mouth (Coll et al., 2020). Depending on the particular tropism of the bacteria, all organs of the human body (individually or jointly) may be affected by infections, including the musculo-skeletal, the urinary tract and the peripheral and central nervous system.[1] Since the discovery of penicillin in 1928 by Alexander Fleming, the main way to treat bacterial infections is by administering antibiotics, although this practice, which has been extended to domestic animals, creates a serious danger of developing resistant strains of bacteria. Vaccination, a technique originally discovered by Pasteur, has been developed for several serious bacterial infections, such as diphtheria, meningococcal disease, pertussis, and tetanus, and is the most effective way to prevent infection.

Fungi are eukaryotic organisms (from Greek *eu* or 'good' and *karyon* for 'kernell'), i.e., having a well-differentiated nucleus containing most of the cell's DNA, unicellular, or multicellular, 20 to 50 times the size of bacterial cells. They come with a rigid chitin wall. According to the morphology of the fungal body cell (called the thallus), they are distinguished in *a)* filamentous fungi or molds, multicellular; *b)* yeasts, unicellular; *c)* dimorphic fungi, which can acquire the appearance of mold or yeast based on environmental characteristics. Fungi can cause diseases in humans known as mycoses.

---

[1] A tropism (from the Greek *tropos* or 'turning') is a term used to describe the particular orientation of certain microorganisms that are primarily located in certain cell types. For example, intestinal tropism, or enterotropism, is the phenomenon by which certain bacteria (choleric vibrio, typhus bacillus, etc.) exhibit elective attraction to cells of the intestinal walls.

Superficial mycoses are infections that affect the skin and mucous membranes and are usually small in size and not life-threatening. Deep mycoses, on the other hand, can affect various organs and are more likely to occur in individuals with compromised immune systems. These infections can be serious and potentially fatal.

Protozoa are single-cell microorganisms that have a eukaryotic cell structure that is more complex than that of a bacterial cell and is more similar to that of an animal cell. Many species have their own motility for the presence of eyelashes or flagella (Flagellates), or for the presence of amoeboid movements (Rizopods); Other species are immobile (Sporozoi). Protozoi are scavengers of soil and water (they need decaying organic matter to survive), but they can also be commensal with animals and plants. Many protozoa are harmless to humans, but there are many others that are capable of causing serious infectious diseases.

Finally, we have viruses. Unlike bacteria, viruses are not considered 'completely alive' because they require a host cell to survive, produce energy, and reproduce. They consist of genetic material, such as DNA or RNA (but not both), surrounded by a protein shell called capsid. Sometimes, this shell is coated with an envelope of fat molecules and proteins from which glycoprotein protuberances, called peplomers, can develop and whose shape could be triangular, pointed, or fungus-like. These protrusions bind only to certain receptors in a host cell and determine the type of host (host cell) that a virus will infect and how contagious that virus will be. To see them, especially powerful microscopes are needed because they are 10 to 100 times smaller than the smallest bacteria.

Viruses are the ultimate parasites, since they rely on a host cell to reproduce and survive, although they can remain latent with the potential to be revived on inanimate surfaces for long periods. Parasites serve this role because they use their genetic material to 'divert' the ribosomes[2] in the host cell to produce proteins (virals) for their own survival, not those that can be useful for the host cell. In addition, they also exploit other components within the host cell, such as ATP (adenosine triphosphate) for energy and amino acids and fats to create new capsids and assemble new viruses. Vaccination is the main way to prevent viral infections; however, a growing number of antiviral pharmacological remedies, not active against bacteria, have been developed to treat several viral infections, such as hepatitis C or HIV.

1.2. Crowding, hygiene, and climate change: the origins of infections

The microorganisms described above are responsible for all infections known today, whose presence persists and increases despite extraordinary advances in biomedical

---

[2] A ribosome is a small organelle that is found in all cells. It is responsible for translating mRNA into proteins.

knowledge. What explains why infections originate and spread are mainly factors having to do with socio-economic conditions and public hygiene rather than with medical and clinical aspects. Phenomena such as high population density, massive urbanization, poor hygiene, climate change, and the spread of anthropophilic vectors (feeding on human blood, such as the anopheles mosquito) through increased travel and trade flows create more and more favorable conditions to bring guests and pathogens into contact.

Although a large proportion of viruses from well-identified mammals are known to be responsible for infections, there are a large number of evolving viruses waiting to infect and adapt. Unfortunately, little is known about these viruses and the very nature of the phenomena often prevents control and prevention. Furthermore, specific geographical regions or interactions between humans, livestock, wildlife, and the environment should be targets for intensive surveillance. In this way, more in-depth research will improve the understanding and ability to predict new pandemics and allow control measures to be designed in advance.

One of the key aspects in the transmission of infections is the relationship between animals and humans that underlies what is technically termed 'zoonosis'. The term zoonosis refers to 'any infectious disease that can be transmitted from animals to humans, or vice versa, either directly, through contact with skin, hair, eggs, blood, or secretions, or indirectly, through other vector organisms or by ingestion of infected food'. The term zoonosis is all-encompassing, but there are others, sometimes used as synonyms, with a more explanatory meaning: 'anthropo-zoonosis' means diseases transmitted to humans by other vertebrates, while 'zoo-anthroponosis' describes diseases transmitted from humans to other vertebrates. Finally, 'amphixenosis' refers to diseases that are present in humans and in other vertebrates and mutually transmitted. Zoonoses therefore give a dangerous character to many activities requiring contact with animals.

In a famous work that appeared in Nature, Jones et al. (2008) claimed that more than 60% of infectious diseases identified since 1940 have zoonotic (animal-borne) origin: two thirds of these are derived from wild animals. More recently, WHO (2018a) estimated that about three-quarters of infectious diseases are the result of zoonotic spillovers. Another, more recent article also in Nature (Carlsson-Szlezak et al., 2020a) documents the increasing likelihood of encountering previously separated species and the consequent risk of cross-special viral transmission determined by climate change.

Of the known viruses that infect humans, about 80% occur naturally in non-human 'basins', mostly in mammals and poultry and, to a lesser extent, in wild animals and arthropods. Animal infectious agents are estimated to make up approximately 60% of known human pathogens and up to 75% of 'emerging' human pathogens. Unfortunately, there is limited knowledge of animal origin (zoonoses) and of the diversity of these viruses in their known ecological niches. Data on some domestic mammals hosting dozens of viral species are limited, and there is insufficient knowledge of wild animals that are estimated to house thousands of viral species. Examples include emerging hu-

---

Box 1: The five stages leading to the development of endemic human diseases

Below is the list of the five stages of the transformation of an animal pathogen into a specialized pathogen in humans. This process is not self-sustaining: at each stage many microbes remain blocked and many of the major pathogens do not reach the last stage.[a]

- **Phase 1**. A microbe present in animals, not detected in humans under natural conditions (thus not voluntarily transferred with blood transfusions, organ transplants or hypodermic needles). Examples: most malarial plasmids are blood parasites that tend to be specific to a host species or a closely related group of host species.
- **Phase 2**. A pathogen of animals that, under natural conditions, was transmitted from animals to humans (primary infection) but was not transmitted among humans (secondary infection). Examples: Anthrax and Tularemia bacilli and Nipah, Rabies, and West Nile viruses.
- **Phase 3**. Animal pathogens that can give rise to only a few cycles of secondary intermission between humans, so that occasional human outbreaks triggered by a primary infection are rapidly extinguished. Examples: Ebola, Marburg, and Monkey pox viruses.
- **Phase 4**. A disease that exists in animals and has a natural (sylvatic) cycle of infection in humans by primary transmission from the animal host, but also undergoes long secondary transmission sequences between humans without the involvement of animal hosts. Examples: yellow fever, dengue fever, influenza A, cholera.
- **Phase 5**. An exclusive human pathogen. Examples: agents causing malaria falciparum, measles, mumps, rubella, smallpox, and syphilis. In principle, these pathogens could be confined to humans in two ways: *i)* an ancestral pathogen already present in the common ancestor of chimpanzees, in which case humans could have co-assumed, when the chimpanzee and human lineages diverged about five million years ago; *ii)* a pathogen could have colonized humans more recently and become a specialized human pathogen.

[a] Adapted from Wolfe et al. (2007).

---

man viruses such as new flu strains, human CoV(h), Hendra virus, Nipah virus, and many others, all of which have been linked to human–animal contacts.[3]

1.2.1 Zoonoses or spillovers: when and how they occur

In the early weeks of the COVID-19 epidemic, the concept of spillovers was widely mentioned in public debates, although it was often misused. For this reason, it is worth clarifying some points: *i)* How do viruses and bacteria spill over? *ii)* To what extent can these events be considered random? *iii)* Do humans have any responsibility in this regard?

Grossi et al. (2020) have clearly explained many of the relevant aspects of spillover. First of all, spillovers have always occurred and will continue to happen as long as there is life on Earth. What needs to be understood is that in the vast majority of cases spillovers remain confined mainly to animal species living in well-diversified ecological

---

[3] For more information on these topics, the interested reader can consult the following bibliographical references: Parvez and Parveen (2017) and Wolfe et al. (2007).

contexts and, in so doing, do not create problems for either humans or domestic animals. Among the most endangered species are certainly 'wild birds that represent the principal reservoir (i.e., the species where the pathogen normally lives and multiplies) of all type A influenzaviruses, as almost all known subtypes (144 combinations of 16 types of hemagglutinins (HA) and 9 types of neuraminidase (NA)) are found in birds, but many subtypes have become endemic in humans, dogs, horses, poultry, and pigs. The evolutionary significance and zoonotic risk of the new subtypes H17N10 and H18N11 discovered in bats remains to be studied. Rodents are the main reservoir of emerging viruses such as hantavirus and arenavirus, which have been responsible for numerous spillover events over the past 50 years due to bites or direct and indirect contact with rodent excreta' (Grossi et al., 2020).

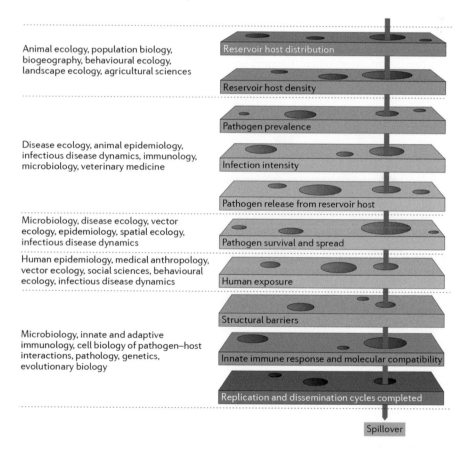

Figure 1.1 Species jumping barriers. *Source*: adapted from Plowright et al. (2017).

These aspects can be better understood by looking at Figs. 1.1 and 1.2. In Fig. 1.1, the so-called 'barriers' to spillover-response relations are shown. The column to the left lists the factors that affect spillovers and are studied by researchers in several disciplines. The column to the right shows how a pathogen has to overcome a series of barriers in order to transmit from one species to another. If one of these barriers is impenetrable, spillover cannot occur. For spillovers to occur, the spaces (represented by holes) in all barriers must be aligned within a narrow window in space and time (indicated by the blue arrow).

Fig. 1.2 shows the so-called spillover bottlenecks. Several barriers allow or limit the flow of pathogens from one species to another. For example, in the case of rabies virus, the species jump may be facilitated by the frequency of the relationship between the animal and the human, but it is limited by the low spread of the pathogen in the animal. The width of the gaps in the barriers should represent the ease with which a pathogen can flow through the barriers and will vary depending on the context. However, Fig. 1.2 is only illustrative. Indeed, there is still insufficient evidence to fully understand the role of these bottlenecks as one or two barriers to each particular system. The question marks in the figure show where barriers are particularly poorly understood and show gaps in knowledge of some globally concerned pathogens (e.g., lack of information on disease dynamics in hosts of Ebola reservoirs).

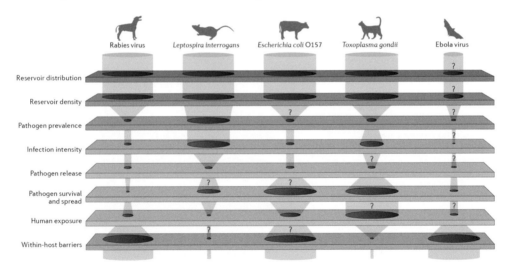

Figure 1.2 Bottlenecks at Species Jumps. *Source*: adapted from Plowright et al. (2017).

In general, spillover transmission is promoted by subsequent processes that allow an animal pathogen to establish infection in humans. The probability of zoonotic spillovers is determined by interactions between different factors, including disease dynamics in the reservoir host, exposure to the pathogen, and some human–related aspects that affect an individual's susceptibility to infection. These factors can be divided into three functional steps that describe the main transmission routes:

- In the first step of the infection process, the pathogenic pressure is determined. This refers to the amount of pathogen (a disease-causing agent such as a virus or bacteria) present in the human host (the person or animal that is being infected) at a particular time and location. This pathogenic pressure is influenced by several factors, including the distribution of the reservoir host (the host in which the pathogen lives and reproduces), the prevalence of the pathogen (the percentage of the population that is infected with the pathogen) and the release of the pathogen from the reservoir host. Pathogenic pressure is important because it determines the likelihood of an infection occurring. The higher the pathogenic pressure, the more likely it is that the infection will start. Once infection has started, the course of the infection will depend on the survival, development, and spread of pathogens outside of the reservoir host. This can include things such as the ability of the pathogen to survive in the new host, the rate at which it reproduces and the ways in which it is transmitted to other hosts.

- In the second step, the exposure to the pathogen is determined. This refers to the likelihood that a person or animal will come into contact with the pathogen, as well as the way in which this contact occurs. Factors that can influence exposure to a pathogen include human behavior (such as whether a person practices good hygiene or follows recommendations to avoid exposure) and the presence of vectors (organisms that can transmit the pathogen from one host to another). The probability of being exposed to a pathogen depends on the level of pathogenic pressure (as discussed in the previous step), as well as the likelihood of coming into contact with the pathogen through various activities. The mode of exposure refers to the way in which the pathogen is transmitted, such as through contact with bodily fluids, inhalation of respiratory droplets, or ingestion of contaminated food or water. The dose of exposure refers to the amount of pathogen encountered during the exposure event. A higher dose of a pathogen can increase the probability of infection.

- In the third step, the genetic, physiological, and immunological characteristics of the human host (the person or animal that is being infected) play a role in determining the likelihood and severity of the infection. These attributes can include things like the host's genetic makeup, the overall functioning of their body and its various systems, and the strength of their immune system. The dose and route of exposure are also important factors in this step. The dose refers to the amount of pathogen encountered during the exposure event, as previously discussed. The route of exposure refers to the way in which the pathogen enters the host's body, such as through the respiratory system, the digestive system, or through a cut in the skin. The combination of host characteristics and dose and route of exposure influences the probability and severity of infection.

For example, a host with a strong immune system may be less likely to become infected after exposure to a pathogen or may experience a milder case of the disease if she becomes infected. On the other hand, a host with a weaker immune system or certain genetic predispositions may be more susceptible to infection or may experience a more severe case of the disease. The dose and route of exposure can also affect the likelihood and severity of infection, with a higher dose or certain routes of exposure more likely to result in infection or a more severe case of the disease.

After defining a spillover and identifying the parties involved, the next steps are to comprehend *i)* the process through which it occurs, and *ii)* how it can escalate into an epidemic or pandemic. A spillover occurs because 'many viruses, especially single-stranded RNA viruses, but also bacteria, fungi, and protozoa are able to cross the species barrier due to mutations in their genome (such as antigenic drift and especially the most decisive antigenic shift)' (Grossi et al., 2020). In this context, the adaptation and evolution process of the virus to better recognize and enter human cells can be compared to a burglar trying different keys to unlock a safe. The virus evolves to create a key, such as a spike protein, that fits well into the cell receptor and allows it to enter the cell. This process can lead to reproductive success for the virus, allowing it to infect more cells and spread further. However, this process can also lead to a dead end, meaning that the virus is ultimately unable to enter the cells and therefore cannot reproduce.

Pathogens with high host plasticity, that is, their genomes are capable of easily adapting to new species, are more likely to succeed in this process. This allows them to 'jump' from one host to another, potentially leading to the spread of the disease to new populations. On the other hand, pathogens with low host plasticity may be less able to adapt to new hosts and may have a harder time establishing an infection.

Upon entering a new host, the virus devises a strategy to move to subsequent hosts before the initial host succumbs. If this were to die before a successful additional infection, it would be impossible to reproduce in other hosts (thus preventing the outbreak of the epidemic). Modes of human-to-human transmission are often the key to the reproductive success of many viruses. As highlighted by Grossi et al. (2020), there are several ways in which pathogens can be transmitted from one host to another. Some pathogens are spread through droplets or aerosols, which are tiny particles of respiratory or salivary secretions that can be emitted when a person talks, coughs, or sneezes. These droplets or aerosols can contain the pathogen and can be inhaled by another person, potentially leading to infection. Other pathogens are transmitted through the oral-fecal route, meaning that they are spread through contaminated food or water that has come into contact with feces. Others are transmitted through direct contact with body fluids, such as blood or saliva. Some pathogens require a vector, or an intermediate host, in order to be transmitted. Vectors can include insects such as mosquitoes or ticks, which can transmit pathogens through their bites.

Many viruses tend to die in the new host after some time, leading to local epidemics that eventually disappear. This can happen when the virus cannot establish a long-term infection in the new host and eventually dies off. This is what happened with the Ebola and Marburg outbreaks in Africa, for example. However, in the case of the SARS-CoV-2 virus, it is believed that the large wet market in Wuhan, China (a city with a population of over 6 million) played a role in the spread of the virus due to the close proximity of live animals and humans and the strong connections to the rest of the world through travel and trade. This allowed the virus to spread more easily and led to a global pandemic.

Virus-induced diseases have been a scourge for humanity and are potentially more dangerous because of their essential property to cause ultimate prey predator relationship across genomes. As such, viruses attack directly the cells of the hosts, with their simple and rapidly evolving DNA or RNA genome surrounded by a protective protein coat. At the borderline between living and inanimate entities, viruses have probably emerged from the primordial soup, perhaps originating from cells or preceding cell formation, and co-evolving over time with a plurality of hosts. Viruses propagate by delivering their DNA or RNA as a complete virus part (called virion) to the cells of the host that provide the biosynthetic machinery for their reproduction, by transcription and translation. The viral genome, often with associated basic proteins, is packaged inside a symmetric protein capsid. More than 30,000 different viruses have been isolated and are classified into more than 3600 species, 164 genera, and 71 families (Gelderblom, 1996).

The coronavirus, of which SARS-CoV-2 is a subfamily, carries its viral genome incapsulated into a protective coat of glycoproteins which merges with a lipid bi-layer derived by the modified host cell. Coronaviruses are characterized by a single-strand RNA genome, which is also one of the largest known genomes for RNA viruses. As in all RNA viruses, the rate of error in transcribing the genetic information is very high, resulting in an equally high rate of mutation. The basic organization of the genome comprises for one third, a set of Non-Structural Proteins involved in proteolysis, transcription, and genome replication. The remaining two thirds include the 'Structural Proteins', a surface or spike glycoprotein, and a protective coating formed by a membrane glycoprotein, an envelope, and a nucleocapsid (Balloux et al., 2022). Viral analysis and vaccine development have focused on the spike glycoprotein, which provides coronaviruses with their characteristic crown or halo. The spike protein is essential to secure viral entry in the host cell, recognition of host receptors, cell attachment, and membrane fusion.

These basic characteristics of SARS-CoV-2 stand out and explain in part the unique features of its diffusion and the uncertainty still surrounding its evolution and relationship to human activities. While it seems likely that the virus 'jumped' from a wild animal to a human, this hypothesis has eluded substantial corroboration so far. More importantly, the evolution of the virus in the past three years has been characterized by several waves of containment, mutations, and resurgence of the diseases, intersecting with non-pharmacological measures of containment, a major research and development effort to discover effective cures and vaccines, and an ongoing vaccination campaign.

The different waves of virus insurgence have coincided with mutations and the generation of several different lineages, in many cases not sufficiently studied because, *inter alia*, the tendency of the different virus strains to rapidly substitute for each other. In general, two tendencies can be recognized. On one hand, mutations appear to successfully increase escapes from vaccine control. On the other hand, pathogenic features seem to become milder and virus lethality to abate, although unsettling questions remain about long-term consequences of even mild infections.

The non-seasonal waves that characterized CoV-2 have not been satisfactorily explained by mutations and the emergence of new virus lineages. Although these phenomena appeared to be responsible for the recrudescence of infections, in selected periods of time, they did not seem to correlate with the different waves, except in a punctuated and approximate way. While competing explanations are still few and unconvincing, the use of predator-prey Lotke-Volterra equations has recently emerged as a promising framework to interpret some of the unique characteristics of the pandemic. A recent article by Dimaschko et al. (2022), for example, claims that the key property that explains the observed cyclical behavior of the epidemic lies in the fact that unlike most other diseases, the main carriers of the infection are the asymptomatic infected. Ordinarily, since the subjects infected are isolated, a population reacts to an infection differently from the individual, in the sense that its degree of infection gradually abates as the infected people are subtracted from social contacts, and thus the probability of infection gradually decreases. In the CoV-2 case, the fact that a substantial part of the population is infected, but keeps having regular social contacts, determines a strong correlation between the course of infection of each individual and the course of the epidemic in the entire population. This model divides the population into two groups: the symptomatic group, which consists of individuals who show symptoms of the disease, and the asymptomatic group, which consists of individuals who do not show symptoms, but are still infected with the pathogen. The asymptomatic group can continue with its regular activities and may unknowingly contribute to the spread of the infection. It is believed that being asymptomatic may also be correlated with a more rapid loss of immunity, meaning that once the infection subsides, asymptomatic individuals may become more susceptible to reinfection and may spread the infection to others more easily. This can lead to new waves of the pandemic, with the asymptomatic group serving as a source of infection for the rest of the population. Although pathogen mutations and the use of vaccines can complicate this picture, they do not change the basic dynamics, which depends on the strong correlation between the state of the environment (whether infected or immunized) and the asymptomatic group.

Despite the many aspects of 'déjà vu', it seems clear that the SARS CoV-2 recurrent epidemics had unprecedented characteristics that made it a totally new experience from two critical points of view. First, even though the coronavirus family is well known and the almost immediate genome sequencing of the COVID-19 virus allowed a rapid vaccine response to the epidemic, the origin of the virus remains unknown. Its design and adaptation to human cells seem to be so peculiar that it may have suggested to many the possibility that the virus was 'manufactured' in a laboratory and its diffusion was the consequence of an accidental escape. Second, the mutagenic characteristics of the virus also appear somewhat novel even for the well-known family of coronaviruses. On the one hand, recombinant RNA mutations seem to be excluded as a source of the multiple lineages of the virus generated so far. On the other hand, the dynamics of mutations reveal a unique capacity to co-evolve with vaccines and immunity. This may depend, *inter alia*, on the fact that the virus is capable of spreading widely to asymptomatic individuals and lingering for long periods of time in immuno-suppressed patients, thus defying both detection and immunological diffusion (Costa Dias et al., 2020). As Zhang et al. (2020) have shown, the virus is also capable of reproducing through cured patients, as these may remain viral carriers for some time, and may be intermittently shedding the virus from some tissues over an even more extended period.

The rapid change of SARS-CoV-2 makes also difficult to establish its key characteristics, especially in a quantitative, nonambiguous fashion. One case in point is its case fatality rate (CFR),[4] which changes with the spread and evolution of the virus. At some point in 2021 America seemed to be the epicenter of the disease, but the primacy of the fatality rate has shifted from one world region to the other as the different variants took hold of the local population at different points in time. The maximum value for the CFR was reached in 2021 with 18.98% in France, while in other countries, such as, for example, Iceland, it was close to zero. Worldwide, CFR was at a peak of 6.06%, in 2021, approaching the value of seasonal influenza, while lower than other coronaviruses, such as SARS-CoV (CFR 10%) and MERs-CoV (35%) (Ghebreyesus, 2021). Evidence has suggested that CFR is further being reduced by the spread of new variants, which can be considered good news, but also testifies to the increasing capacity of the virus to circulate within the world population (Costa et al., 2021).

1.2.2 2019: SARS-CoV-2 infects humans

The origins of the virus have been discussed at various points in the COVID-19 outbreak. In the first months, several conspiracy theories were circulating supporting the idea of an artificial origin or, worse, an accidental error in a laboratory. Former US president Donald Trump, when asked at a press conference if the virus had accidentally left a laboratory in Wuhan, replied: 'We are examining it. Many people are watching this. It seems to make sense'. In an interview with ABC on May 3, 2020, former US Secretary of State Mike Pompeo said there was 'enormous evidence' that the epidemic of the new coronavirus originated in a Chinese laboratory, although he did not

---

[4] The Case Fatality Rate (CFR) is the proportion of individuals diagnosed with a particular disease who die from that disease within a specified time period.

> **Box 2: The main spillovers in the history**
>
> The main spillover events known, arranged in chronological order according to the year of isolation or phylogenetic or documentary reconstruction[a]:
> - 1879: Chlamydophila psittaci, Psittacosi
> - 1908: HiV-1 Group M, AIDS
> - 1930: Coxiella burnetii, Q Fever
> - 1932: McHV-1, Herpes B
> - 1965: Plasmodium knowlesi, Monkey Malaria
> - 1967: Marburg virus
> - 1970: Monkeypox virus, Monkey pox
> - 1975: Borrelia burgdoferi, Lyme disease
> - 1976: Ebola virus
> - 1994: Virus Hendra
> - 1995: Monkey foaming virus
> - 1997: Avian Influenza H5N1
> - 1998: Nipah Virus
> - 2002: SARS-CoV-1, SARS (Severe Acute Respiratory Syndrome)
> - 2009: Swine flu H1N1 pdm09
> - 2012: MERS CoV, MERS (Middle East Respiratory Syndrome)
> - 2019: SARS-CoV-2, COVID-19
>
> [a] Grossi et al. (2020).

provide any of the above-mentioned evidence. In Europe, too, there was some fallout from these statements. German Chancellor Angela Merkel and French President Emmanuel Macron, while distancing themselves from Trump's positions, called China into question, calling on the Beijing government to shed light on some opacities relating especially to the immediate management of contagion after the detection of the first cases in Wuhan. In reality, while the hypothesis of a laboratory accident remains on the table, even in this case, the diffusion of the virus can be considered a byproduct of the growing concern with zoonotic diseases and their emergence as a systemic effect of increasing environmental deterioration.

Following these claims, researchers in various parts of the world collaborated to provide evidence on the true origin of the virus. In an article in Nature published in March 2020, an international research team presented a genetic study of the origins of the virus (Andersen et al., 2020). The authors acknowledge that bat-derived coronaviruses were studied in laboratories around the world, and there have been past documented cases where the virus causing SARS 'fled': a good reason to look at the SARS-CoV-2 case and try to understand what happened. The results of the research appeared to rule out the possibility that the virus had been manipulated or engineered in a laboratory. Two main factors supported this conclusion. First, the section of the genome that codes for a large part of the virus's spike protein, which is responsible for its ability to enter host cells and establish an infection, was not what scientists would have expected to see if the virus had been manipulated by humans. Second, previous research had identified a specific genetic sequence that was believed to optimize the potency of the spike protein, but the sequence found in the real virus did not match this expectation.

These findings suggest that the SARS-CoV-2 virus probably originated naturally, rather than being the result of human manipulation. However, it is still possible that the virus may have undergone some natural mutations or changes after it emerged, which could have affected its properties and behavior.

The conclusion of the study therefore goes in the direction of the thesis that the sequence of the virus was not designed in a laboratory, but is most likely the result of natural selection. However, later evidence has led western governments and many experts to challenge this conclusion, arguing that the group reporting the results lacked qualifications to determine the possibility of a lab leak. The WHO's secretary general, Tedros Adhanom Ghebreyesus, declared that the assessment had not been extensive enough and vowed to conduct more studies in the hope of reaching stronger conclusions. What natural selection has generated SARS-CoV-2 remains to be seen, and scientists have been pondering three possible scenarios regarding its origins:

- natural selection in a host animal (bat or pangolin) before species jump. The new coronavirus probably passed from animals to humans to the Wuhan market. There is a high homology between the SARS-CoV-2 RNA sequence and the bat coronavirus sequence. Two bat coronaviruses share 88% of their genetic sequence with that of SARS-CoV-2. Furthermore, the new coronavirus shares 79% of its genetic sequence with that of SARS and 50% with MERS, the two severe respiratory syndromes that have claimed many lives in recent decades. In addition, the closest virus identified so far, a virus that infects bats of the species Rhinolophus affinis has a 96% sequence identity with the human SARS-CoV-2 virus. The only difference between the SARS-CoV-2 spike protein and that of the bat is the greater capacity of the new coronavirus to penetrate and infect the human body. Pangolin, a species of anteater,[5] has also been found to carry a coronavirus similar to SARS-CoV-2. Thus, this evidence suggests that the spike protein responsible for COVID-19 may have been naturally selected.
- natural selection in humans after the species jump. It is possible that an 'ancestor' coronavirus from SARS-CoV-2 entered the human genome, adapting as it was transmitted from human to human by genetic mutation events. Until then, it would 'fortify' itself to such an extent that it would cause health problems and a pandemic. Only by studying a large number of cases in humans will it be possible to determine when the species jump occurred or the mutation originated. Retroactive serological

[5] The pangolin is a unique mammal known for its protective, overlapping scales made of keratin. Often referred to as 'scaly anteaters,' pangolins are nocturnal creatures that primarily feed on ants and termites.

studies may also be very useful to understand when humans have started to be exposed to SARS-CoV-2.

- SARS-CoV-2 selection during laboratory steps. Basic research on viruses may involve studying them in cell cultures or animal models to learn more about their biology and how they interact with host cells. It is possible that the increased ability of SARS-CoV-2 to cause disease may have been acquired during prolonged cell culture, meaning that it may have become more virulent (able to cause disease) as a result of being grown in the lab. This could have occurred through natural processes such as mutation or through the selection of certain virus strains that were more successful at replicating in cell culture conditions. There have been documented cases of viruses 'getting out of hand' in laboratories, meaning that they have escaped from the lab and caused infections in the surrounding community. While it is not clear how SARS-CoV-2 was transmitted from animals to humans, it is possible that it may have escaped from a laboratory as an unintentional release. However, it is important to note that this is just one possible explanation and further research is needed to determine the exact circumstances surrounding the emergence of the virus.

At the beginning of the pandemic in 2020, the hypothesis of natural origin – that the virus emerged in bats and then transmitted to humans, probably through an intermediary species – was widely accepted. However, as time has passed, scientists have not found a virus in bats or other animals that matches the genetic signature of Sars-CoV-2. Early work based on genetic analysis and comparisons with sequences of other coronaviruses of different animal species suggested that SARS-CoV-2 originates from bats.[6] In particular, SARS-CoV-2 was thought to be closely related to coronaviruses found in some bat species living in Yunnan, about 1600 kilometers from Wuhan, where the first cases were found in humans, although from the beginning there were people infected who had no link to the market in the city. Furthermore, in a letter published in The Lancet on February 15, 2020, an international team of scientists firmly established that the 'virus originated in wildlife', citing numerous genetic analyzes from different countries (Calisher et al., 2020). A paper published in the New England Journal of Medicine in April 2020 indicated that bats probably originated the virus (Morens et al., 2020).

More than three years after the initial documented cases in Wuhan, the source of SARS-CoV-2 has yet to be identified and the search for a direct or intermediate host in nature has so far been unsuccessful. Over the months, with the progression of the pandemic and the collection of new evidence, experts are reconsidering the hypothesis of natural origin, becoming more open to the possibility that the virus may have leaked

---

[6] The note is available at the following address https://www.iss.it/primo-piano/-/asset_publisher/o4oGR9qmvUz9/content/covid-19-most-likely-a-role-per-bats-ma-s-i-search-again-l-host-intermediate.

from a lab conducting bat coronavirus research in Wuhan. Although allegations of a leak from the Wuhan Institute were originally proposed by Donald Trump and were strongly rejected by the Chinese government, little credence was given to this hypothesis until May 2021, when 18 leading scientists sent a letter to the journal Science in which they claimed both spillover and leak theories were equally plausible (Bloom et al., 2014). In that letter, the group of experts also accused a WHO investigation at Wuhan of not giving a balanced consideration to both scenarios. Following the letter, WHO officials reconsidered their position supporting the view that all theories on the origins of COVID-19, including the possibility of laboratory leak, are 'on the table' and urged Chinese scientists to carry out their own investigations and the Chinese government to be more collaborative and transparent.

There is still no definitive answer on this issue, as also stated by the WHO director Ghebreyesus, who agrees that 'the initial investigation by the global health body into Wuhan's virology laboratories did not go far enough'. The strength of the arguments and the reputation of the authors in the letter published by Bloom et al. (2021) was such as to force a global action across different institutions. For example, the authors of the letter stated that '...as scientists with relevant expertise, we agree with the WHO Director General, the United States, and 13 other countries, and the European Union that greater clarity on the origins of this pandemic is necessary and feasible. We must take hypotheses about both natural and laboratory spillovers seriously until we have sufficient data. A proper investigation should be transparent, objective, data-driven, inclusive of broad expertise, subject to independent oversight, and responsibly managed to minimize the impact of conflicts of interest'.[7]

In September 2021 a new article from Segreto et al. (2021) has reported more evidence supporting the possibility of a leak from the Wuhan Institute. These evidence comes from different independent analyses, which tend to support both hypotheses (natural and human) as equally possible. According to the authors, the low binding affinity of SARS-CoV-2 to bat ACE2 studied to date does not support the order of Chiroptera as a direct zoonotic agent.[8] Additionally, using the similarity of the pangolin coronavirus receptor binding domain (RBD) to SARS-CoV-2 as proof for a natural

[7] The WHO director-general's remarks at the Member State Briefing on the report of the international team studying the origins of SARS-CoV-2' is available at the following web address: https://www.who.int/director-general/speeches/detail/who-director-general-s-remarks-at-the-member-state-briefing-on-the-report-of-the-international-team-studying-the-origins-of-sars-cov-2.
The US Department of State, 'Joint statement on the WHO-Convened COVID-19 origins study' is available at the following web address: https://www.state.gov/joint-statement-on-the-who-convened-covid-19-origins-study/. Finally, the statement by the Delegation of the European Union to the UN and other International Organizations in Geneva 'EU statement on the WHO-led COVID-19 origins study' is available at the following web address: https://eeas.europa.eu/delegations/un-geneva/95960/eu-statement-who-led-covid-19-origins-study_en.
[8] Bats are mammals who belong to the order of Chiroptera.

zoonotic transition is questionable. The same analysis suggests that pangolins probably didn't contribute to the origin of SARS–CoV-2, and the idea of recombination is not well-supported. At the same time, genomic analyses pointed out that SARS-CoV-2 exhibits multiple peculiar characteristics not found in other Sarbecoviruses' (Segreto et al., 2021). Moreover, research teams in multiple laboratories have worked to grow and adapt coronaviruses to various cell types, including human airway epithelial cells (Tse et al., 2014; Menachery et al., 2015; Zeng et al., 2016; Jiang et al., 2020). A general conclusion seems to be that while the possibility of a natural origin remains and efforts to find a potential natural host should persist, the unique genetic characteristics of SARS-CoV-2 don't exclude a potential gain–of–function origin (Segreto et al., 2021).

CHAPTER 1 - Take-home messages

- Over 60% of the infectious diseases identified since 1940 are of zoonotic origin, including SARS-CoV-2, as shown by the genetic sequence of the virus analyzed by researchers from all over the world.
- The barriers that prevent the spillover of a virus from a reservoir animal, where the virus can thrive and reproduce, to another animal, including humans, are often insurmountable for microorganisms. However, SARS-CoV-2 managed to bypass these barriers by becoming contagious, though not excessively lethal.
- The contagiousness of SARS-CoV-2 appears to depend critically on the Spike protein, which decorates the surface of the virus like an aura and functions as a key, able to adapt to host cells. The Spike protein (S) is the most characteristic and better studied structural component of the SARS-CoV-2 virus, and is critical for target recognition, cellular entry, and endosomal escape (Huang et al., 2020).
- Socio-economic factors and public hygiene factors, including population density, massive urbanization, and climate change, contribute to making the conditions of viral spread more favorable.

CHAPTER 2

Meaning and dynamics of epidemics

Contents

2.1. What do we mean by epidemics: some basic notions

Before delving into the issues caused by the spread of the COVID-19 pandemic, it is helpful to acquaint ourselves with fundamental concepts and establish a shared vocabulary. This enables us to comprehend what pandemics entail, how they ought to be managed, and their prevalence in our lives. Grasping the epidemiological terms and core principles is crucial for understanding subsequent discussions and statistics on the subject. As a note of methodology: the definitions given in this chapter come from official sources, predominantly the US Centers for Disease Control (CDC) and the World Health Organization (WHO).[1]

2.1.1 The different levels of manifestation of a disease

It is now widely known that COVID-19, the disease caused by the new SARS-CoV-2 coronavirus, is defined as a pandemic. But what is the difference between a pandemic, an epidemic, and an outbreak? And when does a disease become a public health problem? To fully grasp these distinctions, several concepts require clarification. A first notion is the scale of measurement against which an epidemic is defined. This scale represents the proportion of individuals within the total population who are affected by the disease.

---

[1] On the CDC website it is also possible to access a series of information from an online course on 'Principles of Epidemiology in Public Health Practice, Third Edition - An Introduction to Applied Epidemiology and Biostatistics from which is taken part of the material presented in this section https://www.cdc.gov/csels/dsepd/ss1978/lesson1/section11.html. The WHO definition is taken from WHO (2018a).

A possible way to classify diseases is with respect to their 'prevalence', i.e. their level of 'diffusion' in the territory and among populations.[2]

The following definitions of morbid events differ by the different amount of population involved:

Definition 2.1.1 – Sporadic – A sporadic event is an occurrence that is infrequent and random, rather than being part of a pattern or trend. In the context of infectious diseases, a sporadic event refers to a disease that manifests rarely and unevenly, meaning that it is not part of an outbreak or epidemic. An example of a sporadic event in the context of infectious diseases would be an individual contracting a disease like malaria while traveling abroad and returning home with the infection, but not spreading the disease to others. In this case, the infection is not part of a larger outbreak or epidemic and is not spread widely within the community. Sporadic events can occur with any disease, but they are more common in rare or emerging infections without established transmission patterns. It is important to monitor sporadic events and investigate potential cases to identify possible sources of infection and prevent further spread of the disease.

Definition 2.1.2 – Endemic – An endemic occurs when a disease is constant and occurs more often than expected. It refers to the continuous presence of a disease or an infectious agent in a population within a geographical area. Examples are measles or scarlet fever in developed countries today.

Definition 2.1.3 – Hyper-endemics – A hyper-endemics event refers to persistent, elevated levels of onset of disease. This implies that the disease is always present in a given area and that people are infected at a high rate. An example is malaria in some regions of the globe.

Definition 2.1.4 – Epidemic – An epidemic is the widespread occurrence of an infectious disease in a community at a particular time. It is defined as the occurrence of more cases of a disease than what would normally be expected in a specific population within a certain period of time. Epidemics can occur when a disease spreads rapidly to more people than experts expect, often due to the presence of favorable conditions for transmission. An example of an epidemic is the cholera

---

[2] It is important, however, to emphasize that different diseases may have different levels of alert compared to their spread in the territory. The prevalence of a particular disease that is usually present in a community is called the 'basal level' or the 'endemic' of the disease. This is certainly not the ideal level (which would obviously be zero), but rather the 'observed' level at a given time. In the absence of interventions and assuming that the level is not high enough to exhaust the pool of 'susceptible people' (i.e., there is still a basin of 'infectious' people), the disease can continue to manifest at this level indefinitely. Therefore, the baseline level is often considered the 'expected level' of the disease. For example, there are pathologies that are so rare in a given population that an infected individual deserves an epidemiological investigation. This happens in cases linked to diseases such as rabies, plague, or polio. For other diseases, the spread (prevalence) is usually higher. Therefore, only deviations from the norm justify the attention of those who carry out surveillance and, therefore, require the initiation of an investigation.

outbreak in Italy in 1973, in which the disease spread rapidly and affected a large number of people in a short period of time. Epidemics can cause severe public health consequences, leading to a considerable number of infections and fatalities. They may also result in economic and social ramifications for the communities affected.

Related to the definition of an epidemic are as follows:

Outbreak – Like an epidemic, an outbreak refers to the occurrence of a disease in a community or population that is in excess of what is normally expected. It is often used to describe a situation in which a large number of people in a particular area are affected by a disease, especially when the disease is new or has not been seen before in that area. An outbreak can also refer to a single case of a disease that is atypical or unforeseen, such as when an individual contracts an illness not commonly found in their area.

Cluster – A cluster may be identified when there is an unexpected number of cases of a particular disease or health condition within a particular area or population, or when there is a concentration of cases that is higher than would normally be expected.

Definition 2.1.5 – Pandemic – A pandemic is an epidemic that spreads to several countries or continents. It affects more people and causes more deaths than an epidemic. Examples are AIDS-HIV and COVID-19, while past pandemics include a long list of 'scourges of humanity' such as plague, cholera, tuberculosis, and Spanish flu.

2.2. The origins of epidemics

According to the Center for Disease Control and Prevention (CDC) of the United States, epidemics can originate and spread among the population through different channels:

- **The common origin** – An outbreak of common origin is when a group of individuals is exposed to an infectious agent or a toxin of the same source. Within that category, it is possible to further identify three modalities of origin:

  - *Point origin* – If the group is exposed for a short time, so that anyone who becomes sick does so within an incubation period (Cobb et al., 1959), the common-source epidemic is classified as a point-source epidemic. This is, for example, the situation recorded for leukemia cases in Hiroshima following the explosion of the atomic bomb and hepatitis in the 2003 outbreak in Pennsylvania among restaurant customers who ate green onions (Centers for Disease Control and Prevention, 2003).

  - *Continuous origin* – In some outbreaks, patients may have been exposed to the source for a period of days, weeks, or more. In a continuous common source outbreak, the range of exposures and the range of incubation periods tend to flatten and enlarge the peaks of the epidemic curve (see the COVID-19 case).

* *Intermittent origin* – Outbreaks have the same pattern as continuous-source outbreaks, but the effect of extending the epidemic curve is even more marked as it reflects the intermittent nature of exposure.

* **The propagated origin** – A propagated outbreak results from transmission from one person to another. Usually, the transmission occurs by direct contact from person to person (such as syphilis), or by a vehicle (e.g. transmission of hepatitis B or HIV by sharing needles) or vectors (e.g. transmission of yellow fever by mosquitoes).

* **Mixed origin** – Some epidemics have a common and propagated origin. A common source outbreak pattern followed by secondary person-to-person spread is not uncommon. These are called mixed epidemics (Lee et al., 1991). Finally, some epidemics are neither common source in its usual sense nor propagated between individuals. Outbreaks of zoonotic diseases may result from three critical variables reaching a threshold intensity level: infection in host species, presence of vectors, and human-vector interaction (Centers for Disease Control and Prevention, 1999; White et al., 1991).

2.3. The many epidemics of the 21st century: a short history

The SARS-CoV-2 virus, at the origin of the COVID-19 pandemic, is just one of many examples in the long history of animal-to-human diseases (zoonoses). In general, we can think of infectious diseases as developing from agriculture, once a sufficient density of population and animal domestication practices were reached in the course of history. Looking back in time for example, an important spillover that gave rise to what is now the virus responsible for the common cold occurred during the domestication of horses. Later, with the breeding of chickens, the way was opened to diseases such as chickenpox, St. Anthony's fire, and several strains of avian influenza. The origin of influenza can be traced to the breeding of pigs, while measles, cowpox, and tuberculosis come from domesticated cattle. Whenever a spillover from an animal to a human occurs, the so-called 'patient zero' (the first person to become infected with a disease after coming into contact with an infected animal) becomes the human transmission vector of the virus.

In the 21st century, coronavirus spillovers have occurred on three occasions, each time causing deadly epidemics: at the end of 2002 with SARS (Severe Acute Respiratory Syndrome, or Acute Grave Respiratory Syndrome), in 2012 with Middle East Respiratory Syndrome (MERS), and in 2020 with COVID-19 (COronaVIrus Disease 2019).

However, these are not the sole occurrences that warrant our attention. As the WHO Director General, Dr. Tedros Adhanom Ghebreyesus, pointed out, the diseases resulting from zoonotic spillovers are not the only ones that should concern us (WHO,

2018a). A long list of recrudescent well-known communicable diseases can be drawn for the past 20 years or so, beginning with a serious outbreak of plague in the 2017 in Madagascar, which caused a total of at least 2417 confirmed, probable, and suspicious cases, including 209 deaths (among 2417 cases there were hundreds of cases of bubonic plague). Up to nine countries and territories with trade and travel connections to Madagascar were placed on a plague alert. In 2015, the Zika virus, transmitted by the Aedes Aegypti mosquito, triggered a wave of microcephaly in Brazil. This disease causes terrible damage to the brains of unborn children. Almost 70 countries, one after the other, have suffered their own Zika epidemic (and there are likely to be more in the future). Finally, according to the WHO, each year there are approximately 40 outbreaks of cholera.

Table 2.1 Impact of epidemics: outbreak events globally, 2011-2017.

	2011	2012	2013	2014	2015	2016	2017	Total
Yellow fever	17	12	12	2	4	10	4	57
Chikungunya	8	10	10	29	27	14	4	95
Viral hemorrhagic fever						6	4	10
Ebola	1	2	2	11	4	3	1	22
Marburg marburgvirus		1	1	1			2	4
Crimean–Congo hemorrhagic fever	3	5	5	8	7	7	13	49
Rift Valley fever	1	1	1	1	1	4	5	14
Cholera	62	51	51	37	44	42	25	308
Typhoid fever	20	23	23	2	8	3	14	75
Shigellosis	25	24	24	29	4	2	1	113
Paglue	8	7	7	10	7	6	3	47
Lassa fever	2	1	1	3	2	7	6	23
West Nile fever	11	15	15	11	11	18	10	91
Zika				7	19	54	52	137
Meningitis	14	20	20	19	19	23	23	137
MERS–CoV		3	3	17	12	7	8	57
Type A Influence	5	6	6	9	10	5	9	51
Monkeypox					2	2	5	9
Nodding disease		1	1					2
Nipah virus disease	1	1	1	1	1			5

Note: The analysis excluded polio. The following epidemics and pandemics were analyzed: avian influenza A (H5N1), A (H7N9), A (H7N6) A (H10N8), A (H3N2), A (H5N6), A (H9N2), Chikungunya, Cholera, Crimean - Congo haemorrhagic fever, Ebola virus disease, Lassa fever, Marburg virus disease, meningitis, MERS-CoV, monkeypox, Nodding syndrome, Nipah virus infection, plague, Rift Valley fever, shigellosis, typhoid fever, haemorrhagic fever viral, West Nile fever, Yellow fever, Zika virus disease. If a disease has caused more than 1 epidemic event per year in a country, it has been counted once for the year in which it occurred in that country. This includes cases that are imported or locally transmitted.

WHO/IHM data as at 12 January 2018 (note: 2017 data are not complete).
Source: data reported to WHO and by the media (WHO, 2018a).

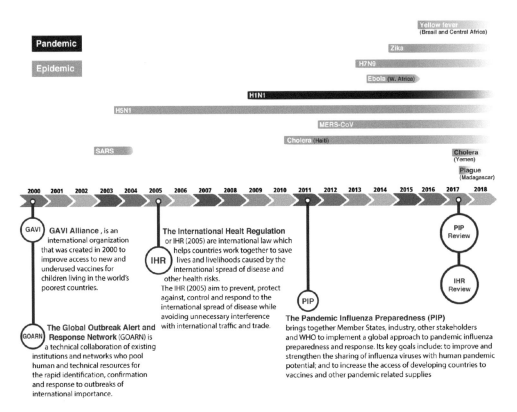

Figure 2.1 Timeline of epidemics around the world. Data reported to WHO and available in the media. *Source*: WHO (2018a).

As can be seen in Table 2.1 and Fig. 2.1, the global situation is far from that of a world free from communicable diseases. In particular, Table 2.1 shows for each year and by type of infectious disease the number of countries that have reported cases of contagion, for a total of 1307 epidemic events in 172 countries. Finally, in Fig. 2.1 the information given in the top section relates to the timeline of the world's major epidemics, while the bottom section highlights the organizational efforts of the WHO to tackle emergencies.

2.4. COVID-19 is only one of the many infectious diseases that spread across the globe

Table 2.1 shows that infectious diseases are increasing globally. This trend has continued after 2018, and while COVID-19 received a lot of media attention in 2019, other deadly diseases have also raised concerns, although they may not have received as much visibility. Based on data from the GIDEON database, which tracks more than 360 infectious diseases worldwide on a daily basis, it is possible to track all existing outbreaks. In 2020, in addition to COVID-19, the following infectious diseases had the most significant impact (in order of the number of cases recorded):

- **Monkeypox** - 4500+ cases, Democratic Republic of Congo (DRC). An ongoing outbreak of Monkeypox has persisted well throughout the COVID-19 pandemic.
- **Lassa Fever** - 5500+ cases, Nigeria. It is estimated that up to 500,000 people are infected in West Africa each year, resulting in 5000 deaths. During the past 50 years, at least 88 travelers have returned home to other countries with Lassa fever – including 11 returning to the United States. The largest recorded outbreak of Lassa fever continues to endanger massive populations. As of August 16th, 2020, 5527 cases (222 fatal) were reported in Nigeria.
- **Hand, Foot, and Mouth Disease** (Enterovirus Infection) - 6000+ cases, Vietnam. Vietnam's largest city, Ho Chi Minh City, has been suffering a prolonged outbreak of HFMD since the beginning of the year, with more than 6000 cases. The highest number of new cases (640) was reported at the end of September 2020, coinciding with the start of the new academic year.
- **Zika virus** - 6000+ cases, Brazil. Arboviral diseases are infectious diseases transmitted by arthropods, such as mosquitoes and ticks. They can have serious health consequences, including birth defects and neurological problems. In the first half of 2020, more than 1.5 million cases of arboviral diseases were reported across the Americas, with the majority of cases occurring in Brazil. Of these cases, more than 6000 were caused by the Zika virus, which is transmitted by mosquitoes.
- **Hepatitis A** - 32,000+ cases, United States. As of September 2020, more than 1000 cases were reported in each of these states: Florida, Georgia, Indiana, Kentucky, Ohio, Tennessee, and West Virginia. The number of cases reported since the arrival of COVID-19 reached 6358 (as of September 26, 2020), and with some of the current state outbreaks dating back to 2016, 2017, or 2018, the total number has now exceeded 32,000 cases.
- **Chikungunya** - Chikungunya is an infectious disease caused by the chikungunya virus, which is transmitted to humans by mosquitoes. In 2020, there have been outbreaks of chikungunya in Bolivia, Brazil, Cambodia, Kenya, Malaysia, and Thailand. Brazil has experienced the largest outbreak, with 78,000 cases and 14 deaths reported in October 2020. Chikungunya virus infection has been a common problem in Brazil since 2014, with a total of 900,000 cases (485 of which were fatal) in eight outbreaks. The disease is characterized by fever and joint pain and can sometimes be mistaken for other diseases transmitted by mosquitoes, such as dengue or Zika.
- **Measles** - 124,000+ cases worldwide. Although Measles vaccines have been available and effective for many years, the World Health Organization (WHO) reports that global vaccination levels had reached only 61% in 2015. Not surprisingly, in

2020, measles outbreaks were reported in 26 countries. The most severe outbreak in 2020 were reported in the Democratic Republic of Congo (DRC), with more than 71,000 reported cases and 1026 deaths. Surprisingly, these numbers represent a significant improvement over 2019, when more than 300,000 cases were reported. During the past 20 years, the DRC has actually increased its coverage of the measles vaccine from 18% to 92%. Sadly, a European country has continued to report an unusually high rate of measles into 2020. Romania has been dealing with a large outbreak, despite having maintained more than 85% vaccination coverage for the past 30 years. In fact, the 2020 outbreak dates back to 2016 and has resulted in almost 20,000 cases, including 64 deaths – far outranking all other European countries.

- **Dengue** – Dengue is an infectious disease caused by the dengue virus, which is transmitted to humans by mosquitoes. It is one of the leading causes of illness and death in many tropical and subtropical countries. In 2020, there have been more than 2.5 million cases of dengue worldwide, with annual peaks in the disease occurring regularly. There were also 49 separate outbreaks of dengue in 2020, the largest outbreak occurring in Brazil, where with more than 1.3 million reported cases and 500 deaths. The incidence of dengue in Brazil was second only to that of COVID-19, making it an important infectious disease to consider in terms of global health.
- **Cholera** – more than 2.5 million cases worldwide. One might forgive the lack of attention to the ongoing cholera pandemic, as this is not a recent occurrence. The current (and seventh) declared pandemic actually started in South Asia in 1961 before spreading to Africa in 1971, then to the Americas in 1991. The impact of cholera was the most severe between 1961 and 1975 but has remained a yearly threat, with millions of cases and tens of thousands of deaths. If not treated, the disease is fatal in 50% of cases. Yemen has endured a cholera outbreak since 2016, amounting to a total of more than two million cases, nearly 4000 of them fatal. Bangladesh, Somalia, Ethiopia, and the DRC have also been severely affected, with nine additional countries reporting smaller outbreaks.

2.5. Are we at the dawn of a reversal in the epidemiological transition?

One of the most celebrated achievements of the 20th century for mankind has been the 'epidemiological transition', a phenomenon in which, at least in the richest countries, the totality of communicable (and preventable) diseases was believed to have disappeared substantially, replaced by the emergence of so-called non-communicable diseases (NCDs). The 'epidemiological transition' model was coined by Abdel Omran in 1971 (Omram, 1971) who transposed the post-war 'demographic transition' model into the public health field (Szreter, 1993). This innovative interpretation of the evolution of diseases was accepted with some enthusiasm by the entire development community, as it

---

Box 3: The main pandemics in the human history

The list of the deadliest pandemics in world history includes:

- The Black Death (Bubonic plague). Experts think that the plague, triggered by bacteria called Yersinia pestis, is the cause of the disease that tore Europe apart in 1347-51 and caused the death of an estimated 25 million people.
- The Spanish flu pandemic of 1918. At least 50 million people worldwide died from the flu during the 1918-19 outbreak. It is often referred to as the 'Spanish flu' not because the virus started there but because Spain was one of the first countries to report cases.[a]
- Smallpox. The smallpox pandemic lasted for hundreds of years. Experts estimate it killed up to 300 million people in the XXth century alone. Thanks to the widespread use of vaccines, smallpox was declared eradicated in 1980.
- HIV and AIDS. The human immunodeficiency virus (HIV), acquired immunodeficiency syndrome (AIDS) and related diseases have killed some 32 million people worldwide.
- The flu pandemic. The flu has also killed millions of people around the world in other pandemics: 1957 (1.1 million), 1968 (1 million), and 2009 (up to 575,000).

[a] Spain was the first to disclose the epidemic as the country was not subject to war censorship (several other countries were involved in World War I). Reading the headlines and articles in the Spanish press, many deduced that the epidemic had begun there. For their part, the Spaniards thought instead that the disease came from France and called it French flu.

---

suggested that health policies (and therefore the role of the WHO) should be re-aligned with demographic policies, which were the prominent goal of all UN agencies and several foundations and national and transnational NGOs (Weisz and Olszynko-Gryn, 2010).

Alleged empirical proofs of this theory at the world level were presented during the 1990s by the Global Burden of Disease (GBD) project, originally launched by the World Bank, and later promoted by the WHO (Disease Burden Unit) and more recently by the Institute for Health Metrics and Evaluation. Thanks to the effort of these institutions, it was possible to track detailed morbidity, disability, and mortality data for more than 200 countries and hundreds of diseases and causes of injury (Keating, 2018). Based on DALYs (disability-adjusted life years) as a measure of years of life lost from premature death (YLL) and years of life lived in ill health and disability (YLD), the GBD project produces statistics that allow us to assess over the years how communicable and non-communicable diseases evolve at world level. The various GBD measures provide indeed support, in absolute numbers, to the thesis of a statistically significant decline of communicable diseases in relation to non-communicable ones, even though this cannot be considered either a definitive or a convincing proof of a lasting achievement.

Epidemiological transitions have been obtained at different moments at different speeds around the world. Richer countries have started the process earlier and have completed it just after the second war world, thanks also to effective sanitation programs, better education, and higher per-capita income levels. A clear example of such a transition in wealthier countries can be observed through the case of Italy. As can be seen from Fig. 2.2, since the unification of Italy in 1861, the country has taken a long journey toward the elimination of infectious diseases from the main causes of mortality. By the early 1970s the epidemiological transition was completed by achieving almost complete disappearance of direct mortality for the major infectious diseases (i.e., tuberculosis, typhoid fever, diphtheria, measles, malaria, scarlet fever, and whooping cough). What remains are fewer deaths caused by gastroenteritis (diseases related to environmental, socio-economic, and climatic conditions), as well as those due to pneumonia, bronchitis, and influenza ('respiratory system' diseases). The decline in these diseases is clearly accompanied by the gradual increase in some chronic diseases (i.e., those related to the cardio-circulatory system and cancer), which go hand in hand with the aging of the population.

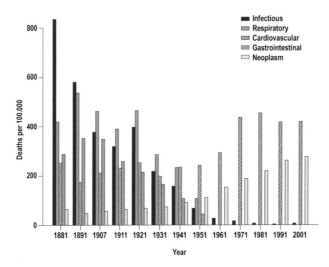

Figure 2.2 The Epidemiological Transition in Italy (1881-2001). *Source*: Atella et al. (2011). Note: The bars in the figure show death rates by cause per 100,000 inhabitants.

In developing countries, at least for the accomplishments already seen for the developed world, transition is still ongoing, though important progress has been made. Fig. 2.3 shows the different transition stages of high, middle, and low income countries at present. In panel (A), we can see that richer countries have completed the transition and are now on a stable path, with non communicable diseases (NCDs) accounting for about 88% of deaths on average between 1990 and 2019. In contrast, in panel (B) we observe the ongoing transition for low and middle income countries. In 1990 communicable diseases (CDs) accounted for 53% of total deaths, while in 2019 the number has almost halved to about 27%. At the same time, deaths due to NCDs have increased from 39% to 65%.

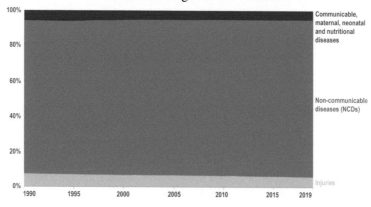

A. World Bank High income countries

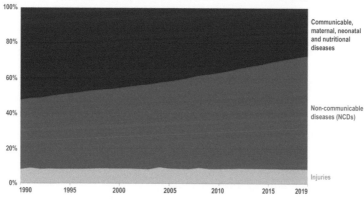

B. World Bank Lower-Middle income countries.

Figure 2.3 Deaths by cause, 1990 to 2019. Note: Non-communicable diseases (NCDs) include cardiovascular disease, cancers, diabetes, and respiratory disease. Communicable diseases (CDs) or infectious diseases include HIV/AIDS, malaria, and tuberculosis together with maternal deaths, neonatal deaths, and deaths from malnutrition. Injuries include road accidents, homicides, conflict deaths, drowning, fire-related accidents, natural disasters, and suicides. *Source*: IMHE, Global Burden of Disease through OurWorldinData (Roser et al., 2021), available at: https://ourworldindata.org/burden-of-disease.

Overall, these trends in both high- and low-medium-income countries seem to suggest that the increasing number of infectious disease outbreaks around the world since year 2000 are not reversing the main historical trends. What is less clear is to what extent these outbreaks have slowed down the epidemiological transition, changing drastically the form and speed of its time trajectory, especially in the low-middle-income countries. Furthermore, this evidence precedes the advent of COVID-19, which will significantly change the smooth patterns seen in Fig. 2.3.

An 'index of relative severity' of COVID-19 mortality (Schellekens and Sourrouille, 2020) recently developed by the World Bank can be used to offer some insight into how COVID-19 could potentially reverse the progress of several years. This index compares COVID-19 mortality with the pre-pandemic mortality profile of individual countries. In the words of an influential World Bank research paper, '...This comparison is useful, as the expression of mortality in relative terms speaks to the fact that countries may have adapted to pandemic shock according to their specific mortality patterns. Deviations from these patterns can create pressure points, for example, in the health system, which lead to different responses and, as a consequence, different mortality outcomes. Comparisons with previous patterns thus give a country-specific flavor to the severity of the COVID-19 pandemic, which could be used to corroborate realities on the ground' (Schellekens and Sourrouille, 2020).' Fig. 2.4 shows the severity index using the same income classification as used in Fig. 2.3. These graphs illustrate how the pandemic may have different effects depending on the type of country and how COVID-19 mortality may have affected rich countries more strongly compared to low–middle-income ones. In 2020, the average COVID-19 mortality in high income countries was comparable to the mortality rate of ischemic heart disease, which is the leading cause of death worldwide. However, in low-to-middle-income countries during the same year, COVID-19 mortality rates were more similar to those of Diabetes Mellitus, which is the 9th leading cause of death. In this way, '...income classification provides a very useful prism for looking at pandemic inequalities because it proxies structural characteristics that, in turn, correlate with pandemic outcomes in terms of infection prevalence and infection fatality' (Schellekens and Sourrouille, 2020). In general, this evidence suggests that COVID-19 may cause a reversal (or at least a stop) in the fight against CDs more in rich countries than in low- and middle-income countries.

2.6. Raising trends in infectious diseases: nothing happens by chance

An interesting question to ask is how spillovers, a phenomenon that used to be rare, have become more and more frequent. Spillovers have always occurred, as seen from the Box 3 list in this chapter. However, in the last century, there has been an increase in the frequency and impact of spillovers on humans and domestic animals. This increase may be due to a variety of factors, including the growth of human populations, increased global travel and trade, and changes in land use and land cover that bring humans and animals into closer contact. In the past 20 years, three coronaviruses have caused successful spillovers: SARS in 2002, which originated from palm civets and was transmitted to humans by horseshoe bats; MERS in 2012, which originated from dromedaries and may have been transmitted to humans by bats at an earlier time; and COVID-19, which is believed to have originated from bats or pangolins. These examples highlight the potential for infectious diseases to emerge and spread from animals to humans and the importance of being prepared to respond to and prevent these types of outbreaks.

A. World Bank High income countries.
Cause #1 in 2019 = Ischemic Heart Disease

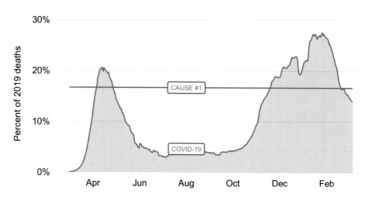

B. World Bank Lower-Middle income countries.
Cause #9 in 2019 = Diabetes mellitus

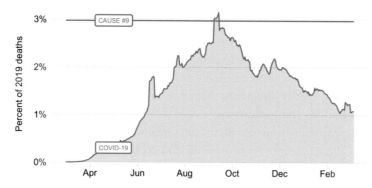

Figure 2.4 Global relative severity index by group of country. Note: Blue line: COVID-19 deaths (7-day avg). Red line: top #n cause deaths in 2019 (yearly avg). Both over total 2019 deaths. 'n' picked from 133 diseases (level 3 of 2019 GBD) to approach COVID-19 peak from below. *Source*: Adapted from Schellekens and Sourrouille (2020). Date: 2021-03-07. Latest: https://blogs.worldbank.org/opendata/relative-severity-covid-19-mortality-new-indicator-world-banks-data-platform.

During the late 1970s, many experts believed that the problem of infectious diseases had been largely solved, due in part to the misinterpretation of the transition theory and the enthusiasm for the advancement of medicine. Paradoxically, the economic growth that occurred since the 1960s, partly due to a healthier population, brought with it several side effects that may have increased the incidence of infections toward the end of the 1990s. According to Bedford et al. (2019), one of these side effects has been the demographic transition, a process by which populations in many countries have shifted from rural to urban areas, often resulting in higher population densities and cohabitation between humans and animals. In 2007, for the first time in human history, the number of people living in urban areas surpassed the number of people living in rural areas. This trend is expected to continue in the coming decades, with the majority of the global population projected to live in urban areas by 2050. Several consequential side effects have emerged alongside significant global changes:

- Intensified agricultural production has transformed human-animal interactions in various habitats.
- Global interconnectivity, via enhanced communication, tourism, and trade, has surged. With affordable and widespread travel, combined with advancements in logistics technology reducing freight costs, there's a heightened human interaction, thereby amplifying disease transmission possibilities.
- The escalating impact of climate change on global ecosystems influences the habitats and migratory patterns of disease vectors like birds, mosquitoes, and ticks. Rising temperatures can extend the range of these vectors, introducing diseases to previously unaffected areas. Moreover, altered rainfall patterns can intensify disease transmission, especially during extreme weather.
- These shifts have augmented the health needs of global populations. However, many regions, especially those with fragile health infrastructures, struggle to address these needs, widening health disparities. This has bred growing skepticism towards national organizations and institutions.
- In recent times, numerous conflicts have arisen in impoverished and vulnerable regions, leading to significant migrations and compromised basic health amenities. The ramifications of such conflicts persist, detrimentally impacting the health and stability of affected communities.

The final result of this complex series of phenomena is that the problem of communicable (infectious) diseases has re-emerged, generating epidemics that have become (and threaten to be) more and more frequent, complex, and difficult to prevent and contain.

The traumatic experience of COVID-19 has shown that human health cannot be considered in isolation from the conditions of living organisms and the environment. This in turn suggests a concept of 'health' that applies to the entire planet as in the Gaia hypothesis (Lovelock, 1972). One of the major problems concerning the management of health concerns the circumstance that health care has hitherto been exercised by targeting only a small range of pathological states and that this has been done essentially through centralized procedures, with both the market and the environment conceived as passive or non-proactive recipients of public health measures. Epidemics caused by viruses such as Ebola, HIV, and avian influenza remind us that 'human, animal, and environmental health are interconnected and that the early response to emerging zoonotic pathogens requires a coordinated, interdisciplinary and cross-sectoral approach. As our world becomes more connected, emerging diseases are a greater threat and require coordination at local, regional, and global levels' (EClinicalMedicine, 2020). Some of these growing trends in connectedness are illustrated in Fig. 2.5. In about 50 years the number of air passengers has increased from about 300 million (1970) to more than 4.5 billion (2020), and the value of goods and services increased passed from less than a trillion dollars to more than 25 trillion dollars.

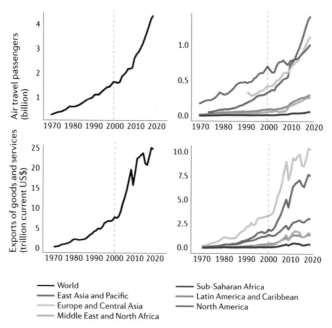

Figure 2.5 Trends in air travel, trade, and urbanization at global (left) and regional (right) level. *Source*: Adapted from Baker et al. (2022). Calculations based on World Bank data.

A further clear demonstration of the tight connections between human health and the ecosystem comes from recognizing that some relevant features of the world population that are exposed to potential spillovers are also changing. According to Baker et al. (2022), for example, 'the elimination of smallpox led to the cessation of smallpox vaccination, which may have enabled the expansion of Monkeypox. More generally, global aging populations may provide an immune landscape that is more at risk of spillover, as aging immune landscapes are less capable of containing infectious agents'. The intersection between declining immunity function at later ages and global aging populations may increase the probability of pathogen emergence, but this remains conjectural and an important area for research. The changing global context may allow existing human pathogens to evolve novel characteristics and expand in scope. Drug resistance selection

---

Box 4: The role of ecosystem in determining the severity of infectious diseases

A striking example of how ecosystems can affect the severity and death toll of infectious diseases on population is represented by what happened in 1918 with the Spanish flu. One of the recurring questions of those who studied that epidemic was to understand what caused the extraordinarily high mortality of the flu. According to Oxford et al. (2002), the most credited cause could be found in the extreme health and stress conditions on the Western Front in the Great War.

According to historians, the virus may have reached the front lines of World War I in 1916, perhaps from as far away as Kansas, and spread globally through the transportation of troops in densely packed ships. The crowded and unsanitary conditions on these ships may have allowed the virus to maintain high levels of lethality and transmissibility. Once the troops arrived at their destinations, the close proximity of humans, hens, ducks, and pigs in army camps may have further facilitated the spread of the virus. In addition, the high concentration of people exposed to the influenza virus, many of whom were in poor nutrition and some of whom had co-existing typhus infections and were under high levels of psychological stress, may have contributed to the lethality of the virus. The extreme conditions and the exhaustion of the European population due to the world conflict may have created the conditions for the virus to thrive and cause widespread illness and death.

---

now occurs throughout the world, and antibiotic resistance has and will evolve repeatedly. As with antibiotic resistance, rapid global spread is commonplace for antimalarial resistance following evolution'.

Wild animals cannot be blamed for the rise in pandemics, either now or historically. Instead, the blame rests squarely on humans who persistently strain ecosystems and the Earth's dwindling biodiversity. Activities such as the trade in wild species, deforestation, and expansion of agriculture in previously undisturbed areas alter the 'normal' circulation of viruses.

The tale of deforestation is extensive and disheartening. We have lost not just forest landscapes and terrains but also the creatures that inhabited them. According to ourwolrdindata.com, since the end of the last great ice age, 10,000 years ago, the world has lost two billion hectares, or one third of its forests. These forests were used to grow crops, raise animals, and make firewood. Most trees have been cut down in the last 300 years (1.5 billion hectares of forest). Fig. 2.6 shows that the demand for agricultural land and wood-based energy caused deforestation to increase step by step in the 20th century. From the 1920s to the 1980s, the amount of forest lost every 10 years nearly quadrupled, reaching almost 120 million hectares, or about the same size as South Africa. Most of this increase was caused by countries in Asia and Latin America following the same path as those in Europe and North America. In the 1980s, 150 million hectares were lost, half the size of India. This was mostly because people cut down forests in the Brazilian Amazon to make room for pastures and farms. Fortunately, since then, the amount of trees that have been cut has steadily decreased, even though the trends are different in different parts of the world. In the 1990s, 78 million hectares were cut down, 52 million at the beginning of the 2000s, and 47 million in the last ten years. While temperate forests surpassed the 'forest transition point' — where reforestation overtakes deforestation — in the 1990s and have consistently grown in coverage, tropical forests have yet to achieve this milestone. Some nations are still a considerable distance from reaching their peak deforestation.

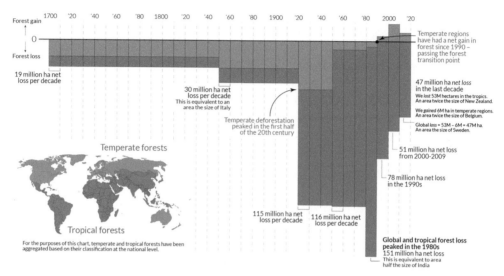

Figure 2.6 Secular trends in deforestation. *Source*: Community Emissions Data System (CEDS). Data retrieved from `Ourworldindata.org` at the following web address: https://ourworldindata.org/deforestation.

These changes can increase the probability of a contact, and hence the contagion of pathogens between wildlife and humans. In other words, as humanity moves closer to places with uncontaminated and wild nature, individuals have a better chance of getting in touch with new pathogens carried by animals. As a result, humanity is at a growing risk of pandemics such as COVID-19. Moreover, climate change, another momentous challenge brought about by human pressure on the environment, amplifies animal migration. This, in turn, heightens the likelihood of interactions with species that carry viruses (Carlson et al., 2022). The problems of pandemics, climate changes, and biodiversity loss are all linked. But so are the solutions.

As Grossi et al. (2020) rightly reported, the synergistic balance of all existing animal species has been broken, exceeding the ecological threshold. Climate change receives the greatest attention, but many other planetary transformations are evident, including interruptions in the nitrogen cycle, ecosystems, biodiversity, and land use (Butler and Soskolne, 2013). The rapid increase in the human population in recent years has led to the expansion of human settlements and the development of intensive agriculture and animal husbandry, which has resulted in the fragmentation of ecosystems and the loss of native biodiversity. As the global population is projected to grow to 8 billion in 2023 and 9 to 10 billion in 2050, the strain on our planet's finite resources will increase. Currently, more than half of the world's population lives in cities, and increasing numbers of younger generations are growing up with little exposure to rural ecosystems (Rockström et al., 2009).[3]

The process of urban sprawl, or the expansion of cities and towns into surrounding areas, has played a significant role in the increase in the number of people living in urban areas.[4] One of the main drivers of urban sprawl is the desire for larger, more affordable housing and the availability of land on the periphery of urban areas. As cities grow, housing prices tend to rise, pushing people to look for cheaper alternatives on the outskirts of the city. This can lead to the development of sprawling suburbs that are often less convenient and less walkable than more densely populated areas. Urban sprawl can have a range of negative consequences, including environmental degradation, increased greenhouse gas emissions from transportation, and a decline in the quality of life of residents who must rely on cars for all of their daily needs. It can also contribute to social and economic inequality, as lower-income residents may not be able to afford to live in more expensive, centrally located areas.

There are several strategies that governments and communities can use to address urban sprawl, including promoting compact, mixed-use development, investing in public transportation, and implementing policies that encourage infill development in existing urban areas. An interesting question is whether urban sprawl is likely to increase as a consequence of the COVID-19 pandemic. Some experts believe that the pandemic may contribute to a shift away from urban living and towards more suburban and rural areas, as people seek more space and less density. Indeed, the pandemic has resulted in widespread remote work and online learning. This could make it easier for individuals to reside in areas beyond major cities while still fulfilling their professional and educational commitments. Conversely, the economic challenges arising from the pandemic might hinder individuals' ability to relocate to pricier suburban or rural locations This could be driving greater urbanization as individuals find more cost-effective housing alternatives in office buildings vacated as a consequence of the diffusion of remote work.[5]

---

[3] In 1960, about 2 billion people lived in rural areas, compared to 1 billion in urban areas. In 2007, for the first time, the number of people living in urban areas was equal to those living in rural areas, around 3.33 billion each. In 2016, the urban population increased to 4 billion while the rural population only slightly increased to 3.4 billion. According to the United Nations, in 2015, 55% of the world's population lived in urban areas and this figure is projected to increase to just over 57% by 2020.

[4] Urban sprawl refers to the spread of urban development beyond a city's boundaries into surrounding areas, typically characterized by low-density, single-use development such as suburban housing subdivisions and strip malls. Urban sprawl often results in a dispersed, car-dependent form of development, with the expansion of roads and highways to accommodate the increased traffic demand.

[5] Another striking example of the rapid increase in human population in highly dense populated cities is represented by the urbanization process around the Mediterranean Sea, where one third of the population

This evidence suggests that urbanization is a new phenomenon for humanity. For most of human history, populations have lived in very low-density rural contexts, and city growth at the current scale is a unique phenomenon that has only occurred in the last two centuries. Moreover, it is predicted that by 2050 about 7 billion people (more than two-thirds of the world's population) will be living in urban areas. This is because people tend to migrate from rural to urban areas as they get richer and because, often, living standards tend to be higher in urban areas. These events have increasingly exposed humanity to a higher risk of spillovers (zoonoses), which is evident in the number of new infection outbreaks witnessed in the last 20 years. Arboviral diseases such as dengue, Zika virus disease, and chikungunya have become more common and spread due to urbanization. These diseases are spread by mosquitoes that do well in cities, such as Aedes aegypti and Aedes albopictus. The population density appears to be linked to the fact that Aedes Aegypti is attracted to human scent and has consequently evolved to bite people, which is how arboviral diseases spread. However, the role of urbanization in the spread of vector-borne diseases is not unidirectional. For example, the fact that the Anopheles vector prefers to live in rural areas may have made malaria less common in the regions that are becoming more urban. Nevertheless, the dense populations and high levels of connectivity in cities make them ideal breeding grounds for viruses and other pathogens. This is why diseases like COVID-19 and SARS can spread so quickly in cities. Cities can also help to speed up the spread of disease globally, as people travel from one city to another.

Another important driver and major source of infections is the trade in wild animals and its relative consumption.[6] Trade and consumption of wild animals are also regarded as ecological anomalies, as they promote relationships between species that would never have met and can have significant negative impacts on the environment and biodiversity, as well as on public health. The trade in wildlife has significant negative impacts on the environment and biodiversity. It can lead to the over-exploitation of certain species, which can result in population declines and even extinction. It can disrupt the balance of ecosystems by introducing non-native species or removing native species from their natural habitats.

The risk of infections in animals rises when they are held in captivity for extended durations under poor conditions, frequently malnourished and stressed. This weakens

is currently concentrated in coastal regions and urban tourism adds to a demographic growth that is especially severe in southern and eastern areas of the Mediterranean rim countries. The population of the 21 riparian countries grew from 285 million in 1970 to 427 million in 2000 and it is estimated to reach 524 million by 2025; the rate of urbanization is expected to grow from 64.3% in the year 2000 to 72.4% by 2025.

[6] Trade in wild animals, also known as the wildlife trade, involves the capture, breeding, and sale of wild animals for a variety of purposes, including as pets, for food, for medicinal use, and for use in traditional cultural practices. Wildlife trade has been identified as a major source of zoonotic diseases, which are diseases that are transmitted from animals to humans.

their immune systems, making them more prone to diseases. Added to this are the poor hygienic conditions of the markets where they are sold, where the fluids of slaughtered animals come into contact with everything that passes in those places (Zhang et al., 2008). And under these conditions, urbanization and a better transport infrastructure only exacerbate the problem.

At the same time, the demand for wildlife products has increased globally, driven by a variety of factors such as cultural and traditional practices, the use of wildlife products for medicinal purposes, and the illegal trade in endangered species. For these reasons, it is important to carefully consider the consequences of such practices and to take steps to mitigate their negative impacts. To explain why certain phenomena consistently originate in the same locations, it's helpful to note that meat consumption in China has increased by more than 30% since 2000, with this unprecedented growth beginning in 2017. At the same time, demand for wildlife products has surged worldwide. The increase in meat consumption in China and the demand for wildlife products around the world is likely related to the phenomenon, as both can be driven by a variety of factors such as economic growth, population growth, and cultural preferences.

The province of Yunnan is one clear example of such a trend. Yunnan is one of the most biodiversity rich provinces in China that witnessed massive urbanization. From 1958 to 2010, Yunnan's population grew from 19 million to 46 million, with massive urbanization disrupting local ecosystems. The deforestation fires and subsequent human settlements destroyed hundreds of thousands of acres of untouched and wild land. Homes, fruit trees, and rubber plantations replaced the tropical rain forest. But urbanization also created a series of food shortages, with many families in the mountain areas of Yunnan beginning to experience hunger for at least part of the year.

The survival of many of these families led to the practice of hunting wild animals for their own consumption or for sale, which exposed them to zoonoses. According to Jabr (2020), 'despite the laws against poaching and the establishment of numerous nature reserves, the collection and hunting of wild species is still common and often represent between 25 and 80% of the income of a rural family'. Not surprisingly, in 2015, when collecting blood samples from a population of 218 in a small village in Yunnan less than four miles from bat caves, a group of researchers found that six of these individuals had coronavirus antibodies linked to SARS-CoV-1, the virus that caused the original SARS outbreak in the early 2000s. None of the six individuals had a known history of SARS or contact with SARS patients, but all had observed bats flying through their villages, suggesting the possibility of direct infection. The isolation of these villages probably meant that there were no previously recorded SARS outbreaks.

For these reasons, many governments and organizations around the world have taken steps to regulate and reduce wildlife trade, through measures such as banning the trade of certain species, establishing protected areas, and implementing conservation and management plans. One of the main ways to reduce the risk of zoonotic disease outbreaks is to implement measures to prevent the transmission of diseases from animals to humans. This can include measures such as regulating wildlife trade, improving the conditions under which animals are kept, and promoting hygienic practices in settings where animals and humans come into close contact. It is also important to invest in research and development to better understand zoonotic diseases and how they are transmitted, as well as to develop vaccines and treatments that can help to contain and control outbreaks when they do occur.

It is evident that eradicating zoonoses is unattainable; however, their containment might be achievable. Zoonotic diseases have always been a part of human history and will continue to be a part of our future. In the next few years, more and more people on Earth will have to share and shed increasingly cramped spaces, making the world a more complex system of connections between the many living creatures (including microorganisms). It may be unimaginable to live in sterile, sealed bubbles that can protect humanity from contagions; however, we can certainly decrease the likelihood of such infections and lessen their impact when they do arise.

This implies implementing measures to prevent the transmission of diseases from animals to humans, including measures such as regulating the wildlife trade, improving the conditions in which animals are kept, and promoting hygienic practices in settings where animals and humans come into close contact. It is also important to invest in research and development to better understand zoonotic diseases and how they are transmitted, as well as to develop vaccines and treatments that can help contain and control outbreaks when they occur. For example, in 2003, after SARS, China had resorted to measures banning the trade in wild animals. Six months later, the ban was lifted, allowing breeding and reproduction facilities to resume operations. On 24 February 2020 China banned the hunting, trading, and transportation of wild terrestrial animals for consumption (except wild animals for fur, leather, and traditional medicine). It remains uncertain whether the ban will persist; however, it seems that the COVID-19 situation has indeed made an impact. There are many signs in China that seem to point to a drastic reduction in the consumption of wild animals. But there is always the risk that what used to happen in the sunlight will now be done illegally and covertly.

However, simply banning the trade and consumption of wild animals is unlikely to be an effective solution to the problem of zoonotic diseases, as it does not address the underlying root causes. According to the FAO, in 2017, more than 150 million families worldwide hunted wild animals, mainly for personal consumption and economic livelihood. These are supplemented by families that consume meat from wild animals as status symbols (especially in urban China). In these cases, total bans are never the correct solution, since they tend to negate the problem without recognizing its roots. In the first place, a ban on wild meat consumption can be made more effective by introducing adequate regulations to ensure better market hygiene and an absolute prohibition on the marketing of wild animals at higher zoonotic risk (bats, rodents, primates, etc.). Second, individual incentives can be provided in the form of alternative sources of livelihood (incomes) to the families that live off these activities today. It's important to remember that the core issue is not necessarily those who hunt wild animals for their livelihood, but rather those who engage in the trade and global transportation of wild animals. Further in this book, we will demonstrate that over the past two decades, international trade and globalization have surged remarkably, contributing significantly to the rise in zoonotic diseases and epidemics. Curtailing these activities, when they are related to produce from wild animals could substantially diminish the occurrence of zoonoses.

All of the above leads to the conclusion that long-term ecological phenomena, such as zoonotic infections, are unlikely to be a pure accidental product of evolution or human progress. This conclusion is further supported by the reasons why spillovers exist and why they have increased in recent years. As recently written by Jabr (2020), 'zoonotic pathogens generally neither seek us out nor stumble upon us by coincidence. When diseases move from animals to humans and vice versa, it is usually because we have reconfigured our shared ecosystems in ways that make the transition much more likely'. Many human activities have contributed to this trend (deforestation, mining, intensive agriculture, and urban expansion), substantially contributing to the destruction of natural habitats and forcing wildlife to 'meet' with human communities.

In addition to regulatory measures, it is important to address the underlying drivers of the trade and consumption of wild animals, such as cultural practices and economic incentives. This could involve educational campaigns to raise awareness about the risks of consuming wild animals, as well as initiatives to support alternative livelihoods for people who currently rely on the trade and consumption of wild animals for their economic survival. In general, a comprehensive approach should be adopted to address the various factors that contribute to the problem of zoonotic diseases and that involve the participation and participation of all relevant stakeholders, including governments, civil society organizations, and local communities.

2.7. We live in a non-linear world!

As people move to once uninhabited areas, there are more opportunities for people and animals to meet. Population growth, the spread of agriculture, rising wealth, and larger property sizes are driving forces behind these interactions and the destruction of habitats they cause. This can happen when people do things that make it more likely for diseases to spread, such as eating wild meat or having more contact between wild and domestic animals. For example, the Nipah virus, a zoonotic virus that is transmitted from animals to humans and can cause severe illness and death, was first identified during an outbreak in Malaysia in 1998-1999, which was primarily among pig farmers. The outbreak was caused by the transmission of the virus from bats, specifically flying foxes, to pigs. Flying foxes are a type of bat that are found in many parts of the world, including Southeast Asia. They are known to be a natural host for the Nipah virus, which they can carry without showing any symptoms. In the late 1990s, the expansion of pig farms into areas inhabited by flying foxes in Malaysia and the intensification of pig farming practices led to a large number of pigs coming into contact with infected bats. The Nipah virus then spread from bats to pigs, leading to a severe outbreak among pig farmers.

The outbreak of Nipah virus in Malaysia is an example of a notable public health incident, after a zoonosis infection, resulting in more than 100 deaths and the culling of more than one million pigs. It also had significant economic consequences, with losses estimated at more than $550 million. The outbreak highlighted the importance of understanding the complex factors that can contribute to outbreaks of zoonotic diseases and the need to implement measures to reduce the risk of transmission. The Nipah virus is believed to have spread from bats to pigs due to three global changes: the spread of pig farms into bat habitats, the intensification of pig farming, which led to a high number of hosts, and international trade, which spread the infection to other pig populations in Malaysia and Singapore.

Increases in population density and urbanization, as well as expansion of agriculture, contribute to the risk of outbreaks of zoonotic diseases. As more people and animals come into close contact with each other, the opportunity for disease transmission increases. Intensive farming practices, such as high-density animal farming, can also increase the risk of zoonotic disease outbreaks by providing a large number of hosts for pathogens to multiply and spread. The use of antibiotics in domestic animals can also contribute to the emergence of antibiotic-resistant bacteria, which can be difficult to treat and can spread to humans through a variety of means. The global demand for meat and the resulting intensification of meat production are also likely to contribute to these processes, as they can lead to the expansion of animal farming into new areas and to an increase in the number of animals raised for meat.

Changes are also happening in the types of people that might be affected by spillover. For example, when smallpox was eradicated, people stopped being vaccinated against it. This may have helped monkeypox[7] spread. Changes in the characteristics of the population can affect the risk of outbreaks of zoonotic diseases. For example, as populations age, their immune systems may become less able to fight infections, making them more susceptible to disease. This could make it more likely that new pathogens will emerge and spread.

The evolution of drug resistance is another factor that can increase the risk of zoonotic disease outbreaks. The extensive utilization of antibiotics and other antimicrobial medications can promote the emergence of bacteria resistant to antibiotics. These

---

[7] Monkeypox is a viral disease that is similar to smallpox, though generally milder.

resistant bacteria can be challenging to manage and can spread to humans through various methods. Similarly, the evolution of resistance to antimalarial drugs can make it more difficult to control the spread of malaria and other infectious diseases.

The risk of pathogen spillover may also be affected by climate change. Changes in the environment can alter the range and number of species, leading to new interactions between species and increasing the risk of zoonotic diseases. The spread of pulmonary hantavirus in 1993 is believed to have been caused by a combination of environmental factors, such as a prolonged drought followed by a lot of rain. In the same way, there is evidence that Australia's black flying fox population, a vital source of the Hendra virus, has moved 100 km south in the last 100 years because of climate changes. This change in range likely caused the Hendra virus to spread to horses in the southern areas of the country, and from horses to people. Alterations are likely occurring in bat populations globally, though they remain largely unexamined. This oversight is of concern, given that bats play a crucial role as reservoir hosts for numerous lethal pathogens.

Finally, as we can see from Fig. 2.7, in recent years, increases in air travel, trade, and urbanization at global (left) and regional (right) scales have accelerated, indicating ever more frequent transport of people and goods between growing urban areas. Increasing trends have been recorded also for several air pollutants as can be seen in Fig. 2.8. The growth rate for many of these pollutants saw a marked shift after World War II, with exponential increases beginning in the 1960s. Another notable trend observed recently is in migration patterns. As depicted in Fig. 2.9, the number of migrants has nearly doubled over 30 years, rising from 150 million in 1990 to approximately 300 million in 2020. Generally, it's essential to understand that these factors intricately influence the likelihood of zoonotic disease outbreaks. Given this complexity, it's crucial to implement a range of strategies to decrease transmission risk and lessen the effects when outbreaks happen.

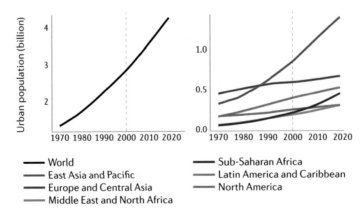

Figure 2.7 Trends in air travel, trade, and urbanization at global (left) and regional (right) level. *Source*: Adapted from Baker et al. (2022). Calculations based on World Bank data.

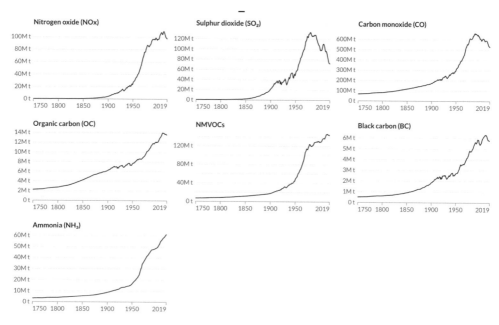

Figure 2.8 Secular trends of air pollutant emissions. *Source*: Community Emissions Data System (CEDS). Data retrieved from Ourworldindata.org.

2.8. The economic and social costs of epidemics

One aspect that should not be overlooked in the analysis of epidemics is the economic impact they can generate. Studies indicate that epidemics impact economies in various sectors, including health, transport, agriculture, and tourism.

Fan et al. (2018) calculated that a severe flu pandemic, akin to the 1918 event, could result in economic losses up to USD 500 billion annually. These losses are attributed to increased mortality rates, reduced workforce size and productivity, heightened absenteeism, and the expenses associated with lockdowns or social distancing measures. These costs would not be equally shared across the population, but would affect the most disadvantaged groups. The same authors calculate that annual losses would vary between countries, with Gross Domestic Product (GDP) of lower-middle-income countries more affected (−1.6%) than high-income countries (−0.3%). In 2019, these figures were revised upwards by a joint report of the WHO and the World Bank Global Preparedness Monitoring Board, which estimated the total cost between 2.2% and 4.8% of the approximately $3 trillion global gross domestic product (Global Preparedness Monitoring Board, 2019b). One of the most concerning factors during a pandemic outbreak is that the majority of the burden falls upon the most vulnerable population groups within countries. People in these groups would therefore risk suffering disproportion-

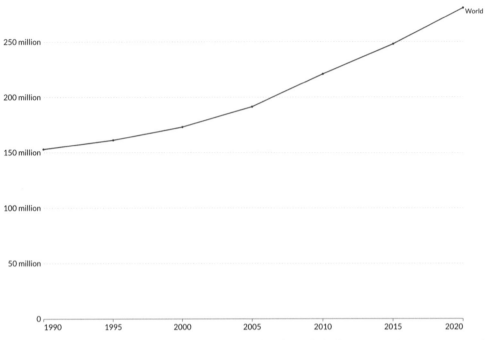

Figure 2.9 Total number of emigrants (people that have left the country). *Source*: United Nations Department of Economic and Social Affairs (UNDESA). Data retrieved from Ourworldindata.org.

ately by often lacking access to healthcare services and having fewer savings to safeguard against a loss of income.

Much of the existing evidence stems from estimates generated by econometric models, rather than direct observations. However, these estimates can still be valuable in understanding the potential consequences of future epidemics and informing policy decisions to mitigate their impact. Past epidemics have caused a substantial reduction in the growth of the private sector, threatened food security due to the decline in agricultural production, and burdened international trade with restrictions on movement, goods, and services (Global Preparedness Monitoring Board, 2019a).

The experience of past epidemics has also shown that economic losses can far exceed prevention costs. For example, according to Barrett (2013), eradication of smallpox is a major public health success, as it has led to significant global benefits with a benefits/costs ratio of 159:1. This means that on average, for every $1 invested in smallpox eradication, there has been a return of $159 in benefits. Yet, it's crucial to acknowledge that the expenses and advantages of eradicating smallpox can differ based on the particular context and period in question. Additionally, the experience of the COVID-19 pandemic has highlighted that the impacts of epidemics can be higher in today's world due to increased travel, trade, and environmental changes. These factors can increase the frequency and severity of outbreaks, leading to more significant economic slowdowns (Pike et al., 2014).

Costs from epidemics are only partially attributable to the treatment and control of the disease itself. Many of the costs will come from over-reaction at all levels (e.g., clogged hospitals, assaulted pharmacies, tilt transportation, etc.). A noteworthy example is the aftermath of the Ebola outbreak in the United States. Although there was only one diagnosed case of Ebola in the country, it triggered a response that led to the allocation of $5.4 billion in federal funding. Of this amount, $1.1 billion was spent on the 'domestic' response, including funds for public health studies, increased laboratory analysis capacity, infrastructure preparedness, and organization of epidemic response (Epstein et al., 2019). Only $119 million was spent on domestic quarantine activities, including screening at five major US airports, medical visits, surveys of potentially ill travelers, and follow-up (Centers for Disease Control and Prevention, 2015).

For the 2003 SARS outbreak, which lasted less than one year, total costs were estimated at between $40 and $54 billion (Jonas, 2014; McKibbin, 2004). According to a World Bank study, the four most affected economies in East Asia (China, Hong Kong, Taiwan, and Singapore) recorded GDP losses of approximately $13 billion (Brahmbhatt and Dutta, 2008). Likewise, Guinea, Liberia, and Sierra Leone experienced a sharp slowdown in their already anemic economic growth. Following an increase in GDP in 2013 in all three countries, Guinea's GDP growth fell from 4% to 0.1% in 2014 and Liberia from 8.7% to 0.7% in 2014, while Sierra Leone's GDP growth fell from 4.6% in 2014 to −21.5% in 2015 (World Bank, 2016a). A significant portion of these economic setbacks might have been prevented had there been a more robust healthcare system in place, designed for outbreak prevention and detection (World Bank, 2015).

To better appreciate the scale of the economic impact of epidemics, consider that between 2000 and 2010, zoonotic diseases caused indirect economic losses estimated at $200 billion (World Bank, 2010). Furthermore, more than 2.5 billion cases and 2.7 million deaths can be attributed to the first 56 zoonoses (Gebreyes et al., 2014). A moderately severe influenza pandemic would result in 700,000 deaths annually and cost around $570 billion (due to loss of income and mortality) worldwide (Fan et al., 2016). The Zika virus, transmitted by Aedes mosquitoes, emerged in Brazil in 2014 and spread rapidly to 26 neighboring countries, infecting up to 1 million people (Petersen et al., 2016). Zika's estimated cost to Latin America and the Caribbean was $3.5 billion in 2016 (World Bank, 2016b). An interesting aspect of the Zika epidemic is that the virus had been eradicated from 18 countries in 1962 and returned due to poor management of public hygiene (Whitman, 2016).

The costs of epidemics are thus amplified by mechanisms for the transmission of effects across economic sectors. For example, during the 2014 Ebola outbreak, high demand for personal protective equipment (PPE) was met with reduced stocks and capacity, and producers struggled to meet the needs. This led many African countries to face the emergency with limited means, leading to additional costs in terms of infections and deaths (UNICEF, 2014). In addition, local political and social conflicts can make epidemics even more expensive. In Nigeria, where the country was declared polio-free in 2015, the conflict with Boko Haram prevented the establishment of adequate surveillance programs. As a result, in 2016, four types of polio viruses were identified, which led to a massive campaign to vaccinate millions of children in West and Central Africa, whose costs have yet to be estimated (Beaubien, 2016).

CHAPTER 2 - Take-home messages
- The essential factor in distinguishing a pandemic from epidemics or endemic diseases is the level of manifestation of the disease. When it is global, the WHO alert is triggered, as happened with COVID-19.
- There are several channels of spread of an epidemic but in all cases there is a common element: patient zero, the first infected human able to start a chain of infections, thus spreading the virus.
- In the 21st century, something has changed in the world: the number and frequency of epidemics have increased. There is a reason and it depends on human beings. The growing pressure on ecosystems brings man into contact with previously unknown pathogens.
- Current and global challenges are hyper-connected. Climate change, access to resources, deforestation, and the spread of infectious agents are related problems, but so are the possible solutions.

PART 2

The COVID-19 crisis management: not an easy task!

CHAPTER 3

How did we manage the COVID-19 pandemic?

Contents

3.1. A contemporary challenge: the political governance and the management of an epidemic

Governance refers to the systems, processes, and structures that dictate how an organization is directed, controlled, and held accountable. It involves setting the organization's strategic goals, establishing policies and plans designed to achieve those goals, and monitoring the progress and performance of the organization. On the other hand, management refers to the process of getting things done. It involves the organization's day-to-day operations, including planning, organizing, coordinating, and controlling resources. Management focuses on implementing the policies and plans established by the governance structure and ensuring that the organization runs efficiently and effectively.

Outbreaks and attacks of bio-terrorism are distinguished from all other natural events or man-made disasters. Unlike bombings, fires, earthquakes, or hurricanes, in which events are limited and well-defined, epidemics can develop slowly and insidiously, and very often, initial traces are lost. When a health crisis disrupts society's normal functioning and seriously affects local communities' responsiveness, local political leaders face both a governance and a management challenge. On the one hand, they must decide on an overall strategy that can include a state of emergency and a request for assistance. On the other hand, they must put in motion a machinery to deal with the daily consequences of the crisis. The challenge in these cases is to make the entire chain of command, composed of local, provincial, regional, and state officials and experts, work as actors who can follow a script perfectly, interfacing with each other while working with other groups, bodies, and institutions outside the chain of command. In these circumstances, the heightened severity of the crisis further escalates the complexity of the response, increasing the risk of inefficiency and confusion.

Evidence and logical reasoning suggest that in addressing a national emergency, the distinction between catastrophe and achievement hinges predominantly on: *i)* having well-established and tested contingency plans, and *ii)* the leadership acumen and expertise of those making decisions. These two factors cannot be disjoined: a good leader will do little without a good plan, and a good plan will do little without someone capable of executing it.

However, it is important to understand the difference between emergency and contingency plans, since this partly reflects the difference between governance and management. A contingency plan is a plan that outlines the steps that should be taken in response to a specific potential event or situation. It is a proactive measure that is put in place in advance, to be prepared for and effectively manage a particular situation if it occurs. Contingency plans are typically developed for various potential events, including natural disasters, technological failures, and epidemics. They are contingent on alternative scenarios and different events within these scenarios. They outline hypothetical actions and assign responsibilities based on hypothetical roles of different institutions.

Nevertheless, an emergency plan can be based on an existing contingent plan, or it may be put together on the basis solely of current information and put into action in the event of an emergency. An emergency is a situation that requires immediate action to prevent or minimize harm to people or property damage. Emergencies can be natural disasters, such as epidemics, or man-made, such as fires or chemical spills. Emergency plans outline the specific steps that should be taken in response to a particular emergency. These plans are generally developed by government agencies and organizations to ensure prompt and effective reactions during emergencies.

In summary, a contingency plan is a proactive measure that outlines the steps to be taken in response to a potential event, while an emergency plan is a reactive measure that is put into action in the event of an actual emergency. Both types of plan are important for effectively managing and responding to different types of situation, and emerging plans can be more effective if they can be prepared and implemented within the framework of existing contingency plans.

Drawing from the lessons of numerous historical emergency experiences, it is also evident that beyond comprehensive emergency and contingency plans, effective crisis management demands robust leadership and transparent communication. This means that a clear strategy must be established with wide participation and consensus on the part of the affected population. At the same time, it is equally important that those in charge of the emergency response have the expertise and experience to make sound decisions and effectively coordinate the efforts of those involved. This includes ensuring a clear chain of command and that personnel know their roles and responsibilities. It is also crucial that there is a system in place to relay information up and down the chain of command so that everyone is kept informed and can carry out their tasks effectively. Additionally, it is important to ensure that all personnel feel motivated and supported during the crisis, as this can help minimize errors and ensure that the response is as effective as possible.

The COVID-19 experience, reaffirming the lessons from many previous crises, highlights that in national emergencies, both the notion of leadership and the configuration of the command chain can evolve as the situation progresses. In general, in the early stages of an epidemic crisis, the leaders of the chain of command are local political leaders and technicians, as these events typically begin in very narrow settings (for example, in hospital emergency rooms or doctors' offices). Leadership problems can mount as the crisis gets worse, as not all local political representatives can be up to the challenge. The best of them can establish good working relationships with public health technicians and physicians, limiting the development of the crisis. In the face of major public health crises, more active political participation is required at the central level, which tends to change both the leadership and the structure of the chain of command. In general, political leaders and their public health advisors find themselves at the center of decision making as the crisis worsens. At that juncture, supported by technicians, the top political leaders define scenarios, identify resources, and establish action plans.

Both evidence and rational thinking indicate that robust emergency and contingency plans are crucial for effective crisis management. These plans should be based on past experiences and updated regularly to take into account new developments and best practices. Plans should outline the steps that must be taken in the event of a crisis, including who is responsible for each task and the timeline to complete those tasks. It is also important that those in charge of the emergency response are able to adapt the plans as needed in case the actual crisis differs from what was anticipated in the plan. This may require a certain level of flexibility and creativity from those in charge as they work to effectively address the unique challenges presented by each crisis.

During ordinary times, developing emergency planning capacity is an activity often neglected by modern organizations, as it appears to address unlikely scenarios that may never materialize. But having well-developed emergency and contingency plans in place and being able to quickly adapt them to the circumstances allows an organization to respond effectively to an emergency and minimize the negative impact on those affected. Planning capacity integrates governance and management abilities. It involves not only having a series of plans in place, but also regularly reviewing and updating them to ensure that they remain relevant and effective. This includes identifying potential weaknesses in the different plans and taking steps to address them, as well as conducting drills and exercises to test the plans and identify any areas that need improvement. Planning capacity is particularly important in the fields of health and civil protection, as these sectors play a crucial role in helping to protect and preserve the lives and well-being of individuals during a crisis.

3.2. The political management of an epidemic: a mix of leadership and expertise

During the early stages of an outbreak, it is common for people to seek medical attention for non-specific symptoms such as fever, cough, and chills. These symptoms can be caused by a variety of different illnesses, and it can be difficult for healthcare providers to accurately diagnose the specific cause without more specific symptoms. This can lead to individuals being sent home with temporary therapies to alleviate their symptoms and being told to return if the symptoms worsen. In fact, this was the case during the 2001 anthrax attacks in the United States. Initially, healthcare providers were unable to accurately diagnose the disease in patients who had gone to the hospital with skin lesions from anthrax. It was only several days after the attack, when a doctor in Florida was able to diagnose the first case of inhaled anthrax, that it became clear that a terrorist attack had occurred. Diagnostic delay refers to the time it takes for a healthcare provider to accurately diagnose the cause of a disease or condition. In the case of the 2001 anthrax attacks, the diagnostic delay was in part due to the initial cases of anthrax infection presented with skin lesions, which further complicated the diagnosis, as it was not immediately clear that the patients were suffering from an inhaled form of the disease.

Drawing from historical lessons, it's evident that diagnostic lags are probable in epidemic situations. Those tasked with handling such crises should anticipate potential delays in detecting and reporting outbreaks. This lag could be exacerbated by the time required to determine the most appropriate response. Furthermore, it's imperative for decision-makers to recognize that the success of interventions largely hinges on a solid foundation in health, medicine, and science, as well as the collaboration between these sectors and policymakers.

The COVID-19 pandemic has highlighted the importance of integrating good governance and effective crisis management with a key role played by leadership, communication, and contingency plans in mitigating the impact of a crisis. At the national and local levels, political leaders had to make difficult decisions about the strategy to embrace to fight the pandemic and how to respond with immediate and concrete measures, including implementing lockdowns, closing schools and businesses, and enforcing social distancing measures. In some cases, these decisions have been met with resistance, and it has been important for political leaders to communicate clearly with their communities and provide information on why these measures were necessary.

At the same time, in several cases, healthcare providers have faced great challenges due to the lack of emergency plans explaining how to treat patients and how effectively protect the health of their communities. This lack has included plans to ensure sufficient medical supplies, such as personal protective equipment, adequate hospital beds, and other resources to handle the influx of patients.

In many cases, political leaders have struggled to respond effectively to the COVID-19 pandemic, for lack of governance, planning capacity, strong leadership, clear communication, or effective crisis management. This has resulted in delays, inadequate actions, serious consequences, and a higher number of infections and deaths. For example, in some cases, political leaders have been slow to implement measures such as lockdowns or social distancing or have been reluctant to enforce these measures, with the consequent higher spread of the virus. In other cases, political leaders have struggled to communicate clearly with their communities about the importance of these measures, which led to confusion and resistance.

Additionally, some political leaders have faced challenges in securing sufficient medical supplies, such as personal protective equipment and vaccines, or ensuring enough hospital beds and other resources to handle the influx of patients. This has resulted in inadequate care for those infected with the virus, which may have led to higher mortality rates.

Even with strong leadership and planning capabilities, the chosen strategies for addressing a crisis and managing the daily response may prove to be inadequate or ineffective. This can be due to a variety of factors, including an incomplete understanding of the crisis, the evolution of the crisis in ways that were not anticipated, or the emergence of unforeseen challenges.

More generally, in a modern society the organization problem is influenced by the political system and its basic choice between a hierarchy and a polyarchy model (Sah and Stiglitz, 1984). A hierarchy is a form of organization favored by authoritarian societies, in which power and authority are distributed among multiple levels, with higher levels having more power and authority than lower levels. In a hierarchy, decisions are typically made at the top levels and then transmitted down through the various levels of the organization. This can be an effective structure for quickly making decisions and implementing them in a coordinated manner, but it can also be inflexible and slow to adapt to changing circumstances.

On the other hand, a polyarchy is a system in which power is decentralized and decision-making is more dispersed. This is a form of organization rooted in open economies and contemporary democracies, that can swiftly adapt to evolving situations. However, it might also present challenges in efficiency and coordination of actions.

In the context of a pandemic, hierarchies and polyarchies can effectively react to the crisis, depending on the specific circumstances. A hierarchy may be more effective in quickly implementing a coordinated response, whereas a polyarchy may be more effective in quickly adapting to changing circumstances. Moreover, polyarchies can be associated with better governance systems, since they depend on consensus and decentralized behavior. Ultimately, the most effective approach will depend on the specific situation and the resources and capabilities of the organization (Sah and Stiglitz, 1985).

An example of human fallibility in the case of a hierarchic system is China's response to the COVID-19 pandemic. While China's leadership implemented some effective measures, such as locking down Wuhan and other affected areas and building new hospitals in weeks, the initial response to the outbreak was also marked by missteps and mistakes. For example, there was a delay in releasing information about the virus to the public, which allowed it to spread further before containment measures could be implemented. Some of the strategies implemented, such as quarantining entire apartment buildings, were criticized as overly harsh and ineffective in preventing the spread of the virus. Lockdowns and other social restrictions were further extended to entire cities and regions and remained in place for many months, and there was a poor focus on vaccination for older individuals at increased risk of severe illness from COVID-19. Eventually, the Chinese authorities had to recognize that the extremely restrictive measures taken were not sustainable and suddenly withdrew them. It is not yet clear what the consequences of this were and will be on the spread and severity of the diffusion of the virus.

The COVID-19 experience also suggests that multiple misunderstandings are likely to arise between political leaders and public health experts at times of health emergencies, and the effectiveness of crisis responses such as epidemics and bio-terrorist attacks can be put at risk. Public health experts need political support for an effective response and politicians need public health support. Historical experience underscores the fact that all crises, especially if they disrupt the normal functioning of society, require governance and not only competent management. Thus, they are all inherently political. The COVID-19 pandemic is no exception.

3.2.1 Control chain models and limits

Substantial scientific evidence exists on emergency management from international studies (Bakewell, 2000; Lee et al., 2011; Lettieri et al., 2009; Saldana, 2009). Much of this literature originates in North America, dealing with emergency preparedness and response, and is based on specific risk and hazard analyses. These analyses encompass countermeasures to terrorism, flood, and earthquake warning systems, or the application of technologies such as geo-referencing systems. On the contrary, there is little or nothing on governance and organizational issues about how to set up a command system that can provide answers in case of accidents and crises. There is also little about disaster learning or performance management, including measuring the effectiveness of responses. Unfortunately, there are gaps in understanding how people face risk and behave in emergencies, and effective tools and strategies have not yet been identified to strengthen community resilience.

Emergency management systems vary globally, and their effectiveness and compatibility remain uncertain. Moreover, a universally accepted method to gauge these systems' performance is lacking. However, what is clear is that numerous cultural, social, and political nuances influence their structure, operation, and efficacy.

According to Kahn (2009), it is possible to identify two 'general' models of leadership in managing a crisis that can easily be traced to two situations that have occurred in the past few years in the United States. The first is the so-called 'Giuliani model', in which the political leader takes decisions with the support of experts; the second is the 'Glendening model', in which the leader of public health makes decisions with the support of the political leader. In the former case, New York's Commissioner of Health, Neal Cohen, during his term of office (January 1998 – January 2002), had preferred to retain the role of advisor to Mayor Rudolph Giuliani. Instead, Dr. Georges Benjamin, Maryland's Secretary of Health from 1999 to 2002, had agreed to take over operations leadership with Governor Parris Glendening's political backing.

Regardless of the type of model adopted, as shown by several examples cited by Kahn (2009), the public health leaders who responded most effectively to crises were those who were able to combine good governance and effective management by working in excellent and well-organized public health and prevention structures and with good working relationships with political leaders. A negative example is the anthrax crisis in New Jersey, where Eddy Di Ferdinando (Acting Commissioner of Public Health in New Jersey) encountered many difficulties in handling the crisis. This was because the State of New Jersey did not have a well-integrated public-private health system, and Di Ferdinando was himself in the unstable and poorly legitimate position of Interim Commissioner. Its role did not allow him to develop a strong relationship with the governor before the crisis, with the result that he made decisions without political support. In that case, Di Ferdinando was even forced to recommend that postal workers obtain antibiotics from their private doctors after CDC officials refused to provide medicines from the national strategic reserve.

3.3. Were we prepared for a crisis of such magnitude?

According to the WHO, a good definition of 'pandemic preparedness' is 'having national response plans, resources, and the capacity to support operations in the event of a pandemic'. As such, preparedness includes several measures ranging from prevention to detection, and from containment to mitigation. For example, they require strong and resilient health systems, with an efficient and pervasive primary care organization capable of facilitating the detection of disease outbreaks, providing essential care, and eventually supporting the deployment of vaccines and other medical countermeasures. They also require surveillance systems and laboratory capacity to detect human and zoonotic outbreaks. Programs must be organized in a way that is capable of responding to problems related to pandemic spread, such as personal protective equipment (PPE) shortages, limited hospital capacity, and acquisition of vaccines and other countermeasures. Since viruses do not respect borders, there are also important cross-national dimensions of preparedness, with regional and sub-regional institutions playing key roles in areas such as regulatory harmonization, standards for reporting and information sharing on disease outbreaks, sharing of key public health assets such as high-complexity laboratories, and pooled procurement. For this reason, preparedness requires the existence of legal frameworks and regulatory instruments to support outbreak prevention and the deployment of countermeasures. All these programs and measures cost money and require effort: for these reasons, low-, middle-, and high-income countries have often sidelined them (Uribe et al., 2021).

Both the OECD (Allain-Dupré et al., 2020) and the Independent Panel for Pandemic Preparedness and Response (PPPR) to prevent future pandemics (Singh et al., 2021) have claimed that at the beginning of the pandemic the level of preparedness throughout the world was very low. In particular, according to Allain-Dupré et al. (2020) 'Most countries, regions, and cities were not well prepared for this pandemic for several reasons: *i)* they underestimated the risk when the outbreak emerged; *ii)* many did not have crisis management plans for pandemics (with the exception of Asian countries that had fought the SARS pandemic, and some others, such as the Nordic countries, where crisis management plans are required by law); *iii)* most countries lacked basic essential equipment, such as masks and ventilators; and *iv)* they had absorbed reduced public expenditure and investment in health care/hospitals. Since the start of the 'Great Recession' launched by the 2008 financial crisis, up until 2018, the number of hospital beds per capita had been reduced in almost all OECD countries, decreasing 0.7% per year, on average'.

It is worthwhile to report extensively the conclusions reached by the Independent Panel for Pandemic Preparedness and Response (IPPR),[1] Singh et al. (2021):

'The panel found critical failings at each step of the COVID-19 response, from preparedness to detection and alert and in both the early and the sustained response. Indicators of how prepared the countries were for pandemics proved inaccurate, having given insufficient weight to the effect of inequities, political leadership, and trust (Baum et al., 2021). International alert systems, including the mechanisms of International Health Regulations (IHR), were not sufficiently rapid (Aguilera et al., 2021). Despite this declaration, many countries were slow to implement comprehensive and coordinated measures to contain and stop the spread of the virus. This delay, along with other factors such as inadequate preparation and response, led to the rapid global spread of the virus and the development of a social and economic crisis. The pandemic has also exacerbated existing in-

---

[1] The IPPR was established by the WHO in May 2020 to evaluate the global response to the COVID-19 pandemic.

equities, with some countries and populations experiencing a disproportionate impact and lack of access to vaccines and other essential resources.

Widespread unpreparedness was at the root of the time lost as SARS-CoV-2 spread. The WHO director general declared the highest alert possible, a public health emergency of international concern (PHEIC), on 30 January 2020 on the recommendation of the second meeting of the IHR emergency committee convened to consider the outbreak. This was a whole month after the spread had first drawn the attention of WHO (Ghebreyesus, 2021). Subsequently, only a few countries have put in place comprehensive and coordinated measures to contain and stop the spread of the virus. Both global and national responses failed to prevent a broad and deep social and economic crisis, with inequities amplified, as evidenced by the maldistribution of vaccines and other essential tools (Sirleaf and Clark, 2021).

Although multiple earlier reports had warned of the dire threat posed by pandemics before COVID-19 emerged, the PPPR concluded that the potential to cause systemic collapse was not appreciated globally (Global Preparedness Monitoring Board, 2019c; Highlevel Panel on the Global Response to Health Crises, 2019). Pandemics and other severe health threats were essentially not on the agendas of heads of state or government in the same way as they continuously appraise threats of war, terrorism, nuclear disaster, and global economic instability. Instead, pandemic preparedness and response were siloed mainly within the health sector. Despite occasional high-level health security discussions and scenario exercises, there was little integration of pandemic threats into government planning, except in countries with recent experience of the health, social, and economic effects of epidemics.

A complex and fragmented international health system resulted in delays and incoherence in response strategies. The IHR proved to be a conservative rather than a proactive instrument of international law. The declaration of a public health emergency was not as fast as it should have been and the response was slow as countries had no specific obligations to act. Financial resources to respond rapidly to the emerging pandemic were not available. The World Bank Pandemic Emergency Financing Facility, established to provide surge funding in 2017, had exhausted the funds available through its cash window for the response to Ebola. Its insurance-based financing proved to be excessively cumbersome and was not triggered until three months after the international public health emergency was declared. (Radin and Eleftheriades, 2021).

Platforms to support the rapid development of therapeutics, diagnostics, and vaccines did not exist and had to be established during the crisis. The most successful element, vaccine development, was built on existing structures, including the Coalition for Epidemic Preparedness Innovations (CEPI). Despite the effectiveness of vaccine development, allocation strategies were not determined in advance and product development pathways were not aligned with global needs. Mechanisms were not available to coordinate across national, regional, and international levels and overcome national competition when the needs for supplies and the essential response workforce increased.

The PPPR's comparative analysis has shown that the most effective responses to COVID-19 were those which established multi-sector coordination reporting to the highest level of government (Haldane et al., 2021). Leadership was lacking, with many national responses dominated by short-term domestic political dynamics that militated against cooperative and solidaristic solutions and the ability to adapt to rapidly changing circumstances. Public health and economic mitigation efforts were not synchronized and mutually supportive at national and international levels, nor were there shared visions for the regulation of trade and intellectual property. The market-driven scramble for tools and supplies, including vaccines, has resulted in inequitable access'.

These unpleasant outcomes occurred even though, over the years, several actions had been taken to ensure that all countries had enough capacity to be prepared for health emergencies. WHO had been particularly active in this direction. For example, in 2010, WHO defined a set of 'capacities' (in terms of prevention, detection, and response) to monitor capacity-building efforts and comply with IHR obligations. Every country was supposed to complete self-assessments of their capacities and then self-report them to WHO. Unfortunately, most of these assessments lacked transparency and accountability and were therefore not considered representative of the true capacity for health security within countries.

Anticipating these problems, during the Ebola epidemic in West Africa in 2013-16, the WHO and the Global Health Security Agenda (GHSA) had developed a new tool based on a voluntary Joint External Evaluation (JEE) process, with the goal of monitoring IHR capacities and assessing the ability of a nation to prevent, detect, and respond to a disease of pandemic potential. Although more transparent and accountable, this new process was adopted only by a fraction of all WHO countries (about 100), with less than 70 countries later developing National Action Plans for Health Security (NAPHS). In 2018, less than half of the WHO member states met their commitments to core IHR capacity, and many lacked even rudimentary surveillance and laboratory capacity to detect outbreaks (Global Preparedness Monitoring Board, 2019c).

In 2019, an additional effort had been promoted that led to the Global Health Security (GHS) Index. This index 'includes important and relevant measures for the current pandemic that go beyond the JEE, such as rapid response to and mitigation of the spread of an epidemic, a robust health system to treat sick people and protect health workers, and adherence to norms' (Bollyky and Patrick, 2020).

Unfortunately, none of these monitoring activities worked as expected. For example, as can be seen in Fig. 3.1, if we look at the GHS index, at the end of July 2020 the nations that scored among the highest on these and other index measures of pandemic preparedness, such as the United States and the United Kingdom, were struggling in their initial COVID-19 response (Center for Health Security, NTI and Economist Intelligence Unit, 2019). On the contrary, countries such as Vietnam, which had relatively low JEE and GHS index scores, appeared to be among the most successful in immediate efforts to contain the coronavirus pandemic. Indeed, at the end of 2020, many countries with higher JEE and GHS index scores had higher death rates, even when adjusting for national differences in population age structure and in the timing of the first COVID-19 death' (Bollyky and Patrick, 2020). In May 2020, a report by the Independent Oversight and Advisory Committee for the WHO Health Emergency Program found no clear relation between JEE scores and country preparedness and response to COVID-19 (Harvey et al., 2020).

Considering the points above, three primary shortcomings in health risk management emerge as foundational reasons for the unpreparedness of many countries to the

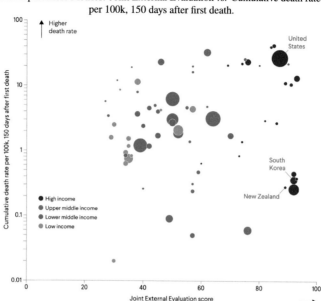

A. Preparedness Metrics: Joint External Evaluation *vs.* Cumulative death rate per 100k, 150 days after first death.

Figure 3.1 Preparedness Metrics Did Not Predict Successful Pandemic Response. Notes: Bubbles are sized by population. Joint External Evaluations (JEE) are voluntary processes developed by WHO to monitor countries' implementation of the core capacities under the International Health Regulations. The 2019 Global Health Security (GHS) Index is an assessment and benchmarking tool that measures a country's capacity to rapidly prevent, detect, and respond to the spread of an epidemic. Universal health coverage (UHC) ensures that all individuals and communities receive necessary health services without suffering financial hardship. The UHC index in this graph is a composite proxy measure produced for the Global Burden of Disease Study 2017. Because COVID-19 is deadlier for older people, an age-standardized death rate is used to account for differences in the average ages across countries. *Source*: Original data are from Johns Hopkins University, World Health Organization, Global Health Security Index, Global Burden of Disease Study, and World Bank compiled by the Institute for Health Metrics and Evaluation. Figure source: https://www.cfr.org/report/pandemic-preparedness-lessons-COVID-19/findings/.

virus. These shortcomings can be categorized under the banners of protection, hedging, and insurance. Understanding these elements is vital not only to discern why current management systems faltered but also to inform enhancements for the future.

Protection includes all actions aimed at directly reducing the threats originating from exogenous events, such as: mitigation, prevention, advance warning, adaptation, and development of capabilities to cope, react, rebound, and adjust. Many of the measures

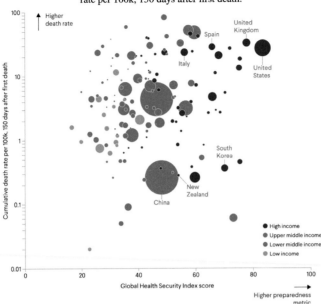

B. Preparedness Metrics: Global Health Security Index *vs.* Cumulative death rate per 100k, 150 days after first death.

Figure 3.1 (*continued*)

initiated by governments prior to the pandemic, including the establishment of special-ized emergency and disaster management departments, go in this direction. However, they largely appeared to be dictated by legitimate, but disjoint, sector worries and typi-cally lacked an integrated policy framework. Because of the inter-connectedness of the sectors involved, they need to be guided by a more cohesive strategy capable to create the capabilities and the options needed to face the many facets of health and environ-mental risks.

The development of a robust tracking system and the increasing emphasis on mon-itoring are important characteristics of a national health governance and infection management capability. A tracking system helps identify and track the spread of in-fectious diseases, while monitoring allows for the early detection of potential outbreaks and the implementation of timely and effective interventions. These capabilities are par-ticularly important in the context of thresholds for 'tipping points', or points at which the spread of disease becomes difficult to control.

The absence of these capabilities in most private and public organizations is likely a contributing factor to the difficulty in responding effectively to the COVID-19 pan-demic. To address this problem, it may be necessary to invest in capacity building, which involves improving the knowledge, skills, and resources of individuals and organizations. This may involve training in disease surveillance, data analysis, and public health policy.

By applying multiple lenses, or different perspectives, to the public health issue, it may be possible to gain insight about the future and better prepare for potential outbreaks.

Hedging is a second missing component of current health risk management, based on risk sharing and productive exchanges between empowered stakeholders. Although the direct objective of risk management is the pursuit of national interest, the fact that health is a global externality has the logical consequence of creating conditions for regional alliances and agreements that go beyond national boundaries. In the case of health, the current relationship of most countries with international organizations (such as the European Union) and multilateral agencies (WHO and the World Bank) already includes various forms of risk sharing, but having a common framework for risk management can promote this and other partnerships in a way that is mutually beneficial to all parties. Hedging need not to be limited to countries of the same political block or to multi-laterals. Various countries and companies can be involved in the protection and prevention of pandemic risks, zoonotic diseases, and other environmental and health disasters through a variety of actions. These can be based, for example, on new technologies, joint ventures or research and development in the field of infectious disease monitoring, and mitigation and adaptation actions to epidemic outbreaks.

Insurance is the final important component of health risk management that is currently missing or not pursued adequately in national and international health plans. Although epidemic risk is still considered to be largely uninsurable, several forms of infectious diseases can be the subject of specific insurance programs. Insurance clauses in the form of exit and/or relief options can also be introduced in contracts (and should be critical elements of future loans for health care investments). Several financial instruments, such as derivatives and catastrophe bonds, are also readily available on international financial markets and can be an effective part of risk management. Some of these instruments can also be particularly effective because they can be the financial basis of insurance programs for key sectors of the economy. A natural form of insurance can be based on the 'One Health' program, which requires developing healthcare, prevention, and monitoring in a diversified and more integrated way. As we will see in more detail later in this book, this implies a holistic focus on the biological equilibrium of all living things, investing in education, infrastructure, technology, and research, and facilitating private investment in a new generation of intelligent health infrastructure.

3.4. A clear example of under-preparedness: the first nine months of the pandemic in Italy

Despite the virus being first detected in France on 24 January 2020 and, days later, in Germany, Italy was the place where it started its pandemic behavior. For months, it was also the hardest hit country in Europe in terms of deaths and cases. As the effects of the pandemic manifested themselves a couple of weeks earlier than any other European country, Italy represented a sort of pilot case to evaluate the effectiveness of policy responses. Its public health system was also believed to be one of the best in the world due to its universal coverage and well-established and effective primary care. However, shuttering prior expectations, in a very little time, the Italian system nearly collapsed, anticipating what was going to happen in many other countries months later. More than three years later, though solid evidence is still scarce, it is helpful to investigate the reasons why things went wrong and why COVID-19 brought the Italian healthcare system close to collapse.

In the next pages, we explore what went wrong and why. In particular, we review the events that occurred during both the first and second waves and to what extent international events and decisions may have affected national ones. Naturally, a significant focus will be on the events of the first wave, a time marked by pronounced unpreparedness. During this phase, as outlined in Bosa et al. (2022), the initial reactions were frequently filled with skepticism and inactivity. Implementing a lockdown, similar to what was done in China, presented unique challenges, as it meant navigating the uncharted territory of balancing measures that infringe upon personal freedoms in a democratic setup against the urgent requirement to curb or at least lessen the virus's spread.

3.4.1 The pandemic timeline in the early months

The first wave of COVID-19 in Italy spans a period of approximately five months (from November 2019 to March 2020). It represents the most critical period to consider to analyze the level of pandemic capacity of the country for the governance and management of a health emergency. It was the period in which the first outbreaks developed in China, the WHO declared a 'Public health emergency of international concern' (PHEIC), and in Italy, the national lockdown period began. Recalling the sequence of events, summarized in Figs. 3.2 and 3.3, helps better understand and compare the times of the epidemic with the response times of the various governments and the WHO. Fig. 3.2 reports the timeline of events over a longer time frame, but with fewer details. In contrast, in Fig. 3.3 the time period is reduced (up to 8 March 2020), but more details are given to better understand the reaction times of the WHO and the Italian government.

As illustrated in Fig. 3.2, between November 2019 and early March 2020, the WHO and the Italian government could have utilized a significant portion of the time more effectively to develop a comprehensive strategy while simultaneously managing swifter responses to the epidemic crisis. In particular, it is possible to identify approximately 30 days of response delay by the WHO, that is, from 31 December 2019 (the date on which the ability of the virus to transmit to humans was made official) to 31 January 2020 (when the WHO declared a public health emergency of international interest)

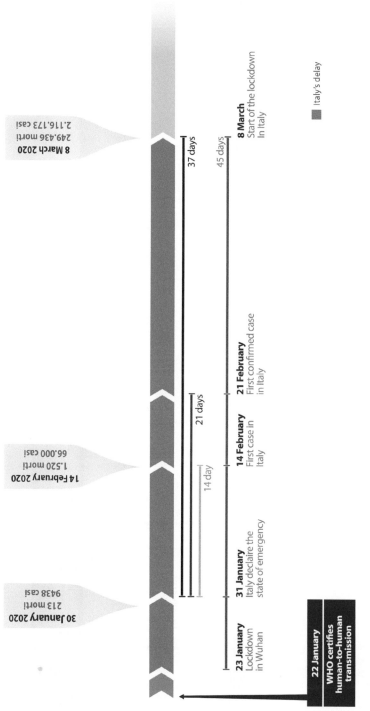

Figure 3.2 Timeline of major COVID-19 events since November 2019.

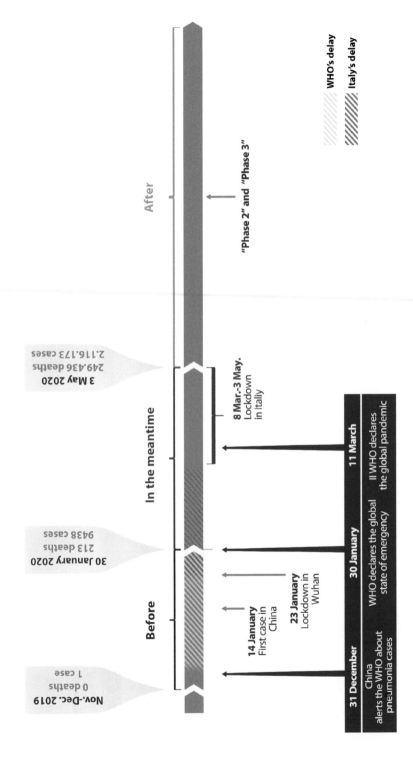

Figure 3.3 Timeline of major COVID-19 events from December 2019 to March 8, 2020.

and around 45 days by the Italian government, that is, from 14 January 2020 (lockdown date in Wuhan) to 8 March 2020 (lockdown date in Italy).

Fig. 3.3 shows also the dates of the three phases ('before', 'during', and 'after') into which we have conventionally divided the first COVID-19 wave. Some pertinent dates have been incorporated into the three phases to more accurately depict the events. Other dates, less relevant but crucial to following the evolution of the crisis, are instead mentioned and discussed in the text.

The timeline conventionally begins at the end of November 2019, as a series of retrospective studies conducted in the first period of the epidemic indicate that the SARS-CoV-2 infection first occurred on an unspecified date of that month (van Dorp et al., 2020). On December 8, 2020, the first case of infection was reported in Wuhan, with a patient with symptoms compatible with a coronavirus infection (the problem was later identified based on a retrospective analysis).[2] The sequencing of this new type of SARS-like coronavirus was first provided between 24 and 27 December 2019 by Vision Medicals in the Guangzhou province. The clinical suspicion of a possible outbreak of infection occurred almost in the same period (27-29 December 2019) at the Hubei Provincial Hospital in Wuhan. On December 31, 2019, the Wuhan health authorities made official the existence of an outbreak of pneumonia of unknown causes. The information arrived through the official WHO and CDC channels in Atlanta, Georgia, on the same day.

On January 1, 2020, the WHO established the Incident Management Support Team (IMST) across the three levels of the organization: headquarters, regional and national headquarters, alerting the organization to tackle the epidemic. From that moment on, a growing volume of news increasingly contrasted with Beijing's official version. The news began on January 4, 2011, when the WHO reported on social media that in Wuhan, Hubei province, there was an outbreak of pneumonia without deaths. The next day, the WHO rushed to publish a technical note for the scientific community and public health experts with more detailed information on the infection caused by the new virus and a first risk assessment and advice on how to behave. In addition, the note contained what China had reported to the WHO on the status of the patients and their public health response.

In Italy, the Ministry of Health was the first institutional entity to deal with the virus. On January 5, 2020, the General Directorate for Health Prevention of the Ministry of Health, relaunching the information received by the WHO, reported that an outbreak of 'pneumonia of unknown cause' had been discovered in China.

---

[2] An interesting overview of the 'backstage' events that occurred before the COVID-19 was made public is available in Farrar and Ahuja (2021). In this book, the author reveals the exchange of emails with various experts and government and institution officials since Christmas 2019, clearly showing that the threat to the world population was already known by experts, but was underestimated by public officials.

Box 5: China vs. WHO: mutual responsibilities and guilty delays

The often ambiguous relations between China and the WHO in the early stages of the pandemic were responsible for several health management problems. That something dangerous was developing between December 2019, and January 2020 was quite evident to all insiders, but few could imagine what degree of danger these relationships could bring about (Farrar and Ahuja, 2021). As illustrated in Fig. 3.2, the WHO declared the emergency nearly a month later, when an epidemic resembling SARS was already spreading throughout China, with all its associated consequences. Officially, the whole story was unknown until the beginning of June 2020, when the Associated Press published an article reconstructing all the background stories of the interactions between China and the WHO in January 2020.[a]

The article provides documentary evidence that during January 2020, China deliberately delayed the sharing of essential data on the virus (in particular its genome, which was available as early as the end of December 2019) and on the evolution of the epidemic. The problem did not seem to be limited to Beijing alone. It was suggested that the WHO had chosen not to put pressure on China, relegating complaints only within its structure, without taking official positions, and even praising the top leaders of the Chinese health organization for the speed of their response and transparency. The evidence was, however, ambiguous, and later records suggest that, contrary to what former President Trump said in April 2020, rather than colluding with China, the WHO had been kept in the dark about the events, receiving only the minimum amount of information required by law.

According to the Associated Press, WHO's behavior was partly dictated by the need to avoid harming Chinese scientists and keep some access to information without forcing their hands. Since the second half of January 2011, as the number of outbreaks increased, the WHO's attitude changed, expecting a repeat of the SARS outbreak in 2002 as an increased risk. Delays in sharing information about the epidemic slowed the recognition of its spread to other countries and the development of tests, drugs, and vaccines. The lack of detailed patient data made it more challenging to determine the rate of spread of the virus, a crucial issue in stopping and containing it.

[a] The Associated Press article is available at the following web address: https://apnews.com/article/ 3c06179497061042b18d5aeaed9fae.

On 10 January 2020, the WHO continued its information work and published online a comprehensive technical assistance package with advice to all countries on how to detect, test, and manage potential cases, based on what was known about the virus up to that point. The guide was shared with the WHO Regional Emergency Directors to further disseminate recommendations to WHO representatives in individual countries. Based on the experience with SARS and MERS and the known mode of transmission of respiratory viruses, guidelines for infection control and prevention to protect healthcare professionals were published.[3]

---

[3] Interestingly, these guidelines mention, for the first time, infection by short-range aerosol transmission in both the public spaces and within healthcare settings, recommending that all precautions be taken in the contact stages during patient care.

The days from January 12 to 20, 2020 recorded the first escalation of the crisis. On 12 January, China released the first gene sequence of the SARS-CoV-2 virus at the time named 2019-nCoV and, on January 14, the WHO admitted that the new virus could be transmitted to humans, given the experience with SARS, MERS, and other respiratory pathogens. In the specific case of Wuhan, transmission between individuals would have occurred between people of the same family. On January 20, after the infection of members of Chinese medical personnel in Guangdong, the China National Health Commission confirmed the human-to-human transmissibility of the virus. Scientists from the China Center for Disease Control (CCDC) identified three 2019-nCoV strains, confirming that the original Wuhan coronavirus had mutated into two additional strains. At that point, China could no longer conceal the situation and Xi Jinping, the General Secretary of the Communist Party of China, officially announced that 'people's life and health should be given top priority and epidemic spread should be resolved firmly' (Kuo, 2020). Until then, Beijing's official version had been a very different one: 'Symptoms reported by patients are common to several respiratory diseases, and pneumonia is a frequent occurrence in winter'.[4]

Given previous experiences with SARS and MERS, discussing these topics with virologists and infectious disease specialists produced evidence that attracted considerable attention. In contrast, in the early stages of the pandemic, the WHO seemed to send reassuring signals, which the Italian Ministry of Health promptly made their own. In particular, on 16 January, the WHO stated that it did not apply 'any restrictions on trade or travel based on the information available' and reiterated the suggestion on 21 January. In those days, press releases from the Italian Ministry of Health were limited to broad assurances that all necessary measures were being taken to prevent the spread of the virus from China by tightening airport controls. An important step was the issuance on 16 January of a statement from the Department of Prevention of the Ministry of Health saying: 'The WHO encourages all countries to continue preparedness activities' and calls for 'stockpiling' (antiviral drugs, masks).[5]

Between January 21 and 31, 2020, the WHO launched a long internal process of official recognition of a potential public health problem. On January 22, the organization certified human-to-human transmission, but deemed it necessary to investigate

---

[4] It is also helpful to point out that on 12 January 2020, despite the news of the first death in Wuhan, Beijing official sources still reported that 'there is no clear evidence of inter-human transmission' and 'since 3 January 2020 no new case were identified'.

[5] Here is the full text: 'The WHO encourages all countries to continue preparedness activities. On January 10, the WHO published information on how to monitor cases, treat patients, prevent additional nosocomial transmission, maintain the necessary stocks, and communicate with the public about SARS-CoV-2. The information included recommendations on the application of hand and respiratory hygiene measures and safe food and market practices. The WHO is still working to prepare and update this information in consultation with the Global Expert Network'. The press release is available at the following Web address: http://www.salute.gov.it/Malinf_Management/10-20.pdf.

further the scope of the transmission and possibly declare the existence of a 'public health emergency of international interest'. During the first meeting held on January 22-23, the WHO expert committee members were unable to reach a consensus based on available evidence and requested a new update within ten days after receiving more information. The consensus was reached on 30 January 2005, the sixth time since the International Health Regulations (IHR) came into force, with the WHO declaring a Public Health Emergency of International Concern (PHEIC). On the same date, the WHO provided a risk assessment, considering it to be 'very high' for China and 'high' globally.

Meanwhile, the first lockdown was launched in Wuhan on 23 January to stop the virus's spread. Road traffic was severely restricted and rail and air services were suspended. At the same time, foreign governments began to develop plans to evacuate citizens from the city. The enemy (the virus) was no longer hidden in the shadows. At the end of January 2020, the world had been warned.

Based on this information, one can better understand how Italy's different and partly contradictory crisis management steps unfolded over time. The first official act that marked the start of the fight against the epidemic was on January 22; the Minister of Health, Roberto Speranza, gathered a task force under his direct control from all ministerial directorates affected by the epidemic crisis and from various indirectly involved institutions. These included institutions linked to the ministry, such as ISS, the Istituto Nazionale per le Malattie Infettive Spallanzani, the Maritime, Air and Border Health Offices, the Italian Medicine Agency (AIFA), the anti-adulteration unit of the Carabinieri corp, and the Agenzia Sanitaria Nazionale (AGENAS). The objective of the task force was to coordinate and oversee the response to the pandemic and to ensure that the necessary resources and measures were in place to contain the spread of the virus and protect the health of the population.

During the same days in China, the total number of cases confirmed by laboratory analysis had increased to 571 and the death toll to 17, and it was clear that the numbers would not stop there. Meanwhile, in Rome, a case involving two Chinese tourists infected with the virus was discovered at the Palatino Hotel. At this point, Minister Speranza decided to take the drastic decision to suspend all direct flights from China, a choice strongly criticized by many.[6]

In those days, infections in China continued to grow and with them the number of deaths. The communication delays by the Chinese authorities and WHO did not help (see also Box 5 in this chapter). In Italy, as in the rest of the world, the emerging scenario appeared extremely complicated to address. By January 31, China had reached

---

[6] The decision to block the flights was criticized in the first place because it created a problem of a diplomatic nature that could compromise a series of economic relations and future investment opportunities. Second, blocking direct flights did not prevent entry to Italy through flights with stopover in other countries, and Chinese citizens could no longer be controlled on arrival.

12,000 cases with 259 deaths, 82 cases were reported in 18 countries, and the WHO had declared a Public Health Emergency of international interest. The Council of Ministers declared a six-month emergency to face the health risks associated with coronavirus infection. It allocated five billion euros in anticipation of the costs to be covered. In addition, the head of the Civil Protection Department was entrusted with coordinating the government actions necessary to respond to the emergency on the national territory.[7]

The machine had started and the chain of command should also have been transparent and effective. However, nothing significant happened in Italy between January 31 and February 14. Those days were punctuated mainly by discussions (often uninformed) among political leaders and experts on the possibility that what was happening in China could happen in Italy. Most agreed that contagion was, above all, a problem for Beijing. Several manifestations of racism and intolerance towards the Chinese population resident in Italy and the few Chinese tourists still in circulation marked those days. Precious time was wasted, ignoring the recommendations of the WHO and the Ministry of Health in mid-January, one of which had been to stockpile equipment to deal with the emergency. Meanwhile, the death and contagion toll in China continued to increase: On February 14, 66,000 cases and 1520 deaths were reported.

On February 14, the medical team of a hospital in Codogno (a small town in the Lombardy region) identified the first Italian patient with coronavirus. It was the first of a long series, with a rapid succession of new cases starting with those registered in another small town, Vo' Euganeo in Veneto. In the following days, until February 21st, information on medical developments was distributed only among insiders and the chain of command was basically idle. In the last days of February, the contagion curve escalated, following a trend that epidemiologists knew well and immediately suggested that a disaster was imminent. In four days, from February 28 to March 2nd, the number of people infected every day tripled from 888 to 2036. The number of daily deaths doubled to 52.

Despite the momentum of the infection, many were still favoring the business-as-usual approach to continue life (and economic activities) and avoid dire economic consequences from lockdowns and other restrictions. But that was not the position of the experts gathered as members of the Scientific Technical Committee. On the contrary, they favored the closure of all activities and a total lockdown. In March 2nd, 2020, an outbreak of infection in Bergamo, a large city in Lombardy, raised further alarm. At that point, it would have been appropriate to immediately apply at least selective restrictions, with more severe lockdowns in areas where the infection seemed

[7] According to the Civil Protection, the Department head's primary focus was to rescue and assist those affected by the contagion, enhance airport and port controls in accordance with urgent measures adopted by the Ministry of Health, facilitate the return of Italian citizens from high-risk countries, and repatriate foreign nationals to their respective countries of origin exposed to the risk.

more threatening (the so-called 'red zones'). Unfortunately, it took at least a week before that happened. The government was considering selected restrictions based on different color-coded zones (e.g., red, orange, yellow) based on the severity of the outbreak in specific regions or areas but no one was ready to propose a generalized closure of all production activities.

In hindsight, easy-to-analyze evidence has emerged: a severe outbreak was occurring in Lombardy. The epidemiological data from March 2nd show that in that region and on that day, there were 1254 positive people, 478 hospitalized, 127 patients in intensive care, and 38 out of 52 registered deaths in Italy. In the adjacent Veneto, on the same day, despite the outbreak of Vo' and the cluster of infections in Padua, Treviso, and Venice, there were only 291 documented infections, 17 patients hospitalized in intensive care, and thousands of infection tests performed. Unlike the Lombardy case, Veneto data suggest that the outbreak was not as widespread in that region, with fewer positive cases and hospitalizations reported.

The disparity in outbreak severity between Lombardy and Veneto could be attributed to various factors, such as differences in population size and demographics, the preparedness and responsiveness of healthcare systems, and the effectiveness of containment measures implemented by local authorities. It is also worth noting that the COVID-19 pandemic was a rapidly evolving situation, and data from March 2 may not reflect the full extent of the outbreak in either region at that time.

Also, on March 3rd, in a climate of great confusion at the government level, all schools were closed (initially for two weeks, 'possibly to be reviewed'). Nevertheless, authorities and experts continued to debate what to do in Lombardy areas affected by the contagion and what kinds of definitions and restrictive measures to use for 'red zones', 'orange zones', or others. Only on March 8th was a government decree (DPCM) issued to impose a lockdown in all Lombardy and the other 14 provinces.

Between March 8th and March 9th, the decision-making process accelerated as never before (unfortunately, it was too late!). Meanwhile, the leakage of confidential information about the draft version of the decree that marked the 'red zones' in Lombardy and the 14 provinces triggered widespread fear in the population. As a result, about 100,000 people moved from these territories, heading mainly to the South, which for some reason seemed to have escaped (though not for long) the contagion curse.

On March 8th, the red zones, defined on the basis of the frequency and extent of the outbreak of the disease, were so large that the government realized that it was no longer possible to block Lombardy alone to stop the contagion. On March 9th, when 97 new deaths were recorded, the government issued the decree 'I stay at home' to impose a lockdown on the country for two months. In March 11th, Italy became a single red zone with travel prohibited outside the municipality of residence, except for work, health, or emergencies. At the same time, all commercial activities were closed, except for supermarkets, pharmacies, and newsstands. However, it took another two weeks until March 23, 2020, when the government extended the shutdown to any production activity considered 'not strictly necessary' to guarantee essential goods and services to the national territory.[8]

3.4.2 The pandemic timeline at the start of the second wave

The May 4th, 2020 set the start of the long-awaited 'Phase 2' in Italy, which allowed some sectors of the economy to reopen and allowed people to meet with relatives. From that date on, some categories of laborers could return to work and it became possible to meet relatives. However, the prime minister reminded the citizens that 'Phase 2' did not imply that all hardship had ended and all had changed for the better. When presenting the new rules, he reminded that the danger of another spike in infection was real, that 'the future of the country will be in our hands' and that 'even more collaboration, civic sense, and respect for the rules by all will be needed'. Unfortunately, that wise advice would go unheeded and soon gave way to the beginning of a new wave of contagion.

Regrettably, experts helped the drift by launching a series of statements about the alleged lost capacity of the virus to infect. In an interview at the end of May, the director of the intensive care unit of San Raffaele Hospital in Milan stated that 'from a clinical point of view, the coronavirus no longer exists'. From then on, he became the leading figure among a wide array of experts who maintained that the weakened virus may not be able to produce a second wave of infections.

Meanwhile, on the economic front, the government issued the 'Decreto rilancio' (Relaunch decree), the largest of its interventions up to that moment (with a value of 55 billion euro, equivalent to two standard financial bills), which introduced new measures and extended those introduced by previous decrees. In consultation with the Member States, the WHO Assembly unanimously gave the green light to the resolution to begin an impartial investigation on the origins of the coronavirus pandemic and the global response to the health emergency. China, which was initially opposed, also voted in favor.

The month of June 2020 was characterized by the beginning of 'Phase 3', with further relaxation of numerous existing restrictions, including the obligation to use outdoor masks and the resumption of the national football championship of the top league ('Serie A'). In addition, playgrounds and summer centers for children up to three years of age, alongside with betting rooms, cinemas and theaters were reopened. However, only outdoor performances were allowed, with some restrictions on the number of seats and access to the public. Immuni, a government-sponsored real-time app that tracks contagion, made its debut (with some delay compared to when it could have been more valuable).

---

[8] During these days, in many empty and apparently lifeless Italian cities, at 6 p.m. of each day, people started to meet on balconies to sing the Italian national anthem (as well as other patriotic songs).

At the same time, the news from the rest of the world was not encouraging, with the situation in South America markedly deteriorating. In China, a new outbreak occurred in Beijing. However, this time, the Chinese authorities had moved to take massive action: 27 city areas were put on lockdown, schools were closed, and 2.3 million people were examined in just a few days to isolate positive cases and counter the spread of the virus. At the same time, the situation in Italy seemed to be heading in the opposite direction. On June 23, 2020, there were only 113 new cases, the lowest number of positive cases in 24 hours since the pandemic outbreak. As a result, intensive care patients (38 cases on 29 July) and the number of deaths (1 death on 29 August) also decreased to their lowest levels in July and August.

With these results in mind and underpinned by the desire for normality, the attention of people to the health emergency waned, as did their compliance with government recommendations. For example, international travel by Italian tourists resumed and peaked in July, with Spain as a preferred destination. These actions led to a new acceleration of infections: In Italy, early signs of resumption of the virus were recorded in Veneto on July 3. Moreover, once re-activated, the engine could not be stopped. The launch of several measures of 'Phase 3', more extensive in terms of relaxation of restrictions (i.e., reopening of discos, authorization for team sports, and people could play cards again in bars and retirement homes), created conditions for a new wave of contagions.

All this happened in the middle of summer. The month of August proved to be the most critical. Although talks about vaccines had intensified (announced by Trump and Putin), new infection outbreaks occurred in Italy, many connected to discos and nightclubs. The contagion curve started to rise again and containment measures reappeared: nightclubs closed and masks began to be mandatory again from 6 pm to 6 am in the 'Movida' places and open spaces.

In September, the situation worsened and infections resumed raging throughout Europe. In Italy, about 1500 new daily cases were reported, far from the peak of 6500 in April, but significantly higher than 200 a day in early July. France and Spain were the countries most affected in Europe, with more daily cases than their spring peaks. On September 7, Spain became the first European country to have half a million patients, of which more than 100,000 were diagnosed in the previous two weeks. However, France was experiencing a significant increase in new infections, exceeding 10,000 daily. Even Germany, an example often cited as a case of excellence in coronavirus response in Europe, saw daily cases increase slowly but continuously, with about 2000 daily infections, doubling since 1 August. In Italy, the situation seemed under control, so much so that on 24 September, an article in the Financial Times praised the country's behavior (no praise ever less deserved!). The government was encouraged by the still positive results and, on 14 September, decided to reopen the schools (extending the closure in some regions to 24 September). Meanwhile, rumors of a possible vaccine became increasingly reassuring: The pharmaceutical company Pfizer had announced that it could establish by the end of October the effectiveness of an anti-COVID vaccine produced together with the German company Biontech. At the same time, Bill Gates also reportedly spoke several times about a vaccine being available soon, raising hopes that it would be possible to return to a daily life similar to that of pre-COVID.

Unfortunately, ignoring the recommendations of the previous months had catapulted Italy into a second pandemic wave. In October, infections increased alarmingly and hospital pressure was felt again. At the end of October, the daily cases exceeded 31,000, much higher than those registered in the first wave. Despite the numbers, the government was sluggish in decision making. On October 19, a new decree imposed restrictive measures on bars, restaurants, and sports in contact. Schools opted for distance education, and several firms for a smart-working organization. Anger also exploded in some Italian cities, with hundreds of people gathering to protest, leading to violent clashes with police officers. At the same time, France began a month-long lockdown at the end of October, and Germany announced restrictive measures for restaurants, bars, theaters, cinemas, and gyms.

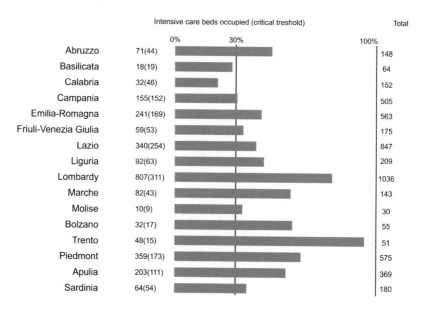

Figure 3.4 Intensive treatment use by region (as of 13 November 2020). *Source*: Adapted from Sole 24 Ore: Lab24, available at: https://lab24.ilsole24ore.com/coronavirus/. *Note*: Bed available at 28 October 2020. Emergency Commissioner.

November saw the situation in Italy worsen on all fronts: contagion, intensive care, and death numbers. Even more restrictive measures were introduced, and the country was divided into three-color zones with different levels of restrictions, which came into force on November 6, 2020. The subdivision assigned yellow, orange, and red colors to different geographical areas of the country, depending on the severity of the local situation. In the red zone, the most at risk, a lockdown occurred, albeit with less stringent measures than in the previous spring. With 40,902 positives, November 13 was the peak day of infections in Italy, and the second wave invested the whole peninsula. As illustrated in Fig. 3.4, hospitals were under stress, and many regions exceeded the critical threshold of intensive care beds. At the same time, good news came to alleviate the increasingly difficult situation: the expectations for effective vaccines became real at the end of November. Pfizer and Moderna announced a few days later that their vaccines were 90% and 94.5% effective, respectively.

3.4.3 Why so many delays: a 'super-chain' with the proliferation of roles, responsibilities, and bureaucracy?

The evidence presented above suggests that the lack of a timely and effective response in managing the COVID-19 emergency was due to a combination of organizational failures at both the international and national levels. One factor may have been the complexity of the administrative structures in place. In market economies and decentralized governments, a number of different organizations and agencies are likely to be involved in responding to a crisis such as COVID-19, which can make it more difficult to coordinate and effectively manage the response. Another factor may have been the failure in the chain of command, that is, the lack of clear lines of authority and decision-making. Polyarchic systems typically feature multiple centers of power and decision-making, which can lead to confusion and delay in responding to a crisis (Sah and Stiglitz, 1984).

The response to the COVID-19 epidemic in polyarchic societies was also hindered by governance issues. Weaker governments with multiple centers of power may have struggled to effectively coordinate and implement a response to the crisis. The lack of a clear and compelling centralized strategy derived from poor planning capacity and strong leadership may have made it difficult to convince other institutions and decision makers to follow guidelines and seek coordination.

Focusing on Italy, in the following, we explore the extent to which both governance and management failures converged in making the chain of command less effective and more prone to errors. To better understand the evolution of the process, it is necessary to go back to January 31, 2020, when Italy, first among Western countries, issued a resolution of the Council of Ministers declaring a state of emergency until 31 July 2020. A state-of-emergency declaration is a measure taken by the Government in exceptional cases. Introduces the 'power of ordinance', which gives the Council of Ministers the ability to similarly empower other bodies and entities. The 'power of ordinance' allows the identified authority to exercise action by derogating from the legislation in force, although in compliance with the general principles of the legal order.[9]

The first step that changed the rules of governance was taken on January 31, 2020, allowing the emergency to be addressed by making fundamental decisions outside the usual regulatory and legislative limits. This step was followed by a very rapid unfolding of events, with the approval of numerous decrees, ordinances, and circulars in an attempt, not consistently successful, to manage the outbreak of COVID-19.

The resolution of January 31 consisted of only three articles. The first declared a state of emergency for six months (up to 31 July 2020); the second defined the first link in the chain of command, identifying the head of the Civil Protection department as responsible for implementing the necessary interventions. The third article allocated 5 billion euros to cover the required expenditure (a sum that would soon prove to be entirely inadequate).[10]

The Civil Protection Department (CPD) - reporting to the Presidency of the Council of Ministers - was one of the institutions entitled with the power of ordinance, thus allowing for exceptional actions derogating from current legislation and procedural prescriptions. CPD was also authorized to identify 'actuating entities', that is, institutional bodies (economic and non-economic public bodies) and private entities acting on specific directives, without additional costs to public finances, and was also allowed to work in derogation of existing legislation.

On 3 February 2020, CPD issued executive order 630, establishing a scientific committee to *i)* coordinate all interventions, and *ii)* list the laws that could be waived by the CPD head and any 'actuating entity' to implement all requested activities.

Given the organization of the Italian national health system, as the outbreak unfolded, it became clear the need to decentralize emergency management by involving the regional governments (henceforth 'the Regions'). Therefore, the governors of the Regions became the third link in the chain (identified as 'other actuators') with the power to implement all anti-COVID measures on their territory. However, in this case, the third link was not hierarchically subordinated to the top links, since the constitutional law identifies health as a joint responsibility of the State and the Regions. The push towards decentralization, with the involvement of the Regions and, at the same time, the centralized outsourcing to technicians and experts, led to multiple conflicts between the various actors. While decentralization gave the Regions the power to deal

---

[9] Most of this section's content relies upon information produced by Openpolis Foundation, which has monitored all legislative and administrative works stemming from the chain of command attribution of powers and responsibilities. The Openpolis Foundation website can be accessed at the following web address: https://www.openpolis.en/.

[10] The value of this allocation may, in part, indicate how unclear the problem was and how ambiguous the actions to be taken were.

with the emergency more autonomously, it also made it more challenging to manage the chain of command and establish a national standard for emergency management.

This initial overlap of powers and responsibilities called for clarification from the Presidency of the Council (in charge of Civil Protection). On 4 March 2020, Council President Giuseppe Conte issued an explanatory note 'defining the chain of command and control, the flow of communications and the procedures triggered by the state of emergency caused by the spread of the virus. A note was issued with instructions on how to coordinate national, regional, provincial and municipal activities. Given the nearly ten days of health crisis where the central government, Civil Protection, and Regions operated without coordination, this action was both necessary and long overdue.

As the outbreak progressed and the problems became complex, the chain of command expanded. A notable flaw in the organizational structure was the provisioning of medical equipment. This oversight was a major error by those overseeing the management of the pandemic crisis. They completely underestimated the role of time in building supply stocks before the pandemic outbreak: a role that was clearly stated in the health emergency plan.[11] The problem was resolved very late by identifying, more than 15 days after the first case of COVID-19 in Italy, a facility that could help address limitations in the provision of health services by hospitals and their ability to cope with the emergency. The delay implied that masks, respirators, and other equipment had to be found in a hurry to allow health facilities and their workers to operate efficiently and safely and, subsequently, to guarantee the availability of personal protective equipment (PPE) to prevent contagion.

Thanks to a Civil Protection Decree, on 2 March 2020, the Managing Director of CONSIP, the government procurement agency, was appointed as the actuating entity, also recognizing special accounting rules for the performance of these tasks, thus providing much power and autonomy for pandemic-related deliberations. Subsequently, on March 17, when the 'Cura Italia' decree was passed, the CEO of Invitalia, an agency responsible for assisting foreign investors in Italy, was tasked with overseeing the boost of supply of protective equipment. Again, as in the case of CONSIP, in his role as a Commissioner, he was authorized to incur costs and 'to provide funding through grants or operating grants, as well as subsidized funding, to companies that manufacture personal and medical protective equipment to ensure that it was adequately provided during the COVID-19 emergency period'.[12]

What we have described covers only the highest part of the chain of command, as granting powers to the lower links led to a proliferation of positions and committees, which in many cases further diluted responsibilities to an unlikely extent. An intuitive

---

[11] Despite the limitations of having an out-of-date document, the existing pandemic plan contained elements that could help ensure proper management of the pandemic.

[12] Based on an article published by the Openpolis Foundation. The full text of the article is available at the following web address: https://www.openpolis.it/esercizi/norme-in-deroga-senz.

way to explain how this system was able to trigger role multiplication is to imagine the structure of a tree that develops both vertically with gradually smaller branches (central state, regions, and municipalities) and horizontally on each unit with myriad twigs and leaves (commissions, task forces, personal assignments in political, health, and civil protection institutions).

The case of Lombardy is emblematic for understanding the undergrowth of regional assignments and positions and how these could have affected the chain of command. Along with the role of 'actuator', as the implementer of central decisions, Civil Protection also asked each region to establish the Regional Crisis Unit (RCU) and to provide for the participation of the Regional Health Contact as a link to the national structure. According to the website of the Lombardy Civil Protection, 'the Regional Crisis Unit is made up of regional and external personnel: technicians from different disciplines who work together to support and resolve event management problems in emergencies'. A total of 154 people participated in the organization, compared to an average of 10 in other regions. In addition to the Crisis Unit, Lombardy also established the COVID-19 task force, composed of only 16 people, directly under the General Welfare Directorate (GWD). Finally, a 'Scientific Technical Committee' was established in GWD on 7 April: 26 members with only scientific medical expertise, of which 4 were also active in the Crisis Unit. All these structures were set up in the same directorate and coordinated by the same person.

The proliferation of centers with delegated functions to respond to the COVID-19 pandemic resulted in a lack of coordination between these different actors. As for the case of Lombardy, this led to different strategies and policies being pursued by the Regions, which caused controversies and conflicts.

The autonomy conferred on the Regions may have exacerbated these issues, as it may have been difficult to determine which rules should take precedence in cases where there were conflicts or discrepancies between the actions of the central and the regional governments. This scenario risked sparking a conflict between the central government and the Regions, with each trying to establish its dominance and follow its individual objectives. The cumulative effect, with Lombardy being just one instance, was an increase in stakeholders operating without synchronization.

The proliferation of delegated powers resulted in increased bureaucracy and conflicting documentation. According to Openpolis data, from January to July 2020, more than 200 acts of various kinds were issued, including ministerial decrees, ordinances, circulars, decrees, and other documents. These documents, often overlapping and disconnected from each other, were issued by a range of institutions, including the government, various ministries, the ISS (Istituto Superiore di Sanità, or National Institute of Health), the National Transplant Center, the 'Commissioner Structure for the Coronavirus Emergency', the Parliament, and others.

All of the above events can be interpreted as manifestations of the specific weaknesses and unpreparedness of Italy. At the same time, they highlight the vulnerability of polyarchies associated with modern democracies, in which decision-making is decentralized and power is distributed among various individuals and groups.

3.5. Before any chain of command: planning and prevention plans

A notable feature of modern democracies, particularly evident in Italy, is to exhibit a reluctance to follow predefined plans and to follow high propensity for improvisation. This may be due to the decentralized nature of polyarchies, in which decision-making is distributed among various individuals and groups, and it may be more difficult to develop and implement long-term, comprehensive plans, as there may be differing opinions and priorities among the various decision-makers. Additionally, the complex and dynamic nature of many organizations and decentralized government may require a degree of flexibility and adaptability to respond to changing circumstances, which may be facilitated by a willingness to improvise and deviate from predefined plans. It may thus be more difficult to strike a balance between the need for planning and the need for adaptability in order to face effectively an unexpected emergency.

As anticipated in the Introduction to the book and reiterated in Chapter 2, it had long been expected that a viral pathogen would eventually spread and cause significant harm to health and society. In 2018, the WHO had said that 'the world must expect a killer influenza epidemic and should always be vigilant and prepared so that we can fight the pandemic that is sure to occur' (WHO, 2018a). As late as 2018, the WHO had developed a practical guide for creating and conducting simulations to test pandemic plans.[13] Moreover, experiences with the other viral epidemics of the last two decades had caused great concern in the scientific community and the public. That is why, after the 2003 SARS epidemic, the WHO forced governments to adopt anti-pandemic plans and, in 2005, laid down guidelines for the actual drafting of the plans. For example, in 2006, the Italian Ministry of Health used these guidelines to draw the 'National Plan for the Preparation and Response to an Influenza Pandemic', which updated the 2002 'Italian Multiphase Plan for an Influenza Pandemic'.[14]

Developing sector prevention plans is a recommendation also contained in the National Prevention Plan (NPP) 2014-2018, whose validity was extended to 2019 by a State-Regions Agreement. In particular, the chapter on infectious diseases states that

[13] The guide is available at the following web address: https://apps.who.int/iris/bitstream/handle/10665/274298/9789241514507-eng.pdf?ua=1.

[14] The plan is available at the following web address: http://www.salute.gov.it/portale/influenza/detailPublicationsInfluence.jsp?lingua=Italian&id=501. According to the WHO, the plan was supposed to be updated every two years. In Italy, it was revised between 2009 and 2010 in response to swine influenza (H1n1). The last update occurred in 2016. The level of update envisaged in the 2016 plan compared to the previous was much debated in a heated media dispute, with legal implications involving the Italian judiciary system.

'with the entry into force in 2013 of the new Decision of the European Commission (No. 1082/2013/EU), Italy is asked to develop a generic plan to prepare for serious cross-border threats to health, both of biological origin (infectious diseases, resistance to antibiotics and nosocomial infections, biotoxins), chemical, environmental or unknown origin, and threats that could constitute an international health emergency under the Sovereign Regulation of Health'.[15] Although not recently updated, the latest available plan already contained several valuable guidelines to be implemented.

According to WHO guidelines, three phases of epidemic evaluation can be distinguished: *i)* when there are no immediate dangers, but there are reports of outbreaks of a potential large epidemic abroad; *ii)* when the first cases occur on the national territory; *iii)* when there are autonomous outbreaks in the country (the most severe scenario).

For each of the three phases, a prevention plan is effective if it provides a list of measures to prevent moving to the next stage. However, in January 2020, all these phases were crossed in a few weeks, gradually at the beginning and more quickly and devastatingly between March and April. It was challenging to come to terms with these events, especially considering that the updates from China had provided ample time to prepare. According to the reconstruction done in Section 3.4 of this chapter, human-to-human transmission was already announced on January 14. At that point, the first part of the National Prevention Plan (NPP), which provides for 'the preparation of appropriate measures to control the transmission of pandemic influenza in hospitals', should have been activated as early as January. This would have led to the implementation of a series of measures to supply hospitals with devices that would allow doctors to work safely and prepare for contagion situations on national territory. It would have been much easier to get hold of masks, ventilators, and all the equipment needed to deal with the virus during that time.

In the 30 days between 14 January and 14 February 2020 (the date of the first verified infection in Italy), the government response to the pandemic clearly failed, since until 22 February the NPP was never considered, at least for what refers to the section suggesting 'all health authorities to carry out a careful monitoring in the areas of competence, identify in time any cases of contagion, and report them immediately to the Ministry of Health and the ISS'. The same plan also stressed the importance of knowing in advance the resources available to deal with a possible pandemic. It recommended 'the definition of the criteria for the suspension of planned hospitalizations and the availability of additional beds'. Furthermore, it gave clear indications on the regulation of relations between the State and the Regions, suggesting that the 'regional operational plans' would stem from the national ones.

---

[15] Taken and translated from the 'Guidance Document for the Implementation of the Central Support Lines to the National Prevention Plan 2014–2018' of the Ministry of Health and available at the following web address: http://www.salute.gov.it/imgs/C_17_pubblicazioni_2477_allegato.pdf.

Another crucial aspect addressed by the NPP was the training of the personnel responsible for managing the crisis, organized on three levels: national/interregional, regional and local. In addition, the plan suggested identifying, at the regional level, figures with specific educational skills that could guarantee the training process throughout the territory. It further emphasized the significance, during outbreaks, of identifying suitable routes for those infected (and those suspected of infection), cataloging isolated beds and negative pressure rooms, and keeping a record of mechanical lung ventilators. The latter is especially relevant for patient management and can be seen as one of the significant failures in epidemic management not only in Italy but in most countries.

Despite the images already seen in Chinese hospitals, in this initial phase of the pandemic, all countries struggled with a lack of experience in dealing with the virus, as well as a lack of clear clinical guidelines and inadequate supplies of personal protective equipment (PPE) and other necessary equipment, such as ventilators. This led to difficult decisions about which patients to prioritize for intensive care and treatment. It was a challenging and unprecedented situation for healthcare workers, and many of them were forced to make difficult choices under extremely difficult circumstances. Hospitals and Long-Term Care facilities (LTCs) turned into major hubs for COVID-19 transmission, gravely endangering the lives of doctors and healthcare workers. These professionals emerged as the unsung heroes of this grim narrative, suffering significant casualties in the line of duty.

3.6. If there were mistakes, who made them?

Are there more general lessons to be learned from the frustrating failures presented by the history of Italian attempts to govern the pandemic? A pandemic is a glaring case for government intervention, where markets and even existing institutions fail to react in ways that ensure that 'things take care of themselves'. During a pandemic, at the same time, citizens turn to the government as the ultimate resource, and their expectations suddenly legitimize an excess of government power. A truly 'Hobbesian' anxiety enters civil society with the attempt to evoke a Leviathan, whose monstrosity is seen as a necessary and even desirable antidote to the plague. Yet this strong delegation of power does not necessarily solve the problem and is generally short-lived. This is because democratic governments are the expression of civil society and are thus torn by the logical and material conflicts between individual and common interests. The result is an ethical failure (the mother of all failures of the market and the state), that notorious 'poverty amid abundance' (Mann, 2015) that affects people, but also the environment and, ultimately, the very survival of humanity.

Governing a pandemic is an epic social problem, since it requires finding a new equilibrium of powers, albeit a transitory one, between the state and the market, as well as within the state and its different sector and regional articulations. For any given governance structure, managing a pandemic is one of the most complex challenges that a society can face. To reduce the social costs of death and economic damage, the whole community must organize itself by managing a wide range of resources and different tools. However, the effectiveness of management may be hindered by cognitive stress on 'beliefs about how to meet needs' and, as a consequence, by various types of emotional disturbances such as anxiety, resentment, or anger, and by the paradigmatic views of agents, which, as Andrew Ross suggests, are often 'absorbed before being chosen' (Ross, 2006). On the other hand, an organized response needs to be well-planned and global because the virus can spread everywhere. Still, an adequate answer also depends mainly on national policies and their implementation at the regional and local levels. Furthermore, saving lives requires excellent attention from those who are involved daily with patients. It also requires vital preparedness by those responsible for foreseeing challenges and creating an environment that enables those dedicated to saving lives to perform at their best. This preparedness should be maintained and updated over the years, even if there may be no natural feeling of a threat for a long time.

The ability to be prepared (preparedness) for the event is crucial. On 22 October 2019, the European Center for Disease Control (ECDC) published an interesting document titled 'Health emergency preparedness for imported cases of high-consequence infectious diseases'. The opening pages read: 'Preparedness planning is essential to respond effectively to epidemics, including individual cases of high-response infectious diseases such as the importation of a case of viral hemorrhagic fever. The preparedness cycle includes planning, identifying, and prioritizing risks; training and simulation exercises; revisions after the action; assessment of lessons learned and implementation of the identified organizational change' (European Center for Disease Prevention and Control, 2019). This checklist was developed for public health planners as an operational tool to review the preparedness system to respond to a possible occurrence of high-response infectious diseases imported into the European Union. The content is based on the work carried out during the Ebola virus epidemic in Africa and on a specific protocol used in the peer review of the health systems of Belgium, Portugal, and Romania. With several bibliographic/resource references, this checklist can complement broader preparedness checklists, such as those available at the WHO.

The publication of this document only predates the COVID-19 emergency by a few months and could have spurred timely action to make everyone ready and prepared to handle such complex and severe events. On the contrary, even in the following months, the document seems to have been thoroughly ignored and never considered or discussed among the insiders.

Lack of attention to health-preparedness calls can be seen as a manifestation of a recurrent neglect on the part of people and governments for seemingly unlikely events, even when the consequences of those events could be severe. Neglecting to be prepared for emergencies may be an example of voluntary, but not atypical, ignorance of negative circumstances possibly affecting the world. This can be due to a variety of factors, including lack of awareness or understanding of the risks involved, lack of resources or funding to allocate towards preparedness efforts, or lack of political will to prioritize these issues. Low probability but high impact events, such as the possibility of a swarm of asteroids striking Earth, evoked by Perrow (2011) are a fact of life. Experience teaches us that humanity, at least as represented by the governments of the world, often ignores events until it is forced to react, even when the price of prevention is negligible. According to Perrow, this attitude represents 'an abdication of responsibility and betrayal of the future'.

However, there are differences between rare events such as the fall of an asteroid, a mass explosion of particles from the sun's corona, or the eruption of a volcano that can temporarily prevent the sun's rays from coming to Earth, and events such as pandemics. Since their discovery, massive solar eruptions have never affected Earth. The last volcanic eruption to impact the climate dates back to 1815 with the explosion of the Tambora volcano. The odds of one such event occurring in the 21st century are estimated at 50%. No government has previously confronted such an overwhelming and catastrophic event, which might account for the oversight. However, it doesn't serve as a justification. One of the primary responsibilities of a state is prevention, and it is imperative for governments to act accordingly.

Unlike undocumented low probability-high impact events, as already discussed in Chapter 2, pandemics are disasters discussed in history books and recent chronicles. In this sense, COVID-19 offers a tragic example. Virologists, epidemiologists, and ecology experts had long warned of the dangers of influenza-like diseases from wild animals. However, when the SARS-CoV-2 virus began to spread, very few countries could put together the winning combination of practical plans, intervention kits required by those plans, and the bureaucratic capacity to implement them. The selected countries that managed to respond according to their established plans, like Taiwan, experienced significant benefits.

In these situations, improvisation is not allowed. In any risk management course, experts teach that crisis management works if we are prepared and have time to think, try, and eventually change. If things seem to have worked in South Korea, it is because the track-test-treat system was not prepared in a short time but had been developed over at least the 15 years since SARS had hit the country. Furthermore, the mechanism had been regularly updated and checked with care and dedication.

As mentioned in Section 3.5, the WHO had been urging governments since 2003 to establish and implement a pandemic plan. Since then, there have been signs that the plan should be considered the bread of many workers.[16] The invitation to prepare had not been lacking over time: the last warning was repeated in 2009 after the latest outbreak of

---

[16] To better comprehend the high number of warnings that occurred in the last 20 years worldwide, Fig. 2.1 in Chapter 2 is useful.

the swine flu pandemic. More importantly, and as mentioned above, in October 2019, the ECDC, the most authoritative body in Europe for infectious diseases, had updated the document with guidelines for preparedness for health emergencies for imported high-impact infectious diseases.

Everyone was alerted, yet it seems those responsible for managing the emergency were overlooked. The effects were not long to come, looking at how the regional emergency was managed: 'The choice of who to test has followed a wide variety of strategies, sometimes leaving people with obvious symptoms and their families undiagnosed; General practitioners had not been prepared and had not received adequate protection. For many days, separate hospital pathways were not set up for COVID-19 patients, and the hospitals themselves became a hotbed of contagion. The inaccessibility of a strategic plan has prevented us from acting proactively, condemning us to remain at the mercy of events'.[17] In many cases, the pandemic seemed to suscitate a standard response of modern bureaucracies to the managing challenges of emergencies: ignore the problem altogether until it is too late, set up a botched response by passing on to ordinary people the responsibility of crisis management, find a culprit.

Although Italy was the first western country on the front line to show this type of unfortunate response, impromptu behavior has been reported in many countries, including those where preventive organization is a tradition. One case for all is the US. In an article published in The Atlantic, James Fallows analyzes how the US administration was tackling the pandemic.[18]

The article describes what would have happened if the National Transportation Safety Board, the body responsible for controlling flight safety procedures in the United States, had studied the response of the United States administration to the coronavirus pandemic. According to Fallows, the answer would have been written using simple, direct, fact-based language, typical of that administration, stating something like: 'There was a flight plan. There was accurate information about what lay ahead. The controllers were ready. The checklists were complete. The aircraft was sound. But the person at the controls was tweeting. Even if the person at the controls had been able to give effective orders, he had laid off people who would carry them out. This was a preventable catastrophe'.

Another article gave a more explicit and stark description of the failures in managing the first phase of the pandemic. 'We were in the best position to respond to the pandemic. However, very little was done to prevent and control it, with much of the initial available resources going to reinforce the hospital and intensive care system, with the

---

[17] Text from an article published in Wired and available at the following web address: https://www.wired.it/actualita/policy/2020/04/20/coronavirus-errors-emergency-government-regions/.

[18] The article is available at the following web address: https://www.theatlantic.com/politics/archive/2020/06/how-white-house-coronavirus-response-went-wrong/613591/.

media spotlight occupied by virologists, vaccine experts, and resuscitation experts'.[19] All this occurred despite WHO providing about 200 million dollars in partnership contributions from 2012 to 2019 to support pandemic influenza preparedness. Unfortunately, in many cases (i.e., Italy), these funds were not used for this purpose.

Returning to the Italian case, according to Ernesto Burgio, an epidemiologist at the European Cancer and Environment Research Institute (ECERI), the errors in managing the pandemic were many and made at different times of the emergency. The first error was underestimating the problem in its early stages, at least since China realized that the situation was worrying and complicated. Many Italian experts and authorities also claimed that the disease was nothing more than a seasonal influence. An immediate consequence was to prevent the tracing of the first cases of infection, making it impossible to understand how many were infected (see also Chapter 4). From this point of view, the absence of protocols (or their non-adoption) played a key role. The lack of experience and experts in this field completed the damage. For months, disputes raged between experts and politicians over whether or not extensive tests should be performed. March 2020 was the prey of absolute confusion that delayed several important choices, first and foremost, the need to impose a larger immediate lockdown than was thought in late February. The Lombardy cases could be mainly attributed to cities like Bergamo and Brescia, where the virus spread quickly without any risk perception and even minimal precautions being taken. This led to the system collapse and the spread of panic. 'However, the main error that led to the most painful consequences of the COVID-19 outbreak in Italy was the inadequate information and protection of health personnel, as well as the inability to adapt the national health system to an emergency that appeared to be only at the beginning' (Burgio, 2020).

Yet, the disorganized response to the coronavirus pandemic in Italy and other countries highlights the vulnerabilities of decentralized governments and open societies to unexpected crises. Decentralized governments, where power is divided among different levels of government and agencies, can be more challenging to coordinate and manage in times of crisis because it is not always clear who is responsible for what, and there may be conflicting priorities and agendas. More generally, open societies, which are characterized by a high degree of freedom and transparency, can be more vulnerable to unexpected crises because they may not have the necessary structures and systems in place to respond quickly and effectively to threats. In the case of the coronavirus pandemic, these vulnerabilities may have contributed to the difficulties that countries faced in responding effectively to the crisis and mitigating its impact. In this sense, the disorganized response in Italy to the coronavirus (COVID-19) pandemic can be considered

---

[19] The article was written by a government medical adviser, Filippo Curtale. Its full text is available at the following web address: https://www.saluteinternazionale.info/2020/04/cera-una-volta-il-piano-pandemico/.

paradigmatic. Its lack of clear responsibility and coordination likely contributed to the difficulties that Italy faced in effectively responding to the crisis. The appointment of multiple commissioners to address different aspects of the crisis, such as the crisis commissioner, the commissioner for improving hospital infrastructure and the commissioner for the supply of health devices, added to the confusion and may have made it more difficult to effectively coordinate the response. Furthermore, the presence of multiple levels of government, including the President of the Council, the Minister of Health, and regional governors and councilors, may have led to communication problems and political infighting, further hindering the response.

3.6.1 Communication in an emergency phase

In a famous article published on his blog, Bill Gates discussed scientific advances we needed to stop COVID-19 (Gates, 2020). The article focuses mainly on technological and medical innovations (vaccines, antiviral drugs, and digital contact tracking apps), limiting the attention on communication to a short sentence: 'It will take a lot of good communication to make people understand the risks and feel comfortable going back to work or school'. Although that article did not address communication issues during pandemic periods, the experience of the pandemic clearly showed that effective communication strategies are essential. This is true both in the short term (for uptake of public health measures like face coverings) and in the long term (allowing people to comply with restrictions or increasing vaccination uptake).

The last decades have left abundant evidence of serious public communication errors made by many local, national, and international government agencies in responding to complex public health emergencies, disseminating inconsistent, incorrect, and contradictory messages (Gamhewage, 2014; Haider, 2005; Taylor-Clark et al., 2010). More generally, communicating clear, reliable, and timely information is essential during pandemics for reasons that go beyond the crisis. For example, a good communication strategy is useful for the sustainability of democracies, as it helps cultivate trust among citizens and their governments and encourages the public to engage proactively with the healthcare system. In the end, managing an epidemic crisis is also a communication problem. As one does not improvise crisis management, one should not improvise crisis communication. A robust communication strategy is essential when enforcing various public health measures. If a flood of disinformation accompanies these measures, a global epidemic of misinformation (an 'infodemic') could ensue, exacerbating the impact of the pandemic.

Despite abundant literature on large-scale health communication campaigns (see for example Nan and Thompson, 2021), COVID-19 institutional communications frequently appeared disorganized and disjointed, particularly during the initial outbreak of the pandemic (Noar and Austin, 2020), reducing public capacity to understand the perceptions of the threat and limiting the potential of the measures to flatten the infection curve. These communication failures originated from the uncertain nature of the phenomenon: for several months, knowledge about the virus was still being developed, outbreaks occurred in unanticipated ways, and there were many uncertainties about the scope of the measures and anticipated results. In such a context, the lack of clear, consistent, and transparent communication triggers mistrust, which is the root of several problems during a crisis (Hafner-Fink and Uhan, 2021).

Several countries have provided clear examples of poor institutional communication. Italy has been one of the most prominent cases, also being the country first hit by the pandemic when no other examples to follow were available. The existence of communication problems was clear from the very beginning. For example, at the end of March 2020, the New York Times criticized the authorities' failure to communicate effectively, stating that 'if the Italian experience shows anything, it is that measures to isolate affected areas and limit people's movement must be rapid, clearly implemented and rigorously applied'.[20]

During the first emergency phase, several problems were due to WHO communication errors, with the spread of non-infrequently contradictory messages that created confusion and fostered fears. Apart from the initial ambiguous relations between WHO and China from December 2019 to January 2020 (see Box 5 in this chapter), on several occasions, WHO changed its position and contradicted previous communications. For example, in March 2020, WHO denied its initial advice to test only symptomatic patients. On the use of the masks, WHO took first a skeptical attitude, recommending the equipment only for medical staff, and then changed its opinion. Later, on 6 June 2020, it acknowledged that masks could help reduce infections, even outdoors. As a final piece of contradictory communication, on June 9, the WHO declared that 'it is rare for asymptomatic patients to infect other people', retracting almost immediately in a hastily convened press conference.

In the absence of clear and accurate information and reliable technical supervision, institutional communication faltered significantly, also driven by hyper-trophic media activity. First, communication from national and international institutions could not ensure the coherence of messages, which is essential to avoid confusion among the public. Instead of speaking with one voice, WHO and most other institutions preferred to speak in chorus, often recording tongue-in-cheek voices. In the first few weeks, all the communication channels were filled with information about COVID-19. This was also when people attempted to 'occupy' seats on TV shows as regular guests, regardless of their roles, whether as experts, politicians, or commentators. These roles often became confusingly intertwined. At that time, with limited knowledge and experience, the media became a battleground for debates where varying opinions clashed from a

---

[20] The full text of the article is available at the following web address: https://www.nytimes.com/it/2020/03/22/world/europe/litalia-pandemia.html.

distance, much like a duel. Notable instances of this communication chaos include the back-and-forth between virologists holding different views on the virus's characteristics. Similar behaviors were taken by often schizophrenic newspapers, which could switch from quiet headlines one day to catastrophic messages to implore rigid lockdowns a few days later.

The erratic nature of the information flow can be partly attributed to institutional communication's failure to take the forefront, provide consistent and trustworthy messages, and curb the adverse effects of political communication, which disseminated conflicting messages precisely as the epidemic curve was on the rise. For example, in Milan, Mayor Giuseppe Sala, in an attempt to protect the city's economy, started a 'Milan does not stop' campaign. In a moment, newspapers began to ride the wave and delusional headlines appeared on several newspapers headlines, like: 'Virus, now it is exaggerated. Let us calm down' (Libero), 'Reopen Milan' (Republica), 'Deaths by COVID-19 in Italy? Zero! (Il Giorno).[21] The ongoing struggle became apparent in the final week of February 2020. During this period, the narrative emerged that the issue was being blown out of proportion, leading people to downplay its severity and overlook the risk of transmission.

With a lack of leadership at the local level, political leaders may have used the available information, however uncertain, to try to shape public opinion and advance their own agendas. This led to a situation where different political leaders presented conflicting information or perspectives, which contributed to confusion and polarization among the public. In this environment, it was difficult for people to distinguish between facts and propaganda, and restoring the truth appeared to be an almost impossible task. The political exploitation of uncertain information during the pandemic thus highlights the importance of accurate and reliable data in times of crisis and the need for responsible leadership and communication from political leaders.

In open societies, characterized by extensive freedom of expression and a robust tradition of independent media, people and the media might be more inclined to question government decisions and display skepticism towards government statements and communications. This tendency can be seen as positive, as it can prompt the government to be more transparent, responsive to the public, and make decisions in the people's best interest. However, during crises, it can also result in a trust deficit between the government and the public if it appears inadequately transparent or unresponsive.

In Italy, as in many Western countries, the lack of clear and transparent communication during the early stages of the coronavirus (COVID-19) pandemic may therefore have contributed to confusion and uncertainty. With conflicting messages being shared in the media, it was difficult for people to understand the actual level of danger posed by the epidemic and to make informed decisions about how to protect themselves.

---

[21] Libero, il Giorno and Repubblica are three major Italian newpapers with nationwide diffusion.

The mistrust between the government and citizens hindered the overall effectiveness of the epidemic response. This made it challenging for people to understand the rationale behind decisions and follow the rules and guidelines to safeguard themselves and others.

The U.S. approach to managing the early phases of the crisis offers another interesting example of poor communication strategy. Since the beginning of the epidemic, President Trump and his officials had repeatedly articulated conflicting opinions about the gravity of the coronavirus and contradictory public health guidelines to states, municipalities, and individuals. During the crisis, the news media often reported tensions between the president's office and the top federal health advisors about how to evaluate and respond to the COVID-19 situation. According to Mathews et al. (2020) 'administration officials delayed and limited outreach from experts at agencies such as the CDC,[22] which should have been front and center in sharing the latest data and offering timely, unvarnished guidance to the American people. Some state and municipal officials also failed to provide consistent messages to the public about the gravity of the threat and the need for evidence-based policy responses'. Often, this behavior leads to social conflicts and disorders and, more generally, to an increase in societal hostilities. These are also the situations where during national emergencies the role of government in unifying and motivating public groups is crucial (Haider, 2005; Seeger et al., 2018). In the words of Mathews et al. (2020) within American society, 'intense ideological divisions often complicated a common understanding among US citizens of the risks of COVID-19 and the most effective strategies to combat it. Individuals and groups retreated to their partisan corners and the pandemic became a political football. Federal, state, and local public health officials were subjected to harassment and personal attacks. Many people questioned the reality of COVID-19 and the value of basic measures such as masks and testing, clearly an unacceptable outcome from scientific and public health perspectives'.

In general, the COVID-19 experience shows the importance that government officials provide citizens with accurate, up-to-date, reliable information on what to expect in an emergency about the risk of infection and the public health measures needed to combat the spread of the disease. This service is essential given the prevalence of misinformation in public discourse, its magnitude on social networks, and the dangers posed by foreign state-sponsored information warfare.

3.6.2 Long term care (LTC) residences: 'horrors' and omissions

According to the European Center for Disease Prevention and Control, in the European Union/European Economic Area (EU/EEA), before December 2019, an estimated 2.9

---

[22] CDC stands for the Centers for Disease Control and Prevention. It is the national public health agency of the United States and is a federal agency under the Department of Health and Human Services.

million residents occupied 43,000 of long-term care facilities (LTCF), representing approximately 0.7% of the total population.[23] From the onset of the COVID-19 pandemic up to November 2021, over 88% of the more than 800,000 COVID-19 related deaths in the EU/EEA involved individuals aged 65 and older. By May 2020, deaths among LTCF residents accounted for 37–66% of all COVID-19-related deaths in EU/EEA countries. These numbers become even more striking if we limit the analysis to March and April 2020, when, according to the WHO Regional Director for Europe, up to half of COVID-19 deaths occurred in long-term care facilities, particularly in nursing homes for the elderly. Surprisingly, this happened even though residents of these institutions were subjected to drastic isolation measures before the general population, a circumstance that unfortunately also negatively impacted their mental health and well-being (OECD Policy Responses to Coronavirus (COVID-19), 2021).

Especially during the first wave, containment strategies made LTC recipients even more socially isolated, particularly in residential facilities, with potentially significant repercussions for residents' mental health. In several cases, this brought about a sharp decline in functional and cognitive abilities, physical health deterioration, severe loneliness, and social isolation. At the same time, as suggested by some evidence, the use of psychotropic medications and physical restrictions also increased (Stall et al., 2020).

The dramatic level of mortality in LTC residences became a public concern at the beginning of the pandemic. A report by the London School of Economics (LSE), first published on 17 April 2020, made it official that the COVID-19 pandemic had thrown a tremendous blow on elderly living in LTC facilities worldwide.[24] Based on reliable data from 19 countries, the LSE report was updated several times. One of the latest updates, published on October 14, 2020, showed that among countries with at least 100 deaths with available official data, the percentage of COVID-19-related deaths among nursing home residents ranged from 23% in Hungary to 81% in Slovenia. In between, the percentage ranged from 53% in Norway, to 48% in Belgium, 33% in France, 63% in Spain, 46% in Sweden and 41% in the United States.

Similar studies confirmed the above estimates for the US. The first outbreak in the United States occurred during the early stages of the spread of COVID-19 when the Kirkland Health Care Center in Washington State became an epicenter of the epidemic. More than three-quarters of the residents of that house contracted the coronavirus and 40 died. At the beginning of April 2020, COVID-19 had spread to up to 163 long-term care facilities in Washington state, resulting in 200 deaths.

[23] This information comes from the following ECDC web page: https://www.ecdc.europa.eu/en/publications-data/surveillance-COVID-19-long-term-care-facilities-EU-EEA.

[24] According to LSE researchers, the term LTC includes a variety of facilities that house older adults and include protected residences, nursing homes, hotel homes, protected homes, housing communities, and other forms. The LSE report is available at the following web address: https://ltccovid.org/wp-content/uploads/2020/10/Mortality-associated-with-COVID-among-people-living-in-care-ho-mes-14-October-2020-5.pdf/.

Detailed data from the Kaiser Family Foundation showed how devastating the first wave had been for elderly residents in US nursing homes. On May 21, 2020, 7732 facilities in 43 states reported COVID-19 infections, with just under 175,000 reported cases and 35,118 associated deaths, representing 42% of all COVID-19-related deaths in the United States at that time. In Minnesota, 81% of all COVID-19 deaths occurred in LTC facilities; the percentage was 78% in Rhode Island and 77% in New Hampshire. Other states with percentages greater than 60% followed, including Connecticut, Delaware, Kentucky, North Carolina, Massachusetts, Pennsylvania, and Mississippi.

Many factors appear to have contributed to high levels of mortality in these institutions (in addition to the fragility of their residents). An initial tragic mistake was not being able to recognize the problem immediately. The chilling picture emerged slowly: for example, it took several weeks before LTC deaths started to be reported among official data in most countries, including France and the United Kingdom, while as late as September 2020, statistics remained incomplete or partial almost everywhere. LTCs' lack of preparedness was also a fundamental problem. Numerous residences faced persistent staffing deficits, a situation exacerbated by private owners prioritizing profits over the quality of care. Workers also expressed concerns about the inadequate supply of personal protective equipment for healthcare staff. Furthermore, a severe lack of epidemiological surveillance was manifested in part due to the lack of coordination between LTC facilities and the hospitals. Overall, it is widely accepted that LTC crisis management was fragmented and chaotic, and caregivers were often left to improvise and without adequate equipment.

LTC residences were also unable to provide adequate health care to their residents, including life-saving treatment and end-of-life care to reduce guest suffering. Many cases of lack of proper patient care were reported in all EU countries, with chaotic management of health needs and many deaths. While this might be understood when hospital beds and emergency medical facilities were saturated, in many other situations, the available beds were not used, as observed in some facilities in Sweden. According to an OECD report, 'many older people receiving home care had to abandon care for fear of infection or were asked to postpone it during July–August 2020 (see Fig. 3.5). One concern regarding continuity of care was access to physiotherapy, rehabilitation, and all types of physical therapies among LTC recipients. Some countries have taken targeted efforts to prevent such concerns [...].' In most countries, a key measure to ensure continuity of care for older people was the expansion of digital technology. More than half of the OECD countries surveyed did not have programs or guidelines on developing telemedicine services in LTC before the outbreak. Since then, 21 OECD countries have expanded telemedicine services to allow remote consultations between patients and the health care sector and keep LTC residents in touch with their relatives outside the facilities' (OECD Policy Responses to Coronavirus (COVID-19), 2021).

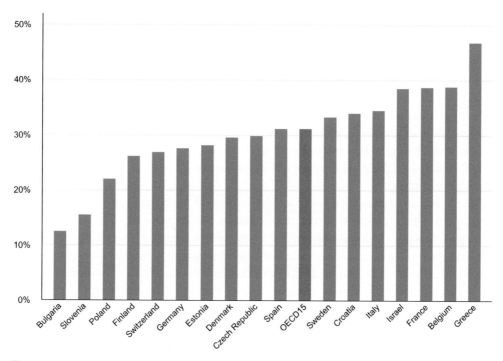

Figure 3.5 Long-term care recipients at home reported forgone care or postponed care. *Source*: SHARE-COVID-19 wave 8 (data refer to June and August 2020). *Note*: LTC recipients refer to those aged 65 years old receiving regularly home care who reported ADL or IADL limitations in SHARE wave 7 (2017). The question on forgone care is, 'Since the outbreak of Corona, did you forgo medical treatment because you were afraid to become infected by the coronavirus?' The question on postponed care is, 'Did you have a medical appointment scheduled that the doctor or medical facility decided to postpone due to Corona?'

Legal actions and civil litigation are among the costly consequences of this state of affairs. In many cases, criminal complaints have been filed against the management of nursing homes. This has already happened in France, Italy and Spain. Violations are often claimed of Article 3 of the European Convention for the Protection of Human Rights and Dignity of the Human Being with respect to the application of biology and medicine (Convention on Human Rights and Biomedicine).[25]

[25] This Convention is the first legally binding international text designed to preserve human dignity, rights, and freedoms through a set of principles and prohibitions against the abuse of biological and medical progress. In particular, Art. 3 concerns fair access to health care and states: 'The parties shall, taking into account the health needs and available resources, take appropriate measures to provide, within their jurisdiction, fair access to adequate quality healthcare'.

Although the LTC residency problem has been widespread, exposing the inadequacy of health care for the elderly, the death toll paid by Italy has been incredibly high during the first wave. While official data is still incomplete, a survey conducted by the Italian National Institute of Health (ISS) revealed that, during the first wave in LTC residencies, a total of 9,154 residents died between February 1st and the questionnaire's completion date (March 26 – May 5).[26] The highest percentage of deaths out of all reported deaths were in Lombardy (41.4%), Piedmont (18.1%), and Veneto (12.4%).

As shown by Fig. 3.6, the number of deaths in Italian LTC residences was generally much higher than the total number of deaths outside these facilities, with Lombardy leading the ranking.

The substantial toll experienced by older individuals and numerous healthcare workers resulted from a combination of various mistakes, some minor and others more severe, ultimately creating a tsunami-like effect. Moving infected patients from hospitals to LTC residences was one of the leading causes of death and was requested by regional governments. For example, the Lombardy Regional Government asked LTC residences to host hospitalized patients (including 'positives') on 8 March 2020, while the infection had already spread to the elderly. This decision came almost twenty days after the first case of COVID-19 was found in Italy and several days after the epidemic spread rapidly. During this critical period, relatives continued to visit older adults in LTC facilities which remained open without protection. On this front, it must be remarked that, on February 23, the Lombardy regional government had taken steps to limit access to one relative per older adult per day. But this measure was useless in the structures that house about a thousand elderly. Relatives represent a valuable workforce for the systems, as they help the elderly eat, change, and move. For this very reason they have close and unprotected contact, so they can easily infect them. On 27 February, the Lombardy Region issued a second directive, requesting that each visitor's temperature be checked and that confirmation of the absence of fever and respiratory symptoms be made. However, in many cases, the institutions ignored the directive or gave it only spotty implementation. They were also not helpful in the process of containing the contagion, giving, for example, the possibility of resuming religious celebrations in the internal church or not closing canteens, bars, and common halls.

The inadequate availability of protective equipment in long-term care facilities is another crucial aspect to take into account. LTC residencies were also largely overlooked

---

[26] The document is available at https://www.epicentro.iss.it/coronavirus/pdf/sars-cov-2-survey-rsa-rapporto-finale.pdf. The survey was carried out by sending a questionnaire to 3417 structures, to which 1356 structures responded, hosting 97,521 elderly persons (34% of the 289,164 beds). For a fair assessment of the data collected, it should be stressed that the response rate from 11 Regions was below 2%, so national validity is conditional on the low participation of the majority of Regions. The survey, among other queries, requested facilities to report the total number of elderly deaths between 1 February and 5 May 2020, distinguishing between confirmed COVID-19 cases and those displaying flu-like symptoms without a confirmed diagnosis.

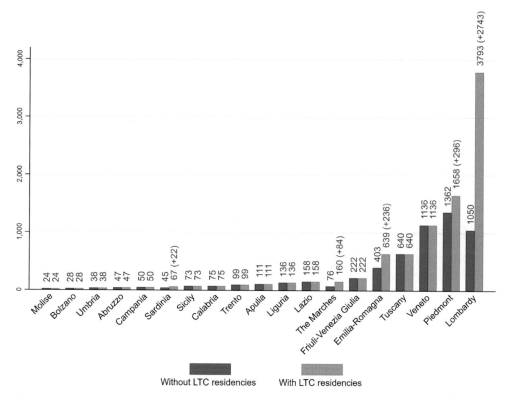

Figure 3.6 Number of COVID-19 deaths by setting (LTC, no LTC) and region in Italy during the first wave. *Source*: Ministry of Health and ISTAT. *Note*: Period March 8 to June 30, 2020. Not all regions have provided information by setting.

in providing protective equipment and emphasizing its need. It is no coincidence that when masks began to be found in hospitals, nursing homes were not considered a priority for supplies. A first supply of about 10,000 masks arrived in Lombardy public LTC residencies only on 19 March 2020, while a substantial supply (122,000 masks) was delivered on 1 April. The lack of action by the regional government was compounded by the lack of action from LTC residences, with the management of some large institutions discouraging the use of available masks to avoid alarming patients.

Finally, outpatient clinics continued to operate until the end of March 2020. Ambulatories were one of the most effective vehicles for transmitting the virus: patients and caregivers arrived without being checked, meeting therapists who assisted residents. Numerous facilities between Bergamo and Milan asked for the clinics to be shut down before 8 March. Yet, the Lombardy Health Agency (Agenzia di Tutela della Salute) advised them to keep the service running.

All this evidence should make it clear that no single factor can explain the increase in mortality in LTC facilities during the COVID-19 outbreak. Rather, the ultimate outcome can be ascribed to a series of events for which responsibility should be distributed among institutions, professionals, and, at times, the general populace. Therefore, while it would be unfair to claim that the decision of 8 March 2020 of the Lombardy Region was the mother of all evils, it is also true that the Lombardy government was effectively unable to carry out effective coordination and to timely understand what was happening. However, it should be recognized that Lombardy and Veneto were the regions where the contagion initially spread. Without past experiences and national directives, operating effectively from the outset was challenging.

3.7. Some final considerations

The Italian experience is paradigmatic in several respects and, as we have observed, tended to be followed to various degrees by many other countries, including the US. The government's reaction to the second wave of the pandemic, widely anticipated in the narrative before and after the first lockdown, is interesting to illustrate the evolution of collective attitudes and politics. Considering the effect of the unexpected events of the first wave, it recalled a strong feeling of 'deja vu'. The surprise of the first wave unfolded in a reaction, which we can define as 'incremental' and marked by three successive attitudes: denial, recognition, and drama. Despite the well-recognized prowess of the health and civil protection systems in Italy, these three phases document the lack of adequate preparation of the public system and the inability to prepare and manage plans and projects concerning unforeseen events and the growing risks of the global economy. That is why one can perhaps understand, but not approve, that the government was trying to make up for this shortcoming through a frantic activity of uncoordinated and progressive intervention measures.

The phase of denial, common to many countries and, between prudence and ambiguity, also to the WHO, consisted in declassifying the pandemic to the category of 'minor risk', from which one could guard with measures widely already included in international health protocols and, therefore, did not require extraordinary actions. In the face of no longer deniable progression of infections and the official declaration of pandemic status by the WHO, the recognition phase involved a reversal of the minimalist narrative of the denial phase and a consequent flood of government interventions. This phase also coincided with the proliferation of scientific committees and task forces and the massive introduction of the medical-epidemiological narrative into the public discourse on the pandemic and its management. In the final phase of national drama, the total lockdown 'closed', at least temporarily, the game.

In the second wave, the same three phases were retraced almost entirely. However, despite *deja vu*, the situation became more worrying than the one developed after the first appearance of the virus because the health crisis was intertwined with an economic and social turmoil that made both emergencies increasingly difficult to govern. The economic rebound in the third quarter of 2020, which had initially sparked hopes for a swift revival of the hardest-hit sectors, suddenly seemed far less promising. However, the international scenario continued to show positive signs, such as a global increase in trade, the return of economic growth in China, and, albeit in a lesser tone, also in the United States, and the overall resilience of financial markets. Therefore, it seemed prudent to expect the fourth quarter to be less positive than expected and with negative implications for the entire next year, but not an economic and financial meltdown. At the same time, the perverse effects of inequality and poverty on lower incomes were amplified by the prospects and by new and uncertain restrictions, more or less general-ized lockdowns, and an expected worsening of the international picture. Furthermore, the burden caused by the restrictions was placed mainly on the less qualified and more precarious employees, on small enterprises in the weaker sectors, such as those of pub-lic services, or more dependent on exports and international flows of travelers, such as tourism and entertainment and culture activities.

According to Italian official statistics, between February and November 2020, there were 83,985 more deaths than the average of previous years, 2015-2019, of which 57,647 were attributed to the pandemic. This means that about 30% of the excess number of deaths were not explicitly attributed to COVID-19, although they were probably associated, in part or in whole, with the health emergency. If we use these numbers to reproportion the mortality figure attributed to COVID-19 (83,157 as of January 18, 2021), we get an excess figure of around 115,000 deaths for the period February 2020-January 2021. These numbers, in themselves dramatic, add a catastrophic dimension to the economic situation that materialized immediately after the second wave of the pandemic. Using the criterion of the Value of Statistical Life (VSL) and the estimates of a recent study by the Italian National Research Institute (CNR), a total loss of about 300 billion euros can be estimated. At the same time, this value quantifies the loss of human capital suffered by the country and the level of health emergency risk, as well as the willingness to pay collectively for the measures that must be put in place to stem its effects.

In the face of these figures, which were also continuously growing, the traditional economic policy instruments used were clearly inadequate, as were the measures taken by the government. The health emergency seemed to dig deep into the inefficiencies of the 'bel paese', exposing the inadequacies of traditional governance and, simultaneously, the inability to overcome them with extraordinary government and public administra-tion actions.

What elements of the health crisis appeared to escape government action to a greater extent? First, there did not seem to be any health governance of pandemic mitigation and adaptation processes. Expert advice was limited to the behavioral restrictions im-posed on the population, but did not seem to concern the behaviors and protocols of the health system as a whole. To a large extent, the latter was left to the management of the regional governments and ordinary administrative structures, according to protocols that varied in time and space and did not seem to respond, except rhetorically, to standard criteria of efficiency and effectiveness. The monitoring of the results and the economic evaluation of the medical, epidemiological, and preventive intervention, from the deployment of resources to the analysis of the effects, were limited to daily bulletins of infections and deaths. The latter were organized in such a way as to provide heterogeneous and statistically inaccurate information, which was difficult to interpret and impossible to analyze using appropriate quantitative methods.

Secondly, the vaccination plan appeared to have been put in place without adequate logistic reserve regarding unforeseen events and inevitable obstacles derived from supplies, difficulties of conservation and distribution, and the heterogeneity and reliability of sources. The timing of vaccinations seemed to follow a do-it-yourself logic by which regional structures competed in execution speed, regardless of the distribution criteria among the categories at risk, susceptibility to side effects and other medical and socioeconomic parameters.

Thirdly, the draft government plan for the extraordinary funds insured by Europe (the so-called Next Generation EU), while trying to cultivate a transformative vision of society and the economy, appeared to spectacularly underestimate the health emergency. On the one hand, it echoed generic proposals for reform based on a territorial network of greater proximity to the sick and their families, but relied on vague technological promises such as telemedicine and digital tools (for example, a frequently evoked and never realized mythical electronic medical record). The dramatic inadequacy of the current network of services in terms of human and social capital was completely ignored. Consequently, even the attempted sizing of the interventions appeared incongruous, both for quantity and for quality of the proposals and projects envisaged. The draft plan ignored the current emergency, as if it could be put in brackets, and was not itself, at least in part, the result of the inadequacy of the health system, the ineffectiveness of prevention measures, the dependence of chronically ill patients on hospital care, and the lack of proximity between medical facilities and the sick.

According to Murphy's law, a popular adage that is often stated as 'anything that can go wrong, will go wrong', five factors determine the universal importance of being prepared to the unexpected for any event or action. These factors are urgency, complexity, importance, skill, and frequency, all present in the critical situation of the pandemic almost everywhere. They were eventually recognized by the national unity government emerging in February 2021 under the leadership of Mario Draghi. They became finally the basis of a well-planned intervention that immediately faced the health emergency with the courage and determination absent in the earlier phases. The new program was an extraordinary effort to mitigate the pandemic through a vaccination campaign, whose logistics was successfully mandated to an army general, and proceeded without stop taking the country out of the health emergency, immediately using all the necessary resources.

One conclusion that can partly explain the Italian case and, more generally, the early reaction to the COVID-19 emergency seemed that democracies may fail to respond timely to extraordinary circumstances because of their polyarchic structures and the high degree of chatter in communication and decision making. This is because decentralized and participatory decision-making cause social actions to be more complex and less timely. In a democracy, power is typically dispersed among multiple branches of government and various levels of administration, and decision-making processes often involve public consultation and debate, with an associated amount of noise and lack of communication.

However, it is important to note that each country has faced unique challenges in responding to the pandemic and that no country was able to display a completely smooth or successful response. In many cases, initial successes and failures were reversed by subsequent events, and over the entire time line of the pandemic performances of most democracies tended to converge. Countries with more autocratic regimes that appeared to be able to implement more centralized and rapid responses to the pandemic also faced challenges and limitations.

Nevertheless, there may be more profound systemic reasons at work. Liberal democracies may be especially unable to cope with emergencies unless they are given sufficient lead time. Governments that have been downsized due to recent neoliberal and populist ideologies, primarily serving as regulators, are ill-prepared and inadequately equipped to govern or manage intricate situations. Emergencies may also expose deeper divisions between short-term individual interests and long-term social goals.

CHAPTER 3 - Take-home messages

- Epidemics often develop insidiously and initial traces of them are lost. To contain them, it is essential to activate a chain of command whose success depends primarily on the existence of appropriate and proven contingency plans, policy strategies, and the expertise and preparation of leaders and decision-makers.
- Policy strategies are based on an analysis of the trade-off between health and the economy, with a major caveat: Different people and countries may have different orientations, and a sound political strategy in one context may not be satisfactory in another.
- Political decision depends on several variables that may change in space and time, but there is a constant: the effectiveness of decisions depends on the competence of policymakers, the leadership capacity, and the robustness and resilience of health systems.
- In all parts of the world, the chain of command has undergone a proliferation of roles that has given rise to an equally large volume of bureaucracy. Unfortunately, the WHO has not set a good example this time. The ambiguous relationship with China at the start of the

epidemic, combined with the WHO's failure to apply sufficient pressure for transparency and its confusing, contradictory communication, diminished the organization's ability to act effectively.

- On top of having a problem of multiple actors, which make the chain of command slow and the centers of responsibility confused, another obstacle to the efficiency of the governance of the epidemic has been the ability to be prepared. While many countries had a prevention plan on record, albeit outdated, it did not receive the due attention it required.
- Effective crisis management necessitates proficient institutional communication. However, this was notably absent: institutional communication was overshadowed by political discourse, leading to a proliferation of opinions rather than dissemination of knowledge, alerts instead of guidance, and confusion rather than clarity and transparency.
- LTC facilities globally witnessed significant outbreaks and fatalities. In specific regions of Italy, this was notably pronounced, and media coverage highlighted various missteps and oversights. Both central and local levels share responsibility for these outcomes.

CHAPTER 4

How did we contain the virus: contact tracing, social distancing, and vaccines

Contents

Preventing the spread of viruses has always been the main objective of all governments during pandemic events. Quarantine measures, social isolation, and other non-pharmaceutical interventions (NPI) were the standard measures adopted when vaccines and drugs were lacking. For example, during the 1918 flu, schools, churches, theaters, and all public spaces were closed. In 2003, during SARS, in China, the central government imposed strict controls. Authority extended to the societal realm, as municipal governments were given the mandate to isolate travelers from affected areas, impose quarantines on those believed to have been in contact with infected individuals, and lock down villages, apartment buildings, and university campuses. All this led to the quarantine of tens of thousands of people. For example, 18,000 people were quarantined in Beijing. In Guangdong, 80 million people were mobilized to clean houses and streets. In the countryside, virtually every village was on SARS alert, with roadside booths to examine all who entered or left. Finally, there was recourse to repressive police measures, which allowed to condemn up to 5 years of jail those who spread 'rumors' about SARS. A century later, the strategy against COVID-19 remained the same, with some heterogeneity in enforcing coercive measures. Still, all measures were relying on NPIs, including bans on social and cultural events and gathering stops to national and international travels, school closures, remote working, national and regional lockdowns, and curfews.

In the absence of vaccines and therapeutic alternatives, NPI-based health measures have proven to be highly effective tools to mitigate the risk and impact of pandemics. However, experience has shown that the right mix of measures depends on the characteristics of the countries and regions of the world. How these measures are imposed may also produce significant changes in human behavior by altering the ordinary course of many social activities. Moreover, they may require considerable logistical and organizational capabilities and entail high private and social costs.

Although helpful, NPIs may not be the most effective measures for containing the pandemic. A comprehensive strategy to control the spread of infectious diseases must incorporate both manual and automatic tracing methods to identify positive patients. Each method has its advantages and disadvantages. For instance, manual tracking provides valuable information, but it relies on patient memory, which may be limited and fallible. On the other hand, automatic tracking can quickly accumulate vast amounts of data, but its usage is constrained by privacy protection principles and laws.

While the arrival of vaccines at the end of 2020 had a decisive effect in limiting the spread of the virus and saving lives, it also highlighted new issues related to the availability and acceptance of vaccines worldwide. Despite the massive success of producing highly effective vaccines in a record time, poor communication strategies allowed misinformation and diffidence into vaccine efficacy and safety and even a 'no-vax' movement to proliferate. This in turn has led to difficulties in achieving high vaccination coverage levels in several countries.

These challenges called all leaders to make crucial decisions about the types of interventions to adopt, with different models of governance, centralized and decentralized management, and ultimate efficacy. The decisions taken often attracted significant criticism, with some people arguing that the government's response was ineffective or too restrictive. A primary point of contention is the widespread belief that there should have been greater public participation in these decisions, with many feeling that the community wasn't adequately consulted in the decision-making process. Additionally, leaders found themselves facing the difficult task of striking a balance between protecting public health and safeguarding individual freedoms and rights. In many instances, emergency measures may have been necessary. However, critics argued that their impact on civil liberties was excessive and the measures implemented were not adequately proportionate, targeted, and subject to review. These decisions often faced significant criticism (refer to Chapter 3). In this chapter, we examine all these aspects.

4.1. NPI interventions: an overview

The SARS-CoV-2 virus is transmitted mainly through contact with people infected while waiting for effective therapies and a sufficient level of vaccination for the entire population. For this reason, avoiding personal contact and maintaining a physical distance of at least 1 or 2 meters represent vital preventive measures. Public health measures that aim to prevent or control virus transmission in the community are called non-pharmaceutical interventions (NPI) and, according to European Center for Disease Prevention and Control (2020c), can be grouped into three main categories:

- individual, such as physical distancing, hand hygiene, respiratory hygiene, or the use of devices such as masks, glasses, and gloves;
- environmental, such as cleaning and ventilation of indoor spaces;
- community, such as the decision to limit the movement and assembly of persons, self-isolation or quarantine, lockdown.

Concerning individual devices, there is no scientific certainty about the distance SARS-CoV-2 can travel when an infected individual breathes, talks, coughs, or sneezes. However, considering other factors such as the concentration of virus particles in droplets or the number of droplets produced, there is evidence that the risk of transmission rapidly decreases with distance from the infectious source (Jones et al., 2020). Additionally, several studies have shown that the droplets created by normal speaking, singing, and breathing are in the range of micro-meters which can quickly settle down, forming a low-concentration droplet-nuclei, called aerosols. Aerosols in turn are in the range of several tens to several hundreds micrometers and can stay suspended in the air for some time. They can travel farther than larger droplets and may remain suspended in the air for longer periods of time, increasing the risk of transmission.[1] The recommended standard of physical distance of one meter or more is correlated with a reduction of about five times the risk of transmission, with a protective effect that increases for each further meter (Chu et al., 2020).

NPIs are also prescribed for hand and respiratory hygiene. Hand hygiene refers to frequent and appropriate hand washing with soap and water or hand-cleaning with solutions, gels, or handkerchiefs. Respiratory hygiene consists of covering the mouth and nose when coughing and sneezing to reduce the transmission of the virus from person to person. Masks, especially surgical masks, are everyday objects. They represent a

---

[1] It's also worth mentioning that various factors impact transmission risk, including the setting (outdoor versus indoor), the length of exposure, and conditions like temperature or humidity.

medical device covering the mouth, nose, and chin to provide a barrier limiting the transmission of an infectious agent. The use of masks may be recommended or imposed when it is impossible to guarantee physical distance in indoor spaces (e.g., supermarkets, public transport) and in crowded outdoor environments, as well as for groups at risk of developing severe complications if infected (e.g., older people or those with underlying diseases). Other NPIs at the individual level include face screens, glasses, or gloves, but it should be noticed that the transmission of the infection via direct contact through the skin has not been documented. In order to avoid introducing infectious material from objects and surfaces into the mouth, nose, or eyes, it is recommended to observe rigorous respiratory and hand hygiene (European Center for Disease Prevention and Control, 2020e). Facial and eye-wear protective equipment is used, mainly by health workers, hairdressers, and aestheticians, in combination with face masks to prevent infectious droplets from reaching the eyes.

Concerning non-pharmaceutical environmental interventions, the evidence shows that SARS-CoV-2 is a virus that can survive in the environment (WHO, 2020e), having been detected on frequently touched surfaces in health facilities (Knibbs et al., 2011; European Center for Disease Prevention and Control, 2020a). As with other respiratory viruses, it is possible that contact with contaminated surfaces and the subsequent transfer of the virus to the nose, mouth, or eyes through the hands is a transmission route. As the virus is susceptible to regular detergents and disinfectants, maintaining good hygiene in both living and working environments is practical for reducing its spread through this route. Other methods for surface disinfection, such as spraying disinfectants (fumigation) outdoors or in large indoor areas (rooms, classrooms, or buildings) or using ultra-violet light radiation, have not been recommended by major international public health organizations due to lack of efficacy, possible damage to the environment and potential exposure of humans to irritant chemicals (WHO, 2020d). Heating, ventilation, and air conditioning systems can play a complementary role in decreasing transmission in indoor spaces (including transport) by increasing the spare rate and decreasing air re-circulation. Poor ventilation in restricted enclosed spaces is associated with increased transmission of respiratory infections (Federation of European Heating, Ventilation and Air Conditioning Associations (REHVA), 2020). Several outbreak investigations showed that virus transmission is exceptionally high in crowded and limited interior spaces such as workplaces (offices, factories) and other indoor environments, including churches, restaurants, gatherings in ski resorts, parties, shopping centers, workers' dormitories, dance courses, cruise ships, and vehicles (Knibbs et al., 2011; Li et al., 2020b; Lu et al., 2020; Rothe et al., 2020; WHO, 2020a). There are also indications that transmission may be linked to specific activities, such as singing in a choir (Hamner et al., 2020) or at church services, which may feature increased droplet and aerosol production.

The last type of non-pharmaceutical interventions to be considered relates to those at the population level. They consist, for example, of isolation of individuals who have tested positive, but do not require hospitalization, quarantine of those who have been in close contact with an infected person, measures to reduce assemblies, work-related interventions up to the closure of schools, non-essential activities, and, ultimately, lockdowns. The first measure to limit the spread of SARS-CoV-2 is the isolation of confirmed COVID-19 cases in dedicated facilities or at home for a specified duration. The aim is to reduce the possibility of potentially contagious individuals contacting others. In the first waves of the pandemic, early identification of patients was considered essential to ensure rapid isolation and contact tracing to prevent further virus spread in the community (European Center for Disease Prevention and Control, 2020b). However, with the successive spread of more infectious variants (such as the two strains named Omega) and the diffusion of rapid home tests, the emphasis on the tracing and quarantine of confirmed cases of COVID-19 has been decreasing. While no rigorous measurements of its impact are available, isolation of individuals with respiratory infection symptoms is likely to reduce the transmission of the disease and limit the spread of the virus in the community (WHO, 2019).

As outlined in the European Center for Disease Prevention and Control (2019) guide, healthy persons who have had high-risk exposure to a confirmed COVID-19 are recommended to be quarantined, i.e., remain in isolation for a specified time depending on the estimated incubation period of the virus. Based on the information available at the end of November 2020, median incubation time for the SARS-CoV-2 virus was estimated to be about five to six days but ranging from 1 to 14 days (Li et al., 2020a; Wei et al., 2020). Generally, 14 days are sufficient to monitor individuals who have had contact with SARS-CoV-2 cases (Backer et al., 2020; Wenxiao et al., 2020). Previous pandemic results suggest that quarantine, if implemented early and combined with other public health measures, is likely to be effective in reducing transmission and preventing new cases or deaths of COVID-19 (European Center for Disease Prevention and Control, 2020e; Nussbaumer-Streit et al., 2020).

As with isolation, implementing measures such as quarantine imply the need to find solutions to logistical, social, and communication challenges. For example, the provision of food must be organized, medical support must be guaranteed to the person infected and to family members caring for those infected. High organizational capacity would also reduce the risk of being unnecessarily quarantined. For elderly patients or those living alone, meticulous management is crucial since an isolated infected individual might not get the necessary care and assistance. On the other hand, if the patient lives with other people and there is not enough space, it may not be possible to enforce quarantine at home. Although a combination of personal and environmental protection measures during isolation may help reduce domestic transmission (WHO, 2019), it appears more appropriate for public health authorities to organize the reception of infected persons in specially designed quarantine facilities (e.g., hotels). Finally, as shown by numerous experiences, including the extreme case of the management of the epidemic in China, quarantine could generate high costs due mainly to the significant number of people forced out of work and the amount of testing required to find contacts. Thus, means must be provided to support the financial, social, physical, and other needs of these patients.

A widely adopted measure by most countries to prevent the spread of the virus has been the prohibition of mass gatherings. This involves limiting the number of participants at both indoor and outdoor meetings and events. In addition, measures such as cancellation, postponement, or reorganization of events were also taken depending on the underlying epidemiological situation. Long-term mass meetings in closed spaces increase the number of close contacts between people and can lead to the introduction of the virus into the community hosting the event and facilitate transmission and spread of infection. The potential role of mass gatherings in spreading SARS-CoV-2 was documented already during the first phase of the COVID-19 pandemic (Ebrahim and Memish, 2020).

Measures such as remote work, flexible work hours, and shifts for employees can be a consequence of lockdowns and quarantines, as well as proactive measures to reduce the spread of COVID-19. In the event of a lockdown or quarantine, it may be necessary to implement these measures in order to keep businesses running while minimizing the number of people who are in close contact with one another. Even in the absence of government-imposed lockdown mandates, if only strong recommendations were given, companies could still implement such strategies to uphold social distancing and other practices aimed at curbing the virus's spread. These measures can also be implemented proactively, as a way to reduce the risk of outbreaks in the workplace.

More generally, smart work, telework, and flexible work schedules are part of a series of actions adopted to reduce the risk of transmission in the workplace and, consequently, in the population. SARS-CoV-2 can be transmitted from person to person in workplaces and other public environments where people gather in enclosed spaces for long periods, as shown by several outbreaks in professional and occupational settings (European Center for Disease Prevention and Control, 2020d). Seasonal influenza and pandemic influenza studies have also shown that measures to reduce contact in the workplace can be partially effective in mitigating an epidemic. Ceasing operations may be necessary in specific scenarios, based on the type of environment and the local epidemiological conditions, especially where more rigorous measures are essential to reduce transmission.

The last stage of non-pharmaceutical interventions involves closures that may affect schools, non-core activities, and, eventually all citizens (lockdown). Schools, school-age children, and nurseries are among the main drivers of infectious agent outbreaks, as was already the case for ordinary seasonal influenza or foot-and-mouth disease.

4.2. The anatomy of contact tracing

4.2.1 Individual freedom and collective safety

The first sections of this chapter discuss the initiatives that have been put in place to control the spread of COVID-19 contagion using NPIs. Social distancing measures, community testing, and the subsequent isolation of positive cases played a key role. As important as it is, this type of activity constitutes, however, a 'reactive' approach to contagion control, simply taking note of who, based on the tests, turns out to be positive and tries to keep people isolated. This approach's limit lies in relying on the voluntary maintenance of social distancing and the choice to undergo tests.

The COVID-19 experience has shown that this form of intervention may not always be practical. For example, the decision to undergo the tests clashes with the waiting times. It is made only by those who think they have come into contact with a positive known to them, leaving entirely outside the perimeter of the checks all those who may have come into contact with people infected but unknown. The classic case is meetings on public transport or in public places, where the probability of contacting infected people may be high. Notification of such meetings may never occur because the people involved do not know each other or are not part of the same knowledge network. Individuals unintentionally coming into contact with the infected person may become instruments of contagion until they develop symptoms (or, worse, for a longer time if asymptomatic).

An alternative method involves proactively addressing the issue. For instance, when there is an infectious individual, a manual or automatic search is initiated for all known or unknown individuals who have come into contact with the infected person within a specific time frame. This process is also referred to as 'contact tracing'. In the case of 'manual' tracking, those who test positive are individually questioned to try to reconstruct, based on their memory, past events of close contact. If the tracing is 'automatic', the process can be conducted via smartphone tracking apps, transmitting anonymous notifications between new index cases and their past contacts. With this approach, the network of contagions can be quickly identified, the people involved can be tested and, if necessary, quarantined to avoid further spreading the disease. These two approaches are complementary: the first is based on the observation and recall of contacts and efficient infrastructure of trained public health officials; the second relies on creating new and well-calibrated technology and population compliance with it. The first approach is a more human-friendly intervention, and a telephone call could be very effective. However, the notification speed of the second approach may be critical given the swift transmission rate of COVID-19 (Kendall et al., 2020; WHO, 2020a,d).

Unfortunately, as already seen in the case of NPIs, even in the case of automatic tracing activities, it is not yet clear whether this type of technology can help contain or even stop an epidemic, as it has never been studied before. Some experts are optimistic that automating contact tracing could help control the spread of the virus. However, everyone agrees that implementing an effective tracking system is only a first step. With today's knowledge, this type of technology is not a silver bullet since it is just part of a more global public health strategy and often the easiest part to manage. Reaching out all people involved in random contacts can be quite simple. However, effective tracing requires a well-run machine that can test everyone, facilities where it is possible to isolate the infected people, and logistics that can meet the needs of those in isolation.

The upcoming sections explore these matters further, probing the advancement of these techniques and assessing their effectiveness in curtailing the contagion's proliferation.

4.2.2 Contact tracing activities

According to Perscheid et al. (2018), contact tracing can be defined as the identification and monitoring of each individual who has been in contact with an infected person. The first attempts at patient tracing date back to the 1930s and were solely manual. The first tracking program was adopted to address the issue of syphilis (Perscheid et al., 2018). In its simplest form, patient identification involves specialized public health professionals interviewing individuals who have been in contact with infected persons. To date, contact tracking is an established part of infectious disease outbreak management, aiming to break infection transmission chains, and has been part of the response to the COVID-19 pandemic in many countries (Jacob and Lawarée, 2021). As such, contact finding and subsequent notification of contact can play an essential role at all stages of the epidemic, especially in the initial stages, before an exponential increase in the outbreak occurs.

Traditionally, contact tracing involves a person who remembers his recent contacts and activities. Individuals who are thought to be at risk of infection (based on contact definitions that may vary by country and change over time) are then approached and asked to take actions to reduce onward transmission, for example, through self-quarantine for a specified period. The ability of any contact tracing system to reduce disease transmission depends on the timeliness with which the detection and subsequent isolation of index cases are carried out and the extent of the perimeter of the analysis. Therefore, the faster and the more thoroughly it can identify the patients infected to advise them to quarantine, the more effective the system will be and the more significant the impact on the culling of the infections.

For COVID-19 specifically, in the initial months following the pandemic's onset, the WHO and several multilateral organizations requested member states to establish tracing systems at the national level. This had a twofold goal: *i)* to halt transmission and *ii)* secondarily, to alter the transmission dynamics.

4.2.2.1 Manual tracking: a key element

Public health authorities usually carry out the manual tracing of cases by interviewing all infected persons. This is a time-consuming and resource-intensive process. In fact, it is necessary to determine the chain of contact for each individual infected within 48 to 72 hours before the onset of symptoms and up to the point of diagnosis or self-isolation. The interviews are instrumental in understanding several critical issues on whether the contacts have caused a new outbreak or if they are entirely harmless. For example, a longer duration of contact and closer proximity mean a higher risk of infection. Likewise, whether a contact occurs in a closed or open environment could make a big difference for the same amount of time and proximity. Of course, if manual tracking can help gain more information, it still has to be handled with care because it relies on a patient's ability to remember what happened in the past 4 or 5 days. Furthermore, if the patient has taken public transport or attended public places (bars, restaurants, shops, etc.), it may be impossible that she knows all her contacts. In any case, producing the people's names and phone numbers (or 'contacts') would be challenging. For all these reasons, the effectiveness of this approach is hampered by:

- the time taken to manually notify contacts, which may delay quarantine, which in turn leads to the poor response by the public health authority (i.e., the time interval due to the manual tracing process) (European Center for Disease Prevention and Control, 2020a; Keeling et al., 2020);
- limited data processing;
- oversights or omissions of the respondent (Ferretti et al., 2020);
- inability to identify individuals in a crowd (Alsdurf et al., 2020; Watts, 2020);
- the need to have many well-trained resources to handle this work (Ferretti et al., 2020).

Despite critical problems, manual contact tracking has a long history of success. For example, it contributed significantly to the eradication of smallpox and the containment of SARS. It has also improved responses to sexually transmitted diseases such as HIV and remains a vital tool for addressing Ebola. The reason for this success is because the benefits of manual tracking go beyond just locating individuals and noting their crucial clinical characteristics. An important role is played by the ability of trained interviewers to assess symptoms, detect asymptomatic carriers, and report other health hazards that automatic tracking systems may have difficulties identifying.

Manual traceability also builds trust within at-risk populations, especially when local norms, values, and concerns are met with respect and empathy. When Ebola was detected in Senegal, for example, the genuine interest of the interviewers in the mental health of the contacts and the provision of a hot line for psychological support helped promote public acceptance of containment measures. Further examples include outbreaks of tuberculosis and HIV in vulnerable populations. Susceptible individuals are more easily persuaded to comply with essential control measures such as behavioral change, data sharing, and acceptance of isolation and treatment in manual tracking systems. This greater propensity to comply may be particularly useful at times when individuals and communities are suspicious of the intentions and interventions of the State, are exposed to misinformation, or are concerned about issues of confidentiality of the information provided or the purposes for which it can be used.

Concerning the current COVID-19 pandemic, an emblematic success is the manual tracking case organized in Japan, where over 450 public health centers have played a crucial role in limiting the spread of the virus across the country. The success of this system relies on the presence of a well-trained team of experts, who were able to trace contacts during the outbreak. Thanks to these individuals, it became possible to swiftly track the connections by instructing newly infected individuals to provide accurate descriptions of their movements, share personal information, and disseminate details about the people they had come into contact with. This enabled the tracing, testing, and isolation of those individuals. The speed with which these activities were carried out enabled a timely monitoring of the outbreaks, making manual tracing effective. This achievement was possible due to the individuals' knowledge of how to address the issue and the existence of a cohesive team with strong communication and collaboration skills.

4.2.2.2 Automatic tracking: a useful complement

Many of the limits in manual tracking systems can be overcome by adopting automatic tracing systems (ATS). Digital tools such as smartphone-related apps with tracking capabilities can support this process. These technologies can help solve the problem of the timeliness of actions and the traceability of contacts outside the usual circle of acquaintances of those affected. Apps allow to identify known and unknown contacts of a confirmed case and may help in their follow-up, particularly in contexts with a large number of cases, where health authorities could be easily overwhelmed in the case of manual tracking. These capabilities can help accelerate the overall process, providing a timely efficiency essential in a pandemic. However, this approach is not free from implementation problems: the practical, technical, legal, and ethical considerations involved are complex.

One major challenge is the diffusion of technology. To be widely adopted by individuals and organizations, ATS is required not only to have the necessary technology available, but also the ability to effectively communicate its benefits and address any concerns that people may have. Privacy is another major concern, because ATS involve the collection of personal information, with related dangers of potential misuse of data. Security may be a challenge, as ATS can be vulnerable to hacking and other forms of cybercrime. Access to tests is also an obstacle to consider, as individuals must be tested before they can be tracked. This requires a significant infrastructure and resources to be in place, including the availability of testing kits and the ability to process and analyze the results. Finally ethical problems are at stake, as ATS can raise questions about fairness and equity, particularly if certain groups are disproportionately impacted.

The first steps to use ATS were taken during the 2014 Ebola outbreak in Guinea. For the first time, efforts were made to convince local public health officials that using apps and other information technologies could help better manage the spread of the disease. In the Guinea case, the ATS adopted was much simpler than the ones used now. Officials would have to locate every person an Ebola patient might have encountered while contagious, and they would have to transfer all this information to the apps to get the data in real-time. In the years since the Ebola outbreak, technology has advanced significantly, and digital contact tracking, mainly via mobile apps, has emerged as a solution to speed up the tracking process. These apps use GPS and other technologies to automatically track individuals' interactions and provide real-time data on potential exposures, making it much easier to quickly trace and isolate individuals who may have been exposed to the virus. This can help to prevent the spread of the virus and control outbreaks more effectively (Kleinman and Merkel, 2020; Oswald and Grace, 2021; Watts, 2020).

In the case of the COVID-19 pandemic, the first operational tracking system was TraceTogether, launched in Singapore on March 20, 2020. According to the Global Pandemic App Watch (2020), at the end of 2020, more than fifty states had implemented (or were planning to launch) a COVID-19 exposure and contact tracking app. However, according to Jacob and Lawarée (2021), 'despite increasing usage, the effectiveness of such apps remains to be demonstrated. A systematic review of the literature finds no evidence of the efficacy of automated tracking apps in reducing infected cases or the number of identified infected contacts (Braithwaite et al., 2020)'.

Nevertheless, there appears to be documented evidence that when the utilization rate is sufficiently high, contact tracking and notification apps significantly reduce the infection rate across the population studied (Alsdurf et al., 2020). The users' number are thus a crucial variable for the effectiveness of digital tracking, and recent research suggests that the utilization rate of an effective digital tracking app should be greater than 60% (Ferretti et al., 2020). In practice, the highest utilization rates recorded are 91% for Ehteraz (Qatar, non-voluntary installation), 62% for BeAware (Bahrain, voluntary installation), and 40% for Covid Tracker (Ireland, voluntary installation). Three-quarters of applications report a usage rate lower than 25% (Global Pandemic App Watch, 2020).

Overall, it may be helpful if manual contact tracing continues to play an essential role in pandemic control in parallel with mobile applications. This thus is partly due to the fact that a fair share of the population (mainly elderly, disabled, and marginalized) may be more vulnerable to infection but less likely to have access to advanced devices and systems on which to spin these tools.

Box 6: How tracing works

The goal of contact tracking apps is quite simple: record all smartphones within a certain distance from a person, and if that person later turns positive for COVID-19, send an alert to any nearby phone.

To ensure that this exchange of information between different types of smartphones occurs without requiring specific skills and knowledge from device owners, a preliminary process of harmonization of data exchange protocols between manufacturers of major operating systems (IOS and Android) installed on mobile phones is necessary. With their operating systems, Apple and Google exert varying degrees of control over nearly all smartphones running on Earth. As a result, the two companies have access to a global network of sensors and computing power of approximately 3.5 billion devices. For this reason, on 10 April 2020, the two IT giants announced their intention to work together to ensure the harmonization program necessary to monitor the COVID-19 pandemic.

In practice, the harmonization process took place by coordinating the operating protocols of the Bluetooth wireless devices. A unifying update joined the two companies' networks to their short-range Bluetooth wireless protocols, which can monitor other phones in an individual's parking area without tracking their location. This procedure also resulted in simplification for developers of contact tracking apps (CTAs), who are now able to operate without specific changes on different platforms.

Under these conditions, CTAs work by transmitting a string of unique numbers and letters from each phone on which they are installed. These transmissions are detectable by any other phone within reach of the blue-tooth device (about nine meters) on which the same app is installed. Simultaneously, the same app installed on another phone will records all the character strings it receives from all phones it has been close to. For security reasons (and because the underlying cryptography protocols of Apple and Google require it), the string of characters transmitted by a phone will change every 15 minutes. Also, at least in the early stages, the records of the strings received will be stored only on the receiving phone, thus making the tracking network a 'decentralized system', whose hacking would be especially difficult.

Suppose a user develops symptoms over a few days from a given date, resulting in positive to COVID-19. In that case, this information will be shared with all mobile phones with which the same user has been in contact within a predetermined time frame (considered valid for attention by authorities). At this point, authorities will transmit several character strings, one for each day the person in question was potentially contagious, to all other apps on the network. These command strings will require apps installed on various mobile phones to search for records that detect signs of proximity to the infected person's phone. If a match is found between the strings sent and the strings present in the cell phone, an alarm is triggered, warning the owner of the cell phone that she had been in contact with an infected person for a sufficiently long time to make it likely that she has contracted the virus.

Later, what happens depends solely on who developed and distributed the app. However, a good response from the system would be to inform the person interested and ask her to contact the health authorities to undergo a test. In this way, infections will be detected quickly: infected individuals will receive helpful information on disease management and possibly quarantine.

4.2.2.3 The limits of contact tracing

Introducing large-scale mobile applications may contribute significantly to the success of contact tracing efforts, allowing health authorities to perform manual tracing in a more targeted way. However, despite the possible low acceptance rates among the population, the implementation of these digital tracking methods has raised many technical and policy issues. Primarily the operation and success of any digital contact tracking app are strongly affected by several technical specifications, including:

- **the proximity measurement mechanism used** (i.e., the wireless communication mechanism). The Bluetooth may have difficulty understanding whether two phones are close enough for their users to become infected. In addition, systems would not be able to monitor contacts of people who have not agreed to use digital tracking devices or people who do not have smartphones (which will not make manual tracking obsolete);
- **the method of data storage** (centralized vs. decentralized). Tracing apps are supposed to be developed and implemented only in close coordination with and under the supervision of the relevant public health authorities. Public health authorities are expected to coordinate the process of finding local contacts following international guidelines, which define which contacts should be followed and what the management of these contacts should be;
- **privacy concerns**. Although the technological and health benefits are provided by digital contact tracing, public decision-makers must consider its privacy impacts (Bengio et al., 2020). Several researchers argue that the adoption of digital contact tracing applications could lead to the economic exploitation of private data and could also create a mass electronic surveillance system (Martinez-Martin et al., 2020; Vaithianathan et al., 2020);
- **the mode of notification** (i.e., the content of alerts sent to users). The content and clarity of the messages sent to the users are crucial for the effectiveness of the service;
- **the relationship with manual tracking** (i.e., how the application is used with manual tracking). Of course, it is easy to see that the app is just a piece of the entire contagion tracking and control system. For example, it will not be worth much unless methods are introduced to test and mass diagnoses people. Without this, there will be no information to be sent to the app's network about who might spread the virus;
- **the existence of the digital divide problem** (i.e., not everyone can use the distributed apps). According to a Counterpoint Research report, about 25% of the 3.4 billion active smartphones cannot support the Bluetooth Low Energy (BLE) standard required by Google and Apple. In the UK, this figure is 12%, but many users have basic mobile phones without access to an app store OS or Android. Many individuals might also find the system difficult to navigate, and a significant number

may belong to low-income groups, lacking the means to afford a phone with the necessary specifications.

All these reasons lead to the conclusion that contact tracing is very complex, and its effectiveness depends on many factors, not least the ability to integrate with all the procedures the public health authorities put in place.

4.2.3 Centralized or decentralized? A question of privacy and democracy

According to their communication protocols, applications for the automatic tracing of infections can be divided into two broad categories: centralized and decentralized. These categories differ primarily in the data storage location and/or how they process COVID-19 arbitrary identifiers and contact data. In the case of centralized apps, the random ephemeral identifiers of all mobile phones in the vicinity of the users are generated, stored, and processed on a central server managed by public health authorities. In contrast, in the case of decentralized apps, the same process is repeated, but arbitrary temporary identifiers are generated, stored, and processed directly on the user's device. The procedure remains the same for both apps, as the updated risk scores are calculated for all interested users. Then, a decision is made as to which ones to inform. When a person is identified as positive for the COVID-19 test by a public health authority, her contact data are loaded on a back-end server. Examples of apps with centralized systems include for example Pan-European Privacy-Preserving Proximity Tracing (PEPP-PT) and OpenTrace/BlueTrace/TraceTogether. In contrast, the DP-3T and TCN protocols and the Google-Apple Exposure Notification API are examples of decentralized systems.

Before addressing the privacy concerns, it is important to acknowledge that the current pandemic situation presents an unprecedented chance for authoritarian regimes to introduce invasive tracking applications as surveillance mechanisms. As Chiusi et al. (2020) emphasized, 'authoritarian countries have made full use of the existing digital surveillance infrastructure and have even added additional equipment and devices to provide automated decision-making (ADM) solutions that prioritize public health and safety issues over individual rights'. The most apparent case in this direction is the Alipay Health Code developed in China with big-data methodologies that allow reaching automatic conclusions on the degree of risk of contagion of a person. According to Chiusi et al. (2020) with this tool, 'citizens must fill out a form with their personal data, which is then exhibited together with a QR code in three colors: a green code allows its holder to move without restriction. Someone with a yellow code could be asked to stay home for seven days. Red means a two-week quarantine. A scan is required to visit office buildings, shopping malls, residential complexes, and metro systems'. Of course, in such a context, the Alipay rating system is mandatory and independently (and opaquely) decides the health status of individuals in a country. The result is that personal rights are defined together with the state of health since people are judged based on algorithmic scoring systems that affect many private and public aspects of their lives.

China is not the only example. Cases of heavy invasion of people's privacy have been reported in Russia, South Korea, India, and Israel. In the latter case, contact tracing was carried out through the location-based tracking app Magen (GPS) and through a digital contact tracking program managed by Shin Bet, the Israeli intelligence agency. According to Chiusi et al. (2020) 'both showed serious flaws: Magen showed recordings of inaccurate location, while the Shin Bet tracking program was revised in July after being widely criticized for forcing people into quarantine by mistake, with complaints often going unanswered'.

Although these extreme control models are more typical of Asia and the Middle East countries, they are not unknown in Europe. In addition to Russia's example, intrusive privacy models were set in Poland (where the 'Kwarantanna domowa' app was mandatory, geo-localization was used, and facial recognition technology ensured that infected people were quarantined), and Hungary (which adopted the same system used in Poland). In Norway, the development of the 'Smittestopp' app was suspended because, according to an Amnesty International investigation, people's privacy had been threatened by highly invasive surveillance tools that went far beyond what could be justified to address COVID-19. In Lithuania, the Norwegian data control authority suspended the app's development for non-compliance with the European General Data Protection Regulation (GDPR).

Against these negative examples, several projects which brought together academics and private individuals implemented methodologies that could be used to develop contact tracing applications in line with European privacy standards. Many of these initiatives are open, and several health authorities have monitored them. A case in point in Europe has been the Pan-European Privacy-Preserving Proximity Tracing (PEPP-PT) action. This is a significant undertaking, partly because it supports the development of national initiatives that pursue a comprehensive privacy protection approach. At the same time, it aims to allow the tracing of chains of infections outside national borders.[2] However, given the presence of private firms in the consortium – PEPP-PT was a coordination effort containing both centralized and decentralized approaches – on 17 April 2020, EPFL and ETH Zurich withdrew from the project, criticizing the PEPP-PT for the lack of transparency and openness and for not respecting enough the principle of people's privacy. Subsequently, institutions including KU Leuven (Katholieke Universiteit Leuven), the CISPA Helmholz Center for Information Security, the European

---

[2] The PEPP-PT (Pan-European Privacy-Preserving Proximity Tracing) initiative, a pan-European project, originated from a collaboration among various European research entities, such as the Fraunhofer Heinrich Hertz Institute and the École Polytechnique Fédérale de Lausanne (EPFL). The goal was to create an app leveraging Bluetooth Low Energy (BLE) technology, circumventing the necessity for invasive state monitoring.

Laboratory for Learning and Intelligent Systems (ELLIS), and the Technical University of Denmark (DTU) also decided to pull out from the project.

The technology agreement between Apple and Google has solved many of the lamented privacy and transparency problems, with a choral effort welcomed and supported by more than 300 experts. Some centralized system developers were pressuring Google and Apple to open up their systems and allow them to acquire more data. On 17 April 2020, the European Parliament gave its support to the decentralized approach, voting by an overwhelming majority on a text stressing 'that [. . .] the data generated are not to be stored in centralized databases, which are prone to potential risks of abuse and loss of trust and may endanger uptake throughout the Union', and demanding 'that all storage of data be decentralized' (CISPA, 2020).

Additionally, on 20 April 2020, Google and Apple defined the jointly developed system as an 'Exposure Notification' rather than a 'Contact Tracking' system, stating that it should be 'at the service of wider efforts of contact tracing by public health authorities' (Russell, 2020). The name change was appropriate since exposure reporting schemes, such as those relying on the Apple-Google procedure, are not 'true' contact tracking systems, since they do not allow public health authorities to identify people who have been in contact with infected individuals (Becker and Starobinski, 2020; Morrison, 2020).

On the same front, the WHO clearly states (in its guidelines published on 10 May 2020) that the adoption of proximity tracing systems should not be mandatory because the data collected can also be used to threaten human rights and fundamental freedoms during and after the COVID-19 pandemic (WHO, 2020a). According to WHO, no advantage should be given to people downloading one of these apps, and no discrimination should result from the decision not to download them. The risk in these cases is crossing the fleeting border, leading to 'normal' mass surveillance using the pretext of an urgent and (presumably) effective solution to the pandemic. In this respect, the WHO document underscores the risk that 'surveillance can rapidly cross that unclear boundary between disease surveillance and population surveillance. [...] Therefore, laws, policies, and supervisory mechanisms are needed to place strict limits on the use of digital proximity tracking technologies and any research using the data generated technologies' (WHO, 2020a). The WHO also reiterated that digital proximity tracking was not essential and should have been considered only to complement, not to replace, manual contact tracing efforts. Digital proximity tracking applications 'can only be effective in providing data to help with the COVID-19 response when these data are fully integrated into an existing public health system and national pandemic response' (WHO, 2020a).

However, if the primary concern is assessing how contact tracing can pose a privacy issue for citizens, we must address a more critical matter: understanding how gathering and controlling the information necessary to manage the pandemic can negatively impact the essence and practice of democracy in various countries. To analyze this complex issue, it's essential to consider that information control can be a vital aspect of effectively managing a pandemic crisis.

In the case of COVID-19, the most evident example is the extent of disinformation ascertained on different occasions and locations. According to Eck and Hatz (2020), 'rumors and falsehoods about origins, symptoms, and remedies for COVID-19 became so widespread that in February 2020, the WHO defined the situation as an 'infodemic', weeks before it declared a pandemic. While governments have taken measures such as censoring online content and criminalizing fake news, it is also true that in some cases, governments contributed to the 'infodemic' by deliberately misrepresenting disease information. These actions are not new: governments have long recognized the value of monitoring information in times of crisis. Scientists may recognize government measures implemented during the COVID-19 pandemic as examples of surveillance, censorship, and information manipulation. Previous research has described how governments use these tactics to contain internal political threats: surveillance enables the detection of dissent and targeted enforcement (Gohdes, 2020); censorship allows states to limit government criticism, stop communication needed for collective action, and silence complaints of abuse (Gunitsky, 2015). Manipulating information can also serve as a tool to distract the public, control the public discourse, and counter citizens' mobilization (Gunitsky, 2015; King et al., 2017). Although a pandemic differs from the controversial policy, we have seen a similar trend across public institutions to increase personal data collection and strengthen their control over information.

As Nabben et al. (2020) rightly emphasize, '[t]he implications for civil liberties of an overt and widespread data collection to the whole population are unpredictable, irrefutable, and permanent. The government's collection of digital contact tracking data as a crisis response measure poses a serious threat to civil liberties due to the potential exploitation of data through hacking, foreign interference, and population manipulation. Such practices threaten individual privacy, public trust, and democratic stability'. For these reasons, the WHO warns against hasty implementation of solutions whose effectiveness is not yet fully demonstrated, and suggests instead providing 'transparency and explicability' of the systems adopted.

To avoid these problems and better monitor the impact and effectiveness of these instruments, it is necessary to address the issue from both a technical and a legal point of view. Nabben et al. (2020) suggest, on the technical side, it would be helpful to adopt the privacy by design approach by considering the protection of privacy during the whole process of setting up the system. This means choosing a decentralized database architecture (blockchain or other peer-to-peer protocols), implementing technical measures for privacy (cryptography and differential privacy), and making all codes open-source. Personal identifiers could be disconnected, relying on non-smartphone-based hardware devices for contact tracking (smart cards or other hardware, data donations in crowdsourcing, or open data initiatives). In addition, third-party audits could make the code more secure in favor of civil society.

From a legal point of view, political responses in line with the principles of human rights, transparent laws, and temporary and proportionate regulations contribute to achieving public health, privacy, and stability results. If data is to be collected, legal guidelines should include clear and transparent measures on how information is gathered, processed, stored, made accessible, and deleted. In addition, the amount of data collected should be minimized. Once the government's ability to collect mass personal information is contained, responsibility for gathering sensitive data for network reconstruction will also be reduced. Maintaining a good relationship of trust between politicians and ordinary citizens will also become easier.

4.3. The use of tracking systems worldwide

Since the early months of the pandemic, there was a worldwide increase in the development of automated tracking applications. These range from basic apps to advanced contact tracing systems capable of identifying and alerting individuals who have come into contact with an infected person. This development involves a diverse array of professionals, including small local programming teams and large-scale efforts by companies like Apple and Google, impacting hundreds of millions globally. As a result, numerous products and services have emerged, many remaining relatively unknown. Additionally, there is limited evidence concerning their societal impact and effectiveness in curbing outbreaks. Even less is understood about the data these systems gather, its future utilization, and if adequate policies exist to prevent potential misuse.

A simple way to better understand the problem is to compare how apps work in Europe and China. In the former case, developers must go through the European privacy watchdog. In contrast, the Chinese system is known to collect extensive information, such as citizens' identities, locations, and even online payment histories. This allows local police to monitor those who who breach quarantine regulations, among other violations.

To gain insight into what was available in the market, its usage, efficacy, and compliance with privacy regulations, a team of experts established the COVID Tracing Tracker. This database collects information on major automated contact tracing efforts worldwide.[3] The initiative aimed to better understand various aspects of tracking, such as relevant and useful monitoring metrics, government records, news and media reports, as well as conducting direct interviews with app developers.

In its first public release, the COVID Tracing Tracker (CTT) initiative gathered information about 25 relevant global automated contact tracking (ACT) projects, with many revealing details of what they were, how they worked, and what policies and processes had been implemented around them (Howell O'Neill et al., 2020). In its release

---

[3] The website is available at this address: https://docs.google.com/spreadsheets/d/1GalIN3Infu_azCG_mNDOljqx2BDZyR234FGd0WK8hYE/edit#gid=0.

in June 2022, the number of initiatives surveyed was 86. As already widely stated in this chapter, ACTs are typically based on apps designed to provide users or public health officials with automated information about who may have been potentially exposed to contagion. The interesting aspect of this project is that the information collected goes far beyond the technical aspects alone. The key questions for each monitored initiative include: Who develops it? Has the app been released yet? Where is it available and on which platforms? What technologies are utilized? The ACT survey also addresses legal concerns such as: Is app usage mandatory? Does it respect privacy? Are citizens' rights safeguarded? How transparent are the developers in their work?

Understanding these questions and others of similar nature is essential to grasp user behavior and the level of adoption. It also sheds light on how different monitoring services can be organized and implemented. In some countries, users willingly engage with apps, while in others, they may be mandated to download and use them. In the case of the Chinese Government, it is understood that, unlike in Europe, law enforcement can use data for purposes other than pandemic control. Other key aspects are the time limits (the use could last longer than the presence of COVID-19 contagion) and the duration of the data use (if the app allows users to delete their data manually or if that task is the responsibility of the service manager). Finally, an important point concerns the transparency of the product/service delivery process, which is usually based on the availability of open-source code. For every one of these characteristics, the CTT system logs a response. If it's positive, a star is marked; otherwise, for negative or uncertain answers, it remains blank. Additionally, the system offers notes to provide context to the data.

Further information on the type of technology used by the apps is available on the CTT site. As reported in Howell O'Neill et al. (2020), the key information listed is:

- **Position**: some apps identify a person's contacts by monitoring the phone's movements (for example, using GPS or triangulation from nearby cell towers) and searching for other phones that have spent time in the exact location.
- **Bluetooth**: some systems utilize 'proximity detection,' where phones exchange encrypted tokens with nearby devices via Bluetooth. These applications more easily maintain data anonymity, so they are considered better for privacy than location monitoring.
- **Google/Apple**: many apps rely on the standard application programming interface (API)[4] that Apple and Google have developed. The API allows iOS and Android phones to communicate via Bluetooth, allowing developers to create a contact-tracking app that will work for both. The two companies have already incorporated the structure to enable tracking directly into their operating systems.

---

[4] API refers to a set of tools, protocols, and definitions that allow software applications to communicate with each other. APIs enable developers to access specific features or data of an application, service, or platform without having to understand the internal workings of that software.

- **DP-3T**: It stands for decentralized proximity tracking that preserves privacy. It is an open-source protocol for Bluetooth-based monitoring where the contact logs of phone are stored only locally, with no central authority, thus having the information on who has been exposed.

Even though contact tracing apps have been launched globally, debates about their appropriate use continue. More than 300 academics and researchers worldwide signed a warning against a possible wrong direction that these technologies may induce or encourage (CISPA, 2020). In their appeal, the drafters say they are 'concerned that some solutions to the [health] crisis, by gradually widening the original objectives, may lead to systems that would allow for unprecedented surveillance of society in general'. In this sense, the drafters warn that '[there] are a number of proposals for contact tracing methods which respect users' privacy, many of which are being actively investigated for deployment by different countries. We urge all countries to rely only on systems that are subject to public scrutiny and that are privacy-preserving by design (instead of there being an expectation that they will be managed by a trustworthy party) as a means to ensure that the citizen's data protection rights are upheld. (CISPA, 2020)' The appeal concludes with four essential principles that must be adhered to by any chosen application for automatic tracking.

- contact tracking apps should only be used to support public health measures for COVID-19 containment. The system must not be able to collect, process, or transmit more data than is necessary to achieve this purpose;
- any solution considered must be completely transparent. Protocols and their implementations, including any sub-components provided by companies, must be available for public analysis. Data processed and whether, how, where, and for how long they are stored must be fully documented. The data collected should be minimal for the given purpose;
- when multiple alternatives exist for implementing a specific component or app functionality, the option that offers heightened privacy protection should be selected. Deviations from this principle are allowed only if necessary to achieve the purpose of the app more effectively and must be justified by temporary provisions;
- the use of contact tracking apps and systems supporting them must be voluntary, used with the user's explicit consent, and the systems must be designed so that they can be turned off and can be deleted at the end of the current crisis.

These four principles address mainly the information-gathering aspect of the tracking system. The system would necessarily be decentralized. This implies that information would not have to be stored on a central server but on individual people's smartphones. Only the code of those who tested positive on the central server would be kept. This code would periodically connect to all other smartphones to check if they met with the phone in question (without transferring confidential information). This is the principle of BLE-based apps[5] that follow the European Commission's guidelines, requiring that it is not possible to browse the code register: and that, once the pandemic is over, all this data will be erased.

4.4. The effectiveness of the various virus contrasting strategies (NPIs): did they work?

A causal effect must be estimated to properly evaluate the efficacy of a virus control strategy. This requires an 'identification approach', which allows for comparing the intervention to a situation where everything is the same except for the intervention being tested, similar to a control group in a clinical trial. In spite of the challenge to obtain proper 'treatment' and 'control' groups, numerous studies have analyzed the impact of these methods in the immediate aftermath of the pandemic. On the whole, they suggest that effectiveness can be achieved, but depends on the quality and the eimeliness of the actions taken by the different countries.

4.4.1 The role of less restrictive NPIs

Regarding masks, confirming earlier results, the studies have collected increasingly solid evidence of the device's ability to prevent virus transmission. In a systematic review, Chu et al. (2020) report that, on average, the use of a mask reduces more than five times the risk of transmission (from 17.4% without masks to 3.1% with mask). Studies on the effectiveness of surgical masks have had mixed results, with some studies failing to find evidence of their effectiveness at both the individual and population level. Nevertheless, a growing body of evidence suggests that masks can have a positive impact in reducing the release of respiratory secretions (Bandiera et al., 2020), as well as in protecting from infections people who wear them correctly (MacIntyre and Chughtai, 2020). For the case of the H1N1 epidemic, Coia et al. (2013) have shown eye protection to be effective in preventing infections in healthcare facilities. However, in most studies, it is not easy to distinguish the effect of eye protection from masking because the two instruments were used together.

On the isolation problem of infected people, the evidence of the COVID-19 experience suggests that it is an effective measure to reduce transmission (Cowling et al., 2020). A simulation study estimated that, in the absence of other measures, a high proportion of cases would need self-isolation, and an equally large proportion of contacts would have to be tracked and quarantined successfully to keep the reproduction number R_0 (the average number of individuals an infected person will transmit the virus to)

---

[5] BLE-based apps refer to applications that use Bluetooth Low Energy (BLE) for communication. BLE is a power-conserving variant of the classic Bluetooth wireless communication protocol. It's especially designed for short-range communication between devices, making it ideal for the Internet of Things (IoT) devices and other applications where power consumption needs to be minimized.

below one (Kucharski et al., 2020). Another study by Chinazzi et al. (2020) suggests that early diagnosis, self-isolation, adequate hand hygiene, and quarantine at home are likely to be more effective than travel restrictions in mitigating this pandemic.

Nussbaumer-Streit et al. (2020) assess the effectiveness of quarantine measures alone or in combination with other non-pharmaceutical interventions. Given the novelty of COVID-19, the authors have included studies on similar viruses (i.e., SARS and MERS) to find as many examples as possible. In quarantine alone, estimates have shown that COVID-19 contagions were reduced from a low 44% to a high 96%. Likewise, estimates of the number of deaths showed a reduction from a minimum of 31% to a maximum of 76%. Combining quarantine with other measures, such as school closures or physical expulsion, may be more effective at reducing the spread of COVID-19 than quarantine alone.

Another critical point to note is that quarantine is more effective and less costly in terms of reducing economic activities than the confinement of citizens, especially when started in the early stages of the virus spread. This, too, is clear evidence of the crucial role that time factors play in managing pandemics.

The effectiveness of policy measures to control the spread of the virus appeared to vary from country to country, depending on the economic situation and the level of government intervention in the economy. This suggests that there is no one-size-fits-all solution, and that a combination of non-pharmaceutical interventions is more effective than individual measures. Furthermore, it is important to note that at some point, additional restrictive measures are likely to have a minimal impact on controlling the spread of the virus. This must be assessed against its often unintended side effects (i.e., the social and economic consequences on communities enduring long periods of extended isolation). Overall this points to the need for decision-makers to continuously monitor outbreaks and assess the implemented measures' impact to maintain the best possible balance of actions. They must also consider that, according to evidence and expert opinions, quarantine alone is essential to epidemic control but has proven insufficient to contain COVID-19.

Regarding the size of the gatherings, an analysis of eight NPIs in 41 countries found the highest reduction in the actual R_t reproduction number when the gatherings were limited to 10 persons or less, compared to 100 or 1,000 persons (Brauner et al., 2020). Based on data from seasonal and pandemic influenza models, mass meeting cancellations before the peak of epidemics or pandemics also appear to reduce the intensity of these epidemics. The same study found that the closure of high-risk businesses (restaurants, bars, clubs, cinemas, and gyms) reduced the actual reproduction of SARS-CoV-2 by 31%, a figure slightly lower than that of the closure of most non-essential activities (40%). A literature analysis also shows that the effectiveness of individual NPIs varies significantly between countries and regions of the world and with human and economic development, as well as different types of governance (Li et al., 2021). The emerging picture is one where there is no solution for all, and no single NPI can bring the R_t indicator below one while, on the other hand, a combination of NPIs is necessary to curb the spread of the virus. According to a study by Haug et al. (2020), the most effective means to reduce the contagion included measures to achieve social distancing, increase the capacity of the healthcare system to cope with the crisis, limit air travel, and improve the communication skills of political and technical decision-makers.

4.4.2 The role of contact tracing strategies

As seen before, contact tracing is used for multiple infectious diseases and only recently has been used massively for COVID-19. However, the number of studies that have explored this disease control strategy is limited and evidence on its effectiveness is scarce. To our knowledge, the most recent reviews are those by Craig et al. (2021) and Hossain et al. (2022).

After analyzing 24 articles that meet designated inclusion standards, Craig et al. (2021) concluded that the evidence suggests that 'contact tracing was mostly evaluated in combination with other non-pharmaceutical and pharmaceutical interventions. Although some degree of effectiveness in decreasing viral disease incidence, transmission, and resulting hospitalizations and mortality was observed, these results were highly dependent on epidemic severity (R_0 value), the number of contacts traced (including pre-symptomatic and asymptomatic cases), timeliness, duration and compliance with combined interventions (e.g., isolation, quarantine, and treatment). Contact tracing effectiveness was particularly limited by logistical challenges associated with increased outbreak size and speed of infection spread'. Similarly Hossain et al. (2022) review 47 articles, of which only six focused on COVID-19. In this case, the authors conclude that 'contact tracing in association with quarantine and isolation is an important public health tool to control outbreaks of infectious diseases. This strategy has been widely implemented during the current COVID-19 pandemic. The effectiveness of this NPI is largely dependent on social interactions within the population and its combination with other interventions. Given the high transmissibility of SARS-CoV-2, short serial intervals, and asymptomatic transmission patterns, the effectiveness of contact tracing for this novel viral agent is largely unknown'.[6]

4.4.3 The role of lockdown as last resort strategy

The strictest non-pharmaceutical interventions (NPIs) to curb the spread of COVID-19 include mandatory home confinement, restrictions on gatherings, shutting down

---

[6] Another significant confounding problem in the evaluation of contact tracing interventions is represented by the time when it is implemented and by the possibility of overlapping with other non-pharmaceutical interventions. All these problems limit the ability of authorities to make informed choices about its deployment.

businesses with high exposure risks, and closing educational institutions like schools and universities. Given the consequences that these policies can determine, it is vital to assess their effects. The most comprehensive collection of studies that have analyzed the effectiveness of lockdowns is presented in the repository created by Marc Bevand (2020). This is a comprehensive list of peer-reviewed and published research articles that evaluate the effectiveness of non-pharmaceutical interventions on COVID-19.[7] According to the repository, articles finding NPIs effective outnumber, by 10 to 1, those finding them ineffective. In June 2022, the list included 78 research articles, of which 63 found that NPIs were generally effective, and nine found inconclusive evidence. Only six of these studies reportedly found that NPIs were ineffective.[8]

An interesting analysis by Bricongne and Meunier (2021) is shown in Table 4.1. It compares eight different studies, all dealing with lockdown-like measures.[9] Although partly mixed, these results seem to convey the idea that these measures work, though not always in a very effective way.

The impact of school closures on reducing SARS-CoV-2 infections appears unclear due to variations in transmission dynamics (Fisher et al., 2020).[10] In the first two phases of the pandemic, only a tiny percentage ($\leq 5\%$) of total COVID-19 cases reported in Europe occurred among minors. When diagnosed, children were found much less likely to be hospitalized or die than adults. Children were also more likely to be infected with mild or asymptomatic forms, which might not be detected or diagnosed. However, when symptomatic, children transmit the virus to adults and infect others just as adults do (Parri et al., 2020; Viner et al., 2020), although it is unknown how infectious asymptomatic children are. In a more recent review by El Jaouhari et al. (2021) the evidence from 24 studies seems to be mixed and varied due to several factors. In particular, most observational studies did not report significant effects. School closures and re-openings did not appear to significantly contribute to COVID-19 transmission when infection prevention and control measures (IPAC) were implemented in schools.

---

[7] The link to the website is available at the following address: https://github.com/mbevand/npi-effectiveness?s=03.

[8] It is worth noticing that the above results are reversed according to Herby et al. (2022). The authors made a systematic review of more than 18,590 studies from which they selected 24 cases for inclusion in a meta-analysis. They concluded that lockdowns reduced mortality by a tiny 0.2%, concluding that they had little or no health effects while having imposed enormous economic and social costs and that these interventions should thus be rejected as a pandemic policy instrument. Following this study, Banholzer et al. (2022) wrote a comment 'raising concerns regarding the subject and conduct of their meta-analysis' and arguing '... that their meta-analysis lacks methodological rigor and should thus not be considered as policy advice'.

[9] On June 2020, six out of the eight studies compared in Bricongne and Meunier (2021) were not present in Marc Bevand's repository.

[10] It is worth noticing that within the repository organized by Marc Boven, only six articles analyze the school closure effect, but none looks at it in isolation.

Table 4.1 Relative impact of various measures on containing infections.

Measure	Deb et al. (2021)	Haug et al. (2020)	Pan et al. (2020)	Jamison et al. (2020)	Li et al. (2021)	Proietti et al. (2021)	Liu et al. (2020)	Askitas et al. (2021)
Workplace closures	Moderate	n.a.	n.a.	Moderate	Moderate	n.a.	n.a.	High
Public transport closures	Weak	Weak	n.a.	Weak	Weak	n.a.	n.a.	Weak
Cancelling of public events	High	Moderate	n.a.	High	High	n.a.	n.a.	High
School closures	Weak	High	Weak	Weak	Moderate	Moderate	Weak	High
Domestic travel restrictions	Weak	Moderate	High	Weak	Weak	High	Weak	Weak
Stay-at-home orders	Moderate	Weak	High	Weak	Weak	n.a.	High	Moderate
Limits on the size of gatherings	Moderate	High	Moderate	Moderate	Weak	Moderate	Moderate	High
International travel restrictions	Moderate	High	High	High	Weak	Weak	n.a.	Weak
Non-essential business closures	n.a.	n.a.	Moderate	n.a.	n.a.	Moderate	Moderate	n.a.

Source: Adapted from Bricongne and Meunier (2021).

'The IPAC measures implemented by the schools were similar across most observational studies and included masks, physical distancing, frequent cleaning, reduced class sizes, and improved hand hygiene. The implementation of these measures in schools has been reported to act as a mediator variable due to reduced transmission and risk of infection. The findings from the ecological studies (19) seem to suggest that 'community transmission were inconsistent, with some studies reporting that school closures were not significantly associated with a reduction in transmission, and other studies reporting a significant reduction in R_t and mortality. Several of these ecological studies reported that other NPIs such as lockdowns, gathering bans, mask mandates, non-essential business closure, and travel restrictions were more effective than school closures in reducing the transmission of COVID-19'. However, it must be noted that ecological studies are considered low-level evidence due to the research design, the multiple confounding factors, and the high degree of variability in the results.

In another study, Lakha et al. (2022) found that 'schools were not the key driver of the latest wave in infections. School re-openings coinciding with returning to work may have accounted for the parallel rise in outbreaks in those settings, suggesting contact points outside school are more likely to seed in-school outbreaks than contact points within the school. Indeed, the wave of outbreaks in all other settings occurred either prior to or simultaneously with the schools' wave'. Looking at the results in Table 4.1, four out of the eight studies confirm moderate to high effects of school closures.

In sum, it is still unclear whether closing schools during certain stages of the COVID-19 pandemic and in different socio-economic contexts effectively reduced community transmission. More generally, it is unknown whether the closure of schools could reduce the burden on the health system, even though it does indeed impose high costs on society and the economy. In the long term, school closures undoubtedly bear negative educational, social, and health consequences, especially for children living in vulnerable settings, marginal communities, or violent families. As shown by the analysis of the economic consequences of these choices for the SARS outbreak (Keogh-Brown et al., 2010), schools and local health departments should weigh the social and psychological consequences of school closures when aiming to shield vulnerable groups from the adverse impacts of epidemics.

The closure of activities considered non-essential (shops, cafes, bars, restaurants, and entertainment venues) while keeping open only those related to the sale of essential goods and services (e.g., food and pharmacies) appears to be an effective intervention. It aids in limiting the contagion by reducing potential human contact while also attempting to lessen the economic fallout of restrictions. On the other hand, measures to force individuals to stay at home are considered a last resort to drastically reduce physical interactions, given their significant impact on society and people. Targeted restrictive measures, both temporally and geographically, would generally be preferable, with priority given to their early implementation to help rapidly reduce the transmission of the virus. On the other hand, the exit phase must be gradual to prevent sudden increases in the contagion and the burden on the health system.

4.4.4 What works

What have we learned from COVID-19 that would be relevant in case of a new pandemic? According to what has been discussed so far in this chapter, we are still learning from our past and present experiences. Although drawing firm conclusions now is challenging, a few tentative lessons can be highlighted.

- First, consistently applying protective measures as early and hard as possible has generally led to better outcomes. The more the intervention is delayed, the higher the risk of worsening health outcomes (deaths in particular) (Buckman et al., 2020; Vinceti et al., 2020) and, at the same time, of economic losses (Eichenbaum et al., 2021; Farboodi et al., 2021). A study by Okell et al. (2020) also found that the dramatic drop in COVID-19 cases recorded in some countries was mainly due to unprecedented government interventions aimed at reducing travel and physical contact between people.
- Second, the early 2020 lockdowns have proved that people can support drastic disease-control measures if they correctly understand their reasons and necessity. Repeated and ineffective lockdowns instead engender widespread rejection. These are seen as government failures to control the pandemic and, therefore, as calls for worthless compliance.
- Thirdly, while drawing comparisons between measures presents statistical difficulties, halting public events has been identified as one of the most effective intervention strategies, as noted by (Bricongne and Meunier, 2021).
- Fourth, lockdowns based on geographical targeting do not seem to produce good results, while better results have been achieved with targeting by age and type of job. Nevertheless, easing lockdown restrictions ought to be undertaken step by step, even as vaccines are being distributed, because preventing a resurgence is dependent on strict health measures.
- Fifth, combined restrictions have the greatest effect, including voluntary responses – such as social distancing – and test and trace. There is no single silver bullet. Moreover, vaccine roll-out should not be considered a good reason to dismantle social distancing measures.

4.5. The role of R&D during the pandemic: where did we do our best?

4.5.1 Cooperation, cooperation, cooperation!

It is broadly recognized that, before the COVID-19 pandemic, countries had not adequately focused on devising a global strategy for serious pandemic prevention, preparedness, and response. No global organization was thus tasked with financing and synchronizing various participants. In hindsight, this condition can be recognized as a form of 'collective failure'. Among other indications, it has highlighted the global need to revamp the existing medical R&D framework, calling for the restructuring of clinical trials and expediting the journey from innovation to production. An other essential outcome of the pandemic has been the highly fragmented nature of the global R&D system. Only a select few countries support the entire process—from basic research to manufacturing and procurement. In this respect, the experience of the vaccine has been emblematic, with only a few regions of the world (North America, Europe, and some countries in Asia) able to develop, manufacture and distribute vaccines, and the rest of the world lagging behind. According to a survey led by National Academy of Medicine President Victor Dzau: 'global research is fragmented, poorly coordinated, under-resourced and cannot manufacture and distribute devices, drugs, and vaccines needed during a health crisis to all countries. Part of the problem is that the traditional linear model of R&D, from discovery and development to application and on to manufacturing, is 'too slow' and needs to change'.[11]

Despite all these problems, the pandemic has generated an unprecedented international effort towards a higher level of collaboration across countries, organizations, and firms. A relevant part of this increase in collaboration comes from the increased mobility of researchers in middle-income countries, which has fostered the rise in collaborative research activities. The knowledge gained from these activities has enabled us to implement crucial and immediate measures at all levels of our society. These actions have successfully limited the spread of COVID-19 by containing new infections, preventing hospitalizations and deaths, and minimizing social and economic disruptions. In this respect, it is worthwhile to report a comprehensive statement from OECD (2021b):

> International scientific cooperation on COVID-19 started through exchanges of data and genetic and viral material, originally from China to other research centers across the world, marking a relatively rapid development compared to previous pandemics. Less than 24 hours had elapsed between the sequencing of the first coronaviruses by Chinese public health laboratories and the full genome data publicly shared in the EpiCoV database of the Global Initiative on Sharing Avian Influenza Data (GISAID), a public-private partnership. Since then, numerous international open data sharing platforms have emerged to provide access to epidemiological, clinical, and genomic data and related studies. Protocols and standards used to collect the data are also being shared with analytical tools.
>
> The COVID-19 Open Research Dataset (CORD-19), created by the Allen Institute for AI in collaboration with the US government and several firms, foundations, and publishers, contains more than 280,000 full-text machine-readable scholarly articles on COVID-19 and related coronaviruses and serves as a basis for applying machine-learning techniques to generate new insights supporting COVID-19 research. Other initiatives include repositories of genome data (such as Nextstrain and GISAID), chemical-structure data (e.g., CAS COVID-19 antiviral candidate compounds dataset),

---

[11] This statement is taken from an interview given by Dr. Dzau to Science|Business during a conference held on 9 September 2021.

and clinical studies (e.g., ClinicalTrials.org for COVID-19-related studies) and data for modeling re-
search (e.g., MIDAS).

The European Commission launched the COVID-19 Data Portal in April 2020 to bring together
relevant data sets for sharing and analysis to accelerate coronavirus research. The portal enables
researchers to upload, access, and analyze COVID-19 related reference data and specialist data
sets as part of the wider European COVID-19 Data Platform. Most scientific journal publishers have
waived traditional access costs related to scientific articles on COVID-19).

This impressive joint effort has allowed the development of safe and effective vaccines in less than ten months from the pandemic's start: a significant achievement, especially compared to previous vaccine experiences that took many years to develop. The vaccine manufacturing and rolling-out phases represent another incredible achievement. In about one year, more than six billion doses of COVID-19 vaccines were administered globally, although 80% of this amount went to high and upper–middle-income countries and only 0.3% to low-income countries.

Despite the enormous success in the development phase, the evidence also suggests that the delivery of COVID-19 vaccines presented several shortcomings, which could have been avoided if there had been better coordination within the R&D value chain. For example, replications of clinical trials could have been circumvented, saving resources and favoring more collaboration. Similarly, the regulatory aspects proved cumbersome and ineffective if we consider that the same vaccines had to be approved at different times in different countries, by various drug agencies, under different regulations. Finally, despite the considerable manufacturing efforts made at the firm level, many developing countries were left unprotected, given that only a few countries have the infrastructure required to manufacture COVID-19 vaccines.

4.5.2 The new vaccine technologies: what are available and how effective are they?

By the end of 2020, scientists were testing about 50 vaccine candidates in clinical trials, and at least 150 were in pre-clinical development, including animal and laboratory tests. The first vaccines approved by the FDA and EMA were the Pfizer/BioNTech vaccine, the Oxford university/AstraZeneca vaccine, and the Moderna vaccine. At the same time, six vaccines in China and Russia received limited or early approval before the completion of Phase 3 clinical trials, which raised some safety concerns.

The vaccine race to protect against SARS-CoV-2 started in January 2020, after the virus's genetic code was deciphered in China and made available to the scientific community. In just nine months, researchers were able to develop very effective vaccines, an unprecedented achievement, given that the average time for vaccine development in the past was about ten years. This was made possible by both recent technological advances and financial efforts.[12] Finally, we should not neglect the logistical efforts. Indeed, while technological advances have simplified product development and data recording, the massive use of social media has made it easier to recruit trial participants, partly because of a strong desire to help. Furthermore, on the logistic side, with the help of public funding, vaccine production was carried out in parallel with clinical trials in the hope that the trials would be successful. It would not have made much sense to arrive at the day of regulatory approval by the FDA and EMA without having the product ready for delivery. Thanks to these record times, the first vaccines started to be administered in the UK and US in mid-December 2020. In the European Union, 27 December 2020 was called vaccination day. As of 21 February, the most advanced countries had used three or more COVID-19 vaccines.

As of 15 February 2022, the WHO's COVID-19 Candidate Vaccine Landscape and Tracker reported 143 vaccines already in use or under human clinical trials and 195 additional vaccines in pre-clinical development (Harris, 2022). The lightning speed with which these vaccines were developed and their effectiveness and plurality demonstrated a response capacity from the medical industry that is unprecedented and unexpected. It runs counter to past experience not only for the speed with which the vaccines have been discovered, authorized, and produced but also for their diversity and competing claims in terms of research, medical control, health industry, and geopolitics.

The vast assortment produced encompasses cutting-edge formulations based on RNA or DNA nucleic acid platforms (making up about 30% of those available to-day). Additionally, there's a substantial group of viral vector vaccines, which are better understood in terms of their immunogenic characteristics and production techniques. In a broad sense, while many other types of immunizations are available for clinical testing and hold promise for the future, the two primary platforms that have produced highly effective and workable vaccines are the viral vector and the RNA messenger methods. These vaccines seem to have very different properties and rather diverse preferential applications for age, sex, and patient condition. The viral vector vaccines are more sta-ble and easily transportable and do not need special facilities or low temperatures to

---

[12] The technological aspect has been crucial. In addition to the record-time decipherment of the virus genome, the vaccine production techniques used were very different worldwide. Vaccines from the University of Oxford/AstraZeneca, Pfizer/BioNTech and Moderna have been developed using tech-nology platforms that involve inserting genetic material from the virus into different delivery vectors (an adenovirus in the AstraZeneca case and an RNA segment in the Pfizer case). Once introduced into the human body, the mechanisms of cellular protein production use this genetic material to churn out the 'spike protein' of the coronavirus, triggering an immune response. These technologies, all discovered and experimented with within the past ten years, not only allow faster vaccine development but also ensure that more is known about the vaccine safety profile from the start. Also, while in the development of traditional vaccines the clinical trial stages were performed sequentially, in the case of COVID-19 they overlapped, making the process much faster.

be stored. They also are known to give immunity, with lower side effects and a higher tolerance for old patients who have underlying pathologies or are otherwise fragile.

The mRNA vaccines are less stable and generally must be stored at very low temperatures. They appear to give more powerful immunity that decays more slowly over time and is more broadly targeted than the viral vector vaccines. All these properties were unknown when they were tested and even after they were authorized. To some extent, they have been discovered during their roll-out in response to unpredictable events, such as the different waves of the pandemic corresponding to the different mutations of the virus. In a broad sense, we can say that we have been, and in a sense we still are, in the middle of a natural experiment, which is unfolding not because of an ex-ante plan but as the consequence of the dynamic interaction of the infecting and the infected. The experiment, conducted over very large samples, must be interpreted in real-time and provides material for thought and scientific understanding in a way that goes much beyond what we have been able to see in any past pandemic experience.

Another aspect to emphasize is that while vaccine data suggests a likely reduction in symptom severity, the impact on transmission chain remains unknown. This lack of knowledge is worrying because, in the absence of vaccines, the more numerous patients showing severe symptoms tend to be excluded from the chain of contagion. At the same time, in a vaccinated population, a greater number of pauci-symptomatic or asymptomatic cases would remain in circulation, thus increasing the spread of the infection.

4.5.3 The vaccine economics: why this time did it work?

What is surprising about these unforeseen and timely outcomes is that they occurred even though economic theory predicts a significant undersupply of vaccines. This prediction is based on the numerous reasons for market failure that characterize vaccine production and consumption. These reasons are broadly summarized by the term 'externality', which economists use to indicate effects of economic activity that are not reflected or are imperfectly integrated into market prices and demand and supply interaction. The most important externalities in vaccine development are on the production side, where R&D costs are unlikely to be repaid since they are often borne without a direct connection to a single product. Even when these R&D costs have been key to vaccine development, when mass vaccination requires public funding and governments' willingness to pay, this externality persists because most of R&D costs will not be recovered by pricing vaccines much above their direct production costs. Furthermore, many technological breakthroughs at the base of a successful vaccine (especially unsuccessful attempts to develop one) cannot be kept secret because of safety regulations or because research results are shared within the scientific community. For these reasons, patents often provide only limited protection and small incentives to investors. Even when they do, they have high social costs since they tend to ration vaccine consumption based on

Box 7: The first COVID-19 vaccines registered

- **Pfizer/BioNTech/Fosun Pharma.** Pfizer partnered with German biotechnology company BioNTech and Chinese drug maker Fosun Pharma to develop a two-dose mRNA vaccine. In mid-August 2020, company officials claimed that the vaccine had produced a 'robust' response in a phase 1/2 clinical trial. The company launched a phase 3 trial in late July, intending to recruit 30,000 people from the United States, Brazil, Argentina, and Germany. Subsequently, plans were announced to increase the number of people recruited to 44,000. In October, the company said it had received approval to enroll 12-year-olds in the trial, the first in the United States to include this age group. By October, the trial had enrolled more than 42,000 people. At the time, the company had not yet carried out an interim analysis of the data in the study, which put it behind the original objective of producing it by September. However, the expectation was that sufficient data would be available in November 2020 to request FDA authorization for emergency use. On 9 November, the company announced the vaccine's effectiveness for over 90% of the participants in the clinical trial. A few days later, company officials said they were applying for an FDA emergency use permit for their vaccine. It was the first regulatory approval in the United States for a COVID-19 vaccine. Officials claimed that the vaccine could be available to high-risk population groups as early as mid-December 2020. On 8 December 2020, the FDA issued documents indicating that the Pfizer vaccine appeared to provide some protection after the first dose and almost complete protection after a second.

- **AstraZeneca/University of Oxford.** At the end of April 2020, the University of Oxford initiated phase 1 clinical trials for a vaccine developed by AstraZeneca. The vaccine is based on a chimpanzee adenovirus, which can carry corona-virus proteins into human cells. In August, AstraZeneca launched phase 3 trials in Brazil, South Africa, and the United States. These studies were interrupted in September when a study volunteer developed a rare inflammatory spinal disorder called transverse myelitis. Rehearsals were resumed a week later in Brazil and the United Kingdom. In late October, the FDA authorized the resumption of the trial in the United States. In mid-November, company officials claimed that their vaccine produced a robust immune response in a clinical trial involving people over 70. Data published on 8 December 2020 indicate that the vaccine appears to be safe and about 70% effective.

- **Moderna**. The company began testing its two-dose messenger RNA (mRNA) vaccine in March 2020 in phase 1 clinical trial, with promising results. In late July, Moderna began phase 3 clinical trials of the vaccine. At the end of August, company officials stated that preliminary data from the Phase 1 study showed that the vaccine had elicited a promising immune response in 10 people aged 56 to 70 and 10 people over 70. After successfully concluding Phase II, the company announced in late October that it had finished recruiting all 30,000 participants in the Phase 3 trial, including more than 7,000 people aged 65 and over 5,000 young people with chronic conditions that increase the risk of severe COVID-19. In early October, company officials announced their vaccine would not be available for wide distribution until spring 2021. Later in the month, Moderna's CEO told investors that the safety monitoring committee could start analyzing the study data in November. In mid-November, Moderna officials reported that their vaccine reached a rate of effective protection of 94% in the initial results of the Phase 3 trial. Experts said that more testing and more information were needed. On 30 November, Moderna started emergency procedures for FDA approval. The vaccine was finally made available at the end of December 2020.

prices, incomes, or both and may thus deprive part of the population of access to a critical public good.

On the consumption side, private willingness to pay for vaccines will generally be lower than their social value because each person's vaccination reduces her probability of being infected and thus reduces the probability of being infected by non-vaccinated persons. Consequently, many people may be inclined to free-ride capturing vaccination benefits from others rather than being vaccinated themselves. Furthermore, people are generally more inclined to access medical services when they are ill rather than to prevent illness, and they may be trying to avoid known or unknown dangers or undesirable side effects of vaccines. The essence of these points is thus that in a typical context, absent a pressing health threat or infection, there would likely be both a shortfall in vaccine supply and a reduced demand for them.

The velocity and efficacy of response in the supply of vaccines and their vast diffusion over the world population, though in the presence of significant inequalities in access, thus seems to have short-circuited the dire predictions coming from economic theory and past experience. How has that been possible? Several explanations can be listed for the success story of the COVID-19 vaccine, which are not mutually exclusive. Here below, we review some of them:

- most of the theoretical arguments on under-supply vanish if the speed of development is sufficiently fast to neutralize the gap between R&D expenditure and vaccine production;
- the scientific community had been working for decades on some of the concepts at the base of the COVID-19 vaccines, although the success had been often meager. However, the lack of success in achieving the goal of manufacturing an effective vaccine also has a silver lining since it is part of a learning-by-doing process that extends to the entire research community. Thus, as a positive externality, new knowledge and new technologies tend to emerge from scientific failure to benefit all scientists working on the same or similar problems. For example, more than 20 years of intensive but unsuccessful attempts at finding an HIV vaccine and also efforts to develop other vaccines for epidemic diseases, such as Ebola and Zika, had a copious positive fallout. By indicating the unyielding paths to discovery, the lessons from the multiple HIV failures contributed to avoiding the efforts and resources for other vaccines going in the wrong direction. Despite their apparent lack of success, the HIV efforts also had the effect of developing and mastering, over a long period, complex technologies, such as those of the adenovirus vector (at the base of the Astra Zeneca vaccine) and the RNA messenger (the mRNA, at the base of both Pfizer and Moderna vaccines);
- unprecedented public funding appears to have been a critical factor in defining a new generation of public-private partnerships for medical research and development. This effort and the richness of past experience are reflected in the present,

highly diversified portfolio of vaccines available for COVID-19. According to one source (Kudlay and Svistunov, 2022), to date, the majority of the vaccines are protein subunit (PS) vaccines, also called peptide vaccines, followed by the ribonucleic acid (RNA) vaccines, the inactivated virus (IV), the non-replicating viral vector (VVnr), and the deoxyribonucleic acid (DNA) vaccines. Other vaccines in clinical development include those containing a viral vector that can replicate (replicating viral vector, VVr), virus-like particles (VLP), VVr with antigen-presenting cells (APC), live attenuated virus, VVnr with APC, or bacterial antigen-spore expression vector. All these vaccines have been researched, tested, and in part produced due to outstanding financial efforts, with unprecedented public sector funding: only for R&D, in the first 15 months of the COVID-19 pandemic, 23$ billion and 70$ billion were reportedly invested in advanced purchase agreements. Similarly, a substantial, though generally undisclosed, private sector investment complemented these figures (only for Moderna, R&D expenditure was reported to be 1.7 billion) (Harris, 2022).

4.6. The arrival of vaccines: the end of a nightmare?

There is no doubt that with the arrival of vaccines, the possibility of seeing a recovery of the entire system (economic, social, and health) and some return to a 'new normalcy' appeared more at hand (Gopinath, 2020). However, achieving global herd immunity was estimated to require several billion vaccine doses. Setting up this massive production and its subsequent distribution presented significant challenges. Persistent efforts are crucial to ensure consistent availability and adaptability to the virus's rapid changes.[13]

Therefore, there is still a long way to go before vaccines are available continuously and in a timely manner in sufficient quantity to immunize the whole world. No wonder, then, that countries rushed to secure supplies of vaccines, even before their safety and effectiveness were established. However, as often recalled by experts, vaccines alone will not succeed in this feat until a considerable greater portion of the world's population has been vaccinated and vaccine updates can be widely available on a continuous basis. The recurrent mutations of the virus with the appearance of vaccine resistant variants may also imply both that new vaccines will have to be developed and that suitable precautions should continue to be exercised by each individual.

4.6.1 The availability of vaccines

As highlighted earlier, making vaccines' development, production, and delivery possible in less than a year was a remarkable accomplishment. According to Science Magazine,

---

[13] The concept of herd immunity will be introduced and discussed later in Chapters 5 and 9. Some technical aspects of how to estimate the threshold are discussed in Appendix A.

developing a vaccine in less than ten months is miraculous and has been defined as the most important discovery of 2020 (Cohen, 2020). According to the OECD (2021a), the rapid development of COVID-19 vaccines results from unprecedented levels of international collaboration, massive public investment in research and development, and production capacity. As we have already emphasized, these outcomes stand out when compared to theoretical predictions of market failures and prior estimates of a vaccine's likelihood to reach clinical trials, which range from 12 to 33% after roughly 7-9 years of development (OECD, 2021a).

However, after the euphoria of the first few weeks of realizing that safe and effective vaccines were on hand, the world became aware that the discovery of an effective vaccine is only a necessary, but not a sufficient condition for its successful deployment in a pandemic. A large number of additional conditions must be satisfied, including authorization, production, distribution, administration, and monitoring. Once the authorizations were granted by FDA and EMA,[14] all other hurdles, as we have seen, have been fraught with political challenges.

According to The Economist, by the date of the appearance of the vaccines, most agreements to secure supplies, known as anticipated market contracts (AMC), had been concluded by wealthy countries. Data from Duke University's Global Health Innovation Center researchers shows that high-income countries accounted for more than half of all confirmed purchases. The United States alone was responsible for nearly one-sixth of AMC, having pre-ordered about 750 million doses from a half-dozen drug manufacturers. That is three doses per person. Canada had purchased ten doses for each citizen (see Fig. 4.1). A group of wealthy countries had agreed to purchase around 600 million doses of Pfizer-BioNTech, almost half of the total quantity drug manufacturers claimed to be able to produce by the end of 2021. The European Commission, with its negative experience with procuring vaccines against H1N1 (swine flu), on behalf of all 27 Member States, organized pre-purchase agreements at a fixed price with key developers of several potential vaccines. These companies received hundreds of millions of euros to set up production plants, even before their vaccines were approved. Approved vaccines were to be distributed by producers in each EU country in proportion to the population as soon as batches of them were available.

This underlines that not every country in the world has been equally lucky. Poorer countries had to rely primarily on the COVAX alliance (see Section 11.4.1 in Chapter 11) to count on the availability of vaccine doses. The COVAX agreement was initially signed to purchase 500 million doses, ensuring equal access to all participating countries, regardless of income. Yet, the coalition ensured that the participating

---

[14] The Food and Drug Administration (FDA) is the US agency responsible for ensuring the safety, efficacy, and security of human and veterinary drugs, biological products, and medical devices. The European Medicines Agency EMA) is a decentralized agency of the European Union (EU) responsible for the scientific evaluation, supervision, and safety monitoring of medicines in the EU.

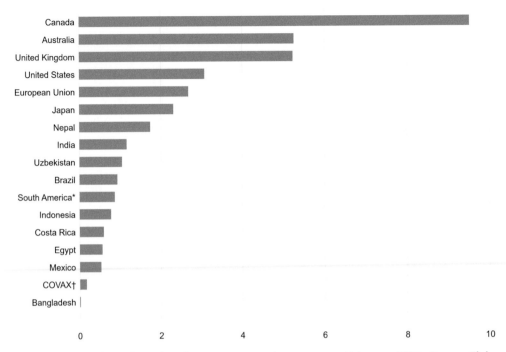

Figure 4.1 Number of vaccine doses guaranteed per person, February 2023. *Source*: Elaborations The Economist on Duke Global Innovation Center, GAVI, and World Bank sources. *Note:* * Brazil is excluded.

countries would get enough doses to immunize only a fifth of their populace. However, there are situations in which high-income countries provided doses to low-income countries. For example, this is the case for Australia, which committed to purchasing five doses per citizen, and signed agreements to provide vaccines for its Pacific neighbors, such as Vanuatu and Kiribati. China, which announced its joining of COVAX on 9 October 2020, also promised to share its vaccines with the poorest countries with close links, including Burma, Cambodia, and the Philippines. In a bold move, Cuba committed to distribute its vaccines free of charge once the research was completed.

According to data collected by the OurWorldinData.org website, by the end of February 2023, 69% of the world population had received at least one dose of the COVID-19 vaccine. In total, 12.27 billion doses had been administered globally, and about 31 million doses were distributed daily. However, only 12% of people in low-income countries received at least one dose. Figs. 4.2 and 4.3 show an international comparison of the percentages of people vaccinated and the total amount of vaccinations administered. Regardless of the non-trivial problems of comparison that may exist in the presence of different vaccination protocols, it is clear the delay with which some countries (especially those with low income) were proceeding with vaccination.

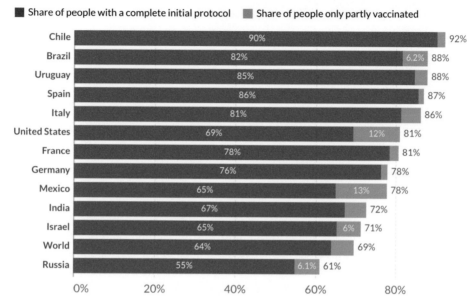

Figure 4.2 Share of people vaccinated against COVID-19, February 2023. *Source*: OurWorldonData.org. *Note*: Alternative definitions of full vaccination are used, e.g., having been infected with SARS-CoV-2 and having received one dose of a 2-dose protocol, are ignored to maximize comparability between countries.

The access and adoption of COVID-19 vaccines, especially in the initial phase, have been heterogeneous and irregular. Countries in North America and Europe started vaccinations earlier, much ahead of regions like Africa and the Middle East. Across all income groups, advanced economies vaccinated a much larger share of their population than emerging and developing economies. The most significant issue appears to be the availability of vaccines for those who require them. The supply of vaccines has been very irregular, and their delivery too slow compared to the challenges of the variants and their uneven diffusion. Production capacity and supply remain crucial factors influencing the pace of national vaccination campaigns. Indeed, as long as the virus remains widespread, there is an increased risk of new, dangerous variants emerging, which can lead to further waves of infection.

Compared to the forecasts and intentions agreed upon before vaccine discovery and approval, an essential tool of the vaccination campaign was the Gavi COVAX facility. This is the only international mechanism aiming to ensure multilateral access to successful vaccines based on a rational distribution of products among governments. It

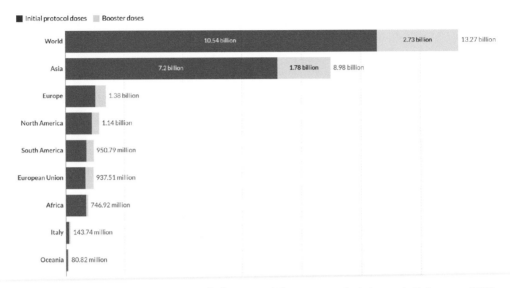

Figure 4.3 COVID-19 vaccine initial doses and boosters administered, February 2023. *Source*: OurWorldinData.org. *Note*: Alternative definitions of full vaccination are used, e.g., having been infected with SARS-CoV-2 and having received one dose of a 2-dose protocol, are ignored to maximize comparability between countries.

subsequently enables access to the vaccine for less affluent countries that might face difficulties in obtaining it.[15]

Throughout the earlier phases of the pandemic, COVAX remained underfunded and continued to compete for supply against bilateral agreements between governments and manufacturers. In 2020, the goal was to raise initial funding of $2 billion for low- and middle-income countries (LMICs), which was achieved by December 2020. However, the money raised afterward was insufficient and too late. It has been estimated that Low- and Middle-Income Countries (LMICs) needed an additional $5 billion in 2022, excluding funds allocated to self-financing nations.[16] This includes $3.7 billion to finance a 600 million dose pandemic vaccine pool as a reserve to help COVAX adapt to uncertainties around the evolution of the coronavirus, including the need for booster shots and vaccines for new variants; $1 billion for supporting readiness among lower-income countries to receive vaccines and deliver them; and $545 million to fund additional rollout costs, such as syringes, transport and no fault compensation insurance.

[15] The Gavi COVAX consortium is a joint procurement program that negotiates supply contracts with manufacturers on behalf of participating countries. The consortium is financed with contributions from various countries.

[16] The full article is available at the following web address: https://www.devex.com/news/we-right-now-are-basically-out-of-money-for-covax-says-gavi-102485.

In the middle of 2022, COVAX hit the milestone of delivering one billion doses of vaccines to 144 countries, of which 90% lower income countries receive donor-supported vaccines. More recently, a paradoxical lack of demand has hampered the attempts to extend vaccination to poorer countries. At the moment (August 2023), the vaccine alliance GAVI has indicated that it will hold onto 60% of the USD 2.7 billion that remain unspent in COVAX to purchase more COVID-19 vaccines, even though demand is low, and most deliveries are currently covered by donations. This leaves Gavi with USD 1 billion in unspent COVID funds.

Global competition or vaccine 'nationalism' can also prevent a COVID-19 vaccine from reaching those most in need. Ineffective assignment of any vaccine could mean that vulnerable people in some countries receive the vaccine after low-risk individuals in other countries, leading to preventable deaths. Investing in vaccine development and fair access would be cost-effective in the long run. Therefore, it is clear that stepping up international cooperation is essential to ensure an adequate supply and delivery of doses worldwide. However, the reality suggests that the success of this operation is limited due to cooperation challenges, even when vital interests are at stake. The pandemic is one indication of how important collaboration is, but the reality remains that of a deep division and the suspicions that inevitably follow. The pandemic will not end until it is over everywhere. As long as few people are susceptible to infection, the whole world will not be safe.

4.6.2 Determinants of vaccinations

Since the onset of the COVID-19 pandemic, countries have been forced to put in place rigorous containment measures to limit the spread of the virus, causing massive economic losses (Carvalho et al., 2020; Coibion et al., 2020; Deb et al., 2021; Caselli et al., 2020; International Monetary Fund, 2021). The development of vaccines has brought increased hope for lifting containment measures swiftly, aiding in the recovery of national economies. However, the mere existence of vaccines cannot guarantee better economic and health conditions. Vaccines must fulfill a series of other requirements to effectively safeguard individuals.

Among these requirements, factors related to demand and supply are crucial determinants for vaccine availability. On the supply side, early procurement, domestic vaccine production, and the required health infrastructure for delivery play critical roles in determining the pace of a vaccination campaign in a country. Fig. 4.4 illustrates the share of the population that completed the initial COVID-19 vaccination protocol in various countries. As evident from the figure, there is a significant difference in growth rates and coverage levels achieved one year after vaccine availability among these countries. For instance, in January 2021, the purchase of vaccines (confirmed and potential agreements) put the US in an advantageous position. In contrast, a two-month delay in vaccine distribution occurred in countries with prolonged negotiations (e.g., France,

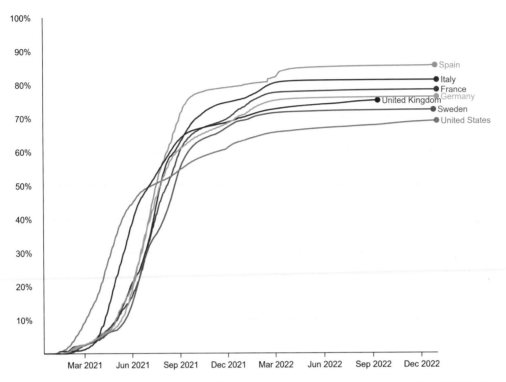

Figure 4.4 Share of people who completed the initial vaccination protocol for COVID-19. *Source*: available at: https://ourworldindata.org/covid-vaccinations. *Note:* Alternative definitions of full vaccination, e.g., having been infected with SARS-CoV-2 and having one dose of a 2-dose protocol, are ignored to maximize comparability between countries. The total number of people who received all the doses prescribed by the initial vaccination protocol was divided by the country's total population.

Germany, and Italy). By March 2021, 20% of the US population had been vaccinated, while this percentage was reached only in June for the three European countries mentioned. The severity of the pandemic may have also had an impact on the population's willingness to accept the vaccine, which was a key factor in demand (see Fig. 4.5).

The significance of these effects on vaccination trends across diverse countries was highlighted in a paper by Deb et al. (2022). The authors used the variation over time and between 46 countries in vaccination rates (as of July 2021) to determine which supply and demand factors explain country differences in the introduction of vaccines (see Fig. 4.5). The results of their study suggest that the severity of the COVID-19 waves in 2020 was the main driving factor (see Fig. 4.4). The most affected countries in terms of cases (e.g., the United States) had higher vaccination rates in the first half of 2021, potentially reflecting the fact that local authorities internalized a greater urgency

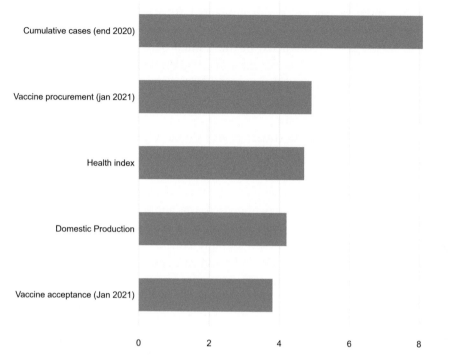

Figure 4.5 Determinants of the introduction of the vaccine. *Source*: (Deb et al., 2022). *Note: The figure reports the impact of a variation in the standard deviation in several factors that may explain the vaccine's introduction on the share of the population vaccinated with at least one dose. Vaccine procurement agreements include confirmed and potential orders by January 2021.*

to vaccinate. The willingness of a population to receive the vaccine also appeared to make a difference (De Figueiredo et al., 2020; Dewatripont et al., 2020).

4.6.3 Who will get vaccinated and who wants to be?

According to several polls, many individuals, including those at high risk of contracting the virus, were reluctant, skeptical, or opposed to taking a vaccine. The reasons are different. Some of these people oppose all vaccines, as they consider them unsafe despite research showing otherwise. In 2021, the WHO declared that rejection and skepticism of vaccines were among the top 10 global health threats.[17] In many countries, hesitation

---

[17] The document is available at the following web address: https://www.who.int/news-room/spotlight/ten-threats-to-global-health-in-2019.

and disinformation about vaccines represent substantial obstacles to achieving coverage and immunity in the community (Lane et al., 2018; Larson et al., 2014).

According to a Gallup poll conducted in the United States in August 2020, one in three Americans would not have accepted a free COVID-19 vaccine. Moreover, a December 2020 Harris poll indicated that roughly 30% of Americans felt they would probably not get vaccinated. Of this group, half stated that they wouldn't accept a vaccine produced outside the United States.

In a survey conducted by IPSOS (International Public Opinion on Social Issues) from October 8-13, 2020, involving over 18,000 adults across 15 countries, 73% conveyed their willingness to receive the COVID-19 vaccine once it was accessible. This is four percentage points lower than a similar survey conducted three months earlier. Among those unwilling to get vaccinated, apprehensions about side effects and doubts over the swift clinical trials were mentioned as reasons. In the survey across 15 countries, 33% of those willing to be vaccinated strongly concurred with these concerns, while 40% somewhat agreed. For the 27% who did not agree, 17% disagreed somehow, and 10% strongly disagreed. The big surprise in the responses to the questionnaire was the variability by country. As can be seen from Fig. 4.6, vaccination acceptance is above average in countries such as India (87%), China (85%), South Korea (83%), Brazil (81%), Australia (79%), the United Kingdom (79%), Mexico (78%), and Canada (76%). It is below average in countries such as the United States (64%), Spain (64%), Italy (65%), South Africa (68%), Japan (69%), Germany (69%), and France (54%). Since August 2020, vaccination intentions have decreased in 10 of the 15 countries, mainly China (down 12 points), Australia (down 9 points), Spain (down 8 points), and Brazil (down 7 points).

These findings are partly the effect of the activity of 'no-vax' groups that had spread their messages on social media before the COVID-19 pandemic, finding many sympathizers, although their claims proved false. However, according to Newhagen and Bucy (2020), anti-vax extremists make up only one-third of respondents who said they would not vaccinate. Moreover, an increasing number of individuals recognize the vaccine's significance but remain unsure and concerned, feeling its rapid development might compromise its safety and effectiveness. Under these conditions, the challenge for public health managers is complicated since they must be able to deliver compelling messages to a range of disparate groups, each holding a unique set of prejudices.

As reported by Berman (2020), the existence of anti-vaccine groups could lead a fair share of the planet to a paradoxical situation so that governments might face the idiosyncratic problem of lack of demand rather than the more structural question of insufficient supply. In part, people's skepticism may also be driven by the excessive public debate about the denial dynamics of the vaccine, which could cause the panic that governments want to avoid. In any case, successful efforts were made in many countries to vaccinate a sufficiently large share of the population. By late 2020, for example,

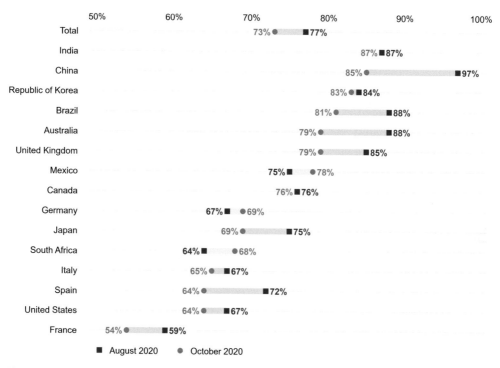

Figure 4.6 Intent on getting vaccinated. *Source*: Processing on IPSOS data. Available at: https://www.ipsos.com/en/global-attitudes-covid-19-vaccine-october-2020. *Note:* The study is based on a sample of 18,526 adults interviewed online, aged between 16 and 74, in 15 countries.

Germany was rushing to set up more than 430 mass vaccination sites and organize vaccination teams for nursing homes and mobile teams that would visit patients at home. Other European countries made similar preparations, though most far behind Germany. Eastern European countries were much slower in organizing the vaccination campaign and, consequently, had lower vaccination rates than Western European countries. This was due to limited vaccine supply, logistical challenges, and vaccine hesitancy. However, by 2021, many Eastern European countries had caught up with the vaccination rate of Western Europe.

4.6.4 The effectiveness of vaccines on health outcomes

Within OECD countries, infection rates began to rise again from June 2021 due to the spread of the Delta variant, and more recently, the highly contagious Omicron variant. Fortunately, the increase in infections was not accompanied by a proportional increase in deaths. For example, in countries with vaccination rates above 65% in mid-October 2021, weekly deaths from COVID-19 decreased by an average of 86% since the end of January 2021, compared to a decrease of 55% for OECD countries with lower vaccination rates. The introduction of COVID-19 vaccines in 2021 thus marked a turning point in global efforts to keep the pandemic under control (OECD, 2021a).

The increases in COVID-19 infections and deaths since June/July 2020 mainly occurred among the unvaccinated, although vaccination rates stabilized at around 60-70% of the population after a rapid initial increase. For example, data from France for the last week of September 2021 indicated that seven-day incidence and mortality rates were eight times higher among unvaccinated than among fully vaccinated. Furthermore, non-vaccinated people accounted for 74% of all hospital admissions to COVID-19 and 77% of all admissions to COVID-19 ICU.

In Italy, according to the ISS Bulletin of February 9, 2022, (Task force COVID-19 del Dipartimento Malattie Infettive e Servizio di Informatica, 2022), the vaccination campaign that began on December 27, 2020 led to the administration of 131,548,249 doses.[18] In the age group 5-11, vaccination began on December 16, 2021, and coverage rates were 13.8% for those who received one dose and 21.5% for those who received two.

As illustrated in Fig. 4.7 and Table 4.2, the vaccination coverage significantly impacted the age-standardized hospitalization rate for the unvaccinated population aged 12 and older. Between December 24, 2021, and January 23, 2022, this rate was roughly six times higher for individuals unvaccinated compared to those who completed their vaccination course within 120 days. Additionally, it was about ten times higher than those who received a booster dose. Similarly, the age-standardized ICU admissions rate for the population aged ≥ 12 years for the unvaccinated was approximately twelve times higher than for those vaccinated with an entire course of ≤ 120 days and about twenty-five times higher than those vaccinated with an add-on/booster dose. Finally, the age-standardized death rate for the population aged ≥ 12 years for the unvaccinated was about nine times higher than for those vaccinated with an entire course of ≤ 120 days and about twenty-three times higher than for those vaccinated with an additional/booster dose.

Moreover, the vaccine's efficacy, measured as a percentage reduction of risk between the vaccinated and unvaccinated, shows that it:
- Reduced the risk of SARS-CoV-2 infection by:
  64% within the first 90 days post-vaccination, 52% between days 91 and 120, and 42% after 120 days post-completion of the vaccine regimen.
  66% for those who received an additional dose or booster.
- Lowered the incidence of severe disease by:
  88% for those vaccinated within the initial 90 days, 90% between days 91 and 120, and 84% for those past 120 days since completing the vaccine series.

---

[18] These are the details: 47,253,208 first doses, 48,642,437 second/single doses, and 35,652,604 third doses; data available from https://github.com/italia/covid19-opendata-vaccini.

Table 4.2 Incidence rate of reported COVID-19 cases, hospitalization, admission to intensive care, death per 100,000, and relative risk by vaccination status and age group in Italy.

Group	Age	Rate (per 100,000)				Relative risk		
		Non vaccinated	Vaccinated with full cycle completed >120 days	Vaccinated with full cycle completed ≤ 120 days	Vaccinated with full cycle + booster	Non vaccinated vs. vaccinated w. full cycle >120 days	Non vaccinated vs. vaccinated w. full cycle ≤ 120 days	Non vaccinated vs. vaccinated w. full cycle + booster
Diagnosis bw. 07/01/2022-. 06/02/2022	12–39	15,413.10	8,403.20	9,865.90	5,247.40	1.8	1.6	2.9
	40–59	14,104.60	7,671.40	7,409.20	4,155.30	1.8	1.9	3.4
	60–79	10,380.50	4,991.70	3,987.00	2,121.90	2.1	2.6	4.9
	80+	10,213.80	3,193.20	3,891.10	2,280.00	3.2	2.6	4.5
	Total	13,268.20	6,860.60	7,052.90	3,842.20	1.9	1.9	3.5
Diagnosis bw. 24/12/2021- 23/01/2022 with hospitalization	12–39	96.5	26.5	28.4	27.1	3.6	3.4	3.6
	40–59	153.7	31.7	30.2	22.4	4.8	5.1	6.9
	60–79	604.6	125.4	89.6	42.9	4.8	6.7	14.1
	80+	1,949.50	433.7	354.7	137.6	4.5	5.5	14.2
	Total	399	87.3	71.8	38.7	4.6	5.6	10.3
Diagnosis bw. 24/12/2021- 23/01/2022 with ICU	12–39	2.6	0.5	0.4	0.6	5.2	6.5	4.3
	40–59	19.6	1.7	1.3	1.2	11.5	15.1	16.3
	60–79	106.9	11.2	7.5	3.0	9.5	14.3	35.6
	80+	69.9	9.5	11.4	3.3	7.4	6.1	21.2
	Total	40.2	4.3	3.4	1.6	9.3	11.8	25.1
Diagnosis bw. 17/12/2021- 16/01/2022 with death	12–39	1.2	0.3	0.0	0.1	4.0	–*	12
	40–59	10.6	1.2	0.7	0.6	8.8	15.1	17.7
	60–79	122.2	15.3	12	5.7	8	10.2	21.4
	80+	815.1	129.5	104	32.2	6.3	7.8	25.3
	Total	103	15.2	12	4.4	6.8	8.6	23.4

* Non-computable estimates for low frequency of events in some strata.
For more details see Methodological note paragraph 3 in Task force COVID-19 del Dipartimento Malattie Infettive e Servizio di Informatica (2022).
For the age group 5–11 it is not yet possible to provide the relative risk estimate as the vaccination of this age group started on 16 December 2020.
The rate relating to the overall population (Total) is equivalent to the standardized rate compared to the ISTAT 2021 population.
Source: Task force COVID-19 del Dipartimento Malattie Infettive e Servizio di Informatica (2022).

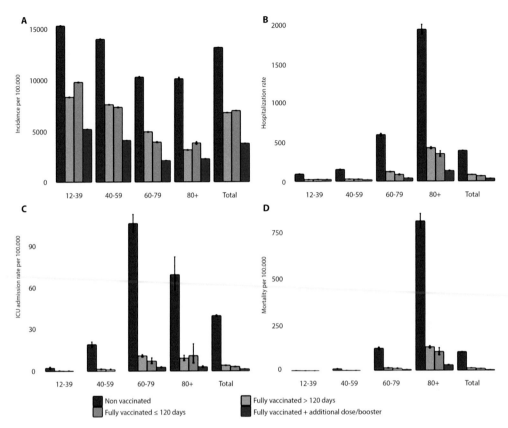

Figure 4.7 Incidence rate of reported COVID-19 cases (A), hospitalization (B), admission to intensive care (C), and death (D) per 100,000 by vaccination status and age group. *Source*: Task force COVID-19 del Dipartimento Malattie Infettive e Servizio di Informatica (2022). *Note:* For more details, see Methodological note paragraph 3 - Non-calculable estimates for the low frequency of events in some strata - The rate for the overall population ('Total') is equivalent to the standardized rate concerning the ISTAT 2021 population (https://demo.istat.it/).

94% for individuals given an extra dose or booster.

Similar outcomes have been observed in the United States where, since the spread of the Delta variant, people unvaccinated faced a five times greater risk of infection, a ten times greater risk of hospitalization, and an eleven times greater risk of death.

To confirm these results from ISS sources, Fig. 4.8 shows the trends in the share of the vaccinated population, the average of new cases per week per million inhabitants, the number of patients in intensive care per million inhabitants, and the deaths per million inhabitants for Italy and the European Union. These numbers show that as

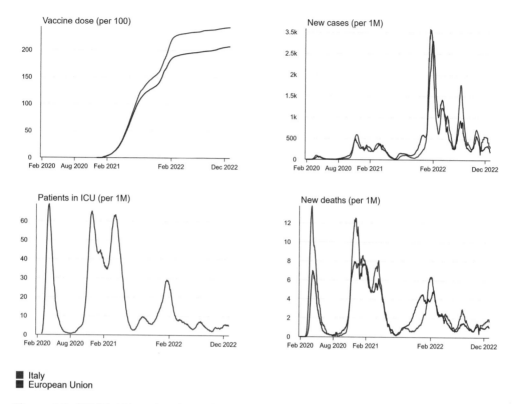

Italy
European Union

Figure 4.8 COVID-19 vaccine doses, Tests, ICU patients, and confirmed deaths. *Source*: Our-WorldonData.org. *Note:* Limited testing and challenges in attributing the cause of death means that case and death counts may not be accurate.

the vaccination campaign progressed, there was a significant reduction in severe and critical cases. This reduction occurred despite the arrival of the Omicron variant which, although less pathogenic, nevertheless infected a significantly higher number of patients. Additionally, Italian statistics show that among those infected with just the Omicron variant, the unvaccinated face substantially higher risk of intensive care treatment and fatalities, age notwithstanding.

It is clear from the above data that vaccination campaigns have helped protect older people and other vulnerable groups. Considering the gradual progress in vaccine distribution and the logistical challenges of swift vaccine implementation, all OECD countries established clear priorities regarding which population groups to immunize first. Although the precise sequence of vaccinations differed from one country to another, the elderly and other vulnerable groups were consistently given a high priority. By October 2021, nearly all OECD countries had made vaccine access universal for adults, with teens also included in most countries' vaccination campaigns.

The impact of vaccination on vulnerable groups was evident. This can be seen from the Italian data reported in Fig. 4.7 and Table 4.2 and from much international evidence. In Austria, for example, infection rates for people aged 80 and over declined since the beginning of 2021 and were close to zero in early July 2021, with nearly 93% of this population group fully vaccinated. The spread of the Delta variant again increased infection rates from around July 2021 across all age groups. However, because the older population group had much higher vaccine protection than the younger groups, the subsequent increase in infection rates – due to the higher transmissibility of the virus variant and the decrease in vaccine efficacy – was much more limited in this age group than in younger people. Similar patterns have been observed in Germany, where data shows a much faster decline in infections among people aged 80 and over than in younger populations since January 2021.

Advances in vaccination coverage have also helped reduce hospital admissions in 2021, particularly among older people. In the United States, for example, hospitalization rates among people aged 85 and over dropped markedly as vaccination campaigns accelerated. By June 2021, hospitalization rates in this most vulnerable age group became very close to the overall hospitalization rate in all age groups. Hospitalization rates increased again since July, including the Delta variant, before peaking in early September. Yet, while the rate of hospitalization for individuals below 50 remained consistent from January to September 2021, the rate for those above 85 was just one-third of what it was in January's peak.

In the context of an international comparison, Deb et al. (2022) demonstrate that the positive effect of vaccines remains valid when the analysis is extended to a large number of countries (46). From Fig. 4.9, it is seen that vaccinations have a large and statistically significant effect on new COVID-19 cases, ICU admissions, and, ultimately, deaths (as a share of the population), as well as the virus reproduction rates. Vaccinations also reduce the number of ICU patients per infected person, thereby improving the resilience of the health system to cope with the spread of the virus. Furthermore, it can be seen that the effect of vaccines varies according to the country's specific conditions.

Interestingly, COVID-19 vaccines were found to be more effective in reducing new COVID-19 infections when implemented alongside stringent containment measures, indicating a synergistic relationship between vaccines and containment policies. Furthermore, an increase in vaccine distribution will likely lead to a more significant decline in new cases if a country is amidst a significant wave of infections. This suggests that vaccines should be channeled wherever possible to countries facing outbreaks.

To sum up, in OECD nations, vaccines have demonstrated their efficacy in lowering the chances of symptomatic COVID-19 infections, as well as hospitalizations and deaths related to the virus.

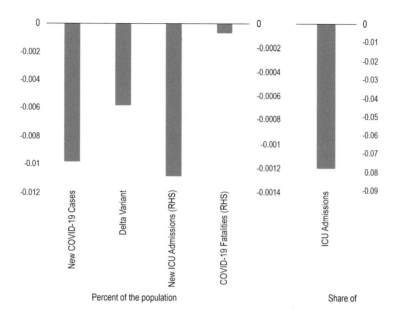

Figure 4.9 Effect of COVID-19 vaccines on health outcomes (percentage of population, share of COVID-19 cases). *Source*: Deb et al. (2022).

4.7. The role of vaccines in the recovery of the economy

Developing safe and effective vaccines against diseases that cause substantial morbidity and mortality has been one of the significant scientific advances of the 21st century. Vaccination, sanitation, and clean water are undeniably essential public health interventions to improve health outcomes around the world. Vaccines are estimated to have prevented 6 million deaths from vaccine-preventable diseases each year. Broadly speaking, the introduction and distribution of a vaccine enable individuals to restart economic tasks that were previously restricted or halted: this includes returning to work, the resumption of recreational activities and social life. People may return to dining indoors at restaurants, traveling, and staying in hotels. Students may return to school, and hospitals can resume more elective procedures linked to prevention and intervention activities to ensure greater levels of health.

However, more than three years after the onset of the epidemic, the availability of vaccines can only be considered a necessary but insufficient condition to bring the economy back to pre-pandemic levels. This is due to several causes, including persistent changes in some habits by individuals, perhaps lowering the demand for some goods and services permanently. According to a survey conducted in several large companies by Willis Towers Watson, it was estimated that 19% of employees would continue to work from home after the pandemic: a value that corresponds to three times that recorded in 2019, but down from the 44% who worked from home in 2020 (Willis Towers Watson,

2021). Those who have been unemployed for an extended period also may face greater challenges in personal relationships, career aspirations, and self-esteem. These challenges might persist and make their recovery difficult, even once the economy stabilizes.

4.8. Some concluding considerations

The first response to the pandemic was based on restrictions and social isolation measures. As these continued and became more stringent, it was expected that some positive signs of the well functioning of the strategy would begin to emerge. At the same time, questions loomed large as to when and how to get out of uncomfortable and costly conditions, which seemed to produce substantial benefits to mitigate the damage and risks of the epidemic, but, at the same time, progressively risked killing the economy.

In a mirror image of the dilemma faced with the option to discontinue collective containment measures, governments around the world faced the choice to exit the strategy of collective isolation, using solutions that contemplated a greater differentiation of protection measures and social distance, the recovery of production and a first re-injection of economic resources into the system. At the same time, this choice was an option to exit from the current situation and invest in a better future. But exercising the exit option presented the risk of reactivating the virulence of the epidemic, with the costs to human lives and associated individual and social suffering.

A possible benefit was the reactivation of the economy and the end of the growing social unease, also dictated, in large part, by the feeling of helplessness in the face of a possible economic catastrophe. In conditions of 'dynamic' uncertainty, that is, of unknown things that only the passage of time can reveal, waiting before acting was what the analogy with financial options suggests. Therefore, in most countries, over and over again, amidst successive waves of infection, the policy chosen was to stick with the measures taken and to wait before taking further actions. Wait until enough new information could be gathered over time to determine if the implemented measures had been effective and successful in the long term, and if it was safe to consider changing the current regime.

On the other hand, this approach carried the risk of being too slow to respond to changing circumstances, which could lead to prolonged lockdowns or other measures after they were no longer necessary or effective. Additionally, it could have led to prolonged economic disruption, as businesses and individuals were forced to abide by excessive and no longer necessary restrictions. By waiting for sufficient information before making changes, countries might have missed opportunities to adjust their policies in a timely manner and mitigate the negative impacts of the pandemic.

By sticking with the measures taken and waiting for more information before making changes, countries might also have been slow to respond to a resurgence of the infection after a new wave of the virus. This could have led to a delay in implementing new measures or adjusting existing ones, allowing the virus to spread more rapidly and increase the number of cases and deaths. This approach might also have led to an over-reliance on the previously implemented measures, which might have lost their effectiveness against the new variants of the virus, leading to a higher number of cases and deaths, as well as prolonging the duration of the pandemic.

The decision to discontinue the measures taken in a previous phase and to adopt new ones to face changing conditions required a difficult balancing act between an ethical and economical choice. The investment in emergency measures had a double cost: on the one hand, social isolation-based containment policies could be replaced by more expensive testing, tracking, protection and insulation measures. On the other hand, the value of these measures depended on the value of human lives that could be lost or saved compared to keeping or discontinuing lockdowns and other social restrictions. Therefore, most countries made different choices based on different estimates of increasing/decreasing infections, hospitalizations, and deaths between the partial or total lockdown regime and the more articulated and proactive alternative.

In countries better equipped and organized, such as Korea and Japan, selective containment measures proved more effective than collective isolation, even though these apparent successes are still poorly documented and remain somewhat disputed. The uncertainty surrounding these effects was also one of the primary sources of the high perception of risks from exiting the social restriction regime. Reflecting this uncertainty, the restrictive interventions were dominated by risk considerations and almost entirely reactive. As such, they tended to discourage any attempt to adopt proactive policies (for example, tracking and isolation) based on an articulated technological and effective set of measures.

Even though broad-based studies are still not available, there is much evidence that the overall social impact of the pandemic has been dismal. Both during the first containment period and the more optimistic phase after the advent of the vaccines, its uneven economic distribution within and across countries resulted from existing inequalities in wealth and human and social capital. Its chaotic unfolding progressively exacerbated social fragilities in the distribution of well-being and wealth worldwide. In many cases, it unleashed selfish and rival behaviors in consumption habits, irrational fears, and unwillingness to cooperate.

The lack of coordination has been a major problem nationally and internationally, and a short-sighted nationalistic policy for procuring and using vaccines has tended to prevail in the most advanced countries at all stages of the pandemic. According to the UN, as of June 2021, the 10 richest countries in the world had effectively monopolized both the consumption and supply of vaccines, with 1 in 4 people vaccinated versus 1 in 500 people in poorer countries. However, the apparent end of the acute pandemic crisis in developed countries, achieved through the success of mass vaccination, soon proved to be a temporary illusion. The pandemic spread to the rest of the world, where the lack of financial resources and international coordination made the supply of vaccines insufficient, poorly distributed, and insecure.

In this context of unsustainable social consequences and urgent international action, the G7 decision to endorse a minimum global tax of 15% on multinational enterprises' profits was good news. In October 2021, the OECD estimated that at least $81 billion in additional tax revenue each year could be raised under the reform. While the tax has been hailed as a conquest of fiscal justice and an aid to the governments' coffers of rich countries under stress from the economic crisis, it could make an immediate and decisive contribution to the international financing of mass vaccination for developing countries.

Given the importance of funding preventive measures and bolstering vaccine production capacity for potential future pandemics, one might wonder: would the revenue from this tax provide a robust framework to preemptively tackle future health crises, especially in underprivileged regions? If applied swiftly and effectively, it appears so. The International Monetary Fund suggests a $50 billion investment to vaccinate 60% of the global populace, a sum easily achievable within a year through even a minimal tax application. Considering the low cost, the IMF projects a return of around $9 trillion by 2025, from combined benefits like vaccinations, diagnostics, and treatments. Notably, over 40% would benefit advanced economies. This approach could lay the groundwork for rapid response mechanisms during future pandemics, ensuring resilient economic recoveries, elevated global demand, and sustained investor and consumer confidence. Furthermore, advanced economies might witness a surge of about $1 trillion in additional tax inflow. Thus, combining a global corporate tax with the IMF's vaccination strategy not only ensures self-sustainability but also serves as a blueprint for high–yield public investments against future health threats.

CHAPTER 4 - Take-home messages
- Non-Pharmaceutical Interventions (NPIs) are widely used containment strategies. They range from individual measures like masks, environmental ones like sanitation, to community-level actions such as lockdowns, which are logistically challenging and costly. While not always deemed the most effective, suspending large gatherings was among the more successful NPIs.
- There is no standard composition of effective measures for each community: The right mix of measures depends on the characteristics of the countries and regions of the world.
- In addition to NPIs, the strategy for containing an epidemic must take a proactive approach by applying manual and automatic tracing methods. In the first case, the network of contacts is reconstructed based on the patient's memory and via interviews. In the second, the contact network is obtained through specific apps that transmit anonymous notifications to those in the proximity of infected people. While manual tracking is essential for getting meaningful information, it presents obstacles due to memory gaps; On the contrary, au-

tomatic tracking can be rapid and allow large amounts of data to be collected. The best strategy is to adopt both methods to mitigate their limitations in relation to each of them.

- A big issue is privacy protection when using automatic tracing. Different countries have adopted different technologies, not all of which provide evidence of protecting users' data. Intellectuals and researchers have made their voices heard worldwide to demand guarantees regarding the use of apps only as a support for public health measures, transparency of protocols and implementations, and non-mandatory adoption of the service by users.
- Creating a vaccine in under ten months was an extraordinary achievement and has been declared the most significant discovery of 2020.
- Even though it has been marked by controversies and unequal access, the availability of vaccines has greatly helped the possibility of seeing a recovery of the entire system (economic, social, and health), and the return to a 'new normalcy' appeared more at hand.

PART 3

The impact of COVID-19 on the economic system

CHAPTER 5

Trade-offs and political economy during pandemics

Contents

One crucial factor during emergencies, capable of significantly altering the course of events, is the leaders' ability to make decisions (refer to Chapter 3). Competence and leadership are two key aspects that enable societies to navigate the adversities imposed by the crisis and emerge unscathed. However, these capabilities hold little value if the decision-maker lacks a clear and comprehensive understanding of potential alternatives. Unfortunately, these alternatives are often vague or may be interpreted subjectively, hindering decision-makers from making accurate choices.

Despite these evident difficulties, decision-makers are forced to make choices. To find a solution to the policy-making problem, economic theory suggests assigning a price (or a 'weight', if the word 'weight' can appear ethically more acceptable) to different alternatives to find the most convenient choice.[1] Assigning a price is beneficial because it enables the establishment of a value scale based on a single unit of measurement. Although this method may seem cold and cynical, it ultimately provides decision-makers with the right conditions to make rational choices. To illustrate how this principle applies to everyday life, one can observe the contrasting decisions made by New York Governor Andrew Cuomo, as opposed to those made by President Trump. Cuomo had vehemently opposed the idea of allocating a dollar value to human life.

---

[1] Timely decisions are critical, especially in pandemics, as widely proven by past experience where, for example, delays in enforcing social distance measures have often resulted in a higher number of casualties.

Even though this seems to deny the trade-off, Cuomo reasoned by the scheme just explained, implicitly attributing an infinite (or larger than any comparable cost) price to life. Although this may seem unreasonable, it may have reflected the feelings of many people in New York State and New York City, particularly due to the high number of deaths during the first months of the pandemic.

But a value to human life, or at least a way of thinking systematically about it, is what leaders need to deal appropriately with the consequences of an epidemic. The real problem with this approach is that the alternatives to be evaluated are not always 'Pareto-improving', which means that one intervention dominates the others so that everyone is happier once the intervention is carried out. In most cases, the choices to be made involve trade-offs or conflicts because they favor one part of the population and harm another, thereby imposing redistributive effects.

Understanding the existence and size of a trade-off between public health objectives (saving lives) and economic objectives (not damaging the economy) is crucial in identifying policies to implement during a pandemic crisis. The best solution would be to find a 'Paretian' policy that maximizes both objectives (saving lives and allowing the economy to grow). But if no such a policy is uncovered, then a compromise would have to be found that might necessarily impose painful choices.

However, during pandemics, optimal policies and their relative trade-offs are dynamic and related to the existing global context at each point in time: a context shaped by several forces: epidemiological trends, economic and social distress, and medical and clinical technological advances. This implies that what is optimal at the very beginning of a pandemic can become suboptimal months later. In this chapter, we summarize the discussions that occurred since the start of the pandemic, highlighting how trade-offs have changed over time, in particular with the advancement of expert knowledge and with the arrival of vaccines, which have profoundly changed the way policymakers see the choices and the compromises to be made to face the pandemic.

The structure of this chapter is organized into three main sections. First, we propose an approach to analyze and evaluate the COVID-19 trade offs, present the different positions of experts on this topic and discuss the validity of such positions. Second, we explore whether and to what extent there have been trade-offs between human life and the economy and how they have changed with the rapidly changing context of the pandemic. In the early phases of the pandemic, lacking preparedness and with a very low level of knowledge about how to handle the virus, the policy responses and outcomes were different from what was observed one year later when vaccines arrived on the market. Finally, we introduce some simple theoretical concepts about the political economy of a pandemic and their implications for policies based on command and control versus policies based on incentives.

5.1. How to evaluate trade-offs? Balance and conflict

According to a consensus by experts, as well summarized in an article published by The Economist in April 2020,[2] in challenging contexts such as those imposed by a pandemic, the best way to make a decision is to rely on three basic principles:

- A single value should be established for one year of life and then used to conduct all evaluations. This amount, whatever it is (it can vary between countries, periods, and events), as in all cost-of-living calculations, is not real money, but an accounting measure that helps to compare very different things like lives, jobs, and the juxta-position of moral and social values in a complex society. When people talk about 'putting a price on life,' they are often referring to two distinct methods, that can also be combined: (a) Productivity measures, based on the potential earnings or economic contributions that an individual might make during their lifetime. (b) In-surance measures, based on how much people are willing to pay to reduce the risk of death or injury. The importance of these measurements increases as the impact of an adverse event becomes more significant;

- help those who suffer losses from the choices made. Since choices are made between alternatives that only rarely can be ordered in terms of 'Pareto' efficiency (i.e., by ranking higher those that are better on all policy objectives, including their distribution across different population groups), decision-makers must have a clear picture of the effects of their actions and their possible creation of different 'winners' and 'losers'. For example, in case of an epidemic, the economic blockade penalizes some people and benefits others (this will be discussed in further detail in Chapters 6 and 7);

- make sure that the decision-making process is as dynamic as the epidemic, which means that the balance between costs and benefits may change as the pandemic evolves. An optimal choice can be made when the outbreak starts, but the same choice is likely to be suboptimal if taken weeks after the outbreak begins. For ex-ample, while imposing a lockdown on a country at the beginning of the epidemic may be effective as it allows it to gain time and organize, but the same lockdown imposed on an ongoing epidemic could generate fewer or no benefits at all.

These problems are also best addressed using tools of welfare economics, the branch of the economy that allows *ex ante* evaluations of possible alternatives and often provides decision makers with objective answers. As Chilton et al. (2020) have argued, welfare economics offers the conceptual framework for thinking about the significant trade-offs

---

[2] The article is available at the following web address: https://www.economist.com/leaders/2020/04/02/covid-19-presents-stark-choices-between-life-death-and-the-economy.

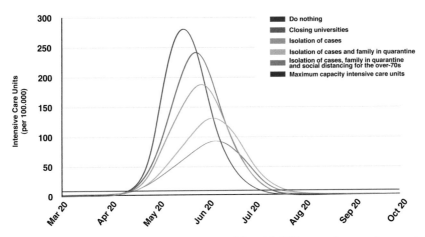

Figure 5.1 Mitigation strategy scenarios compatible with ICU capacity requirements in the UK. *Source*: The chart is an adaptation of Fig. 2 from Ferguson et al. (2020).

between the economy and health that the COVID-19 crisis has brought about. At the same time, it does not neglect the broader social issues related to the other important aspect of trade-offs between equity and efficiency. To understand how these issues relate to the COVID-19 epidemic, it is helpful to start discussing the concept of 'mitigation of the epidemic curve', which allows us to see how increasing lockdown measures can eventually help health systems not to explode under the pressure of intensive care admissions. Based on this concept and looking at Fig. 5.1, the trade-offs essentially arise from the choice between two strategies:

- **Strategy 1 (no containment – dark blue line)**. This strategy could ensure that economic activities are not harmed or stopped in all or in part. However, at the same time, it would likely generate a number of patients far beyond the ability of the health system to treat them, leading many people to die from lack of appropriate care. Its advantage would be its speed of action and return to normal after the epidemic is over, with the possibility of returning to the workplace for people (along with the possibility of having acquired herd immunity (see Box 8). In this scenario, more people could die, but fewer people would lose their jobs or would fall into poverty.
- **Strategy 2 (quarantine containment – red line)**. This strategy would see people stuck at home for relatively longer periods of time and hospitals able to cope with a reduced daily influx of patients. In this scenario, fewer people would die without adequate medical attention. However, the length of the lockdown would lead to the failure of many businesses or the dismissal of a large part of their workers, with consequent income reduction and increased poverty.

The next step requires using economic theory to explain how society makes its choices (Strategy 1 vs. Strategy 2). Although many other approaches are possible, economic theory can help us to understand:

- How a decision can be reached;
- what are the difficulties that must be faced to enable decision makers to make an informed choice that, given certain constraints, can maximize the population's welfare.

To better clarify these concepts, we use the graph in Fig. 5.2, which allows to define the optimal choice in the context of general economic equilibrium and to assess how the population well-being varies with the possible combinations of saved lives and retained incomes/jobs chosen by decision makers.

As detailed in Fig. 5.2, economists define the red curve as a social-welfare indifference curve (a graphical representation of a social-welfare function) that describes all combinations of saved lives and income/job retained that provide a population with the same level of well-being. To plot these indifference curves, we must attribute a monetary worth to human life. Additionally, it's crucial to recognize that it's possible to calculate a 'marginal rate of substitution',or a kind of relative price, which quantifies the willingness to trade lives for jobs/incomes.

To clarify further: the slope of a welfare indifference curve indicates how much a community is willing to trade in terms of lives saved for jobs maintained. If lives saved are on the vertical axis and jobs are on the horizontal axis, then a steeper slope would indicate that the community is more willing to trade lives for jobs, while a flatter slope would indicate that the community is less willing to trade lives for jobs. If the curve had an angle, only one possible ratio could be chosen between these two variables. If the curve were convex as in the figure, it would indicate that, as lives are saved, to maintain the same level of social well-being, society increasingly would prefer jobs/incomes to be saved. Progressives and conservatives may have completely different social preferences, but at any point in time, the curve will have a single shape, reflecting, for a given level of well-being, the preferences of different social groups, with weights depending on the institutional system, the information available, and many other variables.

The blue curve in Fig. 5.2 is the production possibility frontier, which shows the boundary of what is achievable in terms of saved lives and job/income, given the available resources, technologies, and institutions. Therefore, the slope of the red curve, for each level of welfare, represents social preferences as the 'marginal rate of substitution', that is the willingness to trade lives for jobs or incomes and can vary over time and between different groups of people. The production possibility trade-offs between lives saved and job or income losses (the *blue curve*), measured by the slope of the curve at anyone point and known as the 'marginal rates of transformation', are influenced by tangible factors such as available resources and technological limitations. However, the understanding of these factors, as well as the perceived limits of production possibilities,

Box 8: Herd immunity: definition and a brief history

Herd immunity, also known as *indirect protection*, *community immunity*, or *community protection*, is a term used in immunology to describe how a group of people or animals can fight off the spread of a contagious disease. Herd immunity requires that a sufficiently large number of people become immune to a pathogen, making it less likely to spread from infected to healthy subjects (Smith, 2019). In other words, herd immunity makes it harder for contagious diseases to spread from person to person because the chain of transmission is broken, and vulnerable people are less likely to get infected. Herd immunity can be 'natural' (when achieved through exposition to the infection) or 'acquired' (when achieved through vaccination).

In earlier times, people who tried to stop diseases, like Edward Jenner, Louis Pasteur, and William Farr, thought that an infection could be wiped out if enough people got vaccinated. At the start of the 20th century, some veterinarians liked the idea and came up with the term 'herd immunity'. The University of Manchester in England in the 1920s was the first place to study herd immunity to fight epidemics. In particular, in 1923, Topley and Wilson (1923) set up clever studies with hundreds of thousands of mice and pushed the idea into the mainstream. About 15,000 mice a year ran around in a lab where the rooms looked like small moon bases. Each living pod was about a foot wide, and cylindrical tunnels connected them. This made it easy for the rodents to move around the Lilliputian cities. From time to time, the project's leaders would start an epidemic on purpose. The mice in one city would be exposed to bacteria that could kill them, while the mice in another city would be given a vaccine along with the bacteria. The research showed that immunity in part of a population could slow an outbreak and protect mice that weren't immune. The experiment generated the hope that making a small percentage of the population immune could stop a disease from spreading and killing many people. In 1933, Dr. Arthur W. Hedrich, a public health official in Chicago, Illinois, observed that the measles outbreak in Boston, Massachusetts had been halted after 68% of children were infected between 1900 and 1930 (Fine et al., 2011). Also, after the measles vaccine became legal in 1964, the number of cases remained low, and the second dose wasn't given until the late 1980s (Centers for Disease Control (CDC) et al., 1991; McNabb et al., 2007).

There are many examples in human history of blocking or even eliminating infectious diseases through herd immunity (Fine et al., 2011). However, in no case, this result has been achieved through 'natural' herd immunity. Presently, only two infectious diseases have ever been wiped out: smallpox, which was a major killer, and rinderpest, which spread through cattle. All other known diseases, like rabies, leprosy, and the bubonic plague, are not eradicated yet or are still out of control. Smallpox is considered to be among the most deadly infectious diseases human are generally prone to be infected with. Its spread among populations began thousands of years ago, throughout recent human history (Thèves et al., 2016). In 1979, smallpox was officially declared eradicated based on herd immunity achieved by intensive vaccination campaigns (Lane, 2006). Similarly, rinderpest, a highly contagious disease, was eradicated in 2011 through herd immunity in animals (Tounkara and Nwankpa, 2017). Other ubiquitous diseases such as measles, rubella, and pertussis are not eradicated yet, and herd immunity is maintained by keeping a proportion of people immune above some threshold to protect susceptible individuals (Black, 1982; Assaad, 1983; Fine et al., 2011). In the USA, after the whole-cell pertussis vaccines were widely used in routine childhood immunization in the mid-1940s, there was a remarkable reduction in the pertussis incidence, from 150,000 to 260,000 cases annually to just 1010 cases in 1976 (Phadke et al., 2016).

In the early months of the COVID-19 spread, there was a wide discussion wondering if people infected with the SARS-CoV-2 virus would generate enough 'herd immunity' to bring society back to normal. If people counted on natural infection to stop the outbreak, there would be a scary cycle of cases going down and then back up for months or even years. Even if this kind of community-based protection were set up, it would be constantly weakened by the birth of new children and the possibility that those infected would lose their immunity.

are themselves shaped by the views of policymakers and institutions on their existing constraints. These perceptions are also influenced by historical decisions about how to organize various processes, such as healthcare, labor markets, fiscal distribution, and social protection. Point 'A' in the figure is the highest possible level of welfare achievable through a feasible combination of lives and jobs saved. It represents the optimal allocation of resources, given that what is the largest desirable combination of lives and jobs (corresponding to the highest indifference curve) matches the highest feasible supply under current technologies and as a consequence of past policy choices. In practice, due to uncertainty and lack of information, both social preferences and available policies depend on perception, social beliefs, and the adequate support that scientists and technologists can provide to policy makers.

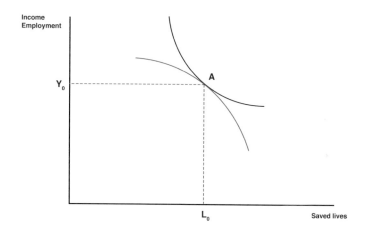

Figure 5.2 A stylized representation of the trade-off between 'lives' and 'income'.

Fig. 5.3 provides some examples that may aid in gaining a better understanding of the role of economic policies in an emergency such as COVID-19. The first scenario might be chosen by a policy maker who decides to go for strategy 2 (lockdown). In this case, the policy takes the country into a situation described in point 'B' where, due to the lockdown, the number of lives saved has increased. Simultaneously, economic losses escalate, and the repercussions of diminished income and job reduction are substantial, leading to a decrease in the country's welfare. This places the country on a lower welfare indifference curve, representing a sub-optimal situation. If these curves accurately reflect the public's preferences and their perception of the feasible choices, then a majority of the country might disagree with such a policy.

The second case again represents a situation in which strategy 2 is chosen. However, this time, we assume that the policy makers have a better degree of understanding and control of the technological and organizational trade-offs so that their perceived

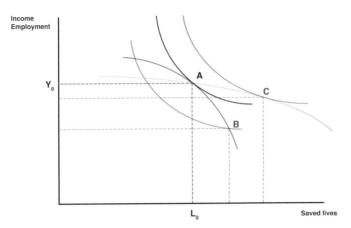

Figure 5.3 A stylized representation of the trade-off between 'lives' and 'income': policy intervention.

production possibility frontier (the blue curve) is flatter (meaning that it is less costly to save lives in terms of jobs/income losses). This makes it convenient to accept a strategy with slightly higher income/job losses in exchange for more lives saved. The results of this policy could now be more acceptable to society, which would see its well-being increase as the optimal balance shifts from 'A' to 'C': society would gain more wealth from saving lives, even at the cost of sacrifices in terms of lower incomes and jobs. A practical example of such policy is presented in Islamaj and Mattoo (2021) and refers to the combination of lockdown and testing measures due to a better understanding of the possibilities offered by these technologies 'to soften the lives vs. livelihoods trade-off, since testing can help attain any desired containment level at a lower economic cost or higher economic benefit'.

Although they refer to highly simplified cases, the proposed examples should help define a conceptual framework for tackling the problems encountered when dealing with such issues. The reader also may appreciate the fact that a framework of this kind can be appropriate for some of the choices or attitudes that different countries can adopt to elaborate responses to epidemics, as well as to other problems. Social preferences (red curve) and production frontiers (blue curve) may differ between countries (and perhaps also between regions of a country). Different levels of civic sense and education, religious belief, type of social relationships, family structures, and overall socioeconomic status shape the social preferences of a community. Similarly, the different levels of perception and information from policy makers on the quality of institutions, technology and available production factors determine the shape of the production possibility frontier (the blue curve). For example, in many countries, increased attention and empathy for the elderly (which may have changed the shape and level of the red curve) revealed new logistical possibilities that were previously overlooked. This may have also changed the perceived trade-off between lives and incomes (the shape and level of the blue curve).

The take-home message of this brief analysis is that it would be helpful to initiate policies that would allow the country to go from point 'A' to point 'C'. Moving the blue curve as far away as possible and flattening it as quickly as possible could also be considered in the realm of feasible policies, even though substantial improvements of the technological trade-offs are more difficult to accomplish in the short run. For example, compared to a complete lockdown policy, targeted quarantine allows governments to implement multiple tests, representing a considerable increase in protecting people's health. This is equivalent to flatten and move outward the production possibility curve, reaching the point 'C'.

5.1.1 'GBD' vs. 'JSM': the two prevalent positions during the pandemic

As the world faced the second wave of SARS-CoV-2 infections in the fall of 2020, the question of which measures to adopt became increasingly pressing. In the first weeks of October 2020, the public debate on the subject intensified due to the polarization produced by the publication of two prominent posters that formulated two alternative strategies to overcome the crisis. The two posters in question pertain to the Great Barrington Declaration (GBD) and the John Snow Memorandum (JSM).[3] The GBD was signed on October 4, 2020, in the town of Great Barrington in Berkshire County, Massachusetts, by Sunetra Gupta, a professor of theoretical biology at Oxford University, Martin Kulldorff, a professor of medicine at Harvard University, Jay Bhattacharya, a professor of medicine and economics at Stanford University, and 35 other experts. This document was followed on October 5, 2020, by the JSM, an open letter signed by 80 international researchers published in The Lancet, one of the most prestigious medical journals.[4] Supporters of GBD, highlighting the enormity of indirect costs in conjunction with more aggressive containment strategies (e.g., failure of businesses and store or failure to treat chronic illnesses), argued that an uncontrolled spread of COVID-19 infections among the low-risk population, e.g., younger people, was desirable and that the immunity thus acquired would be high enough to protect vulnerable groups, which in the meantime should remain confined at home. In contrast, in JSM, the authors warned that a herd immunity approach to COVID-19 management, such as GBD, was 'a dangerous error not supported by scientific evidence'.

The GBD signatories believed that non-vulnerable people should be immediately allowed to resume life normally. This implied that it was desirable to reopen schools and

---

[3] The two declarations are available at the following web addresses: https://gbdeclaration.org/ and https://www.johnsnowmemo.com/.

[4] The name of this memorandum comes from John Snow, an English physician (1813-1858), who is considered the father of epidemiology because he traced a cholera epidemic in London in 1854 in a hospital water source and persuaded the local council to remove the water pump handle.

universities for teaching in presence. All extracurricular activities should be allowed, and work should not cease for working-age adults. GBD also recommended that restaurants, other businesses, arts, music, sports and cultural activities be reopened (or not closed). In addition, it stated that the participation of high-risk population groups in these activities should be voluntary and that simple hygiene measures, such as hand washing and staying at home in case of disease, should be practiced by all with the sole purpose of lowering the threshold for herd immunity. The document did not mention personal protective equipment, such as masks or social distancing.

For the JSM signatories, the GBD recommendations, which in appearance may seem very attractive, concealed enormous dangers. Indeed, the strategy proposed in the GBD could give the impression of reducing the guard against the virus, just as it did in the weeks of July and August 2020. Any unforeseen developments could be disastrous in countries and regions where hospitals were not equipped to handle an uncontrolled wave of infections. Therefore, JSM considered it necessary to communicate the risks posed by COVID-19 and the implications of different law enforcement strategies, especially when winter approached. Furthermore, the JSM acknowledged that in the absence of adequate provisions to manage the pandemic and its socio-economic impact, many countries continued with a lockdown-oriented approach, with dire consequences for several people and leading to a decrease in the overall population confidence index. However, JSM argued that these facts should not be considered a reason to abandon the scientific approach. In fact, at the time of the JSM signature, no evidence of long-term protective immunity against natural infection had been found, and several cases of reinfection had been documented (Iwasaki, 2021), leaving the possibility of recurrent epidemics. Additionally, the long-term effects of COVID-19 made it difficult to identify who was at the highest risk of serious illness or death. All these things being considered, the JSM argued that the best protection for the economic system could only be achieved by controlling the spread of the infection. This conclusion appeared to be corroborated by the fact that in countries such as Japan, Vietnam, and New Zealand, robust public health responses had proven effective in preventing infection transmission, allowing life to return almost to normal in a sufficiently rapid time.

JSM also criticized several other issues of the GBD thesis (see Table 5.1 for details of differences). The first point was that achieving herd immunity through natural infection is not ethically acceptable because economically and socially disadvantaged people are at increased risk of becoming seriously ill. In addition to age, in the case of COVID-19, this risk is significantly increased by both clinical (such as diabetes or hypertension) and socio-economic (poverty, working conditions, and imprisonment) factors. Worldwide, it was clear that COVID-19 deaths was disproportionally affecting minorities and the most disadvantaged individuals. Thus, the strategy of achieving herd immunity would have presented the danger of isolating even more communities already marginalized by society. Second, even if people do not die, they can have serious health problems, such

Table 5.1 Major differences between the 'Great Barrington Declaration' and the 'John Snow Memorandum'.

Main differences	Great Barrington Declaration	John Snow Memorandum
Key points	It's only 540 words long and delineates a focused protection strategy (focused prevention) for the United States. Based on the idea that the risk of death varies drastically with age, it proposes to focus on protecting vulnerable subjects while at the same time allowing young people and other people at low risk to live more normally. This strategy offers some ideas on how to protect nursing homes, which have seen many casualties, and promotes simple measures such as hand washing. It alludes to the idea that the threshold for herd immunity is not a fixed value: it depends on how the populations mix and interact, and on simple measures to take, like better hygiene. It recommends that schools, universities, bars, and restaurants be completely open.	It counts 930 words. The strategy proposed by the signatories of this document provides for continued restrictions until an effective vaccine is available. At the same time, it proposes programs and social services to minimize restrictions. Due to the proposed measures, restrictions would be reduced to very low levels and allowed contact tracking to eliminate outbreaks. It also defines the Great Barrington declaration: a 'dangerous mistake' not supported by 'scientific evidence' since it seeks to obtain immunity through the natural spread of infections.
Risks imposed on society	Low risk for healthy individuals; high risk for vulnerable populations.	Very high risk; society cannot operate normally until the virus is not treated successfully with effective therapies or vaccines.
Prevention	Mainly focus prevention efforts on vulnerable populations.	Concentrate prevention efforts on the whole populations, independently from their susceptibility.
Herd immunity	Over time, immunity acquired through infections will provide increased protection for the vulnerable population.	It is unethical to allow the acquisition of herd immunity through infection in the community. Herd immunity can be built only through vaccination of the majority of the population.
Relative risks	COVID-19 is not a risk for public health sufficient to justify draconian measures that can kill lives and means of subsistence and will damage the future of millions of people worldwide. It should be treated similarly to other respiratory diseases.	COVID-19 is one of the worst health threats that the modern world has ever experienced and therefore requires draconian measures until safe vaccines are available and effective.

continued on next page

Table 5.1 (*continued*)

Main differences	Great Barrington Declaration	John Snow Memorandum
Declared weaknesses	Unethical, unscientific, not supported by a sufficient number of scientists and physicians, with lack of attention to the construction of immune response resilience.	Disproportionate effects on economic and social costs, collateral damage too high, and no certainty that the therapeutic agents or vaccines in phase development will work.
Critical points	The limitations of the GBD include its insufficient guidance on identifying vulnerable individuals, ways they should protect themselves, and the fact that it groups dissimilar entities, such as bars and schools. Open bars can be substituted for drinking beer in the garden with a friend sitting at a distance with little loss of pleasure. However, education, particularly for the poor, is one of the few roads left many countries to a better life, a place to feed poor children, and a way to detect abuse. Another limitation is lack of recognition that temporary measures are likely to be needed in moments of explosive release to prevent, for example, hospitals from overflowing.	Limitations of the JSM include: vagueness on social programs to be created to minimize harm, in the face of growing poverty and exclusion of more and more people. Although GBD authors do not provide a plan to protect the vulnerable population from the virus, the JSM authors also do not explain how they will protect the vulnerable from the harms of the restrictions. Additionally, the JSM does not contain sufficient detail to explain how contact traceability could be achieved in a country like the United States, where many people are reluctant to share information on their contacts and whereabouts.

as hospitalization, long-term symptoms, organ damage, missed work, and high medical bills. Third, at the moment the GBD was published, the number of people infected was still too low (and the death toll already too high) to think of a strategy to achieve herd immunity. These numbers were not even close to the figures required for herd immunity to be activated. Finally, it was not known exactly what 'immunity' means for this specific disease, and it was not even known how long naturally acquired immunity could last (or how common reinfections could be).

5.1.2 Why are the GBD recommendations challenging to implement?

This section provides some simple statistical evidence on how GBD recommendations are difficult to implement. As a reference country, we use Italy, for which we have sufficient information to obtain all relevant statistics. Using prudent assumptions and simple calculations, we conclude that this strategy is almost unfeasible unless it succeeds in confining to their homes at least 50% of the Italian population.

The first step in proceeding with the calculation is to identify fragile people. In 2019 in Italy, 13,783,580 people over 65 lived alone or with other family members, of whom over 8.5 million were married or civilly united. If we also include family members, the number increases to 14,843,119 people. To this number of older adults and their relatives, it is necessary to add those under 65 years of age with significant health problems due to one or more chronic conditions (diabetes, hypertension, respiratory, or cardiovascular disease). In 2019 the number was 8,002,000, although many of the people involved may have suffered from one or more comorbidities. Therefore, we speak of about 20 million people, representing one-third of the Italian population. According to the GBD recommendations, these individuals should have undergone a comprehensive lockdown protocol, similar to the one implemented between March and April 2020. As a first challenge, it would have been necessary to envisage a logistic plan to provide all citizens involved in confinement with the necessities they might have needed and several essential services, first and foremost, health services.

In such a context, it is easy to assume that the isolation of fragile people would not be 100% effective, as there will still be interactions between the most vulnerable and other individuals who, for various reasons, would have to deal with them. For example, vulnerable individuals may need support for medical check-ups, maintaining social contact with family members, or if they live with people who are not strictly isolated and are not considered to be at high risk.

If we use the mortality rates by age group provided by the Italian National Institute of Health (ISS), we can obtain a quick estimate of the mortality impact of such a strategy. At the time GBD was signed, the proportion of people who had contracted the virus in Italy was estimated to be equal to 20% (perhaps overestimated). This would mean that to reach the threshold of herd immunity - established by prudence at 70% - it would have been necessary to infect another 50% of the Italian population (about 30 million people) in good health.[5] Just over 80% Italians, less than 65 years of age in good health, or 36,475,000 people, would have to become infected.

The isolation strategy also presents a high risk of not being 100% effective and causing unnecessary deaths. If we assume that only 5% of the vulnerable population in Italy (around 1 million people) would be infected due to imperfect isolation measures, the projected death toll would be around 70,000, based on the patient count at risk and mortality rates by age group as estimated by the ISS.[6] If the number of fragile infected

[5] As of 31 December 2020, the number of people infected in Italy was just over 2 million. Assuming that the number of people infected was six times higher, it would have caused 12 million people to be infected, representing about 20% of the Italian population.

[6] This value was obtained by multiplying the number of people at risk for the average case fatality rate (CFR) of the Italian population over 50 (7.0%) estimated from the data reported on the website https://lab24.ilsole24ore.com/coronavirus/box_8b under the heading 'Lethality rate per age group' updated on 29 December 2020.

were to increase to 10% or 15%, the number of deaths would increase to 140,000 and 210,000, respectively. Compared to the situation at the end of 2020 (approximately 75,000 deaths), this would have implied a doubling of the number of fatalities at best.

Of course, these numbers are obtained assuming that the healthcare system can function without hindrance (or even collapse) when the situation worsens. We must recount the number of admissions to know how likely this is. According to data from the Civil Protection and Ministry of Health, in Italy, the percentage of hospitalized among the total of infected at the peak of the pandemic was approximately 4.5%. In this case, if 30 million people became infected, at least 1,350,000 'non-fragile' people would have to be hospitalized. To these, we should add fragile people who could not be protected and whose hospitalization rate could be close to 100%. Therefore, the estimated number of hospitalizations could range from 1.42 to 1.56 million, taking into account infections in young people and 5% or 15% of the vulnerable population. Moreover, due to the nature of a pandemic, these cases would not spread evenly over time, but would cluster rapidly, overwhelming hospitals. According to data from the Council presidency, 6458 adult intensive care beds were available in Italy on 9 October 2020.[7] If only 5% of hospitalized patients required intensive care, it is easy to see that hospital beds would not be enough. As a result, the death toll could rise significantly.

These estimated deaths should be added to those of COVID-19 patients who might die because they could not receive treatment and those of non-COVID-19 patients who would die because the health service were busy solving other problems. Other issues concern those who would contract the disease and survive but would develop long-term health problems and irreversible damages (long haulers). Although little is known about these so-called 'long COVID' effects, reports of cases of this kind around the world are increasing.

Finally, it should be considered that a strategy such as that outlined by GBD would be worth considering only if it were established that no vaccine, no rapid mass testing, and no improvement in medical treatment would be available. While no country has adopted the GBD strategy, countries that have had better outcomes, especially during the first two waves of the pandemic, have implemented different approaches. Germany instituted local lockdowns. Japan chose to make great use of masks and disciplined social distance. Korea sacrificed privacy to implement a sound tracking system. It seems better to learn from these experiences than to follow a path that would see healthcare systems overwhelmed and could lead to hundreds of thousands of unnecessary deaths.

---

[7] The source is from the following site: https://www.ilsole24ore.com/art/coronavirus-terapie-intensive-increase-what-regions-are-ready-the-second-wave-ADNUkdv.

5.2. The trade-offs imposed by the COVID-19 pandemic

The COVID-19 crisis represents a textbook case of social choice under emergency, which can be clearly connected with basic economic theory. The first problem policymakers face during a pandemic is defining their general political objectives, which often go far beyond the direct consequences of the pandemic itself. In fact, a broader approach to well-being must be taken, including a wide range of other elements, such as protecting constitutional and human rights while implementing pandemic risk mitigation efforts.

Most of these choices must be made in the first few weeks of the epidemic, often in a state of considerable uncertainty due to the limited information available. Compared to the simplicity of the graphs in Figs. 5.2 and 5.3, real-life choices are more challenging and can be highly heterogeneous between countries and often between regions in the same country. Good examples are the United States, the United Kingdom, and Sweden, which, unlike many other European countries, in the early weeks of the pandemic, strongly supported the view that economic reasons had to prevail over health reasons. Boris Johnson was the pioneer. For several weeks until the beginning of April 2020, and under the advice of the experts of the Nudge Unit, he supported the herd immunity strategy. He surrendered only in the face of a veritable disaster, which led the UK to become first among the countries most affected in terms of deaths per inhabitant.

In partial defense of all policymakers, it must be said that the level of global uncertainty prevailing in the early weeks – also among experts – was such that decisions were often taken following a form of perceived global sentiment. This was especially true when the single COVID-19 outbreak became a pandemic that affected the whole population. At that time, policymakers had to make difficult choices in the face of the scientific community's limited understanding of the virus.

The need to act and the lack of knowledge may partly explain why the choices have been so different across countries. To better understand the varying opinions across different countries, Fig. 5.4 reports the result of a survey conducted in the week of April 15-23 2020 on 13,200 individuals in 11 countries by the Edelman Trust Barometer, where questions were asked whether governments should prioritize saving lives or jobs during the pandemic.[8] About 67% of all respondents agreed that the government should save as many lives as possible, even at the expense of higher costs to the economy; on the contrary, 33% of respondents said it was more important for governments to save jobs and restart the economy than taking all possible precautions to protect people from the virus. The results varied widely between countries. For example, 76% of the

---

[8] The statements on which the respondents were asked to express their preference were: *1)* The government's top priority should be to save as many lives as possible, even if it means that the economy will suffer more damage and recover more slowly; *2)* it is becoming more critical for the government to save jobs and restart the economy than to take all possible precautions to protect people from the virus.

Japanese and 56% of the Chinese respondents considered saving lives more important than economic recovery. The interesting aspect of the survey is that the preferences reported by the respondents at the country level do not seem to be related to the number of infections and deaths that the country was experiencing at the time the analysis was carried out.[9]

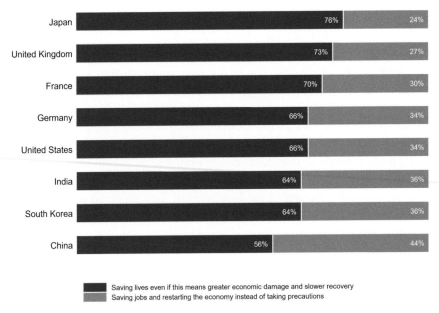

Figure 5.4 Levels of acceptance of the trade-off between health and economy between countries. Percentage of people who agree with their government's priorities between saving lives and saving the economy. *Source*: Trust Barometer Edelman. *Note:* The chart is an adaptation of the Figure of Slides 36 in the Spring Update of Edelman's Trust Barometer. Results obtained from 13,200 interviews in 11 countries in the period 15-23 April 2020.

5.2.1 Identification of a political decision: what were the trade-off?

The trade-off that emerged at the beginning of the pandemic was later articulated by the JSM and the GBD in their manifestos. The majority of virologists, epidemiologists, and a significant group of economists supported the JSM. They even suggested reevaluating the prevailing economic model entirely. The economists argued that state intervention

---

[9] In the week when the survey was conducted, the average death toll was 2653 in the United States, 897 in the United Kingdom, 760 in France and 184 in China. On the other side of the table, Japan recorded ten deaths, South Korea 2.7, and India 34. As these numbers suggest, there appeared to be no correlation between the severity of the situation and the will to act to save lives.

was fundamental in mitigating the problems imposed by the lockdown by protecting jobs or income support for those damaged by the pandemic, focusing on the poorest and most vulnerable. Scientists and economists of this 'rational majority' also supported measures of social distancing and delaying the reopening of the economy. They were convinced that once the first wave of infections had passed, it was possible to imagine a resumption of the epidemic (eventually in the fall) if the threshold of attention towards the virus had fallen in light of the initial successes. This group of people did not see a clear trade-off between the lives to be saved and economic problems and believed that the costs of the epidemic, especially if measured in lost lives, were significantly higher than those associated with a slowdown in the economy.

A different opinion was maintained by a second group of people consisting of representatives from a broad section of the business world, a limited number of virologists and epidemiologists, as well as a group of economists. People in this group felt there was a trade-off, as the lengthening of social distancing measures would reduce health costs but create more severe economic problems. These problems, in turn, would generate economic and social costs that would ultimately be comparable with those of less incisive interventions. Social choices would thus face a significant trade-off between economics and human lives. Moreover, members of the second group saw many of the public support measures advocated by the 'rational majority' as hurting the system's overall efficiency and, in any case, representing a cost to future generations if financed by debt issuance.

The position of this second group was partly shared by the economist and philosopher Amartya Sen, who pointed out that diseases and lack of livelihoods kill people. This view echoed GBD, which states that the trade-off is not between 'lives and money', but between 'lives and lives'. The reasoning behind this position is plain: the longer social distancing is maintained, the more effective the containment of the epidemic will be. At the same time, some productive and commercial establishments, such as hotels, travel agencies, restaurants and bars, airlines, and many small businesses linked, for example, to services to people, culture, and entertainment, will fail. In many cases, inequality will increase: people with low incomes and poorly protected contracts will be the most affected by social distancing measures and could see their health deteriorate.

Similarly, mental health will be another vital concern. Those who have endured extended lockdowns and quarantine phases will carry the emotional toll of their experiences. Individuals will have been isolated from personal interactions during times of profound anxiety and uncertainty. Patients with anxiety or obsessive-compulsive disorder will have more difficult times (with those who live close to them). Older people, already excluded from much public life, will asked to move farther away, increasing their sense of confinement and loneliness. Domestic violence and child abuse incidents will also increase as people are forced to remain confined to sometimes unsafe homes. Children, mostly spared by the virus, may suffer mental trauma that can affect them throughout adulthood.

Finally, we should not forget the problem of an increased probability of post-traumatic burnout and stress, especially in the category of healthcare professionals.[10]

A contrasting perspective comes from Case and Deaton (2020), who contend that a decelerating economy doesn't automatically equate to increased mortality. The results of their study suggest that in the US fewer people die in recessions than in periods of economic growth. During boom time, more people die in car accidents and construction sites. There is more pollution, which is bad for babies and young children. And older people receive less care. Hence, according to Case and Deaton, these groups experience better protection during recessions than they do in boom periods. It is true that, on average, the poor die younger than the wealthy, but their mortality rate does not closely follow the unemployment rate. On the other hand, a long-term decline in employment opportunities for people without a college degree has led to an increase in mortality rates among working-class people, with deaths from drug overdose, suicide, and alcoholism being particularly common.

Case and Deaton cite similar examples from past crises. During the Great Depression of the 1930s, fewer people died compared to the boom period of the '20s; during the Great Recession of 2007-2009, one third of the population in Spain and Greece was unemployed, but mortality rates in both countries decreased. Continuing with similar examples, the authors also address the problem of isolation mentioned by Sen, which could be an important risk factor affecting both physical and mental health (see also Section 8.1.4 in Chapter 8). During lockdowns and periods of social restrictions, the concept of community weakens, leading to significant repercussions. People can't attend church for spiritual comfort, participate in group therapies like Alcoholics Anonymous meetings, or visit their parents for emotional support. This could increase drug and alcohol use and, on the verge, push people to commit suicide. However, Case and Deaton also noted that during the pandemic, the risk of death from COVID-19 has been higher than that of deaths from alcohol or drug use, and the number of suicides does not appear to have been significantly varied. Their conclusion is that the decision should favor maintaining social distancing, at least until, from a health point of view, everything is back to normality.

---

[10] The study of mental health disorders has become a major research focus since 2021, with several studies finding that the prevalence of these disorders has increased. However, the evidence does not yield either convergent or robust results on the causal effect relations. To our knowledge, there are a limited number of studies around the world based on non-subjective data (Dawel et al., 2020; Ettman et al., 2020; Pieh et al., 2020; Sønderskov et al., 2020). More recently, Marazzi et al. (2022) found that purchases of mental health-related drugs have increased with respect to 2019. At the same time, the excess volumes do not match the massive increase in anxiety and depressive disorders found in survey-based studies. Furthermore, while the authors find positive impacts on consumption of anxiolitic drugs in the months corresponding to the introduction of the national lockdown, they do not observe any other significant effect of mobility restrictions.

Despite all the discussions and speculation surrounding these problems, three years after the start of the pandemic, it remains very difficult to predict what will happen in economic terms. The COVID-19 pandemic has little in common with previous crises, as it combines, in an unprecedented manner, negative demand, supply, and financial shocks. In a way, it is as if it were pulling together the problems created by the 1929 Great Depression (demand shock), the oil crisis in 1971 and 1974 (supply shock), and the 2007–2009 Great Recession (financial shock). Moreover, it is difficult to assess the existence of trade-offs due to the time mismatch to observe the benefits and the costs of implementing the two alternatives. The implementation of a lockdown policy, even for a long time, will show its effects soon in terms of contagions and saved lives. At the same time, possible increases in mortality due to economic problems could manifest themselves only much later. Prolonged unemployment and the resulting changes in lifestyle can be harmful, but the full consequences, or what Deaton calls the 'death of despair,' may only become apparent after a long period of time.

5.2.2 The trade-offs in the early phase of the pandemic

In the face of the many discussions that have emerged in the first weeks of the spread of the contagion about how to stop the epidemic without blocking the economy, the accumulated evidence strongly suggests that the trade-off between the economy and health was, in fact, a false problem. The most striking example that supports this conclusion is the experience of northern European countries. For months, Sweden had been taking a non-interventionist stance that was supposed to safeguard the economy, just as the United Kingdom had been trying to do, and insisted on the United States doing so. On the contrary, the other Scandinavian countries had implemented stringent social distancing measures and blocked their economic activities. If the trade-off hypothesis were true, the health and economic performance of the countries concerned would show opposing results. Where one country achieves a great deal in terms of contagion and deaths, the other should fail miserably; on the contrary, countries that fail on the health care front should achieve better economic performance when compared to those that have implemented lockdown measures.

During the first pandemic wave, Authers (2020) carefully compared the economic and health performance of Denmark and Sweden. The results from this study are very interesting and allow us to draw some conclusions about the existence of a trade-off. Fig. 5.5 shows the level of daily outbreaks between March and July 2020 in Denmark and Sweden. The graph clearly demonstrates that the two approaches to contagion control yielded utterly different results between the two countries. Denmark opted for a strict lockdown, whereas Sweden was much more lenient in achieving herd immunity. The results were that, at the end of July 2020, Sweden had a rate of about 90 new cases per day per million inhabitants, and Denmark had less than 10. This led to Sweden having more than three times more total infections (7927 vs. 2369) and more than five times more deaths (568 vs. 106).

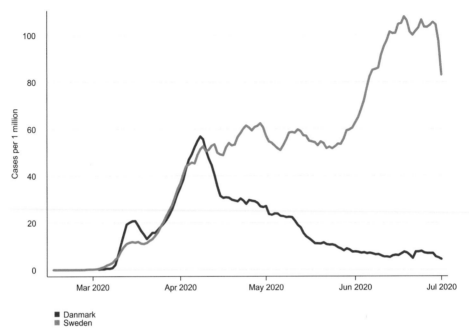

Figure 5.5 Developments in COVID-19 outbreaks in Denmark and Sweden (per million inhabitants, January - July 2020). *Source*: OurWorldInData.org/Coronavirus.

If the trade-off story were true, following these choices, we would have expected a much more prosperous economic situation for Sweden than for Denmark. However, as can be seen from the graph in Fig. 5.6, the numbers on the confidence of the Swedish consumers are not very different from those of the Danes: In both cases, the decrease was significant, on the order of 20 percentage points, with Denmark recovering earlier and faster. Similar considerations apply to the rate of unemployment. Denmark had entered the pandemic at a lower rate than Sweden (3.0% *vs.* 6.5%) and maintained its advantage, since in late July 2020, Denmark's figures were 5.0% and Sweden was 8.5%. Sweden does not seem to have gained any significant benefit from deciding not to block the economy.

The case of Denmark and Sweden is not the only one. As seen in the graph in Fig. 5.7, when comparing the trends in restaurant bookings in Germany and the United States, it is clear that after the intense lockdown imposed in Germany, recovery was faster, even returning to pre-crisis levels. The same is true in the United States, where individual states had implemented different policies. The graph in Fig. 5.8 suggests that

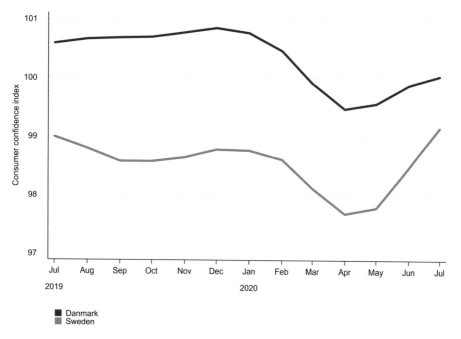

Figure 5.6 Trend in the consumer confidence index in Denmark and Sweden (July 2019 - July 2020). *Source*: Authers (2020).

the US sunbelt states, which refused to implement strict lockdown measures, had a worse performance than the US average.

Further evidence to support the argument that there was no trade-off between health and economy is provided by Chetty et al. (2020), who in a recent US analysis found that the pandemic caused high-income households to isolate themselves and sharply reduce their spending in sectors that require physical interaction. This spending shock led to losses in business revenue and layoffs of low-income workers at businesses that cater to high-income consumers. Because the root cause of the shock was self-isolation due to health concerns, there was limited ability to restore economic activity without addressing the virus itself, at least in the initial months after the pandemic began in mid-March.

These authors conclude that 'state-ordered reopenings of economies had only modest impacts on economic activity; stimulus subsidies increased spending particularly among low-income households, but very little of the additional spending flowed to the companies most affected by COVID; and loans to small businesses had little impact on employment rates'. In general, the suggestion provided by Chetty et al. (2020) is that the most effective approach to mitigating economic difficulties amid a pandemic invariably involves investing in public health measures to restore consumer confidence

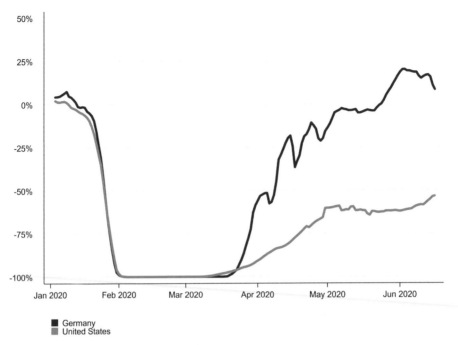

Figure 5.7 Restaurant bookings in Germany and the United States (7-days moving average, 2020 rates of change in 2019). *Source*: www.opentable.com.

and ultimately spending increases. These conclusions are confirmed by data on the development of the savings rate, which in the euro area has approached 17% during the pandemic, a figure never seen before.

A similar analysis has been replicated for a larger number of countries over a longer period (up to December 2020) by Islamaj and Mattoo (2021). According to these authors (see Fig. 5.9), 'cross-country evidence shows that many of the countries that suffered high COVID-related mortality rates were often also those that saw the largest GDP contractions. Thus, saving lives was associated with saving rather than sacrificing livelihoods'. Islamaj and Mattoo (2021) also argue that countries that quickly shifted from lockdowns to a testing, tracing, and isolation (TTI) approach were able to reduce the internal trade-offs between public health and economic activity. This is because TTI can help to control the spread of the virus without the need for widespread lockdowns, which can have a significant negative impact on the economy. These authors also find that mobility restrictions or shutdowns have a strong negative effect on economic growth, while testing is positively (though less robustly) correlated with output growth across countries.

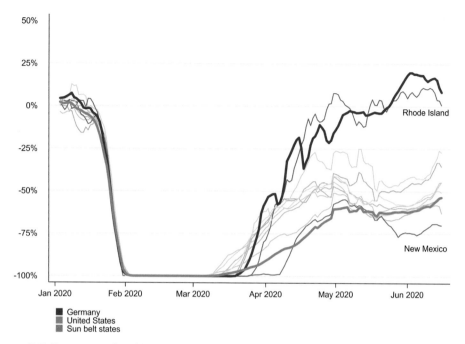

Figure 5.8 Restaurant bookings in Germany, Rhode Island, and the United States 'sun belt' states (7-days moving average, rates of change 2020 on 2019). *Source*: www.opentable. com.

5.2.3 How trade-offs have evolved during the pandemic

The first phase of the pandemic (February–December 2020) was characterized by considerable uncertainty and improvisation for several reasons. On the government side, the COVID-19 outbreak consisted of a sudden and unknown public health threat that required fast policy responses. As we have seen in Chapter 3, public responses often did not comply with traditional channels of democratic decision-making. On the other hand, they imposed several sacrifices on citizens due to the political, social, and economic side effects caused. However, they were accepted with limited opposition everywhere, and citizens of various countries were also highly supportive of the social confinement measures implemented by their national governments (Bol et al., 2021; Sabat et al., 2020).

Opposition increased in the following months when the first wave faded away during the 2020 summer, the accumulation of disease management knowledge progressed, non-pharmaceutical intervention (NPI) shortages vanished, the possibility of new effective vaccines became more than a promise, and it became clear that the economy was severely stalled. At that point, policymakers and citizens began to change the approach to trade-offs. All of this occurred despite the resurgence of contagions and deaths in

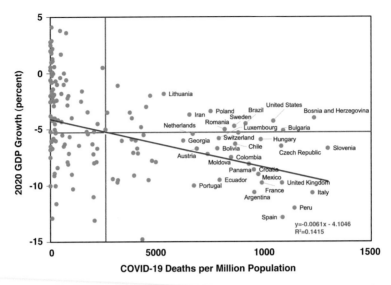

Figure 5.9 Lives vs. Livelihoods: worldwide correlation of growth and mortality in 2020). *Source*: Adapted from Islamaj and Mattoo (2021). Notes: Data obtained from World Bank Economic Monitoring, Global Economic Prospect-January 2021, and Oxford COVID-19 Government Response Tracker 2020. GDP growth (y-axis) are annual 2020 forecast obtained from Global Economic Prospect and World Economic Outlook. The mortality rate (x-axis) is calculated as the number of COVID-19 related deaths reported per one million population in 2020. Quadrants are classified by the mean thresholds of global GDP growth (y-reference line) and mortality rate (x-reference line). Labeled dots represent countries suffering significant mortality (above 500 per million). The red curve (downward sloping) represents the fitted line of the distribution.

the fall of 2020. It also became clear that managing the COVID-19 pandemic could no longer be considered a short-term problem, and the extension of strict lockdowns could not be affordable even for the most powerful economies, with healthier fiscal budgets.

Nonetheless, the transition was rough and fraught with issues. In high-income countries, the death rate per million during the second wave surpassed that of the first wave, as illustrated in Fig. 5.10. Paradoxically, the accumulation of scientific knowledge and its evolving nature became the first cause of disagreement in the decision-making process. As one study noted: 'When there are scientific disagreements, it is not clear which scientific advice should be followed. Even if we are to go with the majority scientific opinion (when there is one), how scientific advice translates into a political decision involves weighing the costs and benefits of different types of interventions' (Bandyopadhyay, 2022).

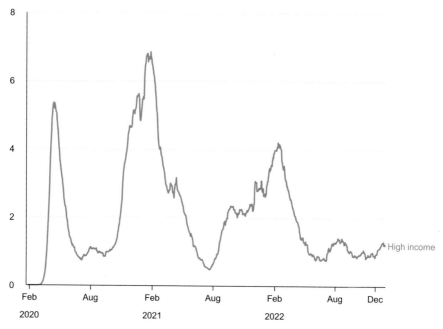

Figure 5.10 Daily new confirmed COVID-19 deaths per million people (Feb 2020 - August 2022). Note: 7-day rolling average. Due to varying protocols and challenges in attributing the cause of death, the number of confirmed deaths may not accurately represent the true number of deaths caused by COVID-19. *Source*: ourworldindata.org.

Over time, the items to be included in the evaluation of the trade-offs increased (i.e., it became necessary to weigh more diverse interests), changing the outcome of the policy makers' decisions. For example, it became clear that stringent lockdowns had undesirable side effects on the economy and population health by delaying the care for other diseases, as was the case, for example, with delayed diagnosis and treatment of cancer. Mental health, especially among young people, became another issue of concern (Adams-Prassl et al., 2022; Dawel et al., 2020; Marazzi et al., 2022; Pierce et al., 2020). School closures with deterioration in educational competencies of children and adolescents were added to the list of costs caused by COVID-19 (Fuchs-Schündeln et al., 2020). Finally, effective tools of protection against COVID-19 (NPI, drugs, and vaccines) and the concurrent evolution of the virus toward less lethal variants drove the trade-off discussion toward less stringent solutions.

However, the importance given to different factors when deciding COVID-19 strategies can vary between policy makers and groups with different interests. For example, vaccines and NPIs can be considered more or less effective and could be made mandatory or optional. Similarly, educational competencies can be seen to be more or less relevant to society, affecting the decision on school opening or closing. Therefore, in a democratic society, the more factors that are considered in the decision-making process, the more likely it is that there will be discussion and debate. This can lead to delays in implementing important and has been a common problem across countries. For example, during the second wave of the pandemic in the fall of 2020, several EU countries delayed reintroducing restrictions by more than a month. This can be seen in Fig. 5.11, where the stringency index is plotted against the date when the death numbers increased again (September 5 - the vertical green line) and when the earliest intervention to increase restriction, which occurred in Germany, was introduced (October 12 - the vertical blue line).[11]

This evidence suggests that managing trade-offs during a pandemic is fraught with difficulties, especially when the duration and severity of its effects are uncertain. During the first phase of the pandemic, much of the discussion focused on implementing lockdowns and the timing, severity, and duration of these measures. The way these aspects are implemented is crucial to understanding the global impact of the pandemic, including its spread and distributional effects. Unfortunately, the discussion around these topics was often affected by inadequate information and, sometimes, biased positions. Ex post, we could claim that gradually opening the economies in the fall of 2020 was probably a good move. However, such a policy evaluation is inherently based on an ethical judgment that implies the assignment of specific weights to lives and the economy (i.e., the trade-off between health and wealth).

In conclusion, the main lessons learned from this discussion can be summarized as follows:

- The virus must be controlled in the early stages, possibly by a mandatory lockdown (even if this violates civil and economic liberties), because this measure manages to contain the behavior of some super-spreaders who are mainly responsible for propagating the infection. This is necessary if more significant economic damage is to be avoided in the medium to long - term;
- the disease must be defeated to allow the economic system to return to its normal level. As long as the virus is not eradicated, there may always be expectations of a resumption of infection and thus a lower than the potential level of consumer confidence;
- the approaches followed by Sweden and the states of the sunbelt in the USA during the first phase failed in their objective;

---

[11] It is worth highlighting that these delays may arise due to the presence of democratic regimes. For example, McCannon and Hall (2021) showed that state-level stay-at-home orders were issued earlier in authoritarian states. Similarly, Geloso et al. (2021) show that higher values of the stringency index of policy responses were significantly associated with lower values of democratization and economic freedom.

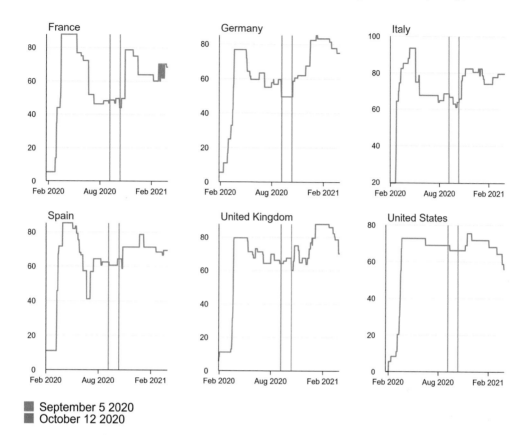

Figure 5.11 Stringency Index during first and second COVID-19 waves in selected countries. *Source*: `ourworldindata.org`. **Note**: The stringency index is a composite measure based on nine response indicators, including school closures, workplace closures, and travel bans, re-scaled to a value from 0 to 100 (100 = strictest).

- following the initial phase, the medical, epidemiological, and socio-economic contexts changed, causing a reformulation of the trade-off policies by all governments in a direction that has privileged a wider opening of the economies and reduced stringency interventions;
- new waves of contagions and new variants may continue to emerge until SARS-CoV-2 is eradicated. This situation leaves governments, individuals, and economies facing a great deal of uncertainty. In terms of economic recovery, it might be reasonable to expect investors to react negatively to news of the spread of infections and to propose their investments mostly in areas where the virus is under control.

5.2.4   COVID-19 trade-offs from a less developing country (LDC) perspective

One of the main arguments presented in the GBD was that lockdowns do not simply result in a trade-off between health and economy (lives and livelihoods), but also between 'lives lost due to the pandemic' and 'lives lost through deprivation'. Although this argument can be criticized if it is applied to all countries indiscriminately, regardless of their development status, it remains a significant consideration when discussing the impact of lockdowns in developing countries.

If the COVID-19 pandemic has affected all countries, the impact on LDCs has been particularly severe due to their reduced resilience and capacity to react to the COVID-19 shock and its aftermath. In many cases, the popular belief that the virus does not discriminate is incorrect. Wealth allows better prevention and care, while poverty does not. This is particularly true when we shift our reasoning to LDCs. According to Loayza (2020), COVID-19 has imposed an immense burden on developing countries, with dire consequences for poverty and welfare. The causes of such a huge impact stem from the diminished demand for exports, with merchandise exports falling by more than 20% in 2020. Due to lockdowns in developed countries, commodity prices fell to record lows, with oil and metal prices falling by 40% and 13% (World Bank, 2020a). International tourism (measured by tourist arrivals and tourist receipts) also decreased by about 25% percent in 2020 (UNWTO, 2020). Remittances, an increasingly important source of income in developing countries, also decreased by approximately 20% (World Bank, 2020b). Finally, external finance escaped from LDCs, recording the largest capital outflow from developing countries ever reported (more than US $80 billion since the start of the crisis) and spreads on sovereign debt increasing by hundreds of basis points (International Monetary Fund, 2022). Unfortunately, the pandemic emerged at a time when development progress was already slow and unsatisfactory, and the combination of these two events highlighted the institutional, economic and social shortcomings of the path of development followed by most of the LDCs. In September 2021, the World Health Organization set an ambitious goal of vaccinating 70% of the global population by mid-2022, with alarmingly low vaccination rates in low-income countries, since only 3% of the population had been vaccinated by the end of 2020. As documented in a careful study by Suárez-Álvarez and López-Menéndez (2022), this difference reflected and was likely to reinforce the conditions of income inequality worldwide. Since then, great progress has been made, but vaccinations in LDCs are still behind, with only 32,2% of the population vaccinated by July 2023, compared to approximately 83,4% of the population in developed countries.

Structural macroeconomic conditions coupled with the pandemic crisis can disproportionately affect low- and middle-income countries because most of them lack the resources and capacity to deal with a systemic shock of this nature. 'Their large informal sectors, limited fiscal space, and poor governance make developing countries particularly vulnerable to the pandemic and the measures to contain it' (Loayza, 2020). With more limited resources and capabilities and younger populations, the policy trade-off approach in these countries should be different from that of high-income countries. In many of these developing countries, the trade-off is not just between lives and the economy; rather, as highlighted in the GBD, the challenge is to preserve lives and avoid destroying livelihoods. For example, the different population age structure between LDCs and developed countries (DCs) allows to approach social distancing and the use of NPIs differently and, in general, to implement more moderate measures. In other words, the goal of saving lives and livelihoods can be achieved with economic and public health policies tailored to the realities of developing countries. Since 'smart' mitigation strategies (such as protecting vulnerable people and identifying and isolating infected people) pose substantial challenges to implementation, 'a combination of ingenuity for adaptation, renewed effort by national authorities and support of the international community is needed' (Loayza, 2020).

Using a mathematical model, Hausmann and Schetter (2022) demonstrate that 'a more stringent lockdown helps to fight the pandemic, but it also deepens the recession, which implies that the poorer members of society find it harder to survive. This reduces their compliance with the lockdown and may cause further deepening of the poverty condition, giving rise to an excruciating trade-off between saving lives from the pandemic and from deprivation. Transfer payments help mitigate this trade-off'. The authors further demonstrate that trade-off policies should differ according to country income levels. In particular, they show that *ceteris paribus*, the optimal lockdown should be stricter in high-income countries. When a country has limited financial resources and a high poverty rate, a milder lockdown is the best option to avoid a higher death toll during a pandemic. Financial aid can help balance the loss of lives and livelihoods. This is especially important in low-income countries, but even wealthy countries can experience a reduction in financial resources during a prolonged outbreak, leading to negative consequences for their populations.

5.2.5 More critical trade-offs will emerge in the next future

The evidence gathered so far suggests that what was initially thought to be a challenging trade-off, ultimately led to a clear choice to intervene to control the spread of the virus. As more information became available, it became increasingly clear which option was the best, and any initial uncertainty dissipated.

Overcoming the mistaken belief that saving lives and saving livelihoods are competing goals does not mean that there will not be other difficult trade-offs that can arise in the future as a result of the pandemic's various impacts. Some of these trade-offs may be even more problematic and challenging than those considered in the early phases. This is because severe crises can lead to significant and lasting changes in people's behavior and preferences, further complicating decision-making. In particular, individuals could develop beliefs and take positions and perspectives opposed to those they had before the crisis. The main elements usually triggering these changes are uncertainty and a feeling of danger, as has been widely documented following periods of crisis (economic, political, social).

As well explained in a wideranging study by Guiso (2020), the economic and financial crises, especially acute and long ones, 'can have very lasting and costly effects because they open up the space to political torments which, far from being the solution to the problems of the crisis, can extend and prolong their effects. The economic uncertainty that weighs on people, the need for protection that follows, the need for immediate solutions that citizens voters complain about, and the impossibility of continuing to meet current needs with their savings, which are hampered by the length and intensity of the crisis, make voters sensitive to illusory promises of immediate solutions'. The consequence of this is that, during the crisis, the fundamental objectives are often 'rethought', leading to different choices than those that would have been made previously. Therefore, changes in these macro-political objectives may lead to various economic and social decisions that may appear inconsistent with earlier ones.

But how does a crisis or an economic recession affect people's behavior? The answer is that people feel and behave differently during a crisis due to various factors. A major reason is related to the so-called 'cognitive biases', which derive from stressful and emotional situations and cause people to deviate from rationality, by resorting to heuristics or more intuitive and simplistic decision-making criteria. This effect is magnified during an emergency because people tend to emulate others, resulting in a cognitive bias known as 'herd' or 'mob mentality', which occurs when individuals are influenced by a group and make decisions that they would not have made individually. In all these cases, people tend not to behave rationally, react to their unconscious emotions, and are more willing to change their habits in times of upheaval. This phenomenon is also related to a recently discovered cognitive bias called the 'Habit Discontinuity Hypothesis' (Bamberg, 2006; Verplanken et al., 2008; Walker et al., 2015). An interesting study by Verplanken and Roy (2016) has shown that when individuals' habits are disturbed, people become more sensitive to new information and adopt a mindset that favors behavioral change. The hypothesis posits that this is because habits are formed and reinforced by repeated actions, and when these actions are interrupted, people are more open to new information and more likely to change their behavior.

Evidence supporting this hypothesis has been found by Giuliano and Spilimbergo (2013). Analyzing data from 37 countries, the two researchers found that people who experienced a recession when they were young (between 18 and 25) believe that success in life depends more on luck than commitment and support more government redistribution (more government intervention is needed to provide safety nets, even if that means higher taxes). They tend to vote for left-wing parties but have less faith in governments. In these cases, the effect of recessions on people's fundamental beliefs is long-lasting.

On the basis of the above, it is possible that during the pandemic crisis, individuals may change their position concerning significant trade-offs, whose effects could potentially be even more disruptive than those discussed in the early months of the outbreak. According to Kessler (2020), the trade-offs on which important changes can be observed are the following:

- *the trade-off between risk and security will be rethought in favor of security.* A pandemic has caused the current crisis, placing the world's population at increased risk of becoming vulnerable. This is a crucial (and new) aspect compared to the standard way of perceiving risk, which by its very definition affects only a limited part of people. This will increase risk aversion, and the least risky option will almost always be chosen over what appears uncertain, random, or capable of producing adverse effects. The impact of these changes will be felt in the financial markets (changed savings and investment choices and increased demands for guarantees and security), in the social sector (increased demand for subsidies and health services), and in politics (adoption of more conservative solutions by politicians for fear of being questioned);

- *the trade-off between local and global will go toward more local.* The balance between local and global priorities may shift towards a greater emphasis on local issues after the crisis. This can be seen as a reaction to problems caused by the fragility of global value chains and the lack of international solidarity. It may lead to a greater focus on national concerns over global ones and regional concerns being prioritized over national ones, in response to worries similar to those observed after the Great Recession, where resentment towards globalization emerged, and people began to prioritize local issues over global ones. Words such as autonomy in public health management, national sovereignty, and strategic interests are now on the agenda in public debate;

- *the trade-off between the short and the long term will shift towards the short term.* This is perhaps one of the most critical aspects to consider. Reducing the time horizon means foregoing significant structural changes instead of living with palliative adjustments that, for example, do not bode well for the growth prospects of developing countries. The underlying mechanism is enduring through repeated emergencies that keep decision-makers busy with short-term issues, neglecting the most important long-term ones. One of the main problems discussed in greater depth in Chapter 7 is the explosion of deficits in public budgets and social security systems or the massive increase in liquidity created by central banks. The latter led initially to a further decline in interest rates, which seemed to make the escalation of public debt relatively painless. However, as the more recent events have shown, interest rates went up in response to an inflation surge, and if they keep increasing, there will be no room for budgetary maneuvers in case of further catastrophes for the weaker countries with higher debt-to-GDP ratios. Economists recognize that trade-offs between present and future costs and benefits can be difficult to manage and often unpopular with the public. Policymakers may be inclined to postpone costs and enjoy current benefits to maintain political support. Unfortunately, future costs are paid with interest and charged to future generations;

- *the trade-off between freedom and responsibility, and between the definition of standards of behavior and the control of that behavior, will shift towards the latter.* The fourth change that could occur is the gradual reduction in the available levels of freedom. In the name of protecting specific individuals and the population in general, many restrictions were imposed during the current crisis (which are standard in wartime but unprecedented in peacetime). The coronavirus pandemic is a textbook case for the concept of endogenous externalities. These occur when one person's behavior may adversely affect others, for example, by transmitting the virus to them. In these cases, it appears legitimate to limit the freedom of persons to move and have social contacts by imposing measures of all kinds, including surveillance of the infected. The advancements in technology have greatly facilitated behavior control as compared to the past (refer to Chapter 4). Control can be exercised in various ways, from compulsory vaccination to monitoring the movement of people. The temptation to tighten control – a fairer term would be 'monitoring' – inevitably reduces individual freedoms and responsibilities. The challenge for policy makers is to find a balance between protecting the community and respecting individual rights, and it is difficult to achieve custom or 'ready-to-use' solutions. Some countries, such as France, have opted for general measures rather than targeted measures. Other countries, such as Germany, have taken more targeted measures toward specific groups, which seem to have been much more effective than the indiscriminate approach;

- *the trade-off between efficiency and inequality will be further exacerbated by the healthcare crisis.* The two main objectives of any policy remain the pursuit of efficiency (achieving growth, raising living standards, increasing incomes, etc.) and equity (tackling disparities in income, wealth, access to health care and education, etc.). Economists aspire to find and prescribe the right balance between these two goals, but this is a more straightforward task to perform in the context of growth than in a recession. Evidence suggests that the pandemic has significantly impacted the pursuit of efficiency and equality. Although the situation now appears to be more diverse, during the early phases of the pandemic, unemployment increased everywhere, incomes and investments fell and, at the same time, inequality increased. As noted by Kessler (2020): 'Each country will face this dilemma and make crucial choices, and this will have important political consequences. In addition to the public health crisis, countries are experiencing an economic crisis, which could lead to social and political crises. The path between these crises is particularly narrow and certainly dangerous'.

5.3. Trade offs and the political economy of a pandemic: a quick review

As mentioned before, the political objectives of policymakers go well beyond the direct consequences of the pandemic itself. They may involve essential aspects of everyday citizen's life, such as safeguarding constitutional and human rights. This points to the importance of analyzing the relationship between epidemic diseases and the State or, equivalently, the existence of a trade-off between public health and freedom in both the short and long run. As epidemics have affected humans since the beginning of history, the problem is old. Each time there has been an epidemic, governments have played a role in controlling it: a control that has been exercised ineffectively until scientific knowledge has developed (starting with the germ theory of diseases), so that the rise of modern states in the 19th century went hand in hand with massive investments in sanitation and publicly funded medical research. The joint presence of these two elements set the foundations for more widespread and effective government interventions during the pandemic periods.

However, as well illustrated by Reisman (1998), as long as there are scientific reasons to introduce restrictions to individual freedom, infectious diseases pose serious challenges to a liberal social order, which should be organized in a way that individuals can make the best decisions for themselves. Adam Smith, John Stuart Mill, Alfred Marshall, Arthur Pigou, Ronal Coase, James Buchanan, and Gordon Tullock, the fathers of liberal thought, tackled this issue and agreed that, in times of emergencies, the 'harm principle'[12] grants the possibility for public interventions. However, 'these exertions of natural liberty of a few individuals that could endanger the security of the entire society are and should be restricted by the laws of all governments, from the most liberal to the most despotic' (Smith, 2010). Similarly, economic theory suggests restricting individual liberties whenever significant non-pecuniary externalities are difficult to observe and measure (see Section 5.3.1). But, according to Buchanan and Tullock (1965), these restrictions should be applied at the lowest level where the externalities in question can be internalized. In the specific case of an infectious disease outbreak, they should be applied using ordinances at the city, provincial, or regional level, whenever the spread of the infection can still be contained by restrictive or other sanitary measures.

Once we have clarified that state intervention is desirable in the presence of epidemic events, it remains to understand how policymakers will react. In Chapter 3, we have seen that policy-makers may have made several errors. The origin of these errors could be attributed to 'information' problems during the first wave (i.e., lack of clear scientific information), and to 'incentives' problems during the remaining waves (i.e., the propensity to take popular decisions that would bring them to be re-elected).

---

[12] The 'harm principle' is a concept in political philosophy and legal theory, first articulated by the British philosopher John Stuart Mill in his work 'On Liberty'. The principle states that individuals should be able to act as they wish, as long as their actions do not cause harm to others.

As discussed in Chapter 3, during the first wave, the choices were mainly determined by the quality of the available scientific information. This explains why the countries reacted so heterogeneously. Good examples can be found throughout the world. In Italy and the UK, significant sporting events were not prohibited, which created a good opportunity for the virus to spread. WHO did not recognize the airborne nature of COVID-19, yet emphasized the role of 'fomites'.[13] Furthermore, in many countries, public authorities recommended not using masks or banned them and, at the same time, prohibited outside activities during the first lockdown in March-May 2020.

On the contrary, during the other waves, decisions were mainly driven by incentives that led policymakers to behave strategically, according to what political economy experts define as the rule of the ruling coalition's median voter. This implies that if policy makers have to choose between controlling a pandemic or mitigating damages to the economy, they tend to choose by the preferences of the median voter of the ruling coalition, since this represents the best strategy for them to gain or remain in power. The final decision could have been restrictive if the restrictions were popular with voters who could influence the governing coalition's support. This could include pensioners who are vulnerable but not financially constrained, or people who can work from home. In contrast, less restrictive policies would have been adopted when the pivotal group comprised young people who could quickly lose their jobs or livelihoods from a restrictive response. A recent study of the political economy of the pandemic persuasively argues: 'It is, of course, impossible to know how the COVID-19 pandemic may have played out had different choices been made by policymakers in early 2020. Given the virus's combination of virulence and contagiousness, it may have been the case that many deaths would have resulted whatever policies were taken. Nevertheless, in 2020, political leaders often lagged rather than led public opinion, and in western democracies, in particular, they were reluctant to take radical measures until the number of confirmed COVID-19 cases was sufficiently high. But that meant that, given the pace at which the disease was incubating and spreading, policies like travel restrictions and lockdowns were imposed too late to prevent COVID-19 from seeding in local populations' (Koyama, 2021).

5.3.1 The externalities of the pandemic and the role of public health

Economists define an externality as the phenomenon that occurs when the consequence of one or more individuals' actions spills over to others. Thus, to the extent that people are selfish, an externality is created whenever individuals behave in a way that ignores the consequences of their actions or inactions on others. The discrepancy between

---

[13] Fomites are inanimate objects or porous/non-porous surfaces which, when contaminated or exposed to pathogenic microorganisms, can transfer an infectious disease to a new host. Examples are dirty clothes, towels, sheets, handkerchiefs, surgical dressings, and contaminated needles.

the choices advocated by these individuals and those desirable to society as a whole defines the distortionary effect of an externality, which is the consequence of an economic transaction that is not reflected in the market price of the goods or services involved. Externalities can be positive or negative and can arise from various activities, including production, consumption, and trade. Positive externalities occur when an individual's actions generate benefits for others, such as when someone's education or innovation benefits others in society. Negative externalities occur when an individual's actions impose costs on others, for example, in the case of pollution. Externalities can be pecuniary, technological, or utility-based. Pecuniary externalities affect the value of resources held by others, such as when innovation makes a previously valuable resource obsolete. Technological externalities physically affect other people, such as when an individual's actions lead to the spread of a communicable disease. Utility-based externalities affect the subjective values of others, such as when another person's illness anguishes or is relieved by healing.

The problem with externalities is that they lead to market failures, since most individuals have no personal incentives to consider the costs and benefits to others when making decisions. Therefore, unless they are altruistic and include the welfare of others in their considerations, they will tend to engage in activities that generate negative externalities, such as pollution or disease transmission, and fail to engage in activities that generate positive externalities, such as innovation or education. To address this problem, mechanisms must be implemented to internalize externalities, which means that the costs and benefits of activities are reflected in the market price. This can be done through various policy mechanisms, such as taxes and subsidies, regulations, and property rights. Mechanisms based on incentives are called 'Pigouvian' taxes (and subsidies) after the Cambridge economist Alfred Pigou (1877-1959), who defined such instruments in 1920 (Pigou, 1920). Two actions to prevent externalities are: *i)* where individuals do less than is desirable from the society's point of view, the classic solution to the externality problem is to subsidize the activity; *ii)* if individuals do more than is desirable by society, then the solution is to tax that activity. In both cases, incentives are created that tend to reduce the formation of externalities. Subsidies and taxes 'internalize' costs and benefits, which means that they become part of an individual's decision-making criteria in a way similar to market-based activities. Another mechanism that can be used to address externalities is based on property rights and is grounded in the 'Coase Theorem' named after the economist Ronald Coase who developed it (Coase, 1960). The Coase Theorem demonstrates that if property rights are well-defined and transaction costs sufficiently low, market failures can be avoided through bargaining and exchange. These may also involve negotiating regulations and legal restrictions through political actions and with the wider participation of the individuals involved.

A clear synthesis of policy instruments and related problems to address externalities in the case of epidemics is provided by Toxvaerd (2020): 'in the context of infection control, there are many ways to help people take externalities into account. For example, the government can offer subsidies for personal protective equipment to encourage people to protect themselves. Other indirect measures include dismissal and statutory disease benefit schemes that encourage people who may be infected or have symptoms to stay home to avoid further spreading the disease. As a last resort, fines may be imposed on those who do not wear face masks in public or do not respect social estrangement restrictions. [...] In addition to providing explicit incentives such as grants and fines, other measures can be used to encourage people to protect each other better. Moreover, many behaviors are driven by expectations and social norms rather than material incentives. Clear communication on the social costs of infections can be used to promote better social norms. Even before the COVID-19 crisis began, it was considered a good norm, as a form of courtesy toward others, to cover your mouth when coughing up in public. Similarly, face masks could become a social norm, even in open spaces where their use is not yet mandatory'.

The mechanisms described may also help solve the trade-off problem between health and economics that has already been discussed in this chapter. Policymakers have faced a key challenge in the coronavirus pandemic (but this applies to any situation involving communicable diseases): a choice between the uncontrolled spread of the virus, which can cost millions of lives in worst-case scenarios, or the imposition of non-pharmaceutical public health interventions, such as social distancing, which harm economic and social activity and may undermine the livelihoods of a much larger number of people. This is a classic case of market failure due to negative externalities (a lockdown imposed to resolve contagion has negative effects on the economy). It is why coordinating health policies and controlling epidemics is a matter of public policy, because infectious diseases involve externalities. When infected individuals engage in social or economic activities, they impose significant externalities on those with whom they interact and place them at risk of infection.

On the other hand, policy makers are not unconstrained social planners but are themselves expressions of their society's institutional context and constitutional rules. Their motivations, goals, and mechanisms are therefore largely endogenous, reflecting the implicit bargaining power and constant renegotiation of the social contract that the various constitutional systems allow. As Allen et al. (2022) clearly noted: '... As such, we can think of the policy decision not through the additive utility lens of a social welfare function (a Pigouvian lens), but rather as political brokering of coalitional exchanges across different risk groups in society (a Coasean lens). In this way, we can shift from what James Buchanan (Buchanan, 1964) referred to as an 'allocation' lens of economic inquiry towards an 'exchange' paradigm. The former emphasizes the allocation of resources by a social planner with relevant information, while the latter focuses on the process of interaction between people within context-specific, varying institutional environments, with more realistic assumptions about information and incentives (Coyne et al., 2021)'.

5.3.2 Incentives vs. constraints: what strategy to adopt in the presence of externalities

In the presence of market failures, policymakers often need to take action by implementing ordinances, decrees, or laws. This was evident during the first wave of infections in March and April 2020, when public health officials worldwide recommended non-pharmaceutical interventions focused on social distancing. Implementing this recommendation required the creation of unprecedented guidelines covering a variety of topics and areas. These ranged from the closure of international borders to the restriction of travel even within regions, to limitations in assemblies, to the closure of all non-essential segments of the economy, together with schools, universities, and places of worship, to lockdowns as orders to stay at home. Even sectors of the economy that were still allowed to operate were subject to a detailed set of regulations that outlined the specific conditions under which they could continue to function.

There has been some debate over whether a regulatory approach is the most effective way to address the negative impacts of the pandemic, with researchers in fields such as behavioral economics and law suggesting that nudging people to take certain actions may be more effective than enforcing them through regulations. Over the past two decades, this method has had significant impact on the field. It has led to re-examining several key policy and legal issues across various disciplines, resulting in a new research paradigm.

In recent years, behavioral science has become increasingly influential in shaping public policy. Governments have used behavioral insights to inform decisions on issues such as education, environmental protection, and nutrition. Some governments have even established specialized units, known as 'nudge units,' to promote evidence-based and behaviorally informed policies. Given this trend, it is not surprising that behavioral scientists are involved in the decision-making process for one of the most significant regulatory challenges in history. They are using their expertise to inform the political discourse and help policymakers make better decisions (see for example Teichman and Underhill, 2020).

In 2010, the UK government established the first Nudge Unit, called the Behavioral Insights Team. This unit is responsible for applying behavioral science to policy making and has been actively involved in UK government decision making. The team played a crucial role during the early stages of the COVID-19 pandemic by advocating for a strategy that focused on changing people's behavior to slow the spread of the virus, rather than implementing strict lockdowns. This approach was aimed at 'flattening the curve' of infections and reducing transmission speed. In that case, the strategy was to make recommendations, the non-mandatory nature of which allowed many discouraging events to continue (the annual Cheltenham horse race attracted 125,000 people between 10 March and 13 March). Soon after that, in light of the escalating infections, the British Government changed its attitude, leading to a significant debate among behavioral analysis experts on the appropriateness of the proposed solutions.[14]

In partial defense of the British government's decision, it should be remembered that in the early months of the pandemic, many researchers, even outside the UK, suggested the nudging strategy. In the influential medical journal 'The Lancet', Betsch et al. (2020) noted that 'a critical element in reducing virus transmission is rapid and widespread behavioral change' and that 'behavioral insights for COVID-19 are crucial'. Similarly, the WHO published a statement stressing that 'behavioral insights are valuable for informing the planning of appropriate pandemic response measures' (Kluge, 2020). This statement by the WHO was the result of years of experience in epidemic containment, mainly when a vaccine or treatment is unavailable for several months or years. According to the WHO, in these cases, epidemic containment depends equally on the behavioral aspects of people as on the medical skills and expertise of professionals. For example, in the case of the Ebola emergency, it was seen how important anthropology was to understand how the disease was transmitted. Similarly, psychology has provided clues on how a social habit, the desire to embrace, could be replaced by elbow touch (as done during the COVID-19 pandemic).

The concept of 'behavioral fatigue' refers to the idea that individuals, over time, may become tired of adhering to behavioral restrictions or guidelines, especially when these are prolonged or perceived as overly restrictive. Some studies have found that people may become less compliant with restrictive measures over time, while others have found no evidence of this phenomenon. The bulk of the evidence suggests that different factors, such as the characteristics of the population, the resources available, and the specific characteristics of the virus, may play a role in determining whether people will comply with restrictive measures over time. In addition, the effectiveness of different strategies will also depend on how they are implemented and communicated.

The approaches taken by the UK and Sweden in their pandemic responses have sparked a broader discussion about the most effective methods for ensuring compliance

---

[14] The Behavioral Insights Team's approach to the early stages of the pandemic was based on the concept of 'behavioral fatigue', which suggests that people would not be able to sustain strict measures such as economic lockdowns, school closures, and travel restrictions for a prolonged period of time. Therefore, the Team argued that implementing these measures would be more effective only when they were most necessary. However, this approach was criticized by some researchers who called attention to the lack of supporting evidence of the behavioral fatigue hypothesis and to the fact that stricter measures should have been implemented earlier. In a highly publicized open letter, more than 600 behavioral economists remarked that they were 'not convinced that enough is known about 'behavioral fatigue' or to what extent these insights apply to current exceptional circumstances' (Chater, 2020; Scientists, 2020).

with regulations. This debate is rooted in the key question of whether policymakers should rely on nudges, which use behavioral insights to preserve choice and encourage people to act in a certain way, or norms, which use sanctions to impose specific behavioral standards. A recent review of relevant studies suggests that the use of behavioral analysis may be beneficial in addressing certain challenges that arise during the emergency phases of a pandemic. However, it is important to remember that different strategies may be more or less effective in different contexts and situations (Teichman and Underhill, 2020).

The main policy recommendation of the review is that 'when people's choices generate massive negative externalities – as in the case of a highly contagious and deadly virus – policymakers should opt for rules backed by command and control policies'. However, as the authors surmise, incentives and nudges can be incorporated into the rules in two situations: *i)* when constraints are difficult to implement due to political or legal constraints; *ii)* when incentives can complement rules and help promote voluntary compliance. This conclusion applies to other contexts with extensive negative externalities, such as climate change and, more generally, environmental protection. For example, while incentives such as incorporating taxes into electricity bills may help reduce the negative environmental impacts of energy consumption, they are unlikely to significantly address global warming by themselves. Behavioral scientists and legal scholars have increasingly come to the agreement that traditional regulatory tools, such as emission restrictions and associated penalties, are essential to change behavior effectively.

5.3.3 The theory of real options as a way of handling trade-offs

The results discussed above have highlighted that the supposed trade-off between health and the economy is a false problem. Its apparent existence is due to the ineffectiveness of strategies that avoid lockdowns, which harm the economy. Additionally, the ongoing circulation of the virus and the lack of understanding of its behavior still create uncertainty. However, these uncertainties can be reduced over time as we gather more information. This suggests that the real options theory could be a useful framework for decision-making in this context.[15] Real options theory is a valuable method for evaluating and managing strategic investments in the context of uncertainty. It provides the theoretical framework for considering opportunities and threats as real options, that is, as 'assets' and 'liabilities' that represent intangible contingent components of wealth. A

---

[15] Real options provide flexibility to adapt decisions as more information becomes available and future events unfold. The idea of real options was laid down in the 1970s and subsequently developed in the 1990s to solve key methodological problems concerning uncertainty and timing, which affected the traditional evaluation of investments, based on the Net Present Value (NPV) and the Internal Rate of Return (IRR) tools.

real option is nothing more than the right (without any obligation) to obtain an advantage from an opportunity in an uncertain environment. Having a real option provides a bonus for waiting, or the postponement of a decision, as it allows to acquire new information. In this way, the future bears the risk of possible threats and positive opportunities: those exposed to uncertainty certainly face the risk of potential losses, but they can also gain greater rewards.[16]

As already seen before, in the presence of an epidemic, there are two polar alternatives for the decision maker: *i)* do nothing and hope to achieve herd immunity quickly, or *ii)* act with interventions, such as the '*hammer and dance*' strategy suggested by Puyeo (2020), through which drastic measures of social distancing are initially applied ('*the hammer*') to reduce the spread of the disease and, subsequently, testing and tracking is used to control the infection until a vaccine is found ('*the dance*').

In the first case, there would be no adverse economic effects, but there may be disadvantages to people's health. In the second scenario, if the disease does not pose significant risk to human lives, it is generally more advisable to opt for the first solution. However, in the COVID-19 case, the effects and severity of the disease were unknown in the first weeks of its spread and remained largely unknown several months after its initial outbreak. And it is precisely the presence of this great uncertainty about one of the alternative strategies that makes the theory of real options the most appropriate theoretical context to tackle the problem.

The value of a real option, as partly reflected in the premium paid to exercise the option, is determined by the level of uncertainty.[17] This uncertainty encompasses potential risks and opportunities related to different possible future outcomes. Choosing the best strategy to respond to COVID-19 implies considering not only the costs and benefits associated with each alternative strategy, as in the traditional evaluation approach, but also the value of the option to defer immediate action and wait for more information. This is bound to be high, given the uncertainty that exists around the effects of COVID-19.[18]

Although the knowledge about the virus and its effects has increased over the months, the level of uncertainty has remained high (Avery et al., 2020a; Garfin et al., 2020; Kresge, 2020; Wu et al., 2020). According to Santos et al. (2020) the uncertainties that must be considered are many and can be summarized as follows:

---

[16] An example of using real options would be an oil company investing in new drilling. This gives the company a real option, or the ability to wait and decide to extract oil later when the market price is higher.

[17] In the context of COVID-19, the premium would be the costs associated with implementing a particular strategy or set of policies. For example, the premium for a strategy that avoids lockdowns and instead relies on voluntary measures could be lower than that of a strategy that includes mandatory lockdowns.

[18] Proponents of the herd immunity strategy tend to overlook these uncertainties.

- people's immune response to the virus. As long as this aspect is not clarified, the herd immunity strategy does not make sense (Bergstrom and Dean, 2020; Iwasaki, 2020);
- mutations of the SARS-CoV-2 virus. If the virus changes over time, this can make it unnecessary and/or impossible to achieve herd immunity;
- uncertainty about the search for treatments and vaccination coverage. Not knowing when the vaccine can guarantee the disappearance of the virus dramatically complicates the calculation of healthcare costs;
- costs of a lockdown. Much of this uncertainty comes from the fact that people are changing their behavior due to the epidemic, and it is not yet known how. Behavioral changes affect people's economic choices regarding demand (i.e., consumption) and supply (i.e., decisions to work). A classic example is the one reported in the previous sections, where it was shown how the Swedes had reacted the same way as the Danes, even though the blockade had only been imposed in Denmark (Andersen et al., 2020). In this way, it becomes difficult to estimate costs;
- costs of not treating other conditions. This crucial topic has begun to be intensely discussed among experts. Since the start of the pandemic, several papers have been published that demonstrate the decrease in utilization of various health services (Baldi et al., 2020; De Filippo et al., 2020). This uncertainty acts through two main channels. The first is that of hospital congestion: If the number of hospitalizations is very high, it can overwhelm the capacity of the health system (Gourinchas, 2020). As a result, the health system may not be able to treat other diseases that will lead to further deaths from those other causes. The second channel is represented by the patient's fear of going to the hospital. In this case, people with other severe conditions may avoid seeking help from the health care system for fear of being infected in hospitals. The higher the stress level to which a health system is subjected due to the epidemic emergency, the greater these effects will be (see the case of Lombardy in Italy).

A simple exercise that allows us to better understand how the current situation can be interpreted in the context of real options theory was conducted by Scandizzo (2020). According to Scandizzo, by not adopting drastic lockdown measures, thus taking advantage of the wait option (see USA and UK), 'you gain valuable information over time for subsequent decisions, but at the same time, you face substantial risks based on the characteristics of the epidemic known at the moment'.

The problem is then understanding to what extent it is helpful to wait before acting (for example, by imposing the lockdown). Researchers at Imperial College London conducted simulations in the early stages of the COVID-19 pandemic, using the basic reproduction number of the virus (R_0) and parameters similar to those used in China. The simulations predicted a scenario in which, without imposing lockdowns, there could have been 20-25 million infections and at least 1 million deaths. However, this approach would have resulted in a lower economic impact, herd immunity, and a younger and more affluent population benefitted (since the elderly and vulnerable would have been disproportionately affected). On the other hand, if a lockdown strategy had been implemented, there would have been a reduction in cases and deaths, but severe short-term damage to the economy and greater difficulty in medium–term recovery.

As long as the costs imposed by inaction (not imposing the lockdown) are lower than the costs of the alternative, it is possible to continue waiting and accumulating information. Scandizzo's (2020) analysis, using a statistical cost of living criterion from Masterman and Viscusi (2018) set at 5.6 million dollars and another based on IS-TAT's (2020a) value of human capital at roughly 400,000 dollars, highlights significant economic implications. With 2020 GDP losses in Italy between 5–10% due to the pandemic, especially its major outbreak in Lombardy, the country had already surpassed the anticipated benefit threshold for implementing isolation measures. That is, the Italian government was making the correct decision based on the underlying economic rationale. In other words, considering the data from epidemiological models, it was the right choice to implement lockdown measures at the early stage of the pandemic.

CHAPTER 5 - Take-home messages

- The debate on soft strategies is open. The GBD and the JSM express two opposing positions. GBD aims to spread contagion among the low-risk population with the concomitant confinement of the most vulnerable to attain herd immunity and the protection of the economy. On the contrary, the JSM emphasizes the absence of scientific evidence to support long-term immunity and argues that the best protection of economic activities is achieved by controlling the spread of the virus. No country has permanently followed the GBD strategy.
- The pandemic has a disproportionate impact on the most vulnerable, both economically and in terms of health. To mitigate this impact, political decisions must balance the health and economic needs of society, including the monetary value of human life.
- Policy strategies are based on an analysis of the trade-off between health and the economy, with a significant caveat: different peoples and countries may have different orientations and a sound political strategy in one context may not be satisfactory in another.
- Research in the epidemiological-macroeconomic field produced interesting results for decision-makers. However, there is still a need for time to refine these models and continue analyzing the effects of policies over the medium to long term.
- The policy choices made to address health crises are contingent on a range of factors that may shift across different times and places. Nonetheless, a consistent element is that the success of these decisions is reliant on the aptitude of the policymakers, their capacity for leadership, and the strength and adaptability of the health systems in place.

- When people's choices generate massive negative externalities - as in the case of a highly contagious and deadly virus - policymakers should opt for rules supported by command and control policies.
- According to the principles of real options theory, epidemiological data offer support to the decision of several governments to exercise the option to implement lockdown measures at the early stage of the pandemic.

CHAPTER 6

COVID-19 macroeconomics: are we using the right toolbox?

Contents

Even after over three years since the beginning of COVID-19, we remain uncertain about whether we have successfully emerged from an unprecedented pandemic, encompassing both biomedical and socioeconomic challenges. Simultaneously, we find ourselves traversing an ambiguous journey towards an elusive and unpredictable state of normalcy. The pandemic has started with an unexpected surge of infections, illnesses, and deaths, with the global economy quickly succumbing to a series of negative supply and demand shocks. These shocks have been the consequence of the lockdowns and other restrictions imposed by governments in the first phase of the pandemic and the fragility and lack of resilience demonstrated by both the market and government structures under its pressure. Disruptions of labor markets and value chains have taken some time to manifest. Even with the relief money abundantly distributed by governments, they have profoundly affected the economy, dramatically impacting the more vulnerable population, firms, and countries. However, the subsequent race for vaccines was won despite its impromptu character and the systemic failure of the pharmaceutical industry to develop them for timely prevention.

The concept of 'waves' of the pandemic is largely metaphorical, and its practical application depends on different definitions and criteria. However, the term 'wave' has been used to describe periods of increased COVID-19 activity, with related sharp increases in cases, hospitalizations, and deaths that characterized the first phase of the pandemic, from late 2019 to early 2021 and led to the appearance of the more virulent but less lethal varieties, such as Delta and Omicron. While dates and time lengths vary according to different definitions and studies, it is generally agreed that the COVID-19

The COVID-19 Disruption and the Global Health Challenge
https://doi.org/10.1016/B978-0-44-318576-2.00019-6

pandemic has had at least three waves (El-Shabasy et al., 2022). The first wave began in late 2019 and early 2020, when the virus emerged and spread rapidly in Wuhan, China, and then to other countries. The second wave began in late summer or early fall 2020, after a decline in cases during the summer months, and the third wave began in late 2020 or early 2021, with a resurgence of COVID-19 cases after a decline in summer and fall 2020.

The three-wave description corresponds to an economic timeline of the pandemic that depends on how the economic downfall and successive recoveries are defined and measured, as well as the specific region or country considered. It is generally acknowledged that most economies experienced a significant contraction in economic activity during the first wave of the pandemic in the first two quarters of 2020, as a consequence of lockdowns and social restrictive measures. After that, depending on the country and the economic sector, some economies started to show signs of recovery. For example, a study by Diebold et al. (2022) suggests that the recession in the United States, one of the shortest in history, began in February 2020, and the economy reached a trough between May and June 2020, also marking the beginning of an expansion that is still continuing. In Europe, although the time path appears to be less sharply defined and the size of the recession and the recovery much more uneven, most countries suffered a sharp contraction in the second quarter of 2020 but started to recover in the third quarter of 2020.

The second wave of the pandemic hit the world economy sometime in the late summer or early fall of 2020, after a decrease in infections during the summer months. The resurgence of illnesses, hospitalizations, and deaths in many countries followed the easing of restrictive measures put in place during the first wave and appeared more severe than the first wave. However, the economic consequences were generally less pronounced than the first wave and the effect was more a slowdown of the economic recovery that was already in action rather than a new recession.

The vaccination campaign started at different times in different countries, depending on regulatory approval and distribution logistics. Nonetheless, by the end of 2020, the most developed countries had started vaccinating high-priority groups such as the elderly and health workers. This seemed to give a new impulse to economic recovery. At the end of 2020, it appeared that a rapid response to government spending and a successful vaccination campaign would allow the economy to bounce back with unsuspected force. In advanced economies the recovery appeared to be back and more robust than expected. In retrospect, the production surge witnessed in developed countries was the result of two exceptional and partly transitory circumstances: *i)* an unprecedented increase in international liquidity and *ii)* a series of concurrent factors, such as the rapid deployment of vaccines, the use of technology and automation in many industries, the rebound of some sectors of the economy, through the expansion of e-Commerce and the use of remote control technologies, and also government policies in support of individuals and small businesses.

The third wave of the pandemic, which started at the beginning of 2021 was mainly based on the appearance of more transmissible virus mutants, which evolved almost simultaneously with the successful vaccine campaign. This long and extended wave of unexpected length somewhat dissipated the illusion that the pandemic might have come to an end and presented several unsettling surprises. On the one hand, vaccine effectiveness appeared to decrease with time with higher than anticipated speed. On the other hand, the virus itself appeared to engage in a mutation race that further eluded vaccination coverage by increasing infectiousness and resistance to vaccines. The returning restrictions, lockdowns, and other emergency measures, combined with disruptions in the supply chain and labor market, acted as bottlenecks in key production and delivery systems and caused an unprecedented lack of supplies of primary goods, labor, and energy and a consequent increase in inflation.

By the middle of 2021, due to the extensive administration of vaccines, developed countries appeared to be in a zone of relative security and hopeful containment of the negative effects due to infections. However, they feared new waves and varieties of the virus. At the same time, most developing countries remained under stress, with the pandemic advancing and the nightmare of a prolonged and unsustainable health emergency. The world economy reflected this dualism, with a rebound in economic activity in richer countries beyond expectations and a fall in incomes and consumption in poorer countries, which also seemed more dramatic than expected. Before the start of the Ukrainian war, the IMF forecasts included an actual increase in world GDP of 5.9% for 2021 and 4.9% in 2022, but the figure for 2022 was revised downward to 3.2%. By December 2022, the IMF expected the global growth forecast to bottom out below 3% for 2023 and to increase again above 3% in 2024. These figures, which suggest moderate optimism, hide the current and potential dramatic differences between rich and developing countries. Furthermore, the prolongation of the emergency and the worrying developments of the pandemic outside the magic circle of advanced economies threaten to undermine further positive developments in recovery.

6.1. Some macroeconomic stylized facts during the pandemic

The dangers on the horizon are not limited to the unsustainable social inequality unleashed by the global spread of the virus. The peculiar characteristics of the possible recovery and the means used to set it in motion create opportunities along with potential threats. The simultaneous expansion of international liquidity and uncertain realization spending programs represents a scenario never experienced before. A phase of global price moderation was followed by a phase of financial bubbles fueled by a rise in asset prices. Commodity prices first contracted, with the uncertain prospect of the energy transition helping to depress investment and capital goods prices. Consumer prices remained stable during the pandemic due to the fall in demand and the government support provided to businesses and workers during lockdown periods. However, the moderation in consumer prices, which was reversed by their turbulent 2022 increases in the US and Europe, had been partly the result of the repression of past inflationary trends fueled by monetary expansion but held back by public intervention and the uncertainty created by the pandemic. At the same time, the lack of investment in crucial sectors such as energy and logistics, in the context of an unexpected war in the heart of Europe, created a perfect storm in which unmet demand and rising costs, slowing growth, and inflation converged.

Even more than the darkest part of the pandemic, the current phase of unequal recovery has revealed some structural knots that need to be solved to relaunch development and growth. First, digitalization and the whole panoply of intangible instruments associated with it can create the dangerous belief of a digital paradise in which generalized smart working allows people to produce value by a collectively fiddling of personal computers, without moving from home except for recreational reasons. To avoid the dangers presented by this illusion, it is important to incorporate digital technologies into physical structures of production and redesign current value chains and logistics systems through an industrial transformation. This will ensure that digitalization is implemented in a sustainable and effective way. Secondly, the observed low growth and prevailing inflation seem to be driven by supply chain disruptions, panic purchasing, and inadequate stockpiling of commodities, primarily energy. These challenges are triggered by the economic revival in advanced nations, impacting both these countries and the rest of the world. This shortage that has already emerged during the peak of the pandemic is the effect of too long and fragile value chains, raw materials concentrated in a few countries, and obsolete logistics systems. It is also the result of the technological and industrial uncertainty caused by the energy transition and climate change mitigation policies, apparently secure and ambitious in the objectives but weak and uncertain in the instruments.

Third, the energy transition remains a big concern, given that the ways and times of its implementation remain to be discovered and are still conditioned by objectives that seem progressively less realistic in light of the results obtained so far. The path to achieving these goals, that is, the actual profile of the transition, is still highly uncertain, and the mix of necessary energy sources is indeterminate. For example, the lack of natural gas, exasperated by the Ukrainian war, which threatened to be a deadly bottleneck for recovery, was simultaneously the result of the ambiguity of the transition and a signal from the governments. It was also a premonition of the difficulties of real progress in the replacement of fossil energy without a way to finance its costs and maximize its benefits. The absence of a transition strategy manifests itself in the lack of credible industrial policies, such as excluding natural gas and nuclear power from the EU's 'Fit for 55' environmental package. The application of the principle of 'do no harm' that the European Commission claims to follow is probably erroneous because it lacks convincing *a priori* reasons and a methodology to assess the consequences of the different choices.

We live in extraordinary times, when the contemporaneous arrival of shocks of different nature (sanitary, economic, and technological) has created the conditions for a perfect storm. How we will respond to these shocks will determine whether and how fast we will recover. However, these are not standard recession times where policymakers had the right tools to deal with problems. Current events may require a different toolbox, which may not be available yet. This is why at the end of this chapter we discuss whether the pandemic is suggesting the need to develop a new macroeconomic theory.

6.2. The global economy shock

6.2.1 A different shock!

For a global economy still in the process of unstable and uneven construction, COVID-19 has caused the largest unexpected shock experienced to date, and it is complicated to predict even the broad lines of recovery that can take place and the additional crises that could appear on the scene. There is no quantitative evidence of similar historical events, except for the narratives of Black Death or Spanish flu. These are not comparable to COVID-19 except for a small part, but they can provide some insight into the transformative nature of pandemics, as they were major turning points in history and had a lasting impact on the way people saw the world and their place in it. The two world wars are the most similar documented historical examples, with virus playing the role of the enemy. Recognizing both the commonalities and distinctions in these scenarios is critical, as they can play a pivotal role in shaping our perspectives and actions.

The COVID-19 crisis began with simultaneous negative supply and demand shocks (Carlsson-Szlezak et al., 2020a,b; Guerrieri et al., 2022). One of the most significant differences was the impact on employment, which was particularly concerning. The effect on financial assets was also distinct; while war destroyed them, the pandemic seemingly created them out of nowhere due to a large and growing money supply, resulting in unpredictable and worrisome overestimations. Moreover, unlike world wars, the pandemic led to two unprecedented global effects: the 'decoupling' effect, which was simultaneously heterogeneous and polarizing across countries, and the disruption of international trade and value chains. In this context, we will limit our discussion to these two phenomena and present the policies implemented thus far to address these challenges.

6.2.2 The decoupling effects

The 'decoupling' effect refers to the heterogeneous impact of the pandemic across different groups of countries, with some countries, sectors, and social groups experiencing a more severe impact than others. This effect can also be described as 'polarization', as it has led to a widening of the gap between those who are faring well and those who are struggling economically. The term 'polarization' implies that the impact is not evenly distributed, but rather concentrated in certain groups or regions. This effect is caused by a combination of factors, including the degree of dependence on international trade and tourism, the level of economic development, and the effectiveness of the policies implemented to mitigate the economic impact of the pandemic. The notion of 'decoupling' in the economy can be elaborated upon by pinpointing sectors like culture, tourism, and commerce that faced significant setbacks during the crisis, while some either sustained stability or experienced growth. This stark contrast between different industries can be attributed to varying vulnerabilities on both supply and demand sides. The interplay of restrictions and relief expenditures during each wave of the pandemic further exacerbated the 'decoupling' effect of demand and supply shocks. For example, using a disaggregated Keynesian model, Baqaee and Farhi (2022) suggest that the effects of negative supply shocks, also reflected in the current recovery phase, tend to be stagflationary. On the contrary, negative demand shocks tend to be deflationary. Thus, business relief policies may be ineffective if related sectors are constrained by demand. In contrast, policies that try to boost demand might lead to supply bottlenecks and generate inflation.

These results depend on the structural characteristics of the sectors affected, including *i)* the process of production of goods and services from procurement to delivery; *ii)* the backward and forward links with other parts of the domestic and external economy; *iii)* the structure of demand for their outputs. A clear example in this direction is represented by the different weights of sectors in an economy in terms of value added and employment.[1] For instance, based on an indicator of 'essentiality' computed from lockdown regulations collected in Germany, Italy, and Spain, Fana et al. (2020) have been able to rank the economic sectors from 'essential' to 'non-essential'. As we can see from Fig. 6.1, within these three countries, there is a large heterogeneity in terms of sector 'essentiality'. Depending on the role sectors play in terms of employment and value-added, this can explain different recovery paths across countries. In particular, based on the previous classification, Zhongming et al. (2021) have estimated the share of employment and value added in 'essential sectors' in the OECD countries. As shown in Fig. 6.2, there exists a significant country heterogeneity, with employment ranging from a high of 42% in Norway to a low of approximately 27% in Italy. Similar differ-

---

[1] With the introduction of lockdown measures, governments have authorized some economic activities, deemed 'essential', to continue operating. For these sectors, restrictions have been weaker, implying that workers were allowed to stay on site, though with stringent safety requirements. In general, an 'essential' sector complies with at least one of the following requirements: *i)* guarantees the health response to the crisis (e.g. health services, R&D, pharmaceutical manufacturing, and retail pharmacies); *ii)* guarantees the supply chain for necessary basic goods and services (e.g., farming, food processing, and grocery retailers); *iii)* guarantees the smooth operation of those critical systems and infrastructures whose incapacity would have a debilitating impact on security, safety or health (e.g., energy production; public administration).

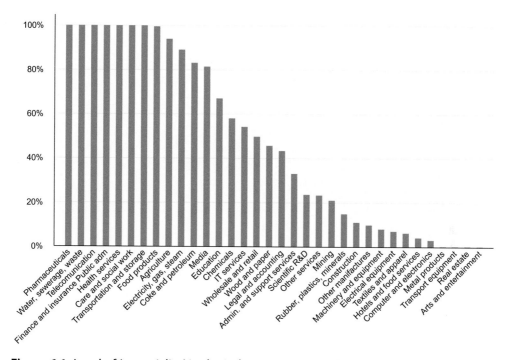

Figure 6.1 Level of 'essentiality' in the industries. *Source*: Adapted from Zhongming et al. (2021). OECD calculations based on Fana et al. (2020). Note: Published NACE two-digit indicators have been aggregated to SNA A38 level for each country based on employment shares taken from the OECD Structural Analysis (STAN) Database. The cross-country median of the resulting country-industry value is then taken as the Essential Industry indicator. This indicator is a continuous variable between zero and one, where higher values indicate higher essentiality, and thus less direct impacts from the initial shocks. The analysis is based on three countries: Germany, Italy, and Spain.

ences can be found for value-added shares, ranging from 43% in Switzerland to about 30% in Israel.

These data suggest a reason for the level of decoupling that the COVID-19 pandemic may have generated in the economy: the higher (lower) the share of essential activities, the stronger (weaker) resilience in the short run. However, the overall response will also depend on the way in which the COVID-19 crisis affects the demand side. This includes the way restrictions may be introduced and lifted, and the frequency at which this may occur, which in turn may have permanent effects on households' income and liquidity, investment behavior and level of risk aversion. In turn, this will send feedbacks across all industries, with those with higher demand elasticity (e.g. manufacturing, luxury retail)

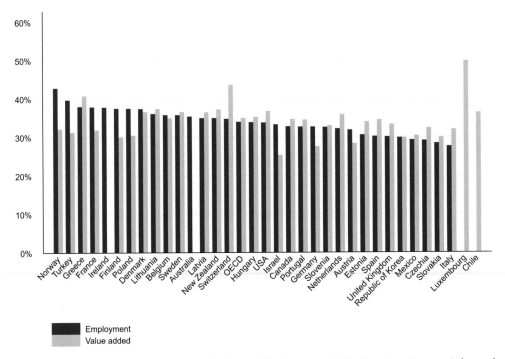

Figure 6.2 Share of employment and value-added in essential industries. *Source*: Adapted from Zhongming et al. (2021). OECD calculations based on data from Structural Analysis (STAN) Database, http://oe.cd/stan. Note: Essential industries are defined as having an index value above 0.8 on the essential industry indicator. These include Agriculture, Food products, Coke and petroleum, Pharmaceuticals, Electricity, gas and steam, Water, sewerage and waste, Transportation, and storage, Finance and insurance, Media, Telecommunications, Public Administration and Defense, Health services, Care, and social work. Average refers to the unweighted average across the reported countries. Data refer to 2018 except for: Australia, France, Germany, Latvia, Lithuania, Luxembourg, Norway, Portugal, Switzerland, and the United Kingdom (2017); Chile, Greece, Israel, and Slovenia (2016); Sweden and Turkey (2015); Canada and Ireland (2014).

more affected in terms of sale declines or fluctuations than those with more inelastic demand (e.g. utility services, healthcare, food).

We have also have witnessed a dramatic decoupling around the world between the rich, who not only suffered less from the crisis, but, at least for a while, saw the value of their assets grow disproportionately in increasingly exuberant financial markets, and the poor, who suffered the consequences of unemployment and the crisis of small and medium enterprises, also most affected by the primary and secondary effects of the crisis. According to Hill et al. (2021) 'the impact of the COVID-19 pandemic has been the largest for the world's poorest. The impact of the crisis on the income of the bottom 40% of the global income distribution has been more than twice as large as the impact on the top 40%. The average incomes of the bottom 40% in 2021 are 6.7% lower than the pre-pandemic projections, compared to 2.8% lower for the top 40%. This large difference comes less from the initial impact of the crisis and more from the fact that the incomes of the poorest 40% have not started to recover yet (indeed they worsened for the very poor in 2021), while the top 40% have recovered more than 45% of their initial income loss'. Similar data comes from World Bank telephone surveys in developing economies. As can be seen in Fig. 6.3, poorer households lost income (and jobs) at slightly higher rates than richer households. This is mainly due to vulnerable groups – women, those with low education, and those informally employed in urban areas – who have been hit particularly hard.

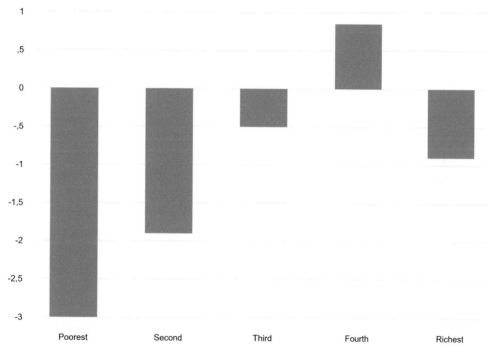

Figure 6.3 Percentage change in income from 2019-2021 by income quintiles. *Source*: Yonzan et al. (2021).

In general, the decrease in income led to an increase of more than 1% point in extreme poverty (from 8.5% in 2019 to 9.4% in 2020), equivalent to approximately 97 million more people living in poverty in 2020 (Mahler et al., 2021), which represents a historically unprecedented surge in global poverty (see Fig. 6.4).

The fall in demand also implies, at least for developed countries, an unprecedented accumulation of savings by lockdown consumers and an equally unprecedented loss of

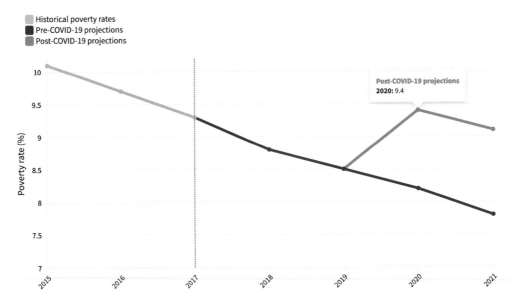

Figure 6.4 Extreme poverty rate, (%). *Source*: Mahler et al. (2021).

receipts from their suppliers (Baqaee and Farhi, 2022). This effect, which leads to an increase in the stock of savings on the one hand and poverty on the other, is equivalent to a massive transfer of wealth. Apart from its iniquity, it is difficult to give an economic interpretation and predict whether, how, and when a sustainable balance between assets and incomes can be restored. Globally, it cannot escape our attention that in the more developed countries, almost a quarter of households appear to be facing an improper tax that reduces their incomes and increases the wealth of the rest of the population. For the poorest countries, the situation is less clear. Still, the polarization between those who accumulate savings by abstaining from consumption in lockdowns and those who see their incomes fall is probably even greater.

This paradoxical situation further exacerbates the decoupling between public capital, which has experienced concerning slowed growth, and private capital. This disparity is further exacerbated by the expanding role of governments in the economy, stemming from the current crisis and the utilization of monetary and fiscal policies that focus on short-term economic cycles. As Clemens and Veuger (2020) show, unlike previous recessions, COVID-19 may have intensified inequality due to a substantial decrease in consumption compared to income levels. Although in previous recessions lower-income consumers could compensate for their income losses by borrowing and using public transfer money, during this pandemic, access to consumer credit has been significantly limited. Simultaneously, government subsidies and charitable donations have declined, making it increasingly challenging for low-income individuals to compensate for their financial setbacks.

6.3. The macroeconomic policy challenges: old tools for new problems?

To address these challenges, policymakers have undertaken the complex task of executing appropriate macroeconomic strategies, incorporating a blend of monetary and fiscal policies. Regrettably, the outcomes have frequently been unpredictable and inconsistent.

In terms of monetary policies, throughout the different waves of the pandemic, different forms of quantitative easing expanded the liquidity of the financial system in many countries through purchases of government bonds by central banks and, in several cases, the purchase of private sector corporate bonds. This great monetary expansion was perhaps inevitable and saved us from a recession. However, it did not solve any structural problems and exacerbated the gap between public capital and private assets. It also had long-term implications and created several macroeconomic risks, including a sudden correction in inflated asset prices due to minimal interest rates and a season of explosive goods and services inflation as economic activity picked up. Even with the optimistic assumption that future inflation will subside and remain low indefinitely, we must be concerned about the explosion of debt.

What about fiscal policies? Globally, the IMF estimated a global fiscal deficit of approximately 14% of the world GDP in 2020, a 10% increase in the deficit compared to the previous year. This was almost three times the deficit after the Great Financial Crisis (GFC), with the significant reduction in deficits from 2020 reflecting an assumed short and deep recession followed by a gradual recovery. Global debt, including private and public liabilities, stood at a record level of around 225% of the global GDP before COVID-19, about 12% more than before the GFC. These increases in debt correspond to reductions in public capital, which, in terms of quantity and quality, can be estimated in almost equal figures: about 15% depreciation at current prices and more than 20% if the data are corrected for lack of maintenance and technological obsolescence.

It is important to distinguish between policies with the nature of fiscal stimulus and those in the form of fiscal relief. The significance of the stimulus is that it corresponds to deliberate budgetary action through higher public spending, tax cuts, more generous social welfare benefits, or payments to revive economic activity compressed by the pandemic. Fiscal stimulus is aimed at the demand side of the economy and should work by increasing total spending to avoid or reverse an unexpected recession. In contrast, fiscal relief measures are primarily directed at the supply side to keep the enterprise system from collapsing and stabilize production by maintaining the payment chain alive. However, these measures can themselves have secondary effects on demand that lead to future gains in output. For example, when corporate tax cuts or investment deductions induce higher investment. Characterizing most business-oriented tax responses as 'Keynesian' is misleading because there is a clear difference between Keynesian defense of direct public spending and immediate purchasing power distributions to stimulate aggregate demand in the economy and fiscal relief measures for businesses to counter an aggregate supply collapse.

In the United States, by mid-2022, Biden's stimulus plan consisted of a \$1900 billion package of interventions. It was based on extending unemployment insurance by \$300 per week until September 6 and the child tax credit for a year. It also invested nearly \$20 billion in COVID-19 vaccinations, \$25 billion in rental and assistance for public services, and \$350 billion in state and local aid. Biden also proposed a \$2 trillion plan for the next eight years to renovate and strengthen the nation's infrastructure, calling it a transformational effort to create 'the most resilient and innovative economy in the world'. This plan would also be extended by 2 trillion over the next ten years to 'rebuild American capitalism' financed by corporate tax increases.

Although the distinction between stimulus and relief is often lost in the mix of different measures, if we add the Inflation Reduction Act approved at the end of 2022, Biden's legislative initiatives on public spending in the past two years point to a decisive shift in economic policy in the direction of public investment. For both scale and purpose, this goes far beyond the mix of short- and long-term policies that characterize Europe and other countries such as the United Kingdom and Australia. The key idea behind Biden's investment plans is that the deterioration in productive public services is the crucial element that explains the general decline in productivity growth and the trend toward stagnation that has recently affected many countries, including the United States. The Biden plan is therefore based on the notion that only a significant relaunch of public investment with massive incentives for the private sector, far beyond the narrow confines of what the EU has been able to concoct so far, can save us from stagnation and respond to global challenges such as climate change, pandemics, and the pervasive systemic risks of an increasingly hyperconnected world. This plan has one danger, though, which is a further decoupling between Europe and the US, with the first falling behind in the scale and scope of its recovery plans and the latter taking advantage once more of the seignorage from the hegemonic role of the dollar in an increasingly unsustainable international monetary order.

6.3.1 Why macro policies are difficult to implement

6.3.1.1 The role of heterogeneous agents and institutions

Modern macroeconomics aims to connect with microeconomic realities by integrating the optimization behavior that economic theory employs to explain market forces through rational economic agents. However, the experience of COVID-19 has shown dramatically that more than rationality and motivation, the heterogeneity of these agents, including individuals and institutions, has been the key driver of aggregate responses to pandemic mitigation and adaptation policies. The main factors of heterogeneity, which differ between countries and between areas of the same country, can be summarized in five main components: *i)* the proportion of young and old in the population, *ii)* the capacity for fiscal intervention, *iii)* the proportion of the shadow economy, *iv)* the frequency of person-to-person contacts (also due to intergenerational coexistence), *v)* the capacity of health care. For all these characteristics, which have a structural character, it appears necessary to address the pandemic with many diversified measures aimed at mitigating its effects on the population and, above all, with immediate investments aimed at removing the underlying causes of fragility.

Population age structure. The first factor has to do with the average age of the population, particularly the proportion of people over 65 years of age. This proportion in the world has an extensive range of values, ranging from a minimum of 5% in Africa to values greater than 25% in some OECD countries. It is not surprising that school closure measures have been deemed necessary in many regions of the developing world, specially if we consider that school facilities are often dilapidated or inadequate from a hygiene, sanitation, and educational point of view. However, school closures in the poorest areas of the world are inequitable because they negatively affect families, especially those with young children and in more precarious economic conditions, and because they withdraw investment in human capital. This latter effect is made more serious, compared to regions with higher incomes, by the digital divide, that is, the lack of network infrastructure, the lower supply of computers, and the available capacity for smart working and distance learning.

Institutions and efficacy of fiscal policy interventions. The second factor, fiscal intervention capacity, presents large cross country variations. While government policies may be similar, their implementation depends on public administrations, and the efficiency and operational capacity of these entities differ widely from country to country. Where public administration is weaker and local governance more problematic and frustrated, as in many developing countries, lockdown policies tend to be heavier and less assisted by social protection measures such as, for example, layoffs or compensatory transfers to households and businesses. In most cases, investing in the management capacity of public administration is a priority, and the emergency conditions of current interventions further underscores the immediacy of this need. The countries that have effectively mitigated the impact of the pandemic owe their success to a series of investments in human capital, training, and equipment, aligned with an organizational model that guarantees the promptness and efficacy of expenditure measures. Similar investments could make an essential contribution to the capacity of local public structures to administer emergency measures.

The shadow economy. The demographic and economic conditions of structural fragility that prevail in the poorest regions of the world are exacerbated by the expansion of the informal economy, which is associated with lower fiscal capacity and local governance. The shadow economy was affected by the lockdown and the consequent contraction of incomes in a dual way, both through the fall in its turnover, which was worse than for the rest of the economy, and by the impossibility of directly receiving subsidies and mitigating provisions reserved to the formal sectors. In the shadow economy, moreover, many operators live daily in conditions of unemployment and dependence, and their state of relative poverty easily tends to degenerate into absolute and, in many cases, desperate poverty.

Person-to-person contact and inter-generational co-existence. The frequency of social relationships, particularly between young and old, is another factor linked to lower economic development and the informal economy. Crowded multi-family housing is typical of the riskiest metropolitan areas and of degraded neighborhoods of the historical centers and suburbs, especially in the South of the world. In these regions, despite the emigration of young people, the informal economy and urban degradation paradoxically cement the daily ties and the relationships of exchange and interdependence between young and old, often co-habiting in extended family groups under one roof. Therefore, the lockdown is more complex and less effective in separating the groups at greater risk. At the same time, contagion can be accelerated by the greater contacts generated by the growing need to obtain income from an economy in crisis.

The healthcare system. The last factor, healthcare capacity, which is also dependent on governance capacity and public administration efficiency, tends to be much lower in developing countries. Although there are some effective local health systems, the past decade has seen a lack of investment in human resources, infrastructure, equipment, and facilities. This has had a disproportionate impact on the southern regions of the world, but the underlying problems have a global impact. Therefore, pandemic containment strategies must urgently address the quantity and quality gaps in a system that is primarily hospital-centric, poorly integrated at the regional level, and varies significantly in terms of effectiveness and efficiency.

COVID-19 has exacerbated world inequality in many ways, including exposing existing intolerable differences in the distribution of health services and imposing egoistic and rival behavior in consumption habits, irrational fears, and willingness to cooperate. Lack of coordination has been a significant problem at national and international levels, and wealthy nations around the world have displayed a more recent nationalistic policy for vaccine procurement and use. According to the Secretary of State of the United Nations, until the end of 2021, the ten richest countries in the world had *de facto* monopolized vaccine consumption and supply. However, the apparent early end of their acute pandemic crisis, achieved through successful mass vaccination, was a temporary illusion since the pandemic threatened to spread incontrollably to the rest of the world, where the lack of financial resources and international coordination rendered the supply of vaccines short, poorly distributed, and insecure. This condition is yet another manifestation of the unsustainable nature of the current development pattern. The challenge of unsustainability arises not only from the emergence of another form of global in-equality and hardship, but also because effectively halting pandemics, whether current or future, depends on achieving widespread vaccination on a global scale in very short timeframes. This urgency is underscored by the potential for viruses to evolve into more contagious variants.

Unfortunately, most of the interventions contained in the current recovery by local governments and multilaterl institutions appear to meet short-term demands and do not address the structural issues underlying the biomedical divide between rich and developing countries. This divide is another manifestation of worldwide inequality and unsustainable social inequities. To a large extent, its existence and the drama created by the pandemic point to a critical structural problem in the provision of health as a global public good.

6.3.1.2 *The unprecedented interweaving of supply and demand shocks*

Although both demand and supply shocks were involved simultaneously, the most sig-nificant impact of the pandemic on the economy was mainly a supply shock, with the reduction in working hours and the increase in private and public health costs being the first cause of the contraction in production. However, as Guerrieri et al. (2022) demon-strate, in a regime of incomplete markets and consumers constrained by liquidity, the COVID-19 supply shock turned into a demand shock, causing lockdowns, layoffs, busi-ness closures and other private and public contractions of aggregate demand. Therefore, the reactions of governments, consumers, and companies resulted in the second round of partly interdependent reductions in supply and demand. The interweaving of sup-ply and demand effects characterized not only the first response to the pandemic, but also the succession of positive and negative feedbacks regarding the subsequent waves of the virus, both from the point of view of preventive and proactive public and pri-vate agents, as well as reactive ones. However, reactive, rather than proactive measures, from lockdowns to qualified restrictions and the various phases of vaccination cam-paigns, dominated and continue to dominate the response of the economic system to the surprises of a new 'ecodemic' reality that appears increasingly unpredictable.

The first supply shock mainly affected sectors with the highest labor density. Never-theless, the immediate intertwining of negative supply and demand shocks, through vol-untary or forced lockdowns, particularly targeted sectors such as international tourism, catering, and hotels. Business lockdowns progressively affected all sectors, but detailed statistics show that the initial shock was strongly asymmetrical. The labor-intensive sec-tors, with little substitutability between labor and capital and low penetration of digital technologies, faced the greatest challenges. Due to consumer reaction, sectors such as remote transport experienced a positive shock, especially when transport replaced direct consumption. Yet, the overall logistic chain for transporting intermediate goods did not have the time and resources to adapt. Business lockdowns caused negative repercussions throughout all production sectors, often dramatically compromising the functionality of national and, above all, international supply chains (see Chapter 7, Section 7.4). Even in these cases, the heterogeneity of businesses and consumers led to additional damages and repercussions, with dramatic differences between and within sectors.

More subtle disruptions in value chains concerned the non–market activities. The most dramatic example is the care sector, formed by a combination of paid and unpaid services for health and assistance for various challenged segments of society: children, the elderly, the poor, and the frail. Much of these activities occur under the radar, so to speak, since they do not appear in national accounts and are only scantily touched by national statistics. They also involve women as the most significant working force for paid and unpaid activities in developed and developing countries. The pandemic caused a substantial positive demand shock for human capital provisioning activities, including education, health, and assistance, with minimal institutional capacity to respond, beyond the intensification of paid and unpaid labor (primarily women) at home and in largely inadequate public and private facilities.

Interpreting what happened in light of the available statistics is not easy. The unprecedented combination of supply and demand shocks that the pandemic has set in motion challenges the capabilities of economic models and the terms of the traditional debate between classical and Keynesian economists. The initial supply shock occurred during prolonged deflationary tendencies in the world economy, which was still metabolizing the aftermath of the Great Financial Crisis. This first shock then created a different condition of excess demand, to which central banks responded with an increase in the money supply. The intertwining of supply and demand shocks made things more complicated, as businesses were often unable to use monetary expansion to offset supply shocks due to shrinking demand caused by lockdowns and changes in consumer behavior. Therefore, the immediate effect was a general increase in substitute financial transactions generated by a wider resource to credit. This increase impacted asset trading, creating a financial bubble that, on the one hand, largely neutralized the objectives of monetary expansion and, on the other, exacerbated its effects on social inequity.

In a world dominated by inequalities, social conflict is an inevitable outcome of economic crises. It is a result of the growing sense of injustice felt by those who are excluded from the benefits of economic growth, the lack of solidarity between different groups in society, and the inadequacy of global and national institutions to address these problems. Social unrest was already simmering under the populist reaction to globalization before the pandemic, but the manifestation of the new and more dramatic inequalities generated by the joint health and economic crisis was relatively slow to appear. Furthermore, the pandemic led to the acquisition of an unprecedented role of authority for the government and health institutions, validated, at least at the beginning of the emergency, by an implicit delegation of power and granting trust by all segments of society. The reason for this widely shared attitude was in part the belief that no other institution or private party could replace public sector action in such a contingency. It was also the consequence of an emergency that required centralized and vigorous governance. Due to these conditions, the first phase of the pandemic was characterized by general compliance with the numerous measures of the government. However, the consensus on their timeliness and rationality was soon destined to decrease. Social conflicts and unrest were not a problem in the first phase of the health emergency and remained somewhat compressed even under subsequent waves of contagion. After the first phase, several unexpected waves of infection and reaction followed, with increasingly severe problems for the poor, elderly, women, and generally the most fragile part of the population. Although vaccines subsequently reduced the burden of pandemics on society, they also became involuntary instruments of discrimination and social exclusion, especially between rich and poor countries and within countries. Finally, amid early symptoms of inflation due to excess demand and cost increases, the Ukrainian crisis fell as an unexpected and ultimate blow to a global economy trying to recover from the pandemic and its social tensions.

6.4. What happened in the real economy?

Overall, with expansionary but ineffective monetary policy, negative supply disruptions dominated demand shocks. Although governments' fiscal policies were implemented later than monetary policies, they also played a significant role in the pandemic response. They gradually became more prominent, especially in the later phases of the pandemic, with a variety of recovery and reconstruction programs. The first response of fiscal policies was aimed at ensuring the survival of enterprises through direct aid and was dependent, by its nature, on the condition that the beneficiary companies continue to remain active. These measures helped avoid the collapse of production chains, even though they did not have large Keynesian multiplicative effects because the beneficiaries mostly used the subsidies to compensate for payments already due. Once the first phase of the pandemic was over and some control appeared to have been gained, reconstruction and recovery plans were produced, although many still remain on paper. At the same time, some expansion of public spending beyond purely relief measures helped to boost aggregate demand.

Despite uncertainties, in most developed countries, before the Ukrainian crisis, credible market sentiment surveys suggested the presence of strong expectations of recovery in public and private consumption, although, for the moment, real investments appeared far from coming. At the same time, the so-called 'green transition' seemed to add to the difficulties for companies to return their production to the risky path of early obsolescence of tangible and intangible capital. This seems even more difficult under current conditions. Yet, the industrial reconstruction of companies and supply chains is undoubtedly a unique opportunity, one not to be missed, to 'build back better' an otherwise likely unsustainable economy.

Inflation, however, is a phenomenon that presents inherent complexities and apparent contradictions. In general, global inflation experiences have historically been associated with asymmetric changes in supply and demand, even though the causal links between the quantities and prices involved are unclear. Past experiences led to the recognition of two distinct components in an inflationary wave, which can be fueled by unexpected increases in income or increases in production costs: i) a tendency to increase price ratios (the so-called relative prices), thus realigning the terms of trade in favor of selected producers and concerning intermediate and final consumers; ii) an increase in a price index, that is, the average price of a basket of goods. In addition to reflecting aggregate supply and demand trends, the increase in a particular price index also reflects relative price movements. For example, in the case of a consumer price index, this records the average price paid by a representative group of consumers and thus indicates the likely impact of changes in current prices on an average consumer. As such, it has a more subtle effect on the economy, as it tends to create expectations of future price increases and thus leads to seeking compensatory behaviors, which can primarily take the form of demands for wage increases.

Inflation becomes a persistent phenomenon at the level of the economy only if the more general price increase is internalized in the market mechanism through a pass-through between expected prices and wage increases. Until this link is established, inflation will not last, and the economy may even experience a temporary boom, as investment plans can be supported by producers' extra profits due to higher sales prices at the expense of steady or only slowly rising wages. The relative price component of inflation, to the extent that it does not become a component of a systematic increase in all prices, can also be advantageous since it can reduce the price misalignments that inevitably accumulate over time due to the various distortions resulting from multiple market failures. These mismatches include a divergence between private incentives and social goals that are particularly dramatic in the case of the energy transition, with international carbon prices much lower than those needed to achieve the minimum targets to avoid a catastrophic evolution of world temperature.

The contradictions and challenges that emerge from this complex picture result from the interweaving of supply and demand shocks. On the one hand, infrastructure and other investments envisaged in the different national recovery plans, while acting more or less promptly on demand, need more time to unfold their effects on supply. On the other hand, companies face unprecedented difficulties in resuming their activities and, simultaneously, in carrying out the transformations of their infrastructure and business models necessary for the required transition. Meanwhile, the demand for final consumption tends to grow relative to that of production inputs and capital goods because of the growing gap between consumption expectations and production capacity. Prices, in turn, reflect this gap and show misalignment and sometimes an increasing divergence between private incentives and public objectives (see, for example, the case of energy).

The expansion of demand by quantity and composition diverges from the supply, with short-term bottlenecks and long-term uncertainties. The result is a growing discrepancy between increasingly unrealistic public investment policies, predictable private behavior, and an unprecedented and difficult-to-govern inflationary environment.

The current upsurge in inflation includes an upward price push, but also a greater realignment of relative prices or, more generally, of the so-called terms of trade. These prices were largely misaligned with the long-term goals of the energy transition, but also with a sustainable distribution of income between countries and within countries. The realignment of energy and energy-dependent goods prices is moving in the right direction, as it is largely associated with an increase in the carbon price. This will accelerate the energy transition, and may reduce the risk of future capital losses from fossil fuel reserves and investment assets. The worrying and structural part of inflation is decoupling, or the danger that value chains may be rebuilt according to a logic of block autarchy, denying the structural interdependencies painstakingly built during globalization.

6.4.1 Is the pandemic suggesting a new macroeconomic theory?

Traditional macroeconomics has gradually embraced the hypothesis of general equilibrium, although its original formulation, especially under Keynesian influence, was based on the attempt to explain why the economy can tend to drift from a good state (perhaps of equilibrium) to a bad one (secular downturns, stagnation, and depression). The so-called 'micro-foundation' school has also led to a reformulation of macroeconomics, assuming that aggregating individual behaviors will result in aggregate outcomes and ignoring the potential errors that can arise from agent heterogeneity. The Great Financial Crisis and the pandemic have shown that both the attempt at micro-foundation and the tendency to absorb the equilibrium notion as a theory of fluctuations around a 'good' equilibrium have led to a disregard of any possibility that internal conditions or exogenous shocks (or both) may lead to bad outcomes, both from an aggregate and a distribution point of view, and that these outcomes may be likely to persist. In other words, general equilibria may still be important focus points for economic analysis, but they must include multiple equilibria as possible outcomes, which can be good and bad from the aggregate point of view and from the different agents involved. Also, as Keynes openly suspected, we have no evidence to argue that the market equilibrium that will be realized at anyone time will be either unique or good. Rather, any given state of the economy may undergo positive and negative changes, and both fluctuations and significant downturns are possible that may or may not be centered on any specific equilibrium. The different agents may have different goals, resources, and constraints, so their individual behaviors may not add up in a predictable way. This is why the so called 'micro-foundation' sought out in economic models may be the source of large errors.

To illustrate how recent crises have challenged current macroeconomic models, let's look at Gechert's (2015) categorization of contemporary macroeconomic models into three groups: *i)* the new classical Real Business Cycle dynamic stochastic general equilibrium (RBC-DSGE) models, *ii)* the New Keynesian (NK-DSGE) models, and *iii)* the structural macro-econometric models. RBC models incorporate the hypothesis that cycles start with shocks in total factor productivity (TFP) (Kydland and Prescott, 1982). These models are based on utility-maximizing representative households operating under fully competitive markets, under the (so-called neo-Ricardian) assumption that they are insensitive to taxes and expenditures because they expect that the government is forced to maintain a balanced budget over time. As a consequence, the RBC models see expansionary fiscal policies in a way that is structurally capable of increasing GDP only through wealth and substitution effects and consequently increasing the labor supply (Baxter and King, 1993). This implies that government expenditure can only expand at the expense of the private sector, with an outcome that will be positive only if the balance between the positive effect on labor supply outweighs the negative effect of crowding out private consumption and investment (Woodford, 2011). These results can be modified by complementarity effects introduced through different assumptions on household utility functions or by crowding-out and tax distortionary effects. Still, they are certainly not consistent with the outcomes observed in the recent financial crisis and, even less, with the downturns and subsequent ongoing recovery witnessed under COVID-19. In both cases, supply shocks have resulted in parallel demand shocks, which expansionary fiscal policies have first moderated and then reversed. Although it may be too early to tell, in the case of the GFC, the effects on aggregate output of increased government expenditure have appeared inadequate to offset the decline in private spending.

New Keynesian (NK) models combine elements of the Real Business Cycle (RBC) framework with certain features of Keynesian macroeconomics by incorporating the idea of sticky prices or wages and a monetary policy rule. This results in a simple mathematical structure that is similar to the famous IS-LM model, which includes an investment-saving IS curve, a Phillips curve, and a rule for the money supply. These components are based on the optimization behavior of representative households and firms. NK models can be seen as a synthesis of Keynesian and Neoclassical economics, where fiscal policies can be used to address gaps created by inadequate or delayed adjustments to unexpected shocks (Azariadis, 2018), but monetary policy is a superior strategy until the zero lower-bound interest rate is reached. The NK models, however, embrace the hypothesis that long-term growth is determined by fundamentals like population and productivity growth rates, as in the stylized neoclassical growth model. Therefore, in the short run, market frictions can cause differences between potential and current output, thereby creating a gap between the levels at which capital is fully employed and labor is willingly supplied, resulting in inflation or deflation. Rather than fiscal policy, this opens the stage for the strategic use of monetary policy to ensure that interest rates align with investment incentives (Tobin Qs) and that demand is sufficient to balance potential supply (Scandizzo et al., 2020).

Unlike the IS-LM model, the functioning of NK models cannot be understood by simply looking at the resulting diagram, but requires a computer simulation. Nevertheless, once this is obtained, the workings of the different assumptions are generally clear. For example, in an experiment with the NK model conducted at the beginning of the COVID-19 shock for the Italian economy, a simulation of a government stimulus in the presence of a significant increase in money supply and a near-zero interest rate showed large multipliers and higher effects of an expenditure package compared to tax cuts (Di Pietro et al., 2020). However, this result was largely the consequence of an increase in employment, partially obtained at the expense of a reduction in consumption and a parallel increase in labor supply. A similar decrease in the discount rate of households reduced the rate of return to capital, resulting into lower private investment. In other words, the fiscal stimulus succeeded in improving the economy *vis a vis* the no fiscal intervention policy, but only at the cost of forcing the sector economy to provide more labor. In retrospect, these indications appear misplaced and largely misleading, since we know that private demand was significantly boosted and determined much of the economy's larger-than-expected rebound in 2021. At the same time, rather than an increase in labor supply, the 2021 rebound generated labor shortages and supply chain disruptions, with the ensuing cost push inflation and repressed consumption.

Fig. 6.5 reports the results of a recent study by Scandizzo and Knudsen (2022) that analyzes the trade-off between private economic costs and social losses from pandemics. The different curves represent private costs as the risk of incurring liabilities by not investing in preventive measures. On the contrary, public costs are represented by the straight line showing how social losses from the spread of the disease increase with private non-compliance. Liabilities depend on the share of private returns (λ) that can be expected to be expropriated by public sanctions. For values of private losses below the social costs (low λ), the system does not have equilibrium since non-compliance is not sufficiently discouraged, and thus the disease is likely to spread without control. For a reasonably significant sanction, different equilibria can be reached according to the degree of uncertainty. For low uncertainty, compliance will be high and social losses (infection rates) low, while for high uncertainty, compliance will be low and social losses high.

As highlighted by Scandizzo et al. (2020), despite their Keynesian claims, all recent macroeconomic models adopt the assumptions of optimal labor supply by the representative household and optimal labor demand by the representative firm. Therefore, they are likely to predict that fiscal policies crowd out consumption and leisure, with households scrambling to find additional employment to finance the increase in government expenditure. But this was not the case either in the aftermath of the GFC or in the provisional recovery of 2021 after the first phase of the pandemic. Under both circumstances, consumption failed to recover sufficiently to return to the pre-crisis path. However, in the GFC case, prices and wages remained depressed in a prolonged deflationary state, while in the COVID-19 case, they were supported by bottlenecks caused

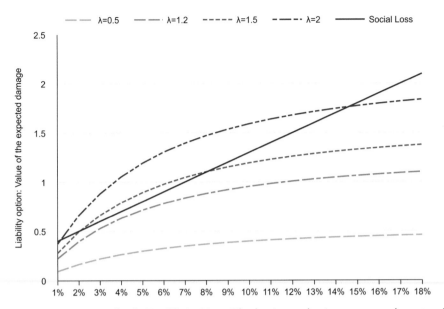

Figure 6.5 COVID-19 Multiple Equilibria. Note: The horizontal axis measures the rate of uncertainty on the evolution of the infection, while the curves are the contingent liabilities faced by businesses operating under the pandemic and λ represents the proportion of private returns that would sanction non-compliance with the regulation. For a range of values of λ multiple equilibria could result. This can be seen from the intersection of the liability curves with the red line, which represents the social loss for each level of uncertainty. *Source*: adapted from Scandizzo and Knudsen (2022).

by a contraction of labor supply and value chain disruptions. Therefore, a neoclassical micro-foundation, softened by the assumption of market imperfections such as monopolistic power or price distortions, is bound to predict, by its very nature, a trade-off between government and private expenditure. Outside the context of the neoclassical framework, on the other hand, the 'new Keynesian benchmark' is provided by a dynamic version of demand-led macro-models – in the presence of 'sticky' prices and wages – if the monetary policy can credibly maintain the interest rate fixed at all times, regardless of the chosen path of government purchases. In the 'benchmark' case illustrated by Woodford (2011), the predicted multiplier (simply defined as the ratio between the increase in GDP and the increase in government expenditure) may be equal to or greater than one, but only in circumstances determined by a combination of recession and extremely accommodating monetary policies. In this case, the multiplier is independent of the degree of resource slackness or the cost of supply, even though such a cost does matter in determining the cost of the monetary policy measures needed to maintain the interest rate constant (cost of present goods in terms of future goods or indexed bonds).

The COVID-19 experience appears to have given a mortal blow to all macro-models that have laboriously tried a 'new synthesis' of neoclassical and Keynesian assumptions in a general equilibrium framework in the past 40 years. The economic models that are used to guide the global economy today, such as those used by the IMF and the European Union, may incorporate some Keynesian ideas. However, they all make the fundamental and likely mistaken neoclassical assumptions that ideal labor supply and demand will eventually be realized. Therefore, they focus on short-term and cyclical fluctuations, predicting that fiscal policies will displace consumption and leisure and that households will scramble to find other jobs to finance the increase in government spending. The effects of the composition of public expenditure and the impact of investment beyond immediate reactions are therefore absent from the horizon of these models, which focus on predicting what will happen immediately, since the markets will still find equilibrium in the long run. But this was not the case either in the aftermath of the GFC or in the rebound of 2021 after the first phase of the pandemic. In both circumstances, consumption failed to recover sufficiently to return to the pre-crisis path. In the GFC case, deflation caused prices and wages to remain low for an extended period of time. In contrast, during the COVID-19 pandemic, supply chain disruptions and a decrease in the labor supply created bottlenecks that supported prices and wages.

More generally, the so-called budgetary rules, which have weighed so heavily on the pre-crisis policies of EU countries, appear to be the brainchild of obsolete currents of thought. They are themselves a product of earlier attempts to save macroeconomics by using microeconomic assumptions to reunify the dichotomy at the very origin of macroeconomics as a separate branch of thought and policy recommendations. These attempts have produced a series of so-called 'micro-founded' models whose hypotheses, although attenuated by the assumption of market imperfections such as monopoly power or price distortions, are destined to produce, by their very nature, a negative relationship between public and private spending. Even the so-called neo-Keynesian versions, which have reintroduced the assumptions of rigid prices and wages downward, have resulted in a reiteration of the power of monetary policy, arguing that only if it can credibly maintain fixed interest rates at all times, regardless of the chosen path of government purchases, fiscal policy may have no effect on the economy.

In addition to their essential short-term orientation, these models lack a critical dimension that the pandemic has brought to the fore: the heterogeneity of agents, reflected not only in income and consumption inequality indexes, but also in the distribution of vulnerability and resilience throughout the population. This heterogeneity and long-term consequences of the pandemic are at the root of the stark choices between different 'bad outcomes' that policymakers can be forced to face, from lockdowns to bankruptcies and personal economic ruin to long-haul effects of infection and death.

For example, the debilitating long-haul effects of COVID-19 and the long-term negative consequences on immune systems and survival rates have been shown to affect disproportionately young adults in their late 30s and early 40s, with women accounting for an estimated 70–75% (Puaschunder and Gelter, 2022). Combined with psychological and delayed learning, these effects can dramatically change any notion of the political economy as a science of choice between short-term aggregate outcomes under the benevolent umbrella of long-term market equilibrium (see also Chapter 5). A new generation of economic models is needed that not only focuses on the long run, on choices between goods and bad or very bad outcomes but also takes into account the stark reality of the disproportionate vulnerability of population groups already disadvantaged in terms of wealth, social conditions, location, and gender.

The crisis of the more widely accepted synthesis of both neoclassical and Keynesian macroeconomics is also one of the roots of the rules invoked to support the credibility of economic policies compromised by the opportunism of governments. Public authorities' commitment to behave in a manner consistent with economic policy choices, it was argued, could not be considered credible due to social pressures and electoral cycles. Hence, the most critical decisions to be accompanied, or even preceded, by a system of rules, with implicit or explicit sanctions, which would guarantee citizens, making it difficult for governments to disregard the commitments made previously. The policies generated by this type of prescription worldwide have included various 'lock-in' measures, including attempts to prohibit specific measures such as devaluation or budget deficits by law. The current consensus of most economists is that they have proven to be largely unsustainable and, like the European rules system, particularly unsuitable for dealing with prolonged periods of recession, even in advanced economies. The COVID-19 pandemic also seems to have conclusively shown that it is unreasonable to try to establish rules that address all possible situations that fiscal policy may face in the future. It also demonstrated that a strong external shock could present an unsolvable trade-off between macroeconomic stabilization and debt sustainability.

In conclusion, the principles of international economic order cannot fail to consider that the theoretical framework that generated them has changed under the mortal blow of the pandemic following the GFC. The macroeconomic paradigm underlying the current rules and prescriptions is obsolete because in the generality of the models adopted, it embraces the assumption that in the long run, growth is determined exclusively by fundamentals such as population and productivity growth rates as in the stylized neoclassical growth model, and that the task of economic policy is to alleviate the effects of market frictions, which can cause differences between potential production (structural production) and current production. The experience of the last 20 years has shown that this approach, essentially reactionary (in the anti-Keynesian sense of the term), is unsuccessful. Perhaps more dramatically, the failure to predict and govern the economic consequences of COVID-19 also suggests that the underlying economic paradigm evades the substantial objective of human development and contributes to the growing unsustainability of the social and environmental system of production, consumption, and distribution of income.

CHAPTER 6 - Take-home messages

- The COVID-19 pandemic has caused a major disruption of the world economic conditions. It has created uncertainty about when and how the global economy, which remains highly vulnerable to shocks, will recover. This uncertainty also makes it difficult to assess the potential impact of future crises.
- We have observed a pronounced 'decoupling' effect, following the diverse impacts of the pandemic among different groups of countries. Some have suffered more intense repercussions than others, whereas some have been unaffected or have even found advantages amid the crisis.
- The COVID-19 epidemic has affected the global production system along several dimensions: it closed production sites, created disruptions in supply chains, and financially impacted businesses and markets. It also changed people's demand and propensity to consume globally.
- The global economy has experienced its largest contraction in GDP ever, but the recovery after the initial nationwide lockdowns has been uneven. As of today, it is still impossible to determine the true cost of the economic crisis.
- The crisis has caused a lot of damage, but it also offers a chance to fix the underlying issues that have been slowing down economic growth. Investing in structural reforms and infrastructure is key to taking advantage of this opportunity.
- Our current macroeconomic paradigm is outdated. It is based on models where long-term growth is driven only by fundamental factors such as population and productivity growth and sees economic policy primarily as a tool to mitigate the effects of market frictions that cause disparities between potential (structural) and actual output. As a result, there is a pressing need for a paradigm shift to effectively address the radical changes in the economic landscape.
- The economic crisis has led to widespread job losses, declining consumer spending, and lower production levels. This is because the pandemic has disrupted the interconnected global supply chains. To overcome these effects, a broad collaborative effort is needed from all over the world.

CHAPTER 7

The effects on the economic systems

Contents

The pandemic has hit the world economy in a serious and perhaps irreversible way through immediate and potentially long-term supply and demand disruptions of goods and services. Perhaps more dramatically, it has highlighted the potential of systemic catastrophes linked to the progressive degradation of the natural environment. This revelation concerns the worrying unfolding of the negative effects of the interaction between man and the environment. More generally, it highlights how, within a pervasive and growing uncertainty over the relationship between the economy and the environment, our social system can be disrupted overnight by unpredictable events and how difficult it is to return to acceptable normality. Conflicts and negative surprises seem to follow one another with an increasing frequency, even within apparently disconnected events. Like the seven plagues of Egypt, we begin to wonder if there is a deeper connection among all the disasters that pour on our planet. More recently, the deadly combination of the pandemic with the invasion of Ukraine has also put the world in front of a no longer avoidable choice between growth and sustainability.

The COVID-19 Disruption and the Global Health Challenge
https://doi.org/10.1016/B978-0-44-318576-2.00020-2

7.1. A pandemic arrived at the wrong time!

The pandemic has arrived on the wings of a series of volatile developments in the world economy. These events, which ranged from repeated financial crises to ever more frequent local conflicts and social tensions, seemed all related to global changes that appeared increasingly difficult to predict and impossible to govern. They included a progressive interdependence across all areas of the world, increased uncertainties, and pervasive risks. They also seem to be related to an ominous trend of economic decline and irreversible degradation of the environment, with secular stagnation and perhaps catastrophic prospects looming large for the future. With its unpredictable dynamics, the pandemic seemed to be just one more 'black swan', as one of a new category of increasingly frequent but paradoxically unexpected and highly improbable events. Below we briefly discuss some macro variables that may have caused the world to suffer from these events.

In 2022, the world's population reached a level of 7.9 billion, an increase of 84% from the level of 4.30 billion in 1978. The world population has increased dramatically in recent centuries, but this growth is actually slowing down. In fact, it is projected that the global population will only reach 11 billion by the year 2100. The United Nations predicts that the real demographic explosion will occur in Africa, which will see its population grow from the present 1.3 billion people to 4.3 billion by 2100. The population of Asia, today of about 4.6 billion, should instead first grow to 5.3 billion by 2050 and then fall in the second half of the century. In summary, the current global population, with 60% concentrated in Asia, is projected to shift towards the African continent following a phase of further growth primarily driven by India. But the disproportion between the 'Western' world and the rest of the planet is already extreme: Europe, the United States, and Australia's total population is roughly the same size as the African population (1.3 billion) and less than 20% of the total world population.

Income and wealth have a distribution that reflects their characteristics in the opposite way to those of the population, that is, following the so-called Pareto law: 15% of the world population controls 66% of income, and more than 70% enjoys an income lower than the average per capita income.[1] Here, too, however, the dynamics are im-

---

[1] The Pareto Principle, also known as the Pareto Rule or the 80/20 Rule, is named after economist Vilfredo Pareto conjectured that 80% of consequences come from 20% of the causes, asserting an unequal relationship between inputs and outputs. The original observation of the Pareto Principle was linked to the relationship between wealth and population. According to what Pareto observed, 80% of the land in Italy was owned by 20% of the population. After surveying a number of other countries, he found that similar proportions applied abroad. The Pareto Principle can be applied in many areas, such as manufacturing, management, and human resources. For instance, the efforts of 20% of a corporation's staff could drive 80% of the firm's profits, or at the individual level that 80% of work-related output could come from only 20% of the individual time spent at work. For the most part, the Pareto Principle observes that things in life are not always distributed evenly. It is worth noticing that, unlike other principles, the Pareto Principle is merely an observational regularity and not a law of nature.

portant. Global gross domestic product (GDP) increased from $26.3 trillion (in 2010 prices) in 1978 to $84.71 trillion in 2020. This means that within 42 years, global GDP has more than tripled, with an increase of 222% and an average annual growth rate of 2.7%. The drivers of this increase are population (+84%), capital accumulation (+92%), and, through technical progress, productivity (+42%). The increase in productivity at an average rate of less than 1% per annum is the result of a dramatic slowdown in the past 20 years, despite the great expectations raised by the progress of digital technologies.

Global primary energy consumption (i.e., energy derived directly from nature) has more than doubled, from 270.5 EJ in 1978 to 556 EJ in 2020.[2] This means that, on average, a percentage increase in world GDP of 1 point has been matched by an increase in primary energy consumption of about 0.7 points. At the same time, GDP per capita grew by 75% from $6117 in 1978 to $10,721 (again at 2010 prices) in 2020, at an average annual rate of 1.34%, while primary energy consumption per capita increased by 12%, from 62.8 GJ in 1978 to 70.38 GJ in 2020 for an average annual growth rate of 0.27%. These figures suggest that energy demand will tend to increase more than proportionally as the population growth rate decreases relative to the increase in income.

The possible balances among these macro variables are many and depend, above all, on the development of new technologies. Current economic models, used to explore alternative scenarios, however, tend to converge on a paradoxical result: unless radical innovations are introduced, the more population growth slows down and income growth increases, the more the sustainability of the world economy will tend to be compromised by excessive growth. The reason for this prediction is that as population growth slows and income increases, the pressure to maintain or increase the growth rate of the economy becomes more intense, and this can lead to more exploitation of resources and increased negative impact on the environment if there are no new and innovative solutions. This outcome, however, can be partly due to the same technology, which tends to prolong, without really solving it, the so-called problem of 'over-reach', that is, the chronic attempt of developed economies to challenge environmental sustainability, by seeking to maintain a high growth rate regardless of the level of well-being achieved. Resulting overall growth may thus be unsustainable, because a minority (less than 20%) does not agree to reduce their growth to allow the majority of the world's population to make the minimum progress necessary to achieve acceptable living standards.

As already mentioned in Chapter 6, the global economy has exhibited a highly dual pattern, with the ghost of 'decoupling' always around the corner. Decoupling is a term that refers to the process of separating or disconnecting different parts of a system, and it has been used to describe a situation where the world is divided between countries or

---

[2] A joule (J) is a unit of energy. It is defined as the amount of energy required to heat one gram of water by one degree Celsius, a GJ or Gigajoule, is a unit of energy equivalent to one billion (10^9) joules, and an EJ or Exajoule is equivalent to one quintillion (10^{18}) joules.

groups of countries that have access to resources and opportunities, and those that do not, leading to increased inequalities, greater political and social divisions, and potential for instability and conflict. Decoupling is a somewhat ambiguous word and has also been extensively used to characterize various risks arising from the differences in income levels and resource constraints around the world. These risks include the danger that economic integration –achieved through international trade and peaceful exchanges of research and technology– may be reversed or even broken down by the attempt of the US and China to cut the node of interdependence that has linked them and the rest of the world so far. More generally, decoupling has taken the meaning of a phenomenon whereby inequalities tend to become extreme, with the world divided between the 'haves' and 'the haves not' along not only economic but also social and political lines. On one side, a little more than a billion people controlling more than 70% of total income and a much larger proportion of global wealth. On the other side, about 7 billion people, growing up in poverty, desperation, and anger.

This form of decoupling has shown its dramatic side during the pandemic, with a biomedical divide becoming apparent through the uneven access to vaccines, the plight of the unemployed and the poor, the lack of medical facilities, the higher burden of the disease on the women, the elderly and the ethnic minorities. Decoupling does not mean only unequal distribution of the costs of the pandemic, but literally a split between those who are protected, by their wealth or their social conditions or just happen to be in a lucky spot, and those who are vulnerable, fragile or are at the wrong place in the wrong moment. For example, while lockdowns have caused inconveniences to all consumers, some of them have continued to earn their incomes, because of smart work or other arrangements, while others have seen their livelihoods vanish. On one hand of the distribution, savings have accumulated, with bank accounts overflowing with unused liquidity. On the other hand, losses and debts have piled up, with bankruptcies and business closures, forced retirements, and sudden losses of family wealth.

Non-pharmaceutical interventions have mainly consisted of lockdowns, behavioral restrictions, and attempts to alleviate economic difficulties by pumping cash into the economy. Relief payments, such as government subsidies and public spending, have been used to address the immediate effects of the pandemic on individuals and businesses. Stimulus measures, such as fiscal and monetary policies, have been implemented to counter the declining demand for goods and services. However, structural economic policies have not been adopted to counter the deeper damages caused by the pandemic to the underlying economic machinery. These damages include persistent disruptions in supply chains, damage to natural capital, and increasing shortages of natural resources. The economic and social disruptions caused by the pandemic, such as job losses and food insecurity, have led to increased tensions and instability in some regions and contributed to determining a global scenario more conducive to growing conflicts, with the war in Ukraine as a major example. The impact of the pandemic was particularly severe for developing countries, where it added to the worsening effects of climate change, poverty, and inequality.

The growing structural weaknesses caused by the 2009 global financial crisis were also a major factor in a series of negative effects that involved virtually all countries. These included collapsing investment, productivity slowdown, unemployment, and poverty increases, combined with rising debt and accelerating natural capital destruction. The pandemic has not only highlighted the existing inequalities in society, but it has also exacerbated them, revealing their severity and drastic nature under pressure. This has created a clear divide between countries and social groups, differentiated by their vulnerability and capacity to recover.

Even for the broader effects of the pandemic, decoupling has been a permanent and present danger, with groups such as women, older people, jobless, and the poor disproportionally affected by the consequences of the infection and the attempts to deal with it, by collapsing investment and domestic violence, the failure of the welfare state and the holes of the health system. The result has been that a part of the population has been hardly affected by the economic downturn and has even benefited from smart working or business reorganization, while another part has seen its livelihood vanish overnight by the severity of the disease or the impact of the restrictions on their business or work. The war in Ukraine has further fueled this dualistic drama by exacerbating the losses of the population of the country invaded, and the destruction of resources, families, and human lives. It has also created a further danger of decoupling, dividing the world between invaded and invaders, winners and losers, fighters and refugees, virtuous and criminal nations.

The ultimate drift of inequality is the replacement of international cooperation with bipolarity, and of international trade with forms of fragmented, semi-autarchical value chains. This trend goes hand in hand with another decoupling: the private sector's enrichment and governments' parallel impoverishment. Over the past 40 years, the share of wealth held by the public sector has become zero or negative in rich countries, while it has grown significantly in the private sector. The impoverishment of governments has been accelerated by the COVID-19 crisis, with the explosion of public debt, and paradoxically contributes to making private businesses more exposed to financial crises, more dependent on the public sector, which in turn is less able to face key 21st-century challenges such as social inequality and climate change. It may also create a dangerous structural inflation environment, in which the expansion of private purchasing power is not matched by a sufficient increase in supply capacity, and private investments fail due to a lack of infrastructure and other public goods.

7.2. The macro-scenario: the winners and the losers

The first effect of the pandemic on the macroeconomic structures was to increase private expectations for government interventions. When these interventions occurred, there was an increase in trust in public institutions, especially at home. The drive for a new form of hyperactive governance seemed to descend directly from the emergency conditions, but it was also the consequence of an unprecedented call for action of all public institutions directly related to the pandemic. The public health system was, of course, at the forefront, but other public centers such as the various ministerial bureaucracies and the government itself were also soon protagonists of a whole series of policy measures, *ad hoc* interventions, pharmaceutical and non-pharmaceutical measures affecting a variety of targets among consumers, businesses and the population at large.

From a macroeconomic point of view, the emergency conditions arising from the pandemic reflected a historical vision that can be indisputably called 'Keynesian', in the sense that virtually all segments of society developed expectations of relief payments, economic stimuli, and fiscal interventions from the government. Perhaps paradoxically, in the first phase of the infection, apart from limited relief packages, public actions were not of fiscal, but of monetary nature, as the central banks, with a lack of restraint reinforced by decades of low inflation, inundated the market with liquidity. The infamous lower bound zero interest rate was reached, and talks of 'helicopter money' began circulating with worries of persistent deflation ('the horse refusing to drink'), filling media economic reports and commentaries. However, the macroeconomic story was never as simple as imagined or depicted by the governments and the media.

Initially and throughout its course, the pandemic acted as a series of uncoordinated demand and supply shocks (see also Chapter 6), which were partly reinforced by some of the policies undertaken to counter them. Monetary expansion and stimulus measures attempted to revive demand, without much success initially, especially on the difficult task of resuscitating investment and convincing the banking system to take the necessary risks. Ineffectiveness derived from economic agents' predictable reaction to the surge in liquidity as a temporary phenomenon led to large increases in households and corporate savings. Except for limited relief measures, very little was done on the supply side, especially from governments and, in general, by the public sector. Aside from temporary measures to increase day-to-day flexibility, this was true even in the case of the supply of health services, where a lack of basic capabilities had been dramatically revealed by the mismatch between hospital bed availability and the flow of patients. Similar degrees of inaction were recorded for transport, logistics, and management of value chains, including protective equipment, medicines, and even vaccines.

Government policies worldwide were inadequate to counter the supply and demand shocks characteristic of the pandemic; they also tended to mimic them and protract their effects. The main reason was the attempt to respond in real-time to the unprecedented, 'wavelike' shape of the pandemic, widely incorporated in the narratives developed between the successive lockdowns, behavioral restrictions, and regulatory interventions. The surprise and the drama of the first wave also determined a pattern of response that tended to repeat itself, with minor changes, in the course of the subsequent waves, even though basic circumstances were rapidly changing with vaccine availability and virus variants. As already discussed in Chapter 3, the first wave established a basic pattern of government response, marked by three successive attitudes: denial, recognition, and drama. Despite the large differences in the health and civil protection systems, these three phases were common to most countries, and document the lack of adequate preparation of the public system, and its inability to prepare and manage plans and projects, to deal proactively with unforeseen events and the growing risks of the global economy.

In the second and third waves of the pandemic, despite the vaccines and the new delta, omicron, and other unspecified mutants, government policies almost entirely retraced the same cyclical pattern of reaction of the first wave. Today, well into what seems the end of the pandemic as a tail of an interminable third wave, further causes of concern are given by the fact that the health emergency is intertwined with an economic and social emergency and, after the Ukraine invasion, with a chaotic geopolitical situation that is increasingly difficult to govern. At the same time, the perverse effects on lower incomes, inequality, and poverty are amplified by the prospect of new and tentative measures of releasing the restrictions still in place, uncertainty about the effects of the release of restrictions and lockdowns, especially in China, and the likely worsening of the international picture because of the Ukraine conflict and the threat of decoupling and regression to cold war conditions.

The burden of the restrictions in the most dramatic phase of the pandemic was placed mainly on the less qualified and more precarious employees, on small enterprises in the weaker sectors, such as those of public services, or more dependent on exports and international flows of travelers, such as tourism and entertainment and culture activities. Faced with the pandemic's new and old uncertainties, the unexpected inflation surge and the geopolitical scenario, the European economic framework has also changed radically as the characteristics and role of the 'Next Generation EU' (NGEU, otherwise called Recovery Fund) wait to be redesigned. In the expectation of an unlikely normalization, which we now know will not be able to find the conditions to take place before the end of the current war emergency, it could be argued that, along with the US recovery plan, the NGEU would offer the opportunity to relaunch growth through a fiscal policy of solidarity, based on supranational strategic choices and on a new ability to plan and manage large public projects and multilateral aid programs.

These expectations have been challenged by the subsequent international developments of the decisions of the governments and the Commission that seem to have reinterpreted the new resources as a particular form of structural funds designed to compensate the countries most affected by the pandemic, saving them from a probable depressive trap, and allowing the financing of the recovery through a critical mass of strategic public investment and structural reforms. But even this interpretation now seems obsolete, because additional resources will be immediately needed, above all, to allow the economies of the EU member countries to emerge from the present phase of combined health and geopolitical emergency without a ruinous collapse of their most fragile productive and social components.

At the time this book was closed, the narrative on the deliverance capacity of recovery programs and 'build back' programs appeared to be compromised by the fact that their approval does not seem to coincide with a real turning point in fiscal policies, despite the radical changes in the international conditions both from the demand and the supply side. Despite the lessons learned through the challenging daily management of the pandemic, it seems that many governments are ill-equipped to handle the uncertainties tied to a potential severe deterioration of the economic climate. Moreover, in the short term, the volatility of the prices of primary commodities, the emergence of diffused bottlenecks and capacity constraints, and the persistence of international inflation cast doubts on the efficacy of any proactive fiscal or monetary policy to support private and public spending. At the same time, it is important to note that the policies that support private and public spending are not the only tools available to governments. For example, in the case of Next Generation EU, the program is designed to provide financial resources through grants and loans, but with conditions to fulfill to access the resources. One of the conditions is that member states must implement structural reforms aligned with the EU's priorities, such as digital transformation, climate and environmental protection, and labor market and social policies. Similarly, many national recovery plans also include provisions for structural reforms intended to improve the country's long-term economic performance and resilience.

All recovery programs imply an expansion of spending leading to higher public debts and deferred payments, against the possible alternative of a prolonged recession or even an unacceptable health, economic, and social collapse. At the same time, a fiscal expansion at levels never experienced in the past creates the risk of dissipating precious resources in a plurality of ineffective interventions, because they lack a perspective of structural improvement with incisive and immediate effects on the expectations of operators. It also seems necessary not to miss the opportunity to use the unexpected extension of the fiscal space and the public resources that this extension allows, for interventions that, in addition to the immediate stimulus, have long-term effects on productivity and sustainable growth. This is the meaning of a genuine 'Keynesian' fiscal policy, in which public intervention has the irreplaceable function of compensating for the fall in private investment, keeping demand for capital goods at a level consistent with future growth and, therefore, sustainable in the long term.

However, to sustain demand, and at the same time, avoid the drift of an uncontrolled expansion of current spending, there is a need for programmatic coherence, timeliness and coordination of spending policies, and constant attention to investment. With the protraction of the pandemic and the government's tendency to proceed with incremental actions based on imperfect and transitory information, reconciling investments and distributive interventions is more difficult, though not impossible. For example, we should consider that health and civil protection need human capital, equipment, plants, logistics, and materials. Many of the investments are urgent because they directly relate to the containment of the virus, and many of them should have already been part of the interventions before the second and the third wave of the virus. Further, the social safety net can be strengthened by immediate actions that create social capital and promote its efficiency when it already exists. In this respect, the most important and difficult measures to calibrate are those concerning families and the most vulnerable groups of the population. Although it is more difficult for these measures to maintain the distinction between current expenditure and investment, it is clear that targeted and incisive income support measures are needed through transfer and assistance programs to the poorest households and those most affected by the economic consequences of the pandemic. Targeted transfers to businesses are also necessary to mitigate the social and economic consequences of the restrictive measures imposed by the government, prevent a ruinous fall in private investment, and broaden the consensus at a time when mutual trust is a critical variable of countries' social cohesion.

7.3. The fiscal policies to support the economies

As widely seen before, the various *lockdown* measures in response to the coronavirus several outbreaks halted economic activity in some sectors and reduced it in others, leading to job losses and bankruptcies. Without massive public interventions, these events could have led to severe social tensions, given the huge number of people involved in Europe and the rest of the world. Economists have widely agreed on the need for massive fiscal and monetary stimuli as the most appropriate response to a 'wartime' economic emergency. This agreement follows from a key concept of public finance theory: higher public debt is the right cushion for the private sector in the face of temporary and unpredictable economic crises (Yared, 2019) since it avoids the distortions that would follow significant changes in marginal tax rates that would otherwise be needed to finance an increase in public spending in the short term. A rare consensus among most economists has thus helped fiscal and monetary policymakers introduce massive stimulus packages almost instantly, in contrast to the much slower response given to previous recessions, including the 2008 financial crisis.

In 2020, several countries adopted a wide range of fiscal measures to mitigate the health and economic impacts of the COVID-19 pandemic. The objectives of these measures fall into three broad groups: *i)* to deal with the health pandemic, *ii)* to support households, and *iii)* to bolster businesses. Governments pursue these objectives using a range of fiscal instruments, including tax and expenditure measures, credits, and guarantees. According to Anderson et al. (2020), fiscal measures adopted by the various countries can be grouped into three categories:

- **Immediate fiscal impulse**: it is additional public spending (such as health care, layoffs, grants for SMEs, and public investments) and revenue waivers (such as the cancellation of some taxes and social security contributions). These measures immediately lead to the deterioration of the budget balance without any subsequent direct compensation.
- **Deferrals**: Several countries have decided to defer certain payments, including taxes and social security contributions, which should, in principle, be repaid later. These measures improve the liquidity positions of individuals and companies, but they do not cancel their obligations. Some of these deferrals lasted a few months and expired in 2020, impacting only some monthly budget balances but not the overall ones. The deferrals expiring in 2021 or later could be expected to cause a deterioration of the balance in 2020 against a later improvement. Some countries also deferred loan services or bill payments, which are important tools for improving the liquidity positions of those affected and deserve inclusion in this category.
- **Other reserves and liquidity guarantees**: these measures include export guarantees, liquidity assistance, and credit lines through national development banks. Some of them improve the liquidity of the private sector. Yet, unlike deferrals, which are automatic and generally apply to specific groups, credit lines require the intervention of the companies concerned. Though credit lines and guarantees did not weaken the budget balance in 2020, they created contingent liabilities that could turn into actual expenditures later. This means that these measures may not have an immediate impact on the budget balance, but they could have a long-term impact on the budget if the companies cannot repay the credit or utilize the guarantees.

According to Lacey et al. (2021), the heterogeneity of types, instruments, and magnitude of the fiscal response during the first year of the pandemic differ significantly by country groupings and regions of the world. As can be seen in Fig. 7.1, in this period, countries adopted a wide-ranging set of fiscal measures: between one to three interventions in support of their health sector, two to five interventions targeted at supporting households, and two to five interventions to support businesses. Later, some of these measures were revised or added to earlier policies. In terms of magnitude, as expected, Advanced economies' (AEs) interventions amounted to more than 9% of GDP in expenditure and revenue measures and another 11% in support through equity and loans, guarantees, and quasi-fiscal activities. Emerging markets and middle-income economies' (EMMIEs) response has been more muted, but they also deployed all instruments for the fiscal response, with about 3.4% of GDP in expenditure. However, the total cost of the fiscal packages shows a weak relationship with the number of policies adopted.

One year later, at the beginning of 2021, the outlook on the evolution of the pandemic was still uncertain due to high levels of COVID-19 cases, perplexity on vaccine rolling out, and the occurrence of new virus variants spreading more easily and more quickly. Under these conditions, policymakers faced difficult choices in an environment where fiscal space was narrowing and additional spending requirements emerged,

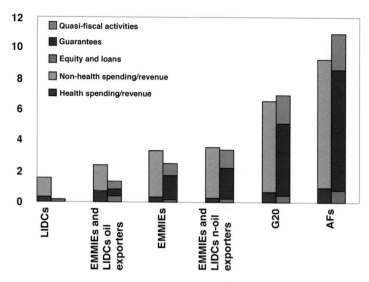

Figure 7.1 Breakdown of Fiscal Support, by Type (percent of GDP). *Source*: IMF Fiscal Monitor (October 2020), available at the web address: https://openknowledge.worldbank.org/bitstream/handle/10986/35904/A-Review-of-Fiscal-Policy-Responses-to-COVID-19.pdf?sequence=1&isAllowed=y.

including those needed for the purchase and distribution of vaccines as well as for measures to support an economic recovery. In 2023, many of these uncertainties are still present and accentuated by inflation, central banks' tightening policies and the disruptions created by the Ukrainian war.

7.3.1 The EU fiscal response to the pandemic

The European Central Bank (ECB) responded swiftly and robustly with monetary policy and supervisory measures introduced as early as March 12 and 18, 2020. In contrast, the policy actions from the EU Commission and other institutions were less decisive and lacked the same level of vigor. After the first phase of lack of common action in the face of uncoordinated nationalistic reactions in the early months of 2020, the EU response gradually moved toward a more assertive cooperative behavior. The measures adopted included several programs which greatly helped EU citizens to cope with the pandemic's adverse effects. For example, the European state aid and tax rules (budget constraint) were suspended, and the RescEU civil protection mechanism was activated. This process took time but strengthened cooperation between the EU countries and six participating states on civil protection to improve prevention, preparedness, and response to disasters. The funding of the initiative was also expanded in the course of time, with the budget leaping from €766.5 million (for 2014-2020) to €772.7 million (for 2021 alone).[3]

In March 2021, the Member States agreed on a far larger budget for health: the restoration of the Health Program as EU4Health, with a yearly budget that grew from around €46 million to €5.1 billion. The budget also allowed expanding the remit and funding of the European Centre for Disease Prevention and Control (ECDC), agreeing on a 'EU Vaccines Strategy' for procurement of COVID-19 vaccines, a pharmaceutical strategy to ensure a supply of pandemic-relevant drugs in the future, and new pandemic preparedness and response organization to be called HERA (Health Emergency Preparedness and Response Authority). These decisions represented a significant improvement in how EU policy had changed compared to a few years before. In particular, the decision to procure vaccines collectively through the EU Vaccines Strategy was perhaps the most important, given that single governments accepted to pool their resources on the most critical issue they faced.

Another important pillar over which EU cooperation has been remarkable is its fiscal governance system. By the time the COVID-19 crisis hit, the existing system was mainly focused on controlling government budget deficits. However, in the face of the pandemic, the Commission invoked the 'general escape clause', which suspended the application of the fiscal governance process. At the same time, it financed the 2021-27 Multi-annual Financial Framework (MFF) with 1210.9 billion euros and the NGEU with an extra 806.9 billion euros. Within the recovery plan (the Next Generation EU or NGEU) the most important action line was the Recovery and Resilience Facility (RRF), a temporary recovery tool that directs funds to the Member States for general budgetary support. It enables the Commission to raise funds to help Member States implement reforms and investments that align with EU priorities and address the challenges identified in the country-specific recommendations under the European Semester for economic and social policy coordination. As detailed in Table 7.1, over a total of € 806.9 billion (at current prices) allocated to the NGEU, the RFF consisted of € 723.8 billion in loans (€ 385.8 billion) and grants (€ 338 billion).

The RRF represented an extremely innovative tool within the EU fiscal policy history. Firstly, unlike the past, where funding was allocated to specific projects or goals, like agricultural policy, the new model focuses on providing support to the overall budget. Second, it comes with conditionality, which means that the Member States must specify the use they will make of it. Finally, these interventions are important as they set a radical change in the commitment to budgetary austerity adopted in the aftermath of the 2008 financial crisis (see Fig. 7.2). Along these lines, the EU is an exception. Its leaders have been able to change quite rapidly the scale and scope of its health policy, civil protection, and fiscal governance by expanding older systems (i.e., the

---

[3] Part of the funding allocated to this program served to secure PPE stockpiling.

Table 7.1 Overview of MFF and NGEU allocations - Current prices in billion of euros.

MFF 2021-27 and NGEU allocations	MFF	NGEU	NGEU breakdown	
1. Single market innovation and digital	149.5	11.5	1. RFF	733.8
2. Cohesion, resilience and values	426.7	776.5	*1.a of which loans*	385.8
3. Natural resources and environment	401.0	18.9	*1.b of which grants*	338.0
4. Migration and border management	25.7		2. ReactEU	50.6
5. Security and defense	14.9		3. Horizon Europe	5.4
6. Neighborhood and the world	110.6		4. InvestEU	6.1
7. European public administration	82.5		5. Rural development	8.1
			6. Just Transition Funds	10.9
			7. RescEU	2.0
Total	**1210.9**	**806.9**	**Total**	**806.9**

Source: https://op.europa.eu/en/publication-detail/-/publication/d3e77637-a963-11eb-9585-01aa75ed71a1/language-en.

RescEU and the ECDC) and creating new ones (i.e., the Vaccines Strategy) (Deruelle, 2022). 'The question now is whether, over the next five years, this newly ambitious and protective EU health policy will convince Member States and others of its utility and value' (Greer et al., 2022).

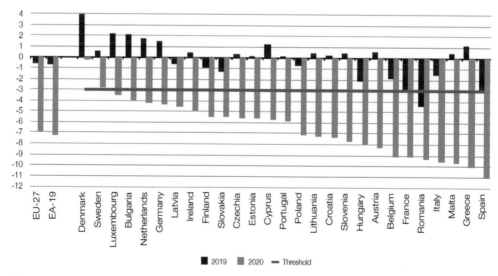

Figure 7.2 Public balance, 2019 and 2020. *Source*: Eurostat, available at the web address: https://ec.europa.eu/eurostat/statistics-explained/index.php?title=Government_finance_statistics.

7.3.2 A future with higher debt levels and little chance of repaying it?

The main objective of the public finance interventions described in the previous sections is to cushion the harmful effects of the pandemic crisis on the economies and to allow them to restart as soon as possible. The actions implemented should produce a series of positive impacts. However, as seen in Fig. 7.2, they will generate a sharp increase in the current deficit and public debt levels.

According to the IMF, at the end of 2019, the public debt was already high compared to the historical standard before the COVID-19 epidemic (IMF, 2019). In almost 90% of the countries described by the IMF as 'Advanced economies', the debt/GDP ratio has been higher in recent years than in the last global recession, which began at the end of 2007 and lasted until mid-2009 (see Fig. 7.3).

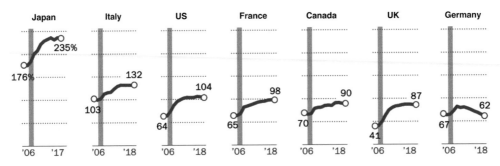

Figure 7.3 Total gross debt in the G7 Group of countries (% of GDP, 2006-2018). *Source*: IMF (2019). Note: Gross debt represents the total liabilities of all levels and units of government - national, state/provincial, and local - minus liabilities held by other levels or units of government unless otherwise specified, by country.

With the arrival of the pandemic, those levels have risen further. According to the IMF,, in 2020, the fiscal deficits of the most advanced countries reached, on average, 11% of GDP, while the debt/GDP ratio moved from an average of 105% in 2019 to 122% in 2020. These changes, which occurred in such a short period of time, are shocking, especially considering that until the beginning of 2020, public finance management was based on strong convictions formed in the aftermath of the 2008 global financial crisis. At that point, many governments focused on what maximum debt-to-GDP ratio would be considered 'safe' to protect them from losing confidence in financial markets. The European Union was not an exception. The euro area budget deficit increased from 0.6% of GDP in 2019 to 8.5% in 2020. In 2020, in the Euro area, eleven countries registered budget surpluses while all other euro area countries registered budget deficits above the reference value of 3% of GDP (see chart (A) in Fig. 7.4). The largest deficits were those of Belgium, Spain, France, and Italy, which were among the countries that entered the pandemic crisis with the highest public debt/GDP ratio.

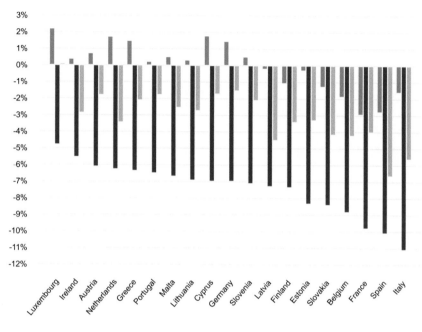

A - General government budget balances (2019-2021).

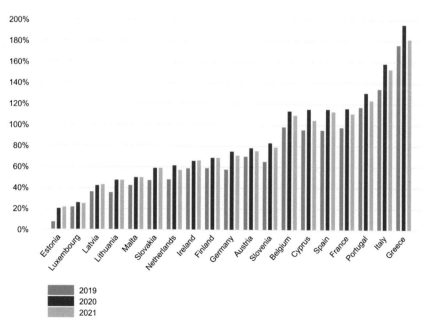

B - Gross debt of general government (2019-2021), % of GDP.

Figure 7.4 Impacts of the COVID-19 crisis on public finance. *Source*: European Commission (AMECO database) and ECB calculations (several years).

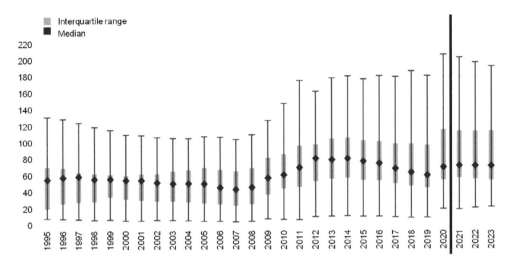

Figure 7.5 Debt-to-GDP ratios of euro area Member States. *Source*: AMECO and ECB staff calculations. Note: The solid vertical line refers to the beginning of the forecast period. These figures represent the latest estimates available at the time the book was closed, and for 2020 and 2021 may not match exactly those presented in panel (B) in Fig. 7.4.

In the Euro area, public debt increased by some 14 percentage points of GDP since 2019, reaching about 100 percent of the euro area GDP in 2020. Looking at graph (B) in Fig. 7.4, we see that countries with debt ratios around 100% at the beginning of the crisis experienced the largest increases. This aggregate number masks significant cross–country heterogeneity, with debt ratios ranging from around 18 percent of GDP to more than 200 percent of GDP (see Fig. 7.5). According to Eurostat, at the end of the fourth quarter of 2021, due to the expenses incurred to mitigate the negative impacts of the crisis, the debt to GDP ratio in the euro area stood at 95.6%, compared with 97.5% at the end of the third quarter of 2021. In the EU, the ratio also decreased from 89.9% to 88.1%. For the euro area, the ratio decrease is due to an increase in GDP and a slight decrease in the nominal debt in absolute terms, while for the EU, the nominal debt continued to increase slightly but was outweighed by the increase in GDP. Compared with the fourth quarter of 2020, the government debt to GDP ratio decreased in the euro area (by about 1.5%) and the EU (about 2.0%). The reductions are due to the GDP recovery outweighing the increase in government debt.

These unfolding events underscore the urgency of a radical revision of the Stability and Growth Pact (SGP). The legacy of higher public debt ratios suggests a natural increase in vulnerability and a greater potential for negative spillovers across member countries. Conversely, a full recovery from the pandemic requires fiscal policy to be sufficiently counter-cyclical. A successful recovery also requires supportive fiscal policies. This highlights the need for effective coordination, increased powers for the Commission, and a focus on growth-enhancing reforms and growth-friendly fiscal policies. An already challenging budgetary agenda has become more challenging due to the pandemic. All member countries must address the macroeconomic and fiscal implications of aging societies. An effective SGP must also acknowledge the urgency of the green transition and digital transformation, which require significant investment to address the decade-long shortfall in public capital, going much beyond the scope of the Recovery and Resilience Facility (RRF).[4]

7.3.3 The public debt problem

The evidence so far points to several important conclusions regarding the role that the fiscal measures implemented over the past few months could play in the future of economic systems. In particular, attention must be focused on the fact that the massive global fiscal stimulus will turn into debt to be repaid in the coming years. As widely reported (see, for example, Kose et al., 2021), the accumulation of debt during the 2020 global recession induced by the pandemic was the largest in decades. This phenomenon applies equally well to the public and private sectors, involves all countries (from advanced to emerging market and developing economies (EMDE)), and applies to both the foreign and domestic sides (see Fig. 7.6). At the world level, in 2020, total global debt reached 263% of the world GDP, and global public debt reached 99% of the world GDP, the highest levels since World War II.

Moreover, the pandemic has amplified the risks associated with debt, as some policy actions, although necessary during the pandemic, may end up weakening fiscal, monetary, and financial policy frameworks. For example, in the face of exceptional fiscal support measures, fiscal rules risk being eroded in some EMDEs. Many governments have also strongly encouraged credit extension and relaxed regulatory policies. While this has prevented a credit crunch during the 2020 recession, private-sector liabilities could eventually migrate to public budgets, both in a financial crisis and indirectly, over a long period of low growth. Unprecedented monetary-policy accommodation was also required to calm financial markets and reduce financial burdens. The war in Ukraine has complicated this scenario, by further expanding debt needs and obligations and by delaying any decisive action to bring the existing debt under control.

The rapid increase in debt levels due to the COVID-19 pandemic was unexpected and challenged the conventional wisdom established after the 2008 financial crisis, with many governments focusing on maintaining low debt-to-GDP ratios to ensure financial market stability and confidence. Before the invasion of Ukraine, it was thought that

---

[4] The Recovery and Resilience Facility (RRF) serves as a temporary mechanism and is the core element of NextGenerationEU, the EU's strategy to recover robustly and resiliently from the ongoing crisis. Via the Facility, the Commission accrues funds by engaging in borrowing activities on capital markets (issuing bonds representing the EU), making them accessible to its Member States for the execution of ambitious reforms and investments.

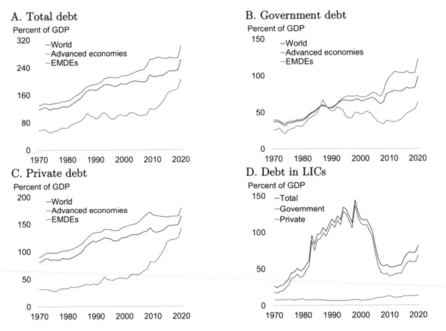

Figure 7.6 Public debt trends by world regions. *Source*: Kose et al. (2021). Note: data are available for up to 192 countries, including 39 advanced economies, 153 EMDE, and 24 LIC. Weighted averages of nominal GDP.

once the pandemic was finally brought under control, governments would focus on reducing their deficits to more sustainable levels. This would enable them to address 'more ordinary' issues, such as aging populations and the relative rise in welfare costs (primarily pensions and health care).

The ongoing war in Ukraine and the resulting increase in commodity prices have added to the factors that conspired to cause a further rise in global interest rates, which have been increasing due to central bank actions. This, combined with the persistence of an inflationary environment, may result in solvency problems for international debtors. The rapid surge in debt coupled with the loosening of fiscal oversight heightens the possibility that not all resources have been directed towards productive ends. The predicted global increase in military spending reinforces the suspicion that a negative quality effect may substantively reduce countries' capabilities to honor their commitments with creditors. Persistent inflation would have mixed effects by relieving some of the burden on the governments in the short-run, but its impact on creditors' willingness to refinance the existing debt could make it ultimately unsustainable.

For countries experiencing debt distress, attaining a successful resolution may be more challenging than before. In particular, future debt restructurings are likely to be more complicated due to a more fragmented creditor base than in the past and a lack of transparency in debt reporting.

The availability of a highly effective new vaccine technology is positive as it suggests that governments' efforts to address the economic impacts of this and future pandemics may be temporary, thereby reducing the long-term pressure on public finances. However, the fact remains that debt levels will still be high after the pandemic, raising concerns about repayability and inter-generational equity.

In a low-interest rate environment, there exists a broad consensus among economists that as long as the debt service (the cost of borrowing) remains below the economy's growth rate ($r < g$), both in actuality and in the operators' expectations, additional public deficit spending (financed by debt) may be considered sustainable. This conclusion depends on the theoretical argument that the sustainability of the national debt depends not so much on the amount, but on its cost in relation to the ability to repay it. In an environment like the one existing until the fall of 2021, with long-term interest rates still at record low levels throughout the developed world, governments can borrow for decades at low cost. For example, Britain's 30-year bonds yielded only 0.9%, making the one-time cost of the pandemic easy to manage for the coming decades. It appeared that there was no better time to address significant budget deficits. Consequently, for wealthy nations, there was no urgency to implement tax cuts, especially if it could potentially hamper economic growth.

Moreover, borrowing is much easier in advanced economies, given the low level of interest rates and the willingness of investors to buy that debt. According to the IMF, interest payments on the sovereign debt of the wealthiest countries have fallen from over 3% of GDP to about 2% over the past 20 years, even though the debt-to-GDP ratio has increased by more than 20 percentage points. Moreover, with much of the newly issued sovereign debt paying negative interest rates, further borrowing will reduce interest charges even more.

Therefore, under these conditions, generating debt was perhaps a good solution. Debt is an appropriate instrument for dealing with such situations: being able to transfer and delay payments until the world economy is in a better state will make the pandemic more manageable and economic scars less profound. The alternative of not contracting debt would have meant mass bankruptcies and unemployment, and, without effective support, workers would have been less inclined to accept the invitation to stay home, contributing more to the spread of the virus and causing more economic damage.

7.3.3.1 What about the future of the debt-to-GDP ratio?

By the end of 2021, interest rate spreads over German yields had increased significantly for the euro-area countries with the highest debt-to-GDP ratios. Although the spread increases did not reach the levels recorded during previous stress episodes (and have declined since the European Central Bank's 15 June 2021 announcement that it is finally working on a new tool to avoid financial fragmentation), they could still represent a risk if *i)* growth will continue to slow because of the energy crisis caused by Ukraine's war, and *ii)* several countries will persist in their historically high debt levels (especially after the highly expansive fiscal policy implemented during the COVID-19 crisis). Moreover, from the monetary policy perspective, the formerly low-interest rate ECB policies have changed to tame the rising inflation in the euro area. This has led to tighter monetary policies. As confirmed after the June 2021 ECB Governing Council meeting, PEPP (Pandemic Emergency Purchase Programme) and PSPP (Public Sector Purchase Programme) net purchases ended respectively in March and June 2022. Additionally, the European Central Bank (ECB) announced a 25 basis point increase in policy rates in July 2022 and a further increase of at least 25 basis points in September 2022, bringing its deposit rate to 0% for the first time since June 2014. By the end of 2022, market participants expected the ECB to raise its policy rate by approximately 225 basis points by the end of 2023.

Despite these alarming trends in interest rate spreads and restrictive ECB monetary policy in 2022 and 2023, debt-to-GDP ratios in advanced countries are expected to continue to fall. This is mainly due to high nominal output growth and implicit rates that should only increase very gradually, given that euro-area countries have locked-in low rates thanks to the combination of low rates for a decade and an increase in the average maturity of their debts. As noted by Claeys and Guetta-Jeanrenaud (2022), 'However, after 2023, the situation might vary across countries. Countries with relatively high primary deficits are not expected to see a further fall in their debt-to-GDP ratios. But countries that will quickly reduce their primary deficits should see their debt-to-GDP ratios continue to fall as long as spreads are kept in check. This shows that current fears about the sustainability of the debt are misguided. It is also important for the ECB to design a new antifragmentation tool that will keep spreads at a level justified by structural differences between countries, and not by self-fulfilling prophecies'.

However, the proposed reasoning is only applicable to rich countries, as it does not take into account the economic realities of poorer nations. Many countries with weaker and less stable economies (Argentina, Belize, Lebanon, Ecuador, Suriname, and Zambia) have relied on much higher-cost, shorter-term debt and either failed to pay their debts or had to restructure them during the pandemic. In the case of emerging and developing economies, other factors such as the efficiency and fairness of expenditure also play a significant role.

This leads to an important conclusion: how debt has been used is more critical than its amount. Historically, we have seen that successful countries can attribute a portion of their economic growth to high-quality social and capital spending. And the most successful countries managed their economies at or near investment-level savings, maintaining minimal current-account deficits. Today, however, many countries, some already heavily indebted pandemic entrants, have not acted as good administrators of public resources, owing to poor project selection and implementation, ineffective targeting of social spending, expensive subsidies, and corruption.

In conclusion, high Debt-to-GDP ratios represent a danger, even if interest rates are low. The key reason is the increased uncertainty of growth prospects that the post-COVID-19 economies will have to face, coupled with the uncertainty about the probability of future large shocks, which are expected to be more frequent than assumed in standard models. In addition, large negative shocks create many more problems when debt is already high. This highlights the crucial question for policy makers: the focus should not be whether high public debt will be a problem for post-COVID-19 economies, but rather how the debt will be utilized. Under favorable low interest rate conditions, proper use of borrowed resources could result in an economic growth rate (g) that exceeds the cost of debt (r), making future debt management manageable and leading to primary surpluses that can reduce the level of debt.

7.4. Global value chains: are we going back to normal or forward to a new order?

The process of globalization of the economy has displayed its effects on all economic activities for more than 600 years, completely changing production processes and consumption patterns. Over the centuries, globalization has induced significant economic changes and social tensions. However, it was not until the early 1980s that the world economy witnessed a significant transformation in the structure of international trade flows, giving rise to the 'Age of Global Value Chains' (Amador and Di Mauro, 2015; World Bank, 2020c).[5] Several factors have fueled such change whose combination has been unique in terms of results produced. First, the information and communication technology (ICT) revolution made it possible to exchange information on a scale and at a speed never witnessed before. Second, there was an incredible acceleration in reducing man-made trade barriers, mainly via dozens of preferential trade agreements and thanks to the highly debated China's accession to the WTO in 2001. Finally, political developments have brought about a remarkable increase in the share of the world population participating in the capitalist system (Antràs, 2016). As Antras and Chor (2021) illustrate, 'these forces worked in tandem to increase the extent to which firms used foreign parts and components in their production processes, as well the extent to which intermediate input producers sold their output internationally rather than to only domestic end users.

---

[5] Global value chains (GVC) refer to the sharing of international production, a phenomenon in which production is divided into activities and tasks done in different countries. One can think of a large-scale extension of the division of labor dating back to the time of Adam Smith. In the famous example attributed to Adam Smith, the production of a pin was divided into a series of distinct operations within a factory, each performed by a dedicated worker. In GVCs, operations are spread across national borders rather than confined to the same hall.

In fact, it has been estimated that trade in intermediate inputs constitutes as much as two-thirds of world trade (Johnson and Noguera, 2012). As a result, the typical 'Made in' labels in consumer goods no longer do justice to the amalgam of nationalities that are represented in the value added embodied in these products'.

7.4.1 International trade and global value chain disruptions

The results of this process over the last 70 years are visible in Fig. 7.7, showing the growth of the total volume of world trade, which has gone from around $62 billion in 1950 to about $19,500 billion USD in 2018. The expansion of international trade, which has reached almost 30% of the world GDP, has been both an effect and one of the leading causes of globalization. The hyper–connection model realized through the network of value chains has been a by-product of exponential trade growth. In recent years, value chains have grown in length and complexity as companies have expanded worldwide in search of margin improvements. From 2000 onward the value of intermediate goods traded globally has tripled to exceed 10,000 billion dollars per

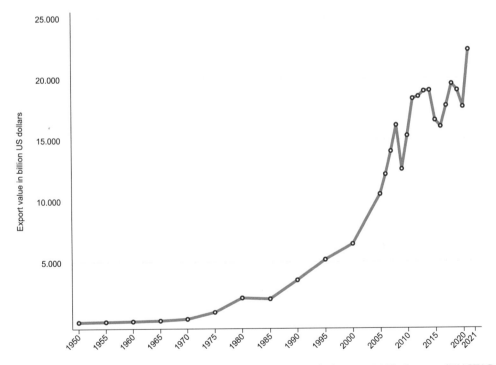

Figure 7.7 Total volumes of international trade in goods (1950-2020). *Source*: (UNCTAD, 2020). Data available at the following web address: https://www.statista.com/statistics/264682/worldwide-export-volume-in-the-trade-since-1950/.

year. Companies that have successfully implemented a streamlined and comprehensive production model have achieved important improvements, such as low inventory levels, timely and complete deliveries, and shorter delivery times.

This meant that at the turn of the year, nearly a fifth of world industry moved from the countries belonging to the G7 to others such as China, India, Korea, Indonesia, Thailand, and Poland. As a result of this new organization, raw materials and intermediate goods are shipped around the world several times and then assembled at different locations. Production is re-exported to final consumers located in both developed and developing markets. For many goods, China is at the center of such GVCs: for example, as a primary producer of high-value products and components; as a large customer of global commodities and industrial products; and as an important consumer market. In addition, China also produces many intermediate inputs and is responsible for machining and assembly operations. A great example is Foxconn, the largest manufacturer of electrical and electronic components for digital equipment manufacturers in the world, which mainly operates under contracts with companies such as Amazon, Apple, Intel, and Sony.

In this scenario, competition has dominated and coordination and cooperation have taken a secondary role in the economy and economic policies. However, even before the 2008 financial crisis and the current emergency, international trade expansion seemed to be reaching levels dominated by diminishing returns. This was partly due to the difficulties in negotiating new mutually beneficial spaces, as well as the rise of global imbalances driven by a few 'superstar' countries with persistent trade surpluses. These imbalances have led to a new form of mercantilism, partly fueled by international capital flows and multinational corporations, where countries are increasingly differentiated based on their comparative advantage in producing value-added, rather than through competition in producing goods. In their participation in international trade, in other words, the different countries have increasingly aimed to specialize in producing final goods or services and semi-finished goods as close as possible to the final goods. The result has been a hierarchical model of international specialization that has accentuated the dependent development of poorer countries and created the dangerous illusion that trade surpluses and deficits can be sustained indefinitely.

Yet, these deep interconnections can occasionally produce unexpected results, and the COVID-19 crisis has brought out a further problem: apparently effective in supplying world markets regularly, international trade networks have proved fragile, and value chains unbalanced and dramatically unreliable in times of emergency. The lack of resilience of many major trade neworks is due to two key factors: first, the development of value chains based on pure comparative advantage and the influence of practices like delocalization, outsourcing, and offshoring. While most of the production of value-added is concentrated in advanced countries, the intermediate goods needed to produce it are mostly the fruit of labor and natural resources located in developing countries. Second, the need for coordination in crucial trade areas, such as medicines and other strategic materials, has been increasingly circumvented by the prevalence of competition over co-operation in multilateral governance. For example, using a world input-output model of international trade, Eppinger et al. (2020) report moderate welfare losses in the majority of countries outside of China, with some experiencing gains due to shifts in trade. In a speculative scenario devoid of GVCs, the median country would see a 40% reduction in welfare loss due to the Covid-19 shock originating from China. These effects are intensified or inverted in several other nations. If the U.S. had chosen to independently repatriate GVCs, it would have faced a significant welfare loss, with negligible alteration to its exposure to the shock.

Complex production networks have been designed for efficiency, cost and proximity to markets, but not necessarily for transparency or resilience. According to Lund et al. (2020), the shock caused by the COVID-19 pandemic has been one of the most damaging to GVC and economic systems, but certainly not the only one recorded in recent years. In 2011, the earthquake and tsunami that occurred in Japan led to the shut down of important manufacturing plants of automotive parts, disrupting assembly lines in the whole world. Also, that same event shut down the world's largest producer of advanced silicon wafers, on which the semiconductor industry relies. A series of floods disrupted the operations of hard disk manufacturers in Thailand, which produced about a quarter of the world's hard disks, potentially creating significant problems for personal computer manufacturers. Furthermore, in 2017, in the United States, Hurricane Harvey, a T4 storm, blocked the operation of some of the most important oil refineries in Texas and Louisiana, creating shortages of plastic materials and key resins for a variety of industries. To this can be added the damage caused in 2019 by about 40 weather disasters, with a cost impact that has exceeded one billion dollars each.[6] Over the years, however, companies have become aware that these interruptions are recurrent events and now expect supply chain disruptions lasting one month or more to occur every 3.7 years, with the most severe events involving a heavy financial cost (exceeding 40% of a year's profits every decade) (Lund et al., 2020).

To these environmental/climate shocks others can be added that are related to geopolitical issues: the ongoing US-China trade conflicts are the clearest example of how clashes between global actors can disrupt the general economic environment. The tensions triggered by these trade confrontations reverberate over the overall functioning of the system, with increased trade disputes, higher tariffs, and broader geopolitical uncertainty. Unfortunately, the outlook for the future is not rosy. On the geopolitical front, the situation is quite complex, and, regardless of the change of the US president,

---

[6] Over the next years, climate change will play an important role on economic growth. It is estimated that, until 2050, the global economy will grow 3% less than the counterfactual scenario obtained without climate change, and the prevalence and severity of natural disasters will increase.

given the interests tied to technological leadership, it's probable that the trade conflict between the world's two largest economies will persist. In addition, the development of computer networks and the flow of information that travels on them daily increases the exposure of supply chains to a wide variety of cyber-attacks, as testified by the number of new ransomware variants doubling from 2018 to 2019.[7]

The COVID-19 pandemic has highlighted the relationship between globalization and the stability of international trade. As we recover from the pandemic, it is important to make significant changes to our economic systems to improve this relationship. This is a chance to make real progress towards a more sustainable future. The reforms needed cover two main fronts: *i)* the global imbalances caused by the pursuit of trade surpluses, *ii)* the reorganization of value chains so that trade networks are more logistically resilient, and the supply of essential goods more reliable. On both these fronts, globalization requires a new commitment from the international community and a new multilateralism based on cooperation and coordination rather than solely on competition.

7.4.2 The problem of global value chains

As globalization intensified, the world's supply chains have become significantly more complex and interconnected. To improve efficiency and achieve greater economies of scale, companies have specialized in increasingly specific segments of the entire process of creating the final product. This has led to two main results. First, it is not uncommon for supply chains to have six or more tiers. For example, in the garment industry the manufacturing process of a single garment includes multiple stages, such as fiber production, fiber spinning, weaving, dyeing, cutting, sewing, and finishing. Furthermore, other parts of the garment (buttons, zippers) are often produced by additional suppliers, while distribution and logistics involve even more actors in a complex network of delivering goods to customers. In other industries, such as automotive, electronics and aerospace, products are even more complex resulting in longer supply chains. Companies, with expertise in highly specialized parts of the manufacturing process, also supply components to numerous firms that ultimately interact with end consumers. In the past, supply chains were thought of as a pyramid with raw material suppliers at the base and consumers at the top. Nowadays, the interconnections between their different parts are so intricate that they can only be visualized as a large network. There are numerous suppliers, consumers, and numerous interconnected value chains, with resulting complexities and lack of transparency.[8]

---

[7] A ransomware is a type of malware that restricts the access of the device it infects, requiring a ransom (ransom in English) to be paid to remove the restriction. Some forms of ransomware block the system and order the user to pay to unlock it, while others encrypt the user's files, asking victims to pay to return the encrypted files in plain.

[8] This is amply demonstrated by what happened following a heavy flood in Thailand in 2011. As already mentioned, at that time, most of the hard drives produced worldwide were manufactured in Thailand.

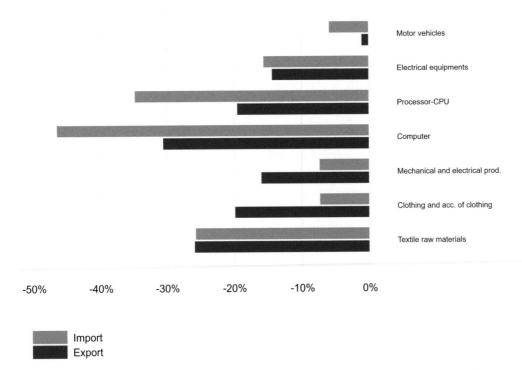

Figure 7.8 Percentage changes in Chinese exports and imports for selected products. *Source:* China General Customs Administration (2020). *Note:* Change in export and import values between Jan/Feb 2019 and Jan/Feb 2020. The selected goods are those for which Chinese involvement in GVC is very high.

The Wuhan pandemic that began in January 2020 caused similar problems, but at a significantly higher level. The first effects of plant closure in China are visible in the import-export trade flows data shown in Fig. 7.8. When comparing the weeks between January and February 2020 with the same period in 2019, Chinese imports and exports decreased by 4% and 17% in value, respectively. In particular, the graph shows significant reductions in imports of products that are used as intermediate inputs in production, such as textiles, electrical, or electronic equipment. Likewise, exports of these goods also suffered sharp decreases. According to Seric et al. (2020), Chinese exports to all regions of the world decreased significantly, with the exception of North America, where trade had already been on the decline for more than a year due to ongoing trade disputes between the United States and China. The collapse of manufacturing activity

Due to the flood, global hard drive production decreased by about 30% and it took several months for companies to realize where these components came from.

at the heart of many GVCs had different implications for producers and consumers in countries placed higher and lower in the product value chain. This meant that, starting from March 2020, vital parts of production began to be lacking in European countries. In Germany, for example, 10% of all imported inputs come from China, and this dependence is particularly strong in the electronics, information technology and textile manufacturing sectors.

Similarly to how a properly functioning set of organs is necessary for human survival, the supply chain plays a vital role in maintaining the health of the global economy. The various lockdowns applied in different countries and the closure of airspace and borders triggered by COVID-19 caused an unprecedented disruption of the production mechanisms of most economies, regardless of their size or stage of development. In particular, the raising of these barriers has strained the world's supply chains, including essential links relating to food and medicines.

7.4.3 Until supply lasts: can globalization be a problem?

As seen before in this book, many phenomena were already underway, and the role of the pandemic was to accelerate their pace; companies had been reviewing how they structure their production activities for years. Fig. 7.9 suggests that the growth trend in GVCs was halted after 2014 since the process had reversed slightly in favor of more domestic production. Although no official statistics are available, it is easy to think that between 2018 and 2019, the trend may have been confirmed, and for 2020, expectations were of a sharp contraction in the production share obtained through GVCs.

A different way to interpret the structural change that has occurred in GVCs is to compare cross-border investment flows in manufacturing activities with trends in GDP and global trade, as represented in Fig. 7.10. The UNCTAD World Investment Report has been monitoring FDI and multinational company activities for three decades. The 2020 report shows that there was a period of strong growth in international production for two decades, followed by a period of slow growth. Since 2009, there has been a slowdown in cross-border investment in manufacturing, although GDP and international trade continued to grow at a lower rate.

To confirm this hypothesis, Lund et al. (2020) pointed out that the crisis triggered by COVID-19 forced many companies to rethink just-in-time supply chains. They also observed that the pressures for a new focus on supply chain resilience and regionalization had been increasing even before the pandemic. The results of Lund et al. (2020) are also confirmed by two reports from Boston Consulting Group (BCG) and AT Kerney (Aylor et al., 2020). These reports suggest that companies are likely to transfer a quarter of their global product supply to new countries in the next five years, as rising threats to supply chains weigh heavily on profits. Lund et al. (2020) estimate that goods worth between $2.9 and $4.6 trillion, or 16-26% of global exports to 2018 could be relocated. For example, by 2019 much of low cost production had already shifted to Mexico and

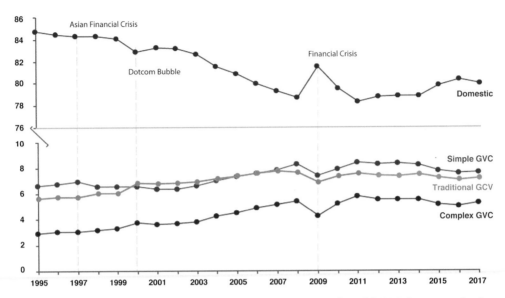

Figure 7.9 Trend in production activities as a percentage of world GDP, by type of value added. *Source*: The 1995-2009 data are based on the GVC indices of the University of International Business and Economics (UIBE) derived from the World Input-Output Table of 2016; the 2010-2017 data are based on UIBE GVC indices derived from the Asian Development Bank (ADB) ICIO Tables 2018. *Note*: The meaning of the variables in the figure is the following: *Pure Domestic* - implies production done in the country and there are no border crossings (Example: haircut); *Traditional trade* - Cross-border trade for consumption (Example: exports of Portuguese wine in exchange for English textiles); *Simple GVCs* - Crossing the border once for production (Example: Chinese steel in US buildings); *Complex GVCs* - Crosses the border at least two times (Example: iPhone/Auto).

Vietnam. In that year, both nations expanded their consumer goods and technology, media, and telecommunications (TMT) markets by 12% and 9% respectively, largely taking from China's share. Furthermore, Vietnam's clothing and smartphone exports grew, along with Mexico's automobile parts and computer shipments.

Moreover, cost considerations and government pressures to become more self-sufficient could cause more than half of pharmaceutical and clothing production move to new countries. According to the BCG, bilateral trade between the United States and China could shrink by about 15% (or about $128 billion) within the 2023 from 2019 levels. Along these lines, AT Kearney's report concluded that the pandemic event would accelerate the 're-evaluation' of supply chains (Gandhi et al., 2020). This report also suggested that technology had diminished the importance of labor arbitrage, while increasing consumer demand for fast delivery was already creating pressure for shorter 'multi-local' supply chains. Along the same lines, S&P Global Ratings reported that

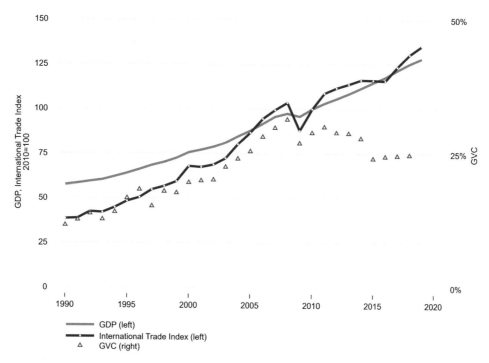

Figure 7.10 GDP, International Trade and GVCs (1990-2020). *Source*: The Eora Global Supply Chain Database, (UNCTAD, 2020; World Bank, 2020d). *Note*: Global exports of goods and services define the international trade index. The share of GVC in international trade is obtained as proxies from the share of foreign added value in exports, starting from data contained in the UNCTAD-Eora GVC database (Casella et al., 2019).

many U.S. companies had already realized that their supply chains had become too long and complex. According to Hedwall (2020), in recent decades, companies have focused mainly on optimizing supply chains through increased efficiency and improved business performance. Yet, recent outcomes indicate a shift in the paradigm, with resilience and adaptability as the guiding principles. The COVID-19 pandemic underscored this, spotlighting vulnerabilities in the existing supply chains. At the same time, S&P Global Ratings warned investors against expecting a rapid overthrow of decades of globalization (Dimitrijevic et al., 2020). It would appear that US producers saw few options to replace their Chinese suppliers. Moreover, beyond the 'potentially prohibitive costs' of seeking alternative sources of production, US companies may be reluctant to risk losing access to the world's second-largest economy. China is also likely to maintain a high level of production, given its domestic demand driven by a population of over 1.3 billion people.

7.4.4 Improving coordination in biomedical supply chains

Supply chain disruptions can occur for many reasons, including natural disasters, acts of war or terrorism, supplier bankruptcy, labor disputes, cyber-attacks, and data breaches. The COVID-19 pandemic added further dimensions to the disruption: *i)* it affected not only the supply but also the demand for products and services, *ii)* it generated a high level of uncertainty on the size and length of the disruption, and *iii)* it had a simultaneous impact on various geographic areas.

For these reasons, the COVID-19 pandemic sheds a specially bright light on the weaknesses in health care supply chains. Over the years, the levels of complexity and opacity of all health supply chains worldwide have increased, including those of critical medicines and medical equipment, creating major vulnerabilities for all countries during pandemics. In recent decades, the problem has been exacerbated, with health supply chains becoming more complex and reliant on offshore manufacturing, often spanning multiple countries and just-in-time modes of production that prioritize quick turnarounds on orders and warehousing as few goods as possible. Often the complexity becomes opacity, generating dangerous situations of upstream monopoly. One of the main causes of this behavior is the constant effort to reduce supply costs, which has pushed many medical manufacturers offshore to take advantage of low-cost labor and tax incentives. Over time, this has led all Western countries to become overly dependent on offshore manufacturing for many essential health care items.

From a pure market perspective, globalized economic geography has guaranteed significant savings and other benefits to producers and consumers at the cost of decreased national autonomy and greater vulnerability to exogenous and geopolitical shocks, with a single broken link capable of interrupting the entire chain. In the end, the pandemic has dramatically shown that all countries lack adequate mechanisms to coordinate their domestic and international activities on supply chain links, vaccine development, and disease surveillance. As a consequence, they are exposed to the risks of dependence on fragile and overextended global supply chains for essential medicines and critical supplies. This implies that they cannot afford to develop and implement domestic preparedness policies and initiatives in isolation, without considering international factors that will help determine their success.

A good example of the implications of value chain vulnerability is the shutdowns that occurred in China in January and February 2020. In a few weeks, decreased production in China and increased worldwide demand for medicines, PPEs, and other critical supplies led to shortages in the United States, European Union, and elsewhere, reducing pertinent countries' capacity to respond to the spread of the disease within their borders, while heightening geopolitical tensions and undermining international coordination. Given the limited knowledge of the industry structure, like in a domino effect scenario, policymakers' misplaced reaction, with frantic attempts to look for alternative production sources, further increased the disruption of entire supply system.

COVID-19 has thus highlighted how both crises and political responses can disrupt supply chains and exacerbate shortages of crucial goods. It has further shown how ineffective can be relying on purely national policies to counter the production problems caused by a pandemic, without a multilateral mechanism to ensure global consistency and equitable distribution.

The absence of a widely supported multilateral mechanism for the equitable distribution of vaccines, treatments, and medical devices during the COVID-19 pandemic is the main challenge to achieve a coordinated global response. Without a global strategy, countries are forced to prioritize their own citizens, which can lead to unequal distribution of resources and an extended crisis. A global plan would ensure that resources are distributed equitably and that the crisis is brought to an end sooner.

However, despite all these problems and following the initial shock, firms and health care organizations have learned about strategies to mitigate disruptions during major emergencies without incurring exorbitant costs. Companies have run risk assessments and implemented business continuity plans. In this way, they have critically reviewed the structure of their supply chain, trying to understand the main bottlenecks and weaknesses. Additionally, many have diversified their product portfolio to respond to changing demands, creating new products based on their existing resources. For example, some apparel manufacturers began to produce PPEs and some distillers began to manufacture hand sanitizers. Finally, several companies have emphasized the need to bring production facilities back onshore or engage in near-shoring. Healthcare institutions had to evaluate which approaches could alleviate supply chain interruptions during significant crises without bearing excessive expenses. For example, holding extensive amounts of safety stocks for a wide variety of health care items and/or reshoring production of a wide array of items would improve resiliency. However, this strategy could be extremely costly and ultimately impractical in most cases. Given that emergency preparedness is an imperative for public health, solutions cannot come solely from the private sector. Federal, state, and local governments need to assess what policy prescriptions they should enact in the wake of this experience as well.

Following Mahmoodi et al. (2021), a series of strategies can be suggested to increase supply chain resilience and mitigate supply chain interruptions.

- **Build redundancy in the supply chain and change compensation programs accordingly**. One way to increase resiliency is by adding redundancy in the structure of the supply process, such as by carrying an extra inventory of essential health care items, holding excess manufacturing capacity for producing critical items, or contracting with backup suppliers. At the same time, compensation programs for managers should be introduced as redundancy is against cost-effectiveness.
- **Utilize technology solutions**. Employing a range of digital and analytic solutions can improve the integrity and resiliency of the supply process at a reasonable cost. For example, the use of cloud-based supply platforms and real-time network visibil-

ity solutions (such as 5G technology and block-chain) can help integrate suppliers and data across the entire supply chain.

- **Gain greater upstream visibility by mapping and monitoring the supply network**. While mapping out the supply network for essential medical products can be time-consuming, it does help organizations to better anticipate how disruptive events will impact their supply chains. It's essential to consistently keep track of the supply chain's present condition by maintaining open dialogue with suppliers. Primary suppliers should also possess thorough risk management strategies in place, such as overseeing their own suppliers and having backup sources for their most vulnerable suppliers.

- **Diversify the supply base**. Some have called for the health care industry to 're-shore' production from Asia (in particular, China). But this approach is no panacea, given the industry's desire to serve the huge Chinese market. Furthermore, reducing dependence on China, which is now the sole source for thousands of items, will take substantial time and investment.

  Health care organizations can improve supply chain resilience by dual sourcing raw materials, whenever possible, and sourcing more of the manufacture of critical medical products such as face masks and shields, respirators, isolation gowns, commonly used medications, and gloves. It is important to realize that diversifying the supply chain is as challenging as it is time-consuming and requires significant investment. Furthermore, healthcare providers with fixed reimbursement income (as opposed to those who are paid on a 'fee for service' basis) may have difficulty absorbing price increases for products that are no longer made in low-cost regions.

- **Improve visibility/transparency of the Strategic National Stockpile (SNS)**. All stakeholders in the system should have a clear understanding of the national repository of antibiotics, vaccines, chemical antidotes, and antitoxins, as well as other critical medical supplies. This was not the case when the pandemic started and gave rise to autonomous ineffective behavior by local health authorities. Healthcare systems and manufacturers should also mutually share their inventory details to prevent backorders and implement contingency strategies.

7.5. An uncertain future?

After many years of sluggish growth, it seemed that inflation had entirely vanished from the world economy. The reasons for its apparent demise were not clear, and several interpretations were provided, including better control of the money supply on the part of central banks, general consumer sobriety, increased import competition, and higher market concentration (see, for example, Heise et al., 2022). Since a moderate degree of inflation was deemed desirable both for theoretical considerations and to avoid the specter of deflation, central banks tended to pursue an inflation target from below, trying to expand the money supply up to a point where inflation was moderately high (roughly 2% in Europe and the US) and yet not growing or falling. This equilibrium point was rationalized as corresponding to a 'natural' rate of unemployment and was believed to be not only a vanishing point for inflation, but also a point where any trade-off between inflation and unemployment would disappear (a vertical Phillips curve). A popular explanation was also that 'inflationary expectations had been broken' or that any passthrough between prices and wages had been eliminated by cutting its vicious circle.

Although no consensus seemed to exist on the reasons for the inflation demise, the possible relationship of this disappearance with the hypothesis of secular stagnation dominated the concerns of many economists and policymakers. The economic environment after the great financial crisis did not seem favorable for taking long-term risks, and investments trended downward everywhere, despite spectacular growth of private profits and multinational companies. Parallel expansion of the money supply did encourage risk taking in financial markets, with quick turn around and short-term bubbles, but this tended to aggravate the so-called 'short-termism' problem generated by the operators' desire to make money and, at the same time, stay liquid to profit from the often frenetic activities in the financial markets.

A more general explanation for the disappearance of inflation was globalization, both as a phenomenon of market integration and as a beneficial and progressive taking over of animal spirits, markets, and endogenous growth over any government attempt to orchestrate the future. After COVID-19, the war in Ukraine, and the surge of inflation and material disarrays that have occurred in the most recent times, we tend to look back at these images of vanishing inflation and boundless global growth more like a temporary illusion and a dangerous dream than anything that had any substantive root in practical reality. The theory of endogenous growth, i.e., growth that is self-supporting, originally arose from the need to explain the lack of convergence of the economic development of different communities and the success at first sight inexplicable of some economies compared to others. Combined with globalization, endogenous growth gave rise to the additional idea that economic development, pursued by individual agents in the name of selfish goals, creates positive externalities. These externalities beneficially influence factor productivity. They include research, education, and individual economic activities that all seem to be able to deploy two key characteristics: they are undertaken for selfish reasons and provide beneficial effects that can be shared by other economic agents.

However, the problem with treating public goods as externalities is that it can lead to the expectation that they will be provided as byproducts of free competition and growth, rather than as investments in public capital.[9] The role of governments and

---

[9] The 'minimal government' or 'minimal state' is a notion found within a particular variant of the limited-government variety of libertarianism. It was introduced by Robert Nozick (1974). Although Nozick

other public institutions both at the national and global levels is thus diminished and tends to vanish. In the long run, the very idea of governance may be undermined by the dangerous illusion that economic and political processes are capable of self-control and that conflicts are only minor disturbances in the fabric of a naturally benevolent global economy.

The endogeneity of growth, on the other hand, also entails the problem of its sustainability. In the same way in which positive external effects can promote growth beyond the limits of pure individual interests, so can negative external effects undermine the private bases of growth, depressing factor productivity and agents' motivation for productive work.

The persistent inequalities and the lack of convergence among the different economies can therefore also be explained by endogenously insufficient development, chronic underemployment, and growing damages to human and natural capital. This can be perpetuated by aggravation factors that prevent the economy from supporting a level of growth compatible with a better quality of life. These factors also tend to render economies more fragile, with pervasive uncertainties and breakdown risks aggravated by their growing interdependence.

In summary, the presence of negative endogenous factors can transform a condition of insufficient economic development into a trap, aggravated by increasing inequality and lack of governance, from which it may be difficult to escape. It is perhaps this trap that characterizes the present crisis of the world model of growth, with polarization and emerging global conflicts, and a tendency to secular stagnation. Some economists have elaborated this concept to explain the slowdown of development and the lack of transformative innovation in many countries and in the world as a whole.

As the still ongoing experiences of COVID-19 and the Ukraine war seem to suggest, the global development trap may be essentially due to the increasing lack of global governance, investment, and public goods that have characterized the globalization phenomenon. While many waves of globalization have crossed the world in the past, what was new with globalization as we have known it in more recent years is that it gave mainstream economics a chance to become the paradigm for a universal practice of free trade, unrestrained competition, and market liberalization. This process went hand in hand with several attempts of economic integration and common trade areas, with Europe being the most prominent example of the application of market-regulated liberalization and virtuous governance rules of diverse economic realities and communities.

As argued by Scandizzo (2018) and Paganetto and Scandizzo (2020), the theory of 'minimal government', combined with a new wave of trust in the power of the private sector, contributed to the abandonment of strategic planning and the demise of public

criticized individualist anarchism, he did hold that the minimal state was the form of government that was morally justifiable.

investment. In the case of health, it contributed to creating the myth of a benevolent evolution of the human condition, with minimal negative feedback from the damages to the environment and largely positive trends in human life spans and infectious diseases. In fact, the main danger from these developments was perceived as an excess of zeal in the medical attention financed by the welfare state and in the dubious utility of ever-increasing financing needs of public health expenditures.

These characteristics contributed to generating several unsustainable patterns of globalization that, by recklessly pursuing an ideal of market absolute freedom, ended up undermining the very principles on which this pursuit was predicated. According to Scandizzo (2018) the unsustainable central feature of these patterns took many forms:

- First, by aggressively asserting the superiority of the free market system in securing sustained growth, it opened itself to dramatic disproof and disappointments against the hard facts of the recurrent economic crises and declining well-being. Far from generating enthusiasm and innovation, it appeared to worsen social injustice and further instability and income inequality;
- Second, by invoking a supposed convergence of economic efficiency with social justice, it let narrow economic targets dominate government policies and concerns, thus exposing itself to major market failures led by a combination of financial bubbles and monetary and fiscal repression;
- Third, it favored an unorderly reallocation of the industrial base. Together with the illusion of the power of austerity and minimal government, this led to a massive restructuring of the value chains, with specialization shifting from the production of final goods to intermediates. Deindustrialization in developed countries followed suit with a broad disenfranchisement of the working middle class;
- Fourth, the 'new' labor policies, austerity, and minimal government, combined with the inability to address economic crises, were instrumental in determining an environment dominated by job insecurity, youth unemployment, and fraught with economic and political risks;
- Fifth, in the process, local social capital from civil society and intermediate bodies were almost dissolved by new institutions, cultural trends, and economic and social policies, without being replaced by any equivalent 'global' infrastructure;
- Sixth, the relationships between health, nutrition, and the environment were compromised by a mixture of unleashed market forces, urban expansion, and terminal threats to wildlife and biodiversity. The medical profession underwent its own dualistic challenges, between widespread humanitarian and personalized precision medicine, a contradiction that was exacerbated by the ever more frequent epidemic crises and the losses of human and natural capital due to pollution and natural disasters.

The upshot of all these features of an increasingly alienating environment is a general process of disillusionment and breakage of trust, which goes to the very root of civil society, and is threatening to destroy the fabric of social capital. The disappearance of intermediate bodies, wiped away by the disintermediation operated through the globalized media, has been especially effective in depriving local communities of their network of mutual obligations based on reciprocity and trust. This process started a long time ago, with the crisis of the traditional political parties, and the demise of locally based cultural and religious groups. The surge of the Internet and social media, with their instant promise of virtual community building, have also been instrumental in accelerating the demise of local communities as we knew them.

In sum, as the COVID-19 crisis has dramatically shown, in spite of globalization, global public goods are strongly under-provided. At the same time, local public goods are also lacking and the capacity to provide them on the part of local communities and governments appears severely depleted.

How do we escape the trap of public policies thorn between largely ungoverned globalization and increasingly ineffective local governments? While no magic bullet is likely to exist, rebuilding confidence in the future appears to be one of the keys. The nature of this challenge is indeed global and a return to nationalism, however responsible, does not seem an adequate answer. Rather, real multilateral cooperation and institution building appears to be necessary, and to an extent that is both practically effective, but also commands an image of vigor and reach, capable to overturn the current expectations of the disgruntled and disillusioned citizens. Rather than turning away from it, the world needs an extension of multilateralism from the narrow financial and economic space where it has been confined by the Bretton Woods partial vision, to a broader array of global institutions. These should receive a clear mandate to act on all economic fronts, including the fiscal, financial, and global investment areas, and a genuine commitment to improving the global practice of societal values to yield inclusiveness, security, innovation, and trust.

7.5.1 Consumption and savings

According to Gardner et al. (2012), on average it takes 66 days for a new behavior to become automatic, with simpler behavioral changes that become more permanent than complex ones. The Italian lockdown, which began on 8 March and ended on 3 June, certainly has the features needed to produce some of these permanent changes. Historically, every big crisis brought about changes, part of which had a temporary character, while others remained forever. For example, women's participation in assembly lines during the Second World War due to men's involvement in wartime activities accelerated the process of women entering the labor market, which would otherwise have taken much longer. The acceleration affected both the production and the social, political, and legal aspects, which quickly adjusted to the need for the transition itself.

Similarly, as the world shifted through the different phases of the COVID-19 crisis, it became clear that the various blockades had profoundly impacted people's ways of life.

Contagions, self-isolation, and the resulting economic uncertainty have changed how consumers behave, in some cases, for years to come. New consumer behavior encompasses all areas of life, from how people work to how they shop or enjoy themselves. These rapid changes have important implications for manufacturers and resellers of consumer goods. Many long-term changes in consumer behavior are still forming, allowing companies to help shape future normality. Behavioral changes are not linear, and their persistence will depend on how much people are satisfied with new experiences.

A consumer is a person who identifies a need or a desire, makes a purchase, and enters the product in the consumption process. Satisfaction, better defined in economics through the concept of a utility function, depends on the consumption of various goods and services, each of which is influenced by different internal and external factors that form the consumer's behavior. Consumer behavior is the result of a decision-making process based on the constant search, purchase, use, and evaluation of products and services, all so that individuals' usefulness or satisfaction by purchasing a good or service is as large as possible, given the constraints of the resources available. The analysis of consumer behavior is a multidisciplinary problem that involves many psychological and sociological aspects in addition to economics. Consumers, not all alike, do not have the same perception of the COVID-19 pandemic or any other crisis. The onset of the pandemic has undeniably ushered in shifts in consumer patterns, heavily influenced by their tangible and perceived sense of risk; with the same level of risk causing different personal attitudes and behaviors, depending on individual characteristics.

The uncertainty associated with the future level of income or its actual reduction due to the economic recession from COVID-19 can lead many people to reduce consumption, increase savings and delay investment and purchases (e.g., cars, appliances, travel, etc.). This could also enhance buyers' attention to the product price and quality regarding reliability and durability. In this respect, the idea of sustainability has the potential to play a more critical role in identifying and setting benchmarks with which consumers will approach the goods and services market. The same uncertainty and the healthcare nature of the COVID-19 crisis also can increase awareness and preference for purchasing goods and services that affect health and safety in everyday life. More attention to content, nutritional details, composition, and provenance are just some attributes of products and services to which consumers could start to give higher consideration.

Mobility restrictions have also encouraged the exploration of local environments, easily accessible near individual residences. To maintain social dislocation, people were more likely to frequent less crowded places, hence smaller businesses. This rediscovery can also alter consumption preferences that in recent decades had shifted toward large global distribution networks and chains.

Moderation in consumption and the climate of uncertainty affected savings since the early months of the pandemic. The ECB (2020) noted that saving attitude reached unprecedented levels in response to COVID-19. The decline in economic activity was partly due to a sharp drop in consumer spending. This was evident in the increase in household bank deposits. However, this increase in savings was not voluntary, but rather a result of the lockdown measures that restricted people's ability to spend. At the same time, according to the same report, the increase in uncertainty also stimulated the propensity to precautionary saving, a type of counter-cyclical saving, which follows the logic of insurance against unfavorable future events and is the greater, the higher the uncertainty people face. The major source of uncertainty for most people was the risk of unemployment. Although government benefits helped to keep unemployment low, many workers and their families still saved money out of fear of losing their jobs or having their incomes reduced. This was especially true in European countries, where the welfare system is more generous.

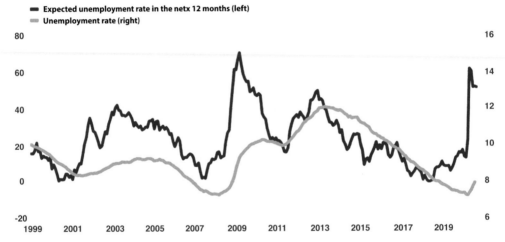

Figure 7.11 Unemployment rate versus unemployment risk perception in the Euro Area. *Source*: European Central Bank calculations based on data DG-ECFIN and Eurostat (ECB, 2020).

Fig. 7.11 shows that regardless of the lower rise in the unemployment rate, by the beginning of 2020 households' concern about potential unemployment had reached levels comparable to those of the 2008–10 financial crisis. In calculating the contribution of the unemployment uncertainty to the saving rate, the ECB (2020) estimated that only slightly less than the 1% of the 15-point change recorded in the second quarter of 2020 (compared to the last quarter of 2019) could be attributed to precautionary reasons, while the remaining 14 points could refer to unintended savings. This gave an impulse to implement fiscal stimulus policies, indicating that liquidity, such as household transfers, would not be blocked on consumers' bank deposits but could circulate in the economy, provided that new lockdowns did not stop buyers. It is hard to predict how long this new equilibrium will last, as it implies knowing the feelings and needs that consumers will have in the post-COVID-19 era. It is true, however, that some spheres of personal life have already undergone radical changes for which it is difficult to imagine a return to the past.

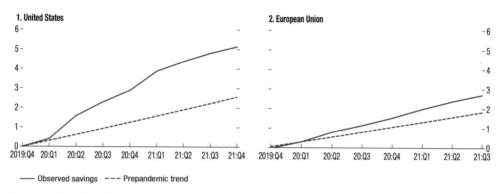

Figure 7.12 Trends in excess gross household savings in Advanced Economies (Trillions national currency, cumulative sum since the fourth quarter of 2019). *Source*: IMF (2022).

According to a study of the International Monetary Fund (IMF, 2022), the increase in excess savings poses important challenges to all governments. As shown in Fig. 7.12, excess savings (above pre-pandemic trends) increased by approximately $2.5 trillion in the United States and $1 trillion in the European Union during 2020-21. This followed from the changes occurred in the propensity to save, as shown in Fig. 7.13 for selected advanced countries. Interestingly, in Italy and the UK, the COVID-19 pandemic has reversed a downward trend in saving propensity, while it has made it more pronounced in the US and Germany. This heterogeneity follows also from different drivers. In the United States, a major role was played by direct government transfers to households in 2020 and early in 2021. In the EU, consumption restraint and excess saving were more protracted up to 2021. In Mexico, where government support was limited, the increase in household savings was driven by larger consumption cuts and personal transfers and remittances from abroad. Managing these excess savings will not be easy, given the potential economic effects they could generate. For example, the International Monetary Fund (IMF) (2022) observed that: 'savings could now help buffer the effect of the higher inflation and lower growth but, in some cases, could add to inflationary pressures if spent quickly. Another challenge relates to the time-bound nature of poverty support programs that can also meet long-term structural needs – when such support ends, poverty rates could rise. This is a risk given the high level of uncertainty and rise in energy and food prices that would disproportionally affect the most vulnerable households'.

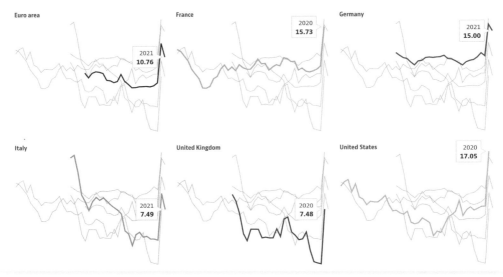

Figure 7.13 Household total savings as % of household disposable income (1980-2021) - Selected countries. *Source*: OECD (2022a), Household savings (indicator). doi: 10.1787/cfc6f499-en (Accessed on 16 July 2022). Data available at the following web address: https://data.oecd.org/hha/household-savings.htm.

7.5.2 The labor market

The consensus of studies on the effects of the pandemic on labor markets is that although they were broadly negative, they were successfully attenuated and eventually even reversed by both relief measures and fiscal policy interventions enacted by virtually all governments (Barišić and Kovač, 2022). The main problem in describing and interpreting these effects concerns their unprecedented nature and their possible persistence over time through some form of hysteresis. Short-term effects on labor were unprecedented since they were the result of a unique combination of demand and supply shocks, and appeared to markedly diverge from those experimented in the previous recessions in both depth and distribution across countries, sectors, and workers' skill class. Concentration of negative impacts in the service sectors, including transportation, and the live entertainment, culture and tourism value chain (Kumar et al., 2021) brought about also the issue of long-term effects.

The hypothesis that a prolonged recession might cause persistent depressive consequences in labor markets was originally raised by Blanchard and Summers (1986) after the economic crisis in the 1970s. As for the time horizon, the complex interrelationship between health crises and fiscal measures to counter them is a challenging issue that has received little attention. The short-run effects of fiscal impulses have been studied intensely while there is only very little evidence on long-run effects. Some evidence of the effects arising from crises (Fall and Fournier, 2015; Fatás and Summers, 2016) suggests that failure to intervene with fiscal and monetary measures on a sufficiently large scale may have negative long-term impacts (Bom and Ligthart, 2014). Both after the oil crisis of the 1970s and the more recent great financial crisis of 2009, unemployment rates failed to return to prior levels and remained high even though the rest of the economy appeared to recover its stance before the crisis.

Although the long-term impacts can only be gauged as potential output losses, history shows that previous health crises often led to sustained drops in GDP per capita and rises in unemployment following such events. According to one IMF study of the more recent epidemics – SARS, H1N1, MERS, Ebola, and Zika for example – '...Job losses cumulate over time, and the employment effects suggest a delayed and prolonged deterioration in the labor market. After five years, the increase of unemployment levels is at about 7 percentage points relative to the pre-pandemic level, a value similar to the estimates of Coibion et al. (2020) for the effect of COVID-19 in the United States' (Emmerling et al., 2021). In a recent study, Verick et al. (2022) compare the short-term effects on the labor market of the COVID-19 pandemic with those observed in the case of the great financial crisis of 2009. In general, the comparison shows that COVID-19 has affected more profoundly and widely the economies of all countries, with negative growth rates in 2020 for both developed and developing countries, and deeper negative impacts in the labor markets.

In general, the pandemic appears to have affected unequally the different sectors, with transport, accommodation, and tourism on the front lines of joblessness and unemployment. During the heights of the COVID-19 crisis, the lockdown and other containment measures have also severely restricted the ability of the jobless to search and to accept employment. Unequal household care for children and the elderly has also caused women to be penalized, with lower participation in the labor market. Spurred by the difficulties created by lockdowns and other restrictions, inactivity levels have been especially high, compared with other economic crises. Against an increase in world unemployment of 33 million people, global employment losses were 114 million, while inactivity rose of 81 million, compared with an increase in unemployment of 33 million (International Labour Office, 2021).

These short-term effects point to possible long-term impacts that also tend to be different from typical employment problems caused by recessions. Based on data from 2022 and 2023, there appears to be a strong resurgence in employment worldwide. However, lingering disparities in inactivity rates across different sectors and genders might still contribute to long-term effects on employment and unemployment levels. Other, more subtle effects are more difficult to fathom. The experience of the pandemic has produced a form of revolution for office workers, by scope of a simple adaptation to the temporary restrictions imposed by lockdowns or other measures of social distance. Not only have digital technologies become ubiquitous in all working arrangements, but their growing use and familiarity have also multiplied the ways in which workers interact within and outside of their relational environment. The environment within which the average worker finds herself immersed in a new multiplicity of tasks and contacts has already profoundly changed, but its effects on the organization of society, which can only be guessed at this time, will come more slowly in the next years.

7.5.3 The redesign of workspace after the pandemic

The unevenness of the short-term impact on the labor markets had also multiple and contrasting effects on the general economic environment, which somewhat abruptly changed during the more recent phases of the pandemic from a deflationary to an inflationary one. In this respect, due to its more flexible characteristics, the US market is an interesting example of more drastic adjustments over time. Here, job losses were at first more pronounced and the recovery more vigorous than in other developed countries in Europe and elsewhere, but market tightness eventually prevailed, as shown by demand side indicators, such as the job vacancy rate, dominating supply–side indicators, such as the prime-age employment-to-population ratio (Domash and Summers, 2022).

The extensive switch to work from home has several dimensions of present and potential changes. First, it has tended to make the divide between 'good jobs' and 'bad jobs' more pronounced, exacerbating the polarization in increasingly segmented labor markets. Thus, jobs requiring lower skills were penalized by the impact of lockdowns and social distancing. Sectors where lower skills prevailed, like personal care, restaurants, transportation, and the entire tourist value chain, suffered from lack of adaptation to smart working, as compared to other service sectors, such as consulting, designing, software, and engineering services. This unequal impact added and is likely to worsen the fragmentation of labor markets and their tendencies to create internal divides as new and increasing sources of exclusion and inequality.

Second, the forcible shift to smart working favored specific classes of workers who were already used to working from home. For these workers, both job satisfaction and productivity may have increased. These effects depend on several factors: worker productivity from flexible hours, better match between workers' preferences and work typology and mode, including hours of the day, intensity of participation, more fluid relation with work hierarchies and command and control structures. In general, it seems fair to conjecture that, confronted with the task of motivating workers under lower supervision, organizational structures within both private and public institutions tended to steer from hierarchies to poliarchies, with much higher weight being given to teamwork and project-based interaction. Therefore, higher-skilled workers within the sectors less affected by the pandemic supply and demand shock saw their status change along several positive dimensions. These dimensions were different across work typologies and skill levels, but they tended all to include a lower weight of institutional hierarchies, and more options to self-organize, exercise creativity, exploit opportunities, and develop connections, beyond the traditional confine of office activities.

Third, the COVID-19 pandemic can be considered a massive natural experiment (Kramer and Kramer, 2020), where workers are redistributed between traditional and smart working modes across a wide range of occupational groups, willingness to adopt digital technologies, skills, and productivity. This experiment is still ongoing. Its outcome is uncertain but what we see so far points to a substantial redistribution of job characteristics, workers' own and society perception of occupational status, as well as job satisfaction across employment patterns and labor markets. Other suggested trends are new employers' and institutional choices on production processes, work organization and space planning. Skilled workers who can effectively work from home will likely receive higher salaries and enjoy a better alignment between their tasks and their preferred modes and types of engagement. They will also enjoy higher flexibility in terms of hours of work, location, and transport. At the same time, the experience of the past two years shows that under the joint push of smart working and hyper-connectivity, the traditional demarcation between the workplace and home may tend to fade, especially for executives and high skilled workers. Because demand on their time may be exercised continuously, it will be likely to expand and become more pressing, with more frequent meetings online and continuous hyper-connectivity. This is also likely to make over-commitment and job fatigue more common. At the same time, higher flexibility and lower constraints in the physical environment will empower many workers to redesign their own material and virtual workspace. While the impact of this new flexibility on work-time and space is still largely unpredictable, the tendency to 'build back better' residential and business places within and outside urban areas can be already observed.

More generally, it seems likely that the new normalcy will be a mixture of old and new arrangements and that this mixture will reflect both challenges and new solutions to the traditional work problems, the relationships between employees and employers, work time and working hours. Because some fundamental parameters of work organization have changed, the work-life balance (WLB), defined by the complex relationship between people's work and private and social life, and which still largely reflects the industrial revolution, is also likely to drastically change.[10] Although the more radical changes appear to directly involve only white-collar office workers, the relationship of individuals to work is bound to undergo substantial changes for everyone. At the same time, as shown by a recent paper on this subject (Scandizzo and Knudsen, 2022), the option to revert to pre-pandemic working patterns will tend to prevail, unless the intensity and/or the duration of the health emergency forces economic agents to invest

[10] WLB is a key concept describing the equilibrium (or the lack of) between labor, leisure and personal and social life. Its formulation can be traced to the Factory Act of 1832 in the United Kingdom, which laid down laws to define maximum working hours specifically for women and children in the manufacturing industry. Following this, in the United States, the Fair Labor Standards Act of 1938 established the work week at 44 hours. In the 1980s, the feminist movement revived interest in this subject, emphasizing the need for more flexible working hours for women, along with maternity leave, enabling them to balance their careers and personal lives effectively.

in new models of organization of public and private spaces, thus committing resources that cannot be easily recovered in short times.

As argued in Addis et al. (2021), to understand the forces that will be unleashed in the reorganization of the economic space as a result of COVID-19, three characteristics of recent developments appear to be particularly relevant. First of all, at this moment in history, the economy is affected by a rising phenomenon of increasing returns to scale from companies' size and imperfect competition. This concerns communications giants such as Apple and Google, but it also depends on the fact that logistics and the Internet have multiplied exchanges between regions and sub-regions progressively in-dependently of their geographical location. In the past, we have gradually moved from the nuclear city to the city-region, with geographical areas also of significant size that develop and live around an urban 'core'. In this center of activity, valuable properties have been concentrated, and with them the services with high value added and the main public goods. Polarization of the center-periphery also results in costs from pollution, social exclusion, and crime. The expansion of urban areas has led to the development of a new 'territorial capital', which is a source of competitiveness for productive settle-ments and residents, including cultural and environmental heritage. The improvement in logistics has created the 'internal market effect', leading to growth in larger, better-equipped areas that benefit from greater accessibility and availability of services (known as 'agglomeration economies'). Before the pandemic these phenomena constituted the backbone of the new economic geography, of the urban macro-regions and of the 'core' citizens and led many scholars to hypothesize forms of further concentration, such as, for example, an exponential growth of urban settlements according to the model of the 'vertical city'.

Finally, and in partial contradiction to the phenomenon of spatial concentration of economic activities, even before the pandemic threatened to upset productive *modus operandi* and lifestyles, digitization had begun to allow companies to dematerialize their presence, separating their activities from their location and allowing for an increasingly less binding management of space. In a certain sense this phenomenon came from afar, since the 90s, the post-Fordist revolution had contributed to determine a separation of the functions of companies in space, allowing a distinct localization of management functions and services with high added value compared to the less valuable ones, through the practices of relocation (e.g., landing on foreign beaches through the so called 'off-shoring'), sub-contracting and out-sourcing. The emerging digital economy amplifies this trend but also shifts its dynamics. It encourages 're-shoring,' or the return of com-panies to their home countries, while simultaneously promoting new types of spatial fragmentation. This is due to innovations that aren't tied to specific locations, such as remote work, robotics, and artificial intelligence.

The ongoing patterns of development come with significant societal implications due to the degradation of the environment and the intensifying spatial disparities. This

includes the overpopulation and overcrowding of central urban regions, and the simultaneous depopulation and deterioration of peripheral or rural areas. Far from discouraging
urban settlements in an attempt to solve the problems of pollution and social degradation, these costs have paradoxically attracted additional technological resources to the
most developed areas. This has caused a further concentration of spatial settlement
around a few economic activities, thus resulting in heterogeneous patterns of territorial specialization. However, these models all have in common the combination of
widespread residential and specialized production areas, often including high-density
city settlements, dedicated to the production and consumption of services. These settlements tend towards a development of the territory around a plurality of centers, with
widespread peripheries and a connective and interstitial model that presents further dangers of environmental and social degradation. Urban and peri-urban centers have thus
grown despite the dis-economies of congestion, due to the digital dematerialization
of companies, the separation of production, distribution, and consumption activities,
and the integration between the production chains of high value-added services in the
fields of finance and administration, training, public administration, tourism, and leisure.
Finally, the fall in public investment due to the increasing financial difficulties of governments has also been accompanied by a greater concentration and apparently higher
productivity of infrastructure and other public interventions in central areas.

COVID-19 has acted as an asymmetric supply shock, selectively determining, first
through the lockdown and then through restrictive measures, a suspension of ordinary
activities that has mainly affected urban regions, with significant repercussions on city
centers and core activities (services of all kinds). Lockdowns have stimulated their consequences continue to generate forms of consumption substitution based on digitalization,
dramatically accelerating some processes, of which they have increased incidence and
persistence. These processes include, in addition to smart working, the logistics of home
delivery and a general tendency to operate digitally from virtually defined individual
spaces, and they are associated with new products that match the changed conditions of
distance working over space and time. These processes and products in large part were
already present before the pandemic but have made an unpredictable leap during the
lockdowns and the social restrictions, whereas before they had advanced with caution
due to the pre-existing infrastructure and the inertia of the behaviors of individuals and
institutions. The tendency to territorial poly-centrism, already in place before the pandemic, but according to forms of strong polarization between centers and peripheries,
appears to be strengthened. However, it follows an organizational model transformed by
the unprecedented enhancement of residential homes, as places of family self-sufficiency
that internalize communication and information services for work and leisure, making
obsolete many current structures of organization of work and consumption within and
outside the city walls. As a consequence, housing is hit by a wave of research, reconstruction and multi functional use reminiscent of similar historical changes, with a foreseeable

boom of demand for self-sufficiency of residences in suburban and even rural areas and the ever-present tendency to transform public services into private services of a society that still aims at mass opulence.

In summary, within an already evolving landscape of territorial development, the pandemic might have either introduced or hastened transformative innovations. These changes have the potential to challenge prior trends and assumptions, paving the way for both new risks and opportunities. On the side of territorial capital, these innovations concern the complementarity between private and public activities and between cultural and environmental heritage, both affected by the emerging changes in the organization of work, production, and lifestyles. These include the transition from traditional mass tourism to newer, more sustainable or regenerative tourism models, as well as the geographic shifts made possible by the rise of telecommuting and changing regional settlement patterns. If, as it is likely, the separation between places of residence and workplaces and the reorganization of production chains are strengthened within a movement towards regional poly-centrism, territorial capital will also be redefined in terms of larger spaces as well as greater geographical spread of production and consumption activities. This will entail a boost to the enhancement of the territorial heritage also through new forms of tourism of longer persistence and greater cultural and environmental awareness.

In this context of risks and opportunities generated by autonomous transformations triggered by the pandemic, the regeneration of spaces that will follow the end of COVID-19 and the European resources that will be available present the challenge of redefining the policy of public investments in relation to the territory. This challenge has particular relevance for labor markets, job polarization, and the supply chains most affected by the economic consequences of the pandemic, such as those of tourism in cities of art and entertainment and culture services. It suggests the importance of investment in shorter supply chains and local labor markets, as well as the maintenance and enhancement of the 'minor' heritage, dispersed on the territory like a historical vegetation largely unknown. It also suggests the need to provide, among other things, for the construction of the context and the adoption by the local communities of a radically new model of planning, maintenance, design, and management of social and human capital.

7.5.4 Economic and social unrest

In a world dominated by inequalities, social conflict is an intrinsic problem of economic crises that exacerbates the sense of injustice from exclusion, the lack of solidarity, and the overall inadequacy of globalized institutions and national government measures. Social unrest was already brewing under the populist reaction to globalization before the pandemic occurred, but the manifestation of the new and more dramatic inequalities engendered by the joint health and economic crisis was relatively slow to come.

From this perspective, the scale of government supports has been responsible to attenuate or exacerbate the uneven effects of the COVID-19 pandemic on households. Since February 2020, all governments have implemented programs and transfers – such as employment subsidies, tax relief, and cash transfers – to enable people to live with containment measures and have prevented a deeper recession (see also Section 7.3). The degree of government support, however, has varied greatly across countries, with distinct effects on household incomes (see Fig. 7.14). Advanced economies, and a few emerging markets, provided the largest support. In some countries, disposable income grew, mainly reflecting governments' direct support to households that more than compensated for the fall in market income (Canada, United States). In other countries, government support was provided indirectly, through job-retention schemes, thereby reducing or preventing a fall in wage incomes. In some cases, it helped keep household incomes broadly stable (France, Germany, United Kingdom), whereas in others it limited their fall (Italy, Spain). Government measures had a limited effect on cushioning the decline in people's income in low-income developing economies, amid large informal sectors and low social protection coverage.

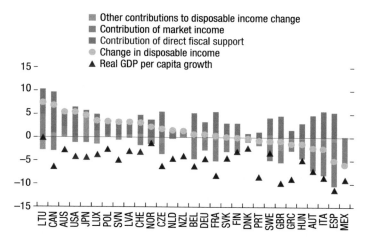

Figure 7.14 Changes in Household Income, 2020 (Percent of 2019 disposable income per capita). *Source*: IMF (2022). Note: Gross disposable household income is reported. Market income includes gross operating surplus, mixed income, compensation of employees, and net property income. Direct fiscal support includes current taxes on income and wealth, social benefits, and social contributions, and does not include support channeled to firms that indirectly supported households such as job retention schemes. Other includes personal current transfers. All quantities are per capita and converted into 2019 prices using the Consumer Price Index. Data labels use International Organization for Standardization (ISO) country codes.

Moreover, the pandemic brought about a role of unprecedented authority for the government and the health institutions, which was met, at least at the beginning of the emergency, with an implicit delegation of power and concession of trust on the part of all segments of society. The reason for this widely shared attitude was in part the conviction that no other institution or private party could possibly substitute the action of the public sector in such a contingency. In part, it was also the consequence of an emergency that clearly required centralized and vigorous governance. Because of these conditions, the first phase of the pandemic was characterized by a general compliance to the governments' many measures, even though the consensus on their timeliness and rationality was soon destined to decline almost everywhere. Nevertheless social conflicts and unrest were not a problem in the first phase of the health emergency and remained somewhat compressed also under the subsequent waves of the infection.

However, as the pandemic evolved and it became clear that it would be a long emergency with uncertain outcomes and duration, the effectiveness of institutional structures and government actions was progressively called into question. In part, this was the consequence of the lack of knowledge and reliable information on the multiple features of the unprecedented pandemic. In part, it was due to the lack of planning and management capacity that appeared to be at the root of the complacent and ineffective bureaucracy of the welfare state. In most developed countries, successive attempts to control the spread of the disease appeared improvised, uncoordinated, and hopelessly too late, in spite of the often frantic usage of executive decrees to launch the various measures. Expanding hospital capacity and flexibility also proved to be a challenge, together with a seemingly too slow vaccination campaign.

7.6. The appearance of new inequalities: what is changing?

All crises may have a devastating impact on the population and, unfortunately, especially on the people who face precarious living conditions. From this point of view, a pandemic has the potential to exacerbate some inequalities in a fairly direct way, affecting those who are most disadvantaged economically, socially, or indirectly, acting on relationships between individuals and on the dynamics of increasing human capital. The notion that the COVID-19 pandemic might have left a lasting imprint on the terrain of social inequalities undoubtedly requires the adoption of policies tailored to meet these emerging challenges.

The stressful conditions caused by the pandemic and the related emergency measures have generated new forms of inequality based on the underlying fragility and vulnerability of different agents, such as the elderly or the chronically sick, or even those that simply reside in places where health care supply is inadequate, no matter what is their income status. Among other causes, this phenomenon has occurred because of significant changes in many people's lifestyles and risky behaviors emerging from containment measures that have been implemented at different times since the outbreak of the pandemic. For all these changes, the pandemic contributed to the existing inequalities and at the same time created new forms of social exclusion and discrimination. Its unfolding showed that inequalities are a dynamic phenomenon, and their underlying causes are not only poverty and income distribution but also capabilities and access to health services and public goods.

One of COVID-19's many alarming characteristics has been its tendency to broaden the socio-economic inequalities linked both to the increased biological vulnerability to the disease and to the lesser ability to mitigate the harmful effects of the pandemic. Heterogeneous biological sensitivity can arise from factors that make an individual weaker against the virus, such as age, immuno-depression, cardiovascular risk, and obesity, all traits confirmed by increased COVID mortality for these clinical profiles. Even if it is not lethal, the infection can have serious consequences, further worsening the clinical picture of some patients in the long term and potentially exposing them to the increased probability of death in the future. Although the temporal arc so far is too brief to fully reveal the long-lasting effects for survivors of the infection, numerous studies show that a significant number of patients, several weeks after the infection, show a number of worrying conditions. These include, for example, cardiac muscle damage, episodes of arrhythmia, hypertension, renal failure, respiratory failure, and numerous neurological damage. It is also common to observe that COVID-19-healed patients show severe fatigue and concentration problems, which, in the best case, lead to reduced quality of life and productivity of work.

A variety of social and economic factors can affect people's ability to cope with the negative effects of all pandemics, including COVID-19. These factors can include age, health status, access to healthcare, socioeconomic status, and living conditions. Specifically, given exposure to the same living conditions imposed by the pandemic, several socio-economic characteristics lead groups of individuals to distance themselves from the rest of the community in the short, medium, or long term, contributing to acute or chronic inequality. Disadvantaged groups may be more exposed to the risk of infection due to various socio-economic conditions inherent in their status. The poor have less ability to exercise social distancing. On the one hand, they have higher levels of employment in 'essential' sector and lower participation in jobs that can benefit from distance work. Furthermore, residential sorting, that is, strategic choice of housing areas, depending on economic status, causes people with more significant socio-economic disadvantages to live in denser places, narrower spaces, and worse sanitary conditions. Alternatively, they live in suburban areas and may be forced to use crowded public transport to go to work. In general, the marginal cost of social distancing is higher for less affluent households whose living conditions have less chance of being altered. Poorer families have higher unemployment rates, lower savings, use greater credit, and face a steeper trade-off in the choice between health protection and financial stability.

7.6.1 Human capital inequalities

During the pandemic, many households' daily activities were conditional on access to the Internet and the availability of devices, such as computers and tablets, without which distance work and distance learning (DL) cannot occur. A big problem with DL concerns the possibility of intensifying socio-economic disparities through parents' different abilities to assist their children in schooling from home. In the economic literature, there is evidence supporting the hypothesis that the children of the most educated parents have the best cognitive and socio-emotional abilities at the beginning of their schooling.

In a well-known study, Cunha and Heckman (2007) explain how human capital is formed through a process driven by a skill production function, characterized by two fundamental properties: self-productivity and dynamic complementarity. The first property means that a higher level of human capital in a given period leads to higher human capital in the subsequent period. The second explains that investment in human capital yields a progressively higher level of human capital. These characteristics imply that the earlier human capital development begins, the best results are produced, and each achievement leads to greater benefits than any other investment in other forms of capital. From the long-term perspective, education obtained in the very early stages of childhood is a crucial determinant of future income. The onset of human capital development at an early stage underscores that even slight disparities between individuals during childhood can result in significant differences in the future. This emphasizes the crucial role of public education in promoting inter-generational mobility. Symmetrically, suppose school fails at crucial stages of development. In that case, some gaps in human capital will also take root and, under the assumption of self-productivity and dynamic complementarity, they will be difficult to bridge in the future. This concern is amplified during lockdowns or localized school closures when distance learning, largely dependent on parental oversight, can lead to inconsistent levels of investment in children's human capital across different households. However, the level of parental education, which is positively correlated with their ability to manage distance learning, plays a significant role. Not only does it influence their proficiency in overseeing the educational process, but it also impacts their perception of the value and effectiveness of their efforts in contributing to their child's learning. This relationship is documented in a study by Fuchs-Schündeln et al. (2020) that highlights the likely severe effects of the COVID-19 crisis for children of parents with low education and limited economic resources.

Moreover, with DL in mind and about managing domestic tasks, it is crucial to focus on the implications of COVID-19 for gender inequalities. From the labor market point of view, the economic sectors most severely affected by the lockdown have been those strongly characterized by female labor. In Italy, for example, women have lower levels of participation in the labor market, but this is not the only characteristic explaining the increased burden of domestic and family commitments that ends up weighing on their shoulders. The time women spend daily in family care is much less related to women's participation in the labor market than men's. Women take care of their children regardless of whether they are employed, while men care much more when they are not working than when they are working. Hupkau and Petrongolo (2020) show how most of the responsibilities related to family care fall on women. In their 2020 study focusing on Italy, Del Boca and colleagues found that women working remotely have taken on a greater share of household responsibilities, irrespective of their partners' workload. In contrast, men were affected in the only cases where their partners belonged to the so-called 'essential' sectors and therefore had to continue to present themselves in the workplace.

Adams-Prassl et al. (2022) show that the market wages are severely penalized by the fragmented working patterns typical of mothers of young children. Even in the case of permanent contracts, the burden of domestic tasks, which require time and energy, weighs on women, resulting, for example, in mothers being forced to reduce their productivity and having a much greater contraction in their career prospects than fathers do. Prolonged inactivity or reduced productivity imposed on women by the COVID-19 pandemic can undermine the achievements in reducing the wage gap in recent decades. Additionally, as the increase in pay linked to an increase in work experience of one year is smaller for women than for men, the gender gap could widen further over time. These negative consequences will be severe for women currently on the labor market and those who were about to enter the market at the beginning of the pandemic.

Reopening schools after the COVID-19 closures was vital for facilitating women's re-entry to the workforce, given that the pandemic had deepened gender disparities, partly due to the mental health strains experienced during periods of school shutdowns. According to Adams-Prassl et al. (2020), the effects of worsening mental health due to the introduction of lockdown measures in the US predominantly affected women, widening the gap in mental health between men and women by 66%. Even at the school level, women were potentially at greater risk of dropping out due to the pandemic, with greater impact on developing countries and those social contexts where gender inequality begins already in the family.

7.6.2 The gender gap

The impact of the pandemic on the gender gap was, at the same time profound and unique as compared to the effects of other economic crises. Rather than primarily on men employed in the manufacturing sectors, the impact on employment was concentrated on sectors such as schools, hotels, restaurants, and other services where women were, on average, more present than men. Child care was especially affected, since schools were closed for long periods of time, affecting both the employment of teachers (women and men, but mostly women), but also causing a veritable crisis in the house-holds, with women often forced to double up as substitute teachers and caretakers.

Women were also in greater demand in all care activities, both formal and informal, from the sick to the elderly, often straining to work longer hours as medical doctors and nurses, and again, doubling up when returning home to bear the main burden of caring for the family.

Like the case of flexible working arrangements, some of the changes induced by long lockdowns and smart-working arrangements will probably vanish when a new normalcy is reached. Some may be more persistent. Remote working and confinement meant that the sanctity of home as a place separated from a workstation and exclusively dedicated to the family was progressively lost, with increased tensions and heavier psychological burdens especially placed on women. The time spent at home by all members of the household tended to increase, 'crowding in' men, women and children that were used to very different patterns of sharing household times and spaces. This tended to increase family stress with reported surges in gender-based violence due to confinement, as family members were forced to face their working tasks, isolated from their colleagues, and often crowded in smaller spaces. However, after the primary lockdowns ended, social distancing and other restrictions reduced commute travel and the greater flexibility and autonomy of workers also suggested some countervailing benefits in the form of greater availability of time for leisure activities and healthier habits. Lockdowns, smart work-ing, and increased childcare needs also favored greater sharing of family tasks formerly largely covered by women.

The impact of COVID-19 on the gender gap was thus uneven, and its long-term implications have yet to fully emerge. While the initial response from the supply side saw a decline of people seeking employment, particularly among women, there are potential indications of emerging patterns of family life that lean towards more equitable sharing models. Reorganization of the health system according to the holistic principles of the 'one health' approach is also likely to further expand the role of women as professional care providers to achieve sustainable and inclusive health development goals.

7.6.3 Inequality in risky behaviors and violence

In terms of risky behaviors, a significant impact was observed on the capacity of many, especially women and children, to escape from abusive households, reach out for exter-nal assistance, or receive proactive support from others. In many countries, this appeared to follow the increased frequency and severity of domestic violence against women and children. As reported by the United Nations (UN Women, 2021), in France, 'official estimates indicate that reports on domestic violence increased by more than 30% in the first ten days of the March 2020 blockade, while reports from Canada, Germany, Spain, the United Kingdom, and the United States indicated that the need for an emer-gency shelter grew during the pandemic as domestic violence increased. In London, the

Metropolitan Police reported that between mid–March and mid–June 2020, domestic abuse increased by 16% among family members and nearly 9% among current partners, but decreased by 9% among former partners' (Suleman et al., 2021). Although we cannot extrapolate data from one or more regions to a whole country, these trends are suggestive of the impact that COVID-19 related movement restrictions have probably had on violence in domestic life.

According to a recent UNICEF report (UNICEF et al., 2021), COVID-19 impact minors has been devastating, jeopardizing decades of progress in crucial childhood challenges such as poverty, health, access to education, nutrition, child protection, and mental well-being. The report states that after two years since the beginning of the pandemic, the widespread impact of COVID-19 continued to worsen, increasing poverty, reinforcing inequality, and threatening children's rights to levels never seen before. As the count of children facing starvation, deprived of education, subjected to mistreatment, living in destitution, or compelled into marriage was escalating, the tally of children with access to necessary healthcare, immunizations, adequate nourishment, and vital services was diminishing. The report estimates that about 100 million more children were living in multidimensional poverty due to the pandemic, a 10% increase from 2019. Furthermore, the number of children in poor households had increased by approximately 60 million compared to before the pandemic. More than 23 million children had lost essential vaccines, an increase of nearly 4 million compared to 2019.

Before the pandemic, it was estimated that around 1 billion children worldwide had suffered at least severe deprivation without access to education, health, housing, food, sanitation, or water. This number is increasing as uneven recovery aggravates the growing divisions between rich and poor children, with the most marginalized and vulnerable suffering the most. But perhaps the most alarming figure comes from school attendance data. During the pandemic, at its peak, more than 1.6 billion students were not attending school due to nationwide closures. Schools closed worldwide for almost 80% of in-person education in the first year of the crisis.

As reported in several systematic reviews, studies have reached mixed conclusions regarding the impact of school closure and subsequent reopening on the spread of COVID-19 (National Collaborating Centre for Methods and Tools, 2021; European Centre for Disease Prevention and Control (ECDC), 2021; Ziauddeen et al., 2020). This inconsistency of the findings may be due to the high risk of bias in most studies. For example, in many regions, the initial decision to close schools for in-person instruction occurred concomitantly with the institution of other non-pharmaceutical interventions (such as masking and distancing), stay-at-home strategies, and economic closures, making it challenging to disentangle the individual impact of school closures. Furthermore, many studies examining SARS-CoV-2 transmission in children have been limited to day-care or summer camps and may not be generalizable to other settings. Understanding the impact associated with school reopening allows the development of additional policies and mitigation strategies to reduce transmission risk (Fitzpatrick et al., 2022).

Lastly, we should not forget the problems associated with early marriages (estimated at more than 10 million in the last ten years) and the number of children involved in child labor which has risen to 160 million worldwide, with an increase of 8.4 million children in the last four years. Another 9 million children are at risk of being pushed into child labor due to the increased poverty triggered by the pandemic. Even in the best-case scenario, it will take seven to eight years to recover and return to the already unacceptably high pre-COVID levels of child poverty.

Finally, the lockdown measures have heightened domestic tensions, generating episodes of violence against women and minors in many contexts (Moawad et al., 2021), as accounts from various countries confirm. For example, the Jianli County police station in Central Hubei Province, China, reported a threefold increase in intimate partner violence cases in February 2020 compared to the same month in 2019 (Wanqing, 2020). In Australia, a survey enrolling 400 frontline workers showed a 40% rise in 'pleas for help' and an increase in complexity of cases by 70% (Lattouf, 2020). In Italy, the national network of shelters for women exposed to gender-based violence (D.i.Re) reported that from 2 March to 5 April 2020, 80 shelters were contacted by 2867 women, reflecting a drastic rise (74.5%) on the usual monthly records of 2018 (Bellizzi et al., 2020). According to ISTAT (2020b), in Italy, the coronavirus epidemic has increased the risk of violence against women, as episodes often take place at home, and the decline in essential social, health, and legal support services has made it more difficult for victims of domestic violence to escape their abusers. Mobility restrictions have further complicated this challenge, as they have made it more difficult for victims to leave their homes and seek help. As the statistics gathered by ISTAT (2020b) show, the regions where the growth of violent acts has been most pronounced have been Lazio, Tuscany, and Liguria. Between 2019 and 2020, stress calls in Lazio went from 6.8 for every 100,000 residents to 12.4, in Tuscany from 4.8 to 8.5, and in Liguria from 4.1 to 7.2. The detection period is still too short to determine the overall effect of the pandemic, also because problems related to mental illness tend to emerge gradually and, coupled with the context of domestic isolation, can be a dangerous ally of future incidents of violence.

7.6.4 Inequalities in access to vaccination and social disadvantages

The most emblematic form of the new inequalities revealed by the pandemic was the vaccination divide between countries, and created by a combination of pre-existing conditions and opportunistic behavior. Unequal access to vaccination has been dramatic and a phenomenon that went much beyond what could be expected on *a priori* grounds of pre-existing wealth inequalities. Vaccines are a form of contingent wealth in the sense that they give uneven and uncertain access to prevention, but also in the sense that their efficacy depends on the underlying distribution of health as a critical capability of human capital.

According to the UN, by the end of 2021, 147.4 single and multiple vaccinations per 100 people had been administered in developed countries, against 23.4 per 100 in the least developed countries. Inequalities in vaccination numbers were compounded by the widely believed lower efficacy of vaccines, such as Sputnik, Sinopharm, and Janssen, mostly used in developing countries, compared to the Pfizer and Moderna products used more predominantly in higher-income locations. Within countries, lower access and higher hesitancy in vaccination have been reported for disadvantaged groups, including the poor as well as the ethnically or racially discriminated (Ndugga et al., 2021; Painter et al., 2021). The latitude and the persistence of the vaccine divide have revealed a form of massive market and government failure in the provision of a most basic public good, with profound epidemiological, social, and economic implications (Bollyky et al., 2022; Schellekens and Wadhwa, 2021). Nine months later, at the end of August 2022, data collected by `OurWorldinData.org` showed that 68% of the world's population had received at least one dose of COVID-19 vaccine. A total of 12.57 billion doses had been administered globally, and about 30 million were administered daily. However, only 10.6% of people in low-income countries had received at least one dose. Fig. 7.15 shows an international comparison of the rates of vaccination. Regardless of the non-trivial comparative problems that may exist in the presence of different vaccine protocols, it is clear that some countries (especially those with low incomes) have been proceeding with a large delay in vaccination. By contrast, richer countries have reached in record times levels close to or above 90% of the population.

The uneven distribution of vaccination reflected, to some extent, the division between rich and poor countries, even though the resulting divide was for many aspects surprising. While China appeared at the head of a successful effort to control the spread of the virus since its very beginning, the Western countries, whose management of the pandemic appeared broadly failing, stroke a critical success with the vaccines. The development, production and delivery of vaccines in less than a year was indeed a remarkable feat. It was combined with the discovery of an entirely new technology (the RNA messenger) based on genetic research, that also suggested, as a silver lining of their seemingly chaotic market and governance systems, that the Western countries were still ahead in the technological race.[11]

Nevertheless, as already illustrated in Chapter 4, having a vaccine is only one necessary step in addressing the challenges posed by the pandemic. While the development

---

[11] According to Scarpetta et al. (2021), the rapid development of COVID-19 vaccines is the result not only of unprecedented levels of international collaboration, but also of massive public investment in research and development and production capacity. These results are particularly significant when compared to previous estimates of the probability of approval of a vaccine reaching the clinical trial phase: 12 to 33% after about 7-9 years of development (Scarpetta et al., 2021).

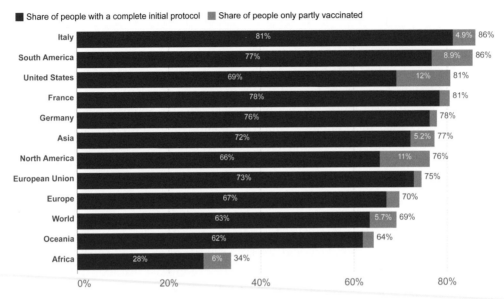

Figure 7.15 Percentage of people vaccinated against COVID-19, August 2022. *Source*: Our-WorldinData.org. Note: Alternative definitions of full vaccination are used, e.g. having been infected with SARS-CoV-2 and having received 1 dose of a 2-dose protocol, they are ignored to maximize comparability between countries.

of vaccines was a major achievement, it was just one of many things that needed to be done to overcome the crisis. There are a number of necessary and sufficient conditions, widely different for different areas of the world, different countries, and different population groups. These differences range from incomes and status to production, distribution, administration, and monitoring capacities. As we have seen, all these phases are fraught with political challenges.

The access and adoption of COVID-19 vaccines, especially in the early phases of the vaccination campaign, were heterogeneous and irregular. As previously depicted in Chapter 4, Nations in North America and Europe initiated vaccination of their citizens earlier, maintaining a considerable lead over countries in other areas, including Africa and the Middle East. Across all economic spectrums, advanced nations, on average, have vaccinated a significantly larger proportion of their populations compared to emerging and developing economies. Compared to the predictions and intentions agreed upon before the discovery and approval of vaccines, a very important tool that worked only imperfectly in the vaccine campaign was the Gavi COVAX infrastructure. This was the only mechanism aiming to ensure multilateral access to successful vaccines according to a rational distribution plan across countries. If adequately financed, COVAX would have also allowed access to the vaccine for less wealthy countries that lacked the resources to promptly purchase and distribute it.

However, COVAX, even though it was the only global initiative aimed to ensure equitable access to COVID-19 vaccines, faced persistent underfunding and competition from direct supply agreements between governments and vaccine manufacturers. In 2020, the target was to raise initial funding of $2 billion for middle- and low-income countries (LMICs), and that target was met by December 2020. An additional $5 billion were needed for LMICs in 2021, excluding funds for self-funded countries. The goal of providing 1.8 billion doses of vaccine in 2021 was not met, with the same goal reached only 8 months later, mainly due to due a reduction in vaccine roll-outs in AEs. Overall, it is estimated that by the end of 2023, it will still not be possible to vaccinate populations in parts of South America, Africa, and Asia.

Unequal access to vaccines, however, is not a simple consequence of pre-existing inequalities, since it reflects deeper drivers of social imbalance and fragility. As such, it is both the effect and the cause of the growing differences between countries' ability to respond to health crises through their care and social systems, institutional capacity, tax revenues, and employment. More generally, the vaccination divide reveals the precarious foundations of the world health systems on essentially private management of health care and a market or quasi-market approach. Under this approach, in most countries, and by implication at the global level, the public sector appears to be only playing the role of an ineffective safety network with no serious prevention and non-discriminatory capacity to provide public goods for society at large.

A recent study of Rydland et al. (2022) draws on two social science theories, Fundamental Causes Theory (FCT) (Link and Phelan, 1995) and Diffusion of Innovations Theory (DoI) (Rogers, 2004) – to argue that the unequal vaccination phenomenon reflects structural issues that go much beyond the mere differences in vaccine distribution or even in social wealth. FCT explains the skewness of health and education within and between countries through the underlying endowment of capabilities and wealth, which generally disfavors the most disadvantaged, regardless of the plague or risk that society faces at any particular moment. The key variable, however, is not monetary income, but access to goods and services, healthcare and education, and all public goods whose distribution is overwhelmingly skewed between and within countries against the poor, the elderly, the women, and the other most fragile components of the population. This means that the intensity and diffusion of COVID-19 is higher for the most disadvantaged to begin with, both in high- and low-income countries, resulting in a syndemic pandemic, where the infection interacts with existing social inequalities in disease and the social determinants of health (Bambra et al., 2020). In the case of COVID-19, poorer people are at increased risk due to their higher exposure to infectious encounters and practices, lower capacity to understand and practice prevention measures, higher hesitancy to accept vaccines, and more precarious health conditions. Once infected, they are more vulnerable and are more likely to suffer more serious health conditions, as well as long-term health and economic consequences of the disease.

FCT, however, does not limit itself to interpreting inequality as a reflection of underlying social conditions, but it is a theory of social change that tries to explain why inequalities persist. While the social context may be very dynamic and cause previous causes of inequalities to disappear, if the fundamental disadvantages affecting the most vulnerable part of the population remain, new inequalities will be continuously created. In the case of diseases, for example, the higher incidence of diphtheria and tuberculosis among the poor was originally due to low quality housing and hygienic conditions. Once these variables were improved, however, the poor proved to be victims of higher incidence of cardiovascular diseases and cancer, which had previously been a prerogative of wealthier people. Even though, overall, life expectancy and morbidity improved for the poor, inequality did not, as a new social gradient emerged from the same fundamental characteristics of poverty, including unhealthy lifestyles, and working conditions.

The DoI theory, on the other hand, takes a completely different angle to explain inequality as the result of the diffusion of innovations, including for example the diffusion of hybrid corn to farms or beta-blockers to hypertension patients. The theory was first formulated by Rogers and Agarwala-Rogers (1976) and has undergone a number of modifications and refinements since then. It is based on the idea that individuals are heterogeneous in their acceptance of innovation and that diffusion of new technologies tends to follow a series of alternatively reluctant and eager waves of adoptions (Scandizzo and Savastano, 2010) or, in other versions, a first phase of early adopters followed by an intermediate phase of moderate and a final phase of laggard adopters. The resulting inequality tends first to increase, as a consequence of the rise of the innovators, to stabilize and then to decrease. This theory appears relevant to the recent vaccination events for several reasons. First, the discovery of vaccines for COVID-19 marked a significant series of innovations, for the unprecedented speed and the path-breaking nature of some of the vaccines discovered. Second, the radical nature of these innovations, had also the effect of delaying the adoption of vaccines among the most disadvantaged population groups. As predicted by DoI, the first wave of adopters was thus followed by an increased reluctance to use the vaccine from the less informed and more vulnerable part of the general public. Third, the new and more effective vaccines, based on the innovative technique of RNA messenger were monopolized by the richer countries, which were also able to launch more extensive and successful mass vaccination campaigns. Finally, inequality tended eventually to decrease as poorer countries caught up on vaccination rates, even with less effective vaccines, and were able to refrain from larger and economically more exacting measures of prolonged lockdowns. However, vaccine nationalism and lack of international coordination delayed the beneficial effect of a greater diffusion of vaccines and a reduction of the medical divide.

In conclusion, we should celebrate the huge progress made so far. At the same time, the large vaccination gaps around the world present clear vulnerabilities. One of these is represented by the continuous appearance of new variants that are heightening worries worldwide and should serve as a wake-up call, independently of the severity they will show. As of today, the world remains wholly under-prepared against the downside contingency of the next variants. If a more transmissible and more lethal variant emerges a year from now, what will our strategy be and how different will it be from the several ones we have used over the past?

CHAPTER 7 - Take-home messages

- The pandemic has inflicted substantial damage on the world economy, underscoring the potential for systemic disasters connected to environmental degradation. This highlights the adverse effects of human interactions with the environment. It illustrates how these interactions can cause abrupt, unpredictable events that may profoundly disrupt our societal structures.
- The various lockdown measures in response to the coronavirus have halted economic activity in some sectors and reduced it in others, leading to job losses and bankruptcies.
- The pandemic has caused a decoupling effect, a phenomenon whereby inequalities tend to become extreme, with the world divided between the 'haves' and 'the haves not'. It has had a devastating impact on the population, especially on those who face precarious living conditions. It has also created new forms of inequality based on the underlying fragility and vulnerability of different agents, such as the elderly or the chronically sick.
- Policymakers are more likely than economists to agree that substantial fiscal and monetary stimuli are the best way to respond to an economic emergency that is similar to a 'wartime' situation. They may also view elevated public debt as an acceptable buffer for the private sector to help it deal with temporary and unforeseen economic downturns.
- The COVID-19 pandemic has exacerbated socioeconomic inequalities by increasing people's biological vulnerability and reducing their ability to mitigate the harmful effects of the pandemic. Pre-existing structural conditions have aggravated these effects.

CHAPTER 8

The effects on healthcare systems and health status

Contents

8.1. Three years after the pandemic's start, what happened to our healthcare system?

8.1.1 Evolution of contagions

According to `ourworldindata.org`, by mid–February 2022, there were approximately 675 million reported cases of COVID-19 worldwide, of which nearly 247 million were in Europe. One year after the pandemic outbreak, those numbers were about one-sixth, reaching around 112 million infected around the world, of which about 33 million were in Europe. Fig. 8.1 shows in all its drama how the waves that followed the first in 2020 were more explosive. But the drama of these numbers is partially mitigated by the aggregation process at the national level. In fact, the spread of infection was highly uneven, and the impact on sub-national territories (regions and municipalities), in terms of reported cases and related deaths, diverse. According to a study on the first phase of the pandemic by Allain-Dupré et al. (2020), 'In China, 83% of confirmed cases were concentrated in Hubei province. In Italy, the northern part was the hardest hit, and one of the richest regions in Europe, Lombardy, recorded the highest number of cases (47% in November). In France, the Île-de-France and Grand Est regions were the most affected, with 34% and 15% of national cases, respectively. In the United States, New York had the highest share of federal cases (14.6%), followed by Texas (8%). In Canada, the provinces of Quebec and Ontario represented 61% and 31% of the total cases in November. In Chile, the Metropolitan of Santiago represented 70% of the cases in November. In Brazil, Sao Paulo recorded 25% of cases in November. In India,

Maharashtra reported 21% of confirmed cases, while in Russia, Moscow represented 24% of the total cases in November'.

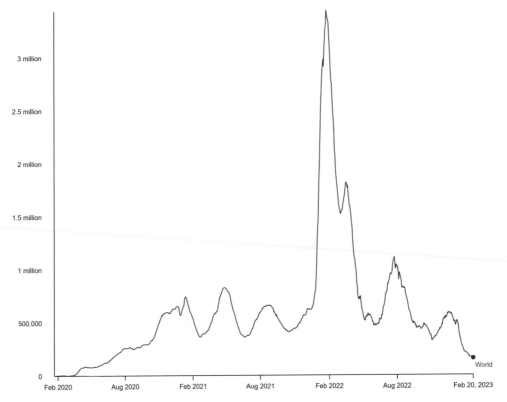

Figure 8.1 Evolution of new daily COVID-19 cases worldwide. *Source*: Ourworldindata.org, available at the web address: https://www.ourworldindata.org/. Note: 7-days moving average. The number of confirmed cases is less than the number of actual cases as the number of tests is limited.

The timing of the contamination was also mixed. As seen in Fig. 8.2, the evolution of infections was highly different across continents. With all the limitations of measuring such a phenomenon: in August 2020, the peak was in South America, in November 2020, it was in Europe, in January 2021, in North America, in April 2021 again in South America, and then moved to Asia during summer 2021. With the arrival of the waves driven by new variants, the cycle across continents was repeated. The wave at the end of 2021 triggered in part by the variant 'Omicron' started in Europe, leading to about 1000 new daily cases per million inhabitants, followed in this dynamic by a few weeks delay in the other continents. Similar misalignments in the timing of the contagion can also be found within continents. Fig. 8.3 shows the same data for selected AEs: as we can see, both the magnitude of the contagion and the timing are very heterogeneous, and this depends on several variables, notably the presence of new variants, vaccination rates, social behavior, and stringency measures adopted. In July 2022, according to the World Health Organization, the number of new coronavirus cases was increasing again, with more than 4.1 million cases reported worldwide only in the last week of July 2022. Infections increased by approximately 32% in Europe and Southeast Asia and about 14% in the Americas. However, the number of deaths remained similar to the week before, with some increases only in the Middle East, Southeast Asia, and the Americas. The cases increased in 110 countries, mainly driven by BA.4 and BA.5. omicron variants.

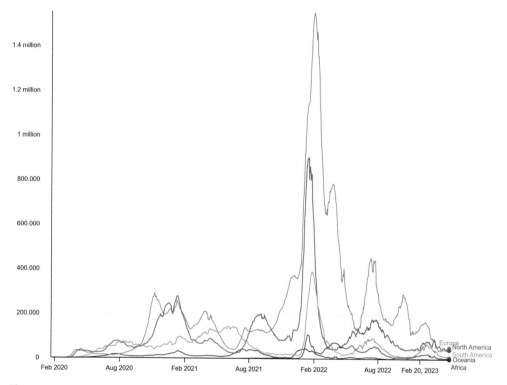

Figure 8.2 Evolution of new daily COVID-19 cases per million people by continent. *Source*: Ourworldindata.org, available at the web address: https://www.ourworldindata. org/. Note: 7-days moving average. The number of confirmed cases is less than the number of actual cases as the number of tests is limited.

8.1.2 Evolution in COVID-19 hospital admissions

In the phase of the spiraling dynamic of infection, the COVID-19 pandemic has shown us how important it is to have a robust healthcare system with enough hospital beds to care for patients who need them. The pandemic has also highlighted the scarcity of these beds, placing immense pressure on health care systems in the last few years.

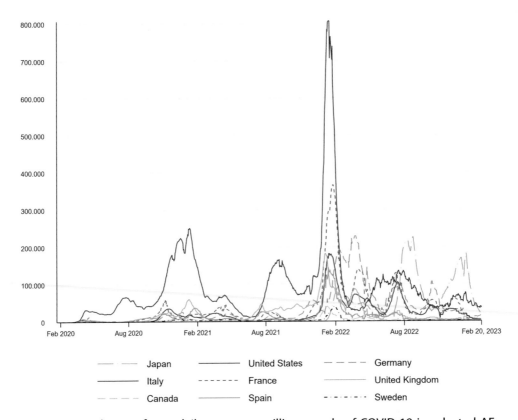

Figure 8.3 Evolution of new daily cases per million people of COVID-19 in selected AEs. *Source*: OurWorldinData.org, available at the web address: https://www.ourworldindata.org/. Note: 7-days moving average. The number of confirmed cases is less than the number of actual cases as the number of tests is limited.

According to OECD (2022b), since 2009, the number of beds per capita has decreased in almost all OECD countries. Part of the decrease can be attributed to advances in medical technology, allowing more surgery to be performed on the same day, or as part of a broader policy strategy to reduce the number of hospital admissions. For a given number of available hospital beds, its rate of occupancy (the number of beds occupied over the total number of beds) provides crucial information to assess the adequacy of hospital capacity. During the early days of the COVID-19 pandemic, many countries experienced high occupancy rates, which reflected the strain on their healthcare systems.

The level of occupancy rate within the OECD countries was already problematically high before the pandemic started. Although there is no consensus on the 'optimal' occupancy rate, according to NICE,[1] a threshold of 85% of occupancy was considered adequate (NICE, 2021). While the literature on best practice recommended having some spare bed capacity to absorb unexpected surges in patients requiring hospitalization, in 2019, the bed occupancy rate was higher than 85% in four of 27 OECD countries with comparable data: Canada, Israel, Ireland, and Costa Rica (see Fig. 8.4). In contrast, occupancy rates were relatively low in the United States, Hungary, and The Netherlands (less than 65%). In 2019, around half of the OECD countries had bed occupancy rates of 70–80%, and the OECD average was 76%.[2]

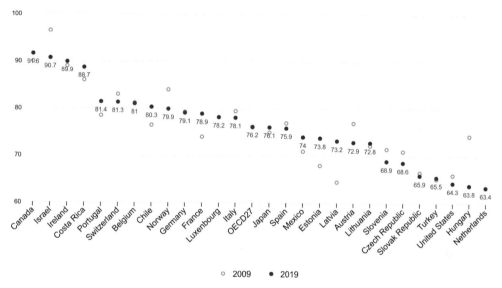

Figure 8.4 Occupancy rate of curative (acute) care beds, 2009 and 2019 (or nearest year). *Source*: OECD Health Statistics 2021.

An essential part of the total number of hospital beds is represented by the intensive care units (ICU). The necessity of beds in these units became apparent during the severe stages of COVID-19, as the survival of numerous critically ill patients hinged on their accessibility. As shown in Fig. 8.5, neglecting the definitional differences between countries, in 2019, the average number of ICU beds in the OECD was 14.1 per 100,000 people. When it comes to hospital beds, however, a significant disparity exists across countries e.g., between the Czech Republic, which has 43 beds per 100,000 individuals, and Costa Rica, where the figure is a mere 2.9 beds per 100,000 inhabitants. Amid

---

[1] NICE stands for: National Institute for Health and Care Excellence, a UK organization that provides guidance and advice to improve health and social care.

[2] Five of the nine countries analyzed in this report show heterogeneous occupancy rates, ranging from a high 81.4% in Portugal to a low 65.9% in Slovakia, and in any case, below the 85% threshold suggested by NICE.

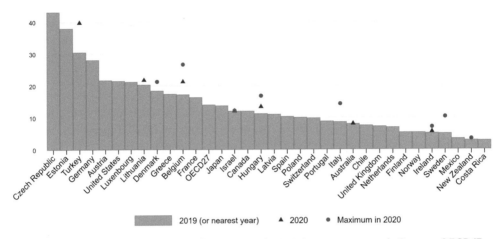

Figure 8.5 Adult intensive care beds, 2019 and 2020 (or nearest year). *Source*: OECD/Eurostat/WHO Regional Office for Europe Joint Questionnaire on Non-Monetary Health Care Statistics 2021 (unpublished data); Country Health Profiles 2021; Health at a Glance: Latin America and the Caribbean 2020; national sources. Note: Neonatal and paediatric ICU beds included. 2. Data cover critical care beds only. 3. Data refers to England only.

the pandemic, countries also undertook numerous policy measures aimed at enhancing their hospital capacity. For example, during 2020, Italy was able to increase ICU's capacity value from 8,7 to 14,3 beds per 100,000 people (one of the most significant increases recorded in the world). Most of this additional bed availability was obtained by transforming other clinical wards into ICUs, creating field hospitals with ICU units, and transferring patients to localities with spare ICU capacity (sometimes also to other countries).

Hospitalizations (including those in intensive care) closely correlated with the number of confirmed cases of COVID-19. This correlation was only broken following improvements in vaccination coverage, which led to a reduction in hospital admissions since 2021, especially among the elderly. For example, in the United States, hospitalization rates among people 85 and older dropped markedly with more intensive vaccination campaigns. A particular resurgence of hospitalizations began with the arrival of the Delta variant. However, as can be seen in Figs. 8.6 and 8.7, the availability of the vaccine reduced hospitalizations even in the presence of the different new variants that rapidly follow each other, with the greatest effect on the most fragile people (> 50 years) and, particularly, on the over 80s.

8.1.3 The evolution of COVID-19 mortality

As the pandemic spread throughout the world, people began to familiarize themselves with the grim statistics of mortality. According to the website ourworldindata.org,

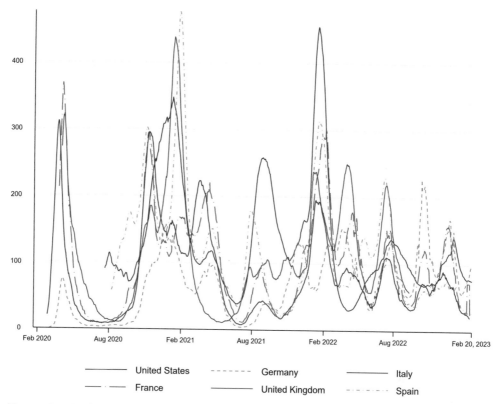

Figure 8.6 Evolution of new weekly COVID-19 hospitalizations per million people in selected AEs (Advanced Economies). *Source*: Ourworldindata.org, available at the web address: https://ourworldindata.org. Note: Total new entries compared to the previous week.

there were approximately 6.9 million deaths from COVID-19 worldwide at the end of February 2022 (an estimated 800 deaths per million inhabitants). Europe alone had just over 2 million, but the value per million inhabitants was almost three times higher (2725). The differences in the evolution of new infections and deaths from COVID-19 reflect demographic differences, variations in containment and mitigation strategies, and the timing of their implementation, as well as differences in the ability of health systems to treat patients with COVID-19 and adapt to current challenges. In fact, the OECD reports that mortality rates in most OECD countries had declined to around 1-2% by early October 2021 (OECD, 2022b).

Several arguments can be invoked to explain these differences. On the one hand, there has been an increase in case detection over time. Furthermore, vaccination campaigns and better disease management and capacity building in the health system have significantly reduced mortality rates. However, factors beyond the immediate control of policymakers, such as geography, population demographics, and the prevalence of some

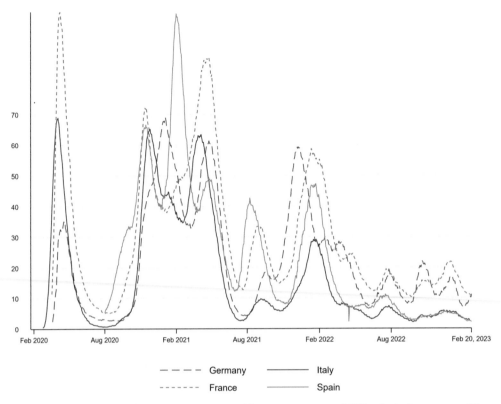

Figure 8.7 Evolution of new COVID-19 weekly intensive care (ICU) admissions per million people in some AEs. *Source*: Ourworldindata.org, available at the web address: https://ourworldindata.org. Note: For countries where the number of ICU patients is not reported, we display the closest metric (patients ventilated or in critical condition).

risk factors such as obesity, have made some countries more susceptible to higher rates of infection and mortality.

At the European and global levels, the evolution of mortality is shown in Fig. 8.8. As can be seen, three major mortality peaks can be distinguished in Europe: the first in the initial months of the pandemic, the second during the winter months of 2020/21, and the third during the winter months of 2021/22. Europe was hit more than the rest of the world, throughout the entire pandemic period, with an average of approximately three/four daily deaths per million people versus approximately one daily death worldwide (including Europe).

However, the European numbers hide considerable differences and some countries record much larger numbers. In Fig. 8.9, the same phenomenon appears to occur for selected AEs. Although we still observe the same three waves, the timing has been somewhat different between countries, and the death rates are much higher in some of

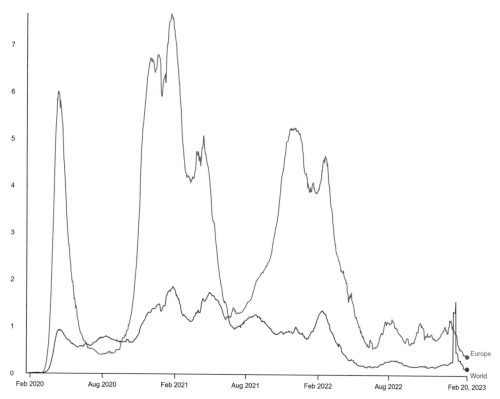

Figure 8.8 Evolution of COVID-19 daily deaths per million people in Europe and worldwide. *Source*: Ourworldindata.org, available at the web address: https://ourworldindata.org. Note: 7-days moving average. For some countries the number of confirmed deaths may be lower than the actual number of deaths. This is due to limited testing and challenges in attributing the cause of death.

them. Higher death rates were a problem that was limited to France, Italy, and Spain during the first wave, with the other countries being only slightly affected. Croatia, Czechia, Greece, Poland, and Slovakia recorded fewer than one daily death per million people. The situation changed once the two other waves hit Europe during the winter periods: although with some time differences, all countries were affected. In terms of magnitude, all of these countries have recorded peaks of close to 20 daily deaths per million people (with Portugal reaching a high of 28.6 in February 2021).

Unfortunately, measuring mortality trends based on COVID-19 deaths can be misleading, as published statistics may not accurately reflect the true causes of death. This is due to the numerous factors that can contribute to an individual's demise, with not all of them being recorded in official documents.

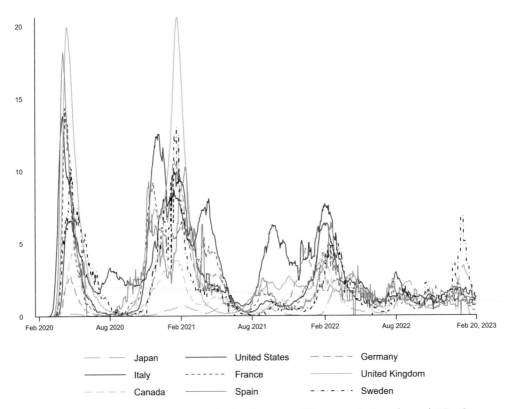

Figure 8.9 Evolution of COVID-19 daily deaths per million people in selected AEs. *Source*: Ourworldindata.org, available at the web address: https://ourworldindata.org. Note: 7-days moving average. For some countries the number of confirmed deaths may be lower than the actual number of deaths. This is due to limited testing and challenges in attributing the cause of death.

To avoid such problems, the Economist Expert Group has developed a specific methodology, which consists of counting all deaths – rather than trying to distinguish between types of death – and comparing them with those of a 'reference' period defined as 'normal'. The difference between the two values provides a fairly accurate estimate of how many people died in a given region during a given period of time, regardless of cause, and how many deaths there would have been had no particular circumstance occurred (such as a natural disaster or an epidemic). Nonetheless, this 'indirect' method only offers a ballpark figure, as it includes individuals who succumbed indirectly due to COVID-19, such as those who passed away due to inadequate care. This was due primarily to the fact that hospital capacities were exceeded, leaving no room for everyone, and ambulance services were not able to promptly attend to the ill. Based on this methodology, at the end of July 2022, the Economist estimated an actual death toll of 20.5 million people (which is 95% likely to be between 16.7 and 27.3 million excess deaths), although the official death numbers for COVID-19 were only around 6.5 million.

These data clarify that COVID-19 has caused more deaths than the official statistics suggest. Fig. 8.10 shows the estimated global mortality overkill in the three years of the epidemic. Of course, the difference between confirmed and estimated data depends greatly on the type of data collected in individual countries. For example, if the excess mortality rate is applied to population numbers, many of the world's worst-affected countries are in Latin America. Even in Russia, the death toll suggests that the country has been hit fairly hard by COVID-19, and India's death toll is estimated to be in fact in millions, rather than in hundreds of thousands. For a handful of countries fewer people died during the pandemic than in previous years, as shown by Table 8.1 below. In these cases, the main reason may be attributed to specific factors, such as, for example, changes in lifestyle that reduced the burden of other causes (such as influenza).

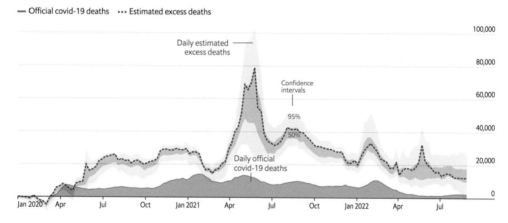

Figure 8.10 Estimated global excess mortality and official deaths from COVID-19. *Source*: The Economist, available at the web address: https://www.economist.com/graphic-detail/coronavirus-excess-deaths-estimates?fsrc=core-app-economist. Note: The model estimation methodology is available at web address: https://www.economist.com/graphic-detail/2021/05/13/how-we-estimated-the-true-death-toll-of-the-pandemic.

Another important feature of the excess mortality indicator is to allow credible comparisons between countries. Table 8.1 reports the official mortality data (total and per 100,000 inhabitants) and the confidence intervals of the estimated excess deaths over the period 27 January 2020 – 31 July 2022 by continent and for a selected sample of EU countries. The data show that Asia had the highest number of excess deaths (from 6.5 to 16 million). However, when we look at the same data per 100,000 people, the picture changes, with Europe being the most exposed continent. Within Europe, Poland,

Table 8.1 Cumulative excess deaths by country in July 2022.

Countries	Official COVID-19 Deaths	Per 100,000	Estimated Excess Deaths	Per 100,000	Estimate vs. Official
Asia	1,470,427	31.3	6.5m to 16m	140 to 340	+800%
Europe	1,923,142	256.9	3.2m to 3.4m	430 to 460	+70%
Africa	256,507	18.4	1.3m to 3.3m	93 to 240	+1000%
Lat. Am. and Carib.	1,730,977	264.2	2.5m to 2.8m	380 to 420	+50%
European Union	1,136,146	253.8	1.4m to 1.4m	310 to 320	+20%
North America	1,088,391	290.1	1.2m to 1.4m	330 to 370	+20%
Oceania	18,496	41.6	13k to 44k	30 to 100	+70%
Croatia	16,05	393.2	22k to 24k	550 to 580	+40%
Czech Republic	40,313	375.9	44k to 47k	410 to 440	+10%
France	149,386	221.6	110k to 120k	160 to 180	−30%
Greece	30,178	291.0	33k to 36k	320 to 340	+10%
Italy	168,102	278.5	210k to 230k	350 to 370	+30%
Poland	116,417	308.0	180k to 190k	480 to 500	+60%
Portugal	24,013	236.2	27k to 30k	270 to 290	+20%
Slovakia	20,142	369.6	28k to 30k	520 to 550	+40%
Spain	107,799	230.6	120k to 130k	250 to 270	+10%

Source: The Economist. Data are available at the following web address: https://www.economist.com/graphic-detail/coronavirus-excess-deaths-estimates?fsrc=core-app-economist.

Note: The model estimation methodology is available at the following web address: https://www.economist.com/graphic-detail/2021/05/13/how-we-estimated-the-true-death-toll-of-the-pandemic.

Croatia, and Slovakia are the countries most affected, where the estimated excess mortality is higher than the official mortality by 40% or more. Italy follows with an excess death toll that is close to 30%, while all other countries have lower estimated excess death rates, and France recording a remarkable −30%.

Finally, it is useful to emphasize that national data often conceal large differences at a more local level, which can tell very different stories. To this end, the information available on the website 'Tracking the coronavirus across Europe' is very useful, as it can help to understand the scale of the phenomenon in an international comparison context. In particular, a comparison of eight sub-national regions is made in Fig. 8.11, looking at infections and deaths.[3] To facilitate cross–country comparisons, the available data was standardized using the seven-day moving average and all series rescaled to 100 for the highest infection and death rates in each region. Therefore, it is possible to see that Lombardy is by far the European region with the highest death toll per 100,000 inhabitants (350 per 100,000) and that the majority of those deaths occurred during

[3] The data shown in Fig. 8.11 report the most updated estimates produced by The Economist using the EuroMOMO model.

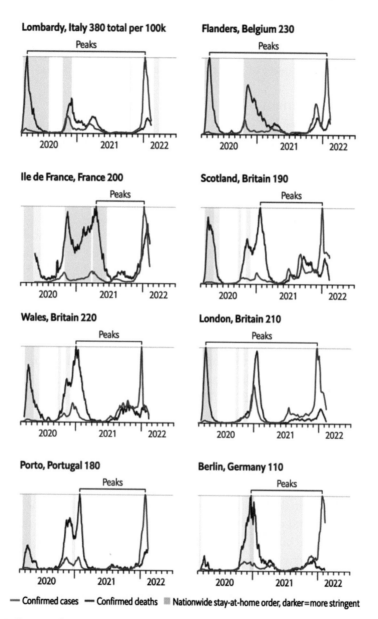

Figure 8.11 Regional COVID-19 cases and deaths per 100,000 people. *Source*: The Economist, available at the web address: https://www.economist.com/graphic-detail/coronavirus-excess-deaths-estimates?fsrc=core-app-economist.

the first wave, even though the second wave also exacted a heavy toll. It can also be observed that since the beginning of 2021, thanks to the introduction of vaccines, the situation greatly improved everywhere (see the next section). The only exception was the region of the Ile de France, where the increase in cases at the end of 2021 seems to be accompanied by an increase in mortality, the only case among the eight regions considered. Finally, the timing of deaths varies greatly between regions.

8.1.4 The effects on mental health

The epidemic of COVID-19, caused by the novel coronavirus SARS-COV-2, originated in Wuhan, China, at the end of 2019 and spread rapidly throughout the world in the following months. In the absence of effective treatment and vaccination, to prevent contagion, countries resorted to various non-pharmaceutical interventions (NPIs). School closures, travel restrictions, national lockdowns and various social distancing measures affected people's lives from different perspectives.

Italy was the first advanced economy to experience a major outbreak of the virus, with the first infected patient isolated on February 18th, 2020. From March 9th, 2020, Italy declared a complete national lockdown that lasted until May 18th, 2020 (referred to as 'phase 1'), along with other strict mobility restrictions that continued until June 3rd ('phase 2'). In parallel, the World Health Organization (WHO) declared COVID-19 a global pandemic on March 11th, 2020, leading to the introduction of large-scale non-pharmaceutical interventions (NPIs) in many other countries. The events of the following months caused widespread and dramatic disruption in daily life, with an emotional, social, and economic burden that remains to be understood.

According to the OECD (OECD, 2021c), the prevalence trends of mental health[4] were stable until the outbreak of COVID-19 in 2020, when the rates of depressive and anxiety disorders increased in many countries. In particular, according to several surveys, self-reported prevalence more than doubled in Belgium, France, Italy, Mexico, New Zealand, the United Kingdom, and the United States in the case of anxiety; and in Australia, Belgium, Canada, France, the Czech Republic, Mexico, Sweden, the United Kingdom, and the United States in the case of depression. Despite the limited scope for representativeness and cross-country comparability of these surveys due to different designs and sampling techniques, the evidence clearly points to a significant increase in the prevalence of subjective anxiety and depression.

A recent WHO report (WHO, 2022a) provides a comprehensive overview of the mental health effects of the COVID-19 pandemic, based on the Global Burden of Disease (GBD) data and other research, including a framework of systematic reviews and meta-analyses (John et al., 2020; WHO, 2020c, 2022b). According to GBD (Santomauro et al., 2021), the COVID-19 pandemic has led to a 27.6% increase in the case of major depressive disorder (MDD) and a 25.6% in the case of anxiety disorders

---

[4] Prevalence is a statistical concept referring to the total number of cases of a health-related event (in this case, mental health conditions) in a population at a given time. It is often reported as the number of cases per 1,000 or 100,000 people.

(AD) worldwide in 2020. In general, the pandemic is estimated to have caused 137.1 additional disability adjusted life years (DALYs) per 100,000 population for MDD and 116.1 for AD. The report suggests that the increase in mental health disorders has disproportionately affected young people, women, and patients with pre-existing health conditions.

The results of both the OECD and WHO reviews are far from conclusive. On the one hand, longitudinal studies are necessary to assess how sustained are these new trends. However, as the WHO highlights, 'most of the eligible meta-analyses reviewed were rated as low quality and with a high risk of bias, ..., which makes rates across studies difficult to interpret' (WHO, 2022a). This is partly due to the types of survey design used, which in most cases were non-representative and frequently conducted on on-line social media platforms, imposing several forms of statistical and methodological bias. Moreover, most of the studies were purely descriptive and did not provide the means to identify and measure the real effects and mechanisms at play.

The results of longitudinal data on mental health trajectories suggest more optimistic conclusions: Fancourt et al. (2021) and Saunders et al. (2021) use the results of a weekly survey administered in the UK and find, after a sharp increase at the beginning of the lockdown, a common declining trajectory of depression and anxiety symptoms in the weeks after the beginning of the mobility restrictions. They also highlight how individuals with previous diagnoses of mental health conditions, although reporting higher anxiety and depression in the early phases of the lockdown compared to the rest of the sample, did not show higher levels of emotional reactivity, possibly because they had already experienced coping strategies in stressful situations.

In general, despite a significant increase in studies on the link between COVID-19 and mental health, the evidence appears inconsistent and inconclusive, at least from the limited number of studies based on objective data up to the end of 2020 (Dawel et al., 2020; Ettman et al., 2020; Pieh et al., 2020; Sønderskov et al., 2020).

For the Italian case, studies documenting the mental health effects of COVID-19 have been mainly, if not entirely, based on online surveys and self-reported mental health conditions, and several of them suffer from limited representativeness of the data and/or small sample sizes. Mazza et al. (2020); Cellini et al. (2020); Amendola et al. (2021); Gualano et al. (2021) mention that several studies reported a higher prevalence of depressive and anxiety symptoms in the Italian population during the lockdown. This appeared to have occurred against a background of reduced voluntary admissions in the 40 days after the beginning of the COVID-19 epidemic, a non-significant increase in outpatient pharmaceutical consumption in March of antidepressants, and a significant one of anxiolytics. In general, these studies find a consistent deterioration in mental health conditions (Amerio et al., 2021; Rossi et al., 2020; Amendola et al., 2021), in particular for women, young people, and patients with pre-existing conditions and a worse socioeconomic status (Delmastro and Zamariola, 2020; Fiorillo et al., 2020). The prevalence of anxiety (self-reported) and diagnosed depressive symptoms, which range between 17.6% and 41.5% for the first and between 12.4% and 33.2%, is strikingly above the prevalence of depressive symptoms reported by the Italian National Health Institute (ISS) for the pre-pandemic period (6% for the period 2016-2019).

The Italian Observatory for Drug Consumption (OSMED) led by the Italian Medicines Agency (AIFA) (The Medicines Utilisation Monitoring Centre, 2021) documents a significant increase in the use of anxiolytics and sedatives in recent years due to a recent surge in stress disorders. In particular, in Italy, the consumption of benzodiazepine increased from 47.4 DDD/1000 inhabitants in 2014 to 55.0 in 2020, while that of antidepressants increased from 39.2 in 2014 to 43.6 in 2020. According to AIFA, this increase is symptomatic of a consolidated habit of easily resorting to pharmaceutical therapies, which calls for more controlled prescribing practices. Additionally, the same report also mentions an increase in the prevalence of depression, with a consequent increase in the attention and management skills of physicians. However, only one in three patients with depressive symptoms receives pharmacological treatment. In general, the report does not find an excessive increase in consumption in these drug categories compared to historical trends.

A recurrent weakness of the literature presented above is inherent in sample selection bias and, as a consequence, lack of representativeness. Individuals who readily take surveys on the Internet are likely to represent a non-generalizable sample of the population. The correlations between mental health measures and sample selection criteria may result in inflated or deflated results. Furthermore, studying isolated snapshots of these mental health measures without establishing baseline conditions offers limited scope. Instead, longitudinal studies that monitor the evolution of mental health in response to the pandemic would provide more informative insights. Cross-sectional observations can also capture short-term fluctuations in mental health responses to the pandemic, with such an effect heightened by the subjective nature of the evaluations. The self-administration of the survey could be temporarily affected by social isolation, intolerance to uncertainty, or loneliness, influencing the well-being of people during the outbreak, without a persistent effect on mental health. This is the case, for example, of the analysis of subjective well being through social media data, such as Twitter (Carpi et al., 2021; Monzani et al., 2021). However, some of the studies available point to a significant reduction in well-being in Japan and Italy (−8.3% and −11.7%, respectively) in the first nine months of 2020 (Carpi et al., 2021), due to prolonged mobility restrictions, flu and COVID-like symptoms, economic uncertainty, social distancing, and news of the pandemic. In Italy, researchers also found that an increase in daily deaths and new daily cases due to COVID-19 caused negative emotions and somatosensory words,[5] often linked to traumatic events and symptoms of PTSD[6] (Monzani et al., 2021).

Using a more robust approach, Marazzi et al. (2022) provide a clinician-based reading of the mental health consequences of COVID-19 in Italy in the first year of the pandemic. To avoid the limited generalizability issue that pervades other studies, they use detailed data on the universe of purchases of anxiolytics and antidepressants (measured at three-level ATC codes[7]) during the COVID-19 pandemic and quantify the eventual increase in the consumption of anxiolytics and antidepressants sold in Italy in 2020 compared to 2019. One of the main advantages of these data is the absence of selection bias, because of their equivalence to administrative data. Importantly, to tease out the channels through which the COVID-19 pandemic could affect mental health in Italy, the authors of the study combine drug purchases with granular mobile phone data on mobility within and across Italian provinces. Subsequently, by adopting a multivariate regression approach, they test whether there is an interplay between the use of drugs for mental health disorders and the COVID-19 pandemic, controlling for heterogeneous local mobility restrictions, the threat of COVID contagion and the effect of mortality.

The final findings of the study is that purchases of mental health-related drugs have increased compared to 2019, but excess volumes do not match the massive increase in anxiety and depressive disorders found in survey-based studies. Furthermore, while the authors find incremental effects on anxiolytics consumption in the months corresponding to the introduction of national lockdown, they do not observe any further significant effect of mobility restrictions. The divergence of these results from those based on self-administered survey studies can be explained according to three main hypotheses. First, the mismatch is likely to be inherent in the milder nature of self-reported psychological distress with respect to conditions that require pharmaceutical interventions. Second, since these studies observed an increase in the consumption of anxiolytics and antidepressants in the last part of 2020, an important share of cases of mental health disorders could have been overlooked during the first year of the pandemic, possibly leading to the onset of more severe conditions in the longer term. Third, individuals affected by pandemic distress might have exhibited differential mental health responses due to their differential (optimal) investments in defensive expenditures, coping mechanisms or

---

[5] These are words that refer to the body and its sensations. Some examples of somatosensory words include pain, ache, hurt, and numb.

[6] Post-traumatic stress disorder (PTSD) is a mental health condition that can develop after someone experiences or witnesses a life-threatening event, such as war, sexual assault, or a natural disaster. Symptoms of PTSD can include flashbacks, nightmares, avoidance of reminders of the event, and increased anxiety and fear.

[7] ATC codes are a system for classifying pharmaceutical substances. They are used to track the use of medications and to identify patterns of prescribing and dispensing. The three-level ATC codes refer to the third level of the ATC classification system. This level is the most specific level, and it is used to identify individual medications.

compensatory behaviors, such as mild non-specific drugs, support groups or specialist psychotherapy. Therefore, estimates of the marginal effect of the COVID-19 pandemic on mental health are likely to measure the net effect, whose welfare and policy design implications highlight an important role for economic incentives in determining such defensive spending.

8.2. Changes in life expectancy

Data on mortality and excess mortality are considered reliable indicators for understanding the impact of the COVID-19 pandemic on health systems and human health because they are less sensitive to coding errors, competing risks, and mis-classification potential and, as such, allow cross-country comparisons. Although these indicators are useful and are used regularly, their metric has the disadvantage of not considering the age at which individuals die. In fact, with the same number of deaths, if the person who dies is old, the number of years lost in a population is less than if the people who died were young. To address this problem, indicators of 'life expectancy' and/or 'lost years of life' are used, which explain the demographic composition of the deceased and thus can provide impact estimates of increased mortality due to the pandemic in society as a whole.

Life expectancy is a measure of how long people can expect to live on average. It is calculated by taking the age-specific mortality rates for a given year and assuming that these rates will remain constant for the rest of people's lives. In contrast, the lost-life-years indicator considers the distribution by mortality age, giving greater weight to deaths occurring at a younger age. There is an important difference between life expectancy and lost years of life. Although life expectancy is a standardized measure based on a hypothetical life table cohort, the indicator of lost life years is calculated from the number of deaths observed in real populations. Therefore, while life expectancy depends solely on mortality, the years lost depend on both mortality and the population's age structure.

According to the latest official OECD data (OECD, 2022b), life expectancy has increased in all OECD countries over the past 50 years, although progress has slowed over the past decade. This trend was reversed in 2020, with the COVID-19 pandemic causing a widespread decline in life expectancy in most of the world. Although in 2019 life expectancy at birth in the OECD countries averaged 81 years, more than ten years longer than in 1970, at the end of 2020, it showed a decline almost everywhere (see Fig. 8.12). These negative developments are primarily due to the exceptionally high number of deaths caused by the COVID-19 pandemic (OECD countries have recorded approximately 2.0 million excess deaths). As shown in Fig. 8.12, in 2020, the annual reduction in life expectancy reached one year or more in nine countries and was particularly large in the United States (−1.6 years), Spain (−1.5 years), Poland (−1.3

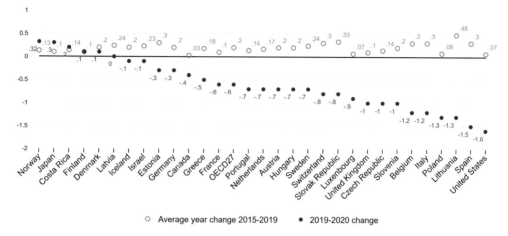

Figure 8.12 Reduction of life expectancy during the pandemic. *Source*: OECD Health Statistics 2021. Note:2020 figures are provisional for some countries.

years), Italy (−1.2 years), and the Czech Republic (−1.0 years). The only countries that have recorded an increase in life expectancy are Norway, Japan, Costa Rica, Denmark, Finland, and Latvia.

8.3. Changes in lifestyles and risky behaviors

The COVID-19 pandemic has significantly changed the lives and attitudes towards risk of many people, and this is a factor that must be taken into account. Some of these changes are due to containment measures that have been implemented at different times since the outbreak of the pandemic.

According to the OECD (OECD, 2022b), significant changes in lifestyles are indicated by the following findings:

- In four of the five OECD countries with available data, the impact of the pandemic led to an increase in the proportion of people with higher frequency of alcohol consumption, in particular among women, parents of young children, people with higher incomes and those with anxious and depressive symptoms;
- a reduction, albeit temporary, of physical activity and an increase in sedentary behavior during lockdowns;
- changes in smoking habits, with some individuals increasing daily cigarette consumption, and others, particularly older people, such as in France and Japan, reducing consumption, possibly due to the association between smoking and the risk of contracting the virus.

8.4. The lack of continuity of healthcare services during COVID-19

During a pandemic period, the maintenance of essential health services is crucial. In the specific case of vulnerable people, disruption of services such as health promotion, disease prevention, diagnosis, treatment, rehabilitation, and palliative care can cause serious adverse problems to the population and individual health, sometimes more than the pandemic itself. Unfortunately, as also reported by Xiao et al. (2021) and Bodilsen et al. (2021), the restrictive measures imposed during the pandemic period have caused a decline in access to care, surgery, and other territorial, specialist, and hospital care services.

As discussed, among several other studies, by Mogharab et al. (2022), the literature finds a reduction in emergency department (ED) visits during the pandemic period. Some studies also reported delayed emergency medical care in the case of prehospital services, such as response to outpatient cardiac arrest (Baldi et al., 2020). Others showed that untimely and improper management of emergency medical needs contributed to increase morbidity and mortality in patients without COVID-19 during the pandemic (Czeisler et al., 2020; Lazzerini et al., 2020; Maringe et al., 2020; Santi et al., 2021). Emergency departments also appeared to give comparatively less priority to non COVID patients. Diagnostic delays caused by COVID-19 have been reported to cause a significant increase in the incidence of preventable cancer deaths in England (Lazzerini et al., 2020).

An earlier study reported that 41% of individuals in the United States postponed or avoided medical care, including urgent (12%) or non-urgent care (32%) (Czeisler et al., 2020). Avoiding healthcare is a type of patient disengagement that leads them to delay seeking medical care (Byrne, 2008). During the pandemic, some people who were experiencing urgent medical emergencies also avoided seeking healthcare services because they were afraid of contracting the virus. This reduction in overall utilization of healthcare services could have worsened health outcomes for patients with other chronic diseases or acute medical emergencies (Santi et al., 2021). Although in some countries we could have expected such outcomes, where data are available, the size of the problem is emerging larger than forecasted, especially for individuals living with long-term health conditions. For these people, COVID-19 significantly affected their access to routine medical care, with many chronically ill patients experiencing severe discontinuations in in-person care during the pandemic.

To better understand the role of the COVID-19 pandemic in disrupting essential health services worldwide, since 2021 the WHO has been monitoring the global situation through its 'Global Pulse Survey', a rapid key informant survey on the continuation of essential health services during the COVID-19 pandemic (WHO et al., 2022). The third round of the survey was launched in November – December 2021, involving 223 countries, territories, and areas defined in various ways. These surveys provide a rapid and timely assessment of how the pandemic impacted over time and worldwide in terms of 'disruptions and reversals in services and responses, mitigation strategies, and bottlenecks to the implementation of essential COVID-19 tools. [...] The surveys also aim to capture the challenges health systems face in ensuring continued access to essential COVID-19 services and tools (including COVID-19 diagnostics, COVID-19 therapeutics, COVID-19 vaccines and PPE) and how countries are responding to mitigate challenges and recover services. The findings can be used to support evidence-informed planning and implementation of mitigation strategies in countries. The results may also be used to monitor the progress of multiple WHO and other response-related plans' (WHO et al., 2022).

In the same 2022 Global Pulse Survey, we read that, more than two years into the pandemic, '...worldwide health systems are still not recovering or transitioning beyond the acute phase of the pandemic, and COVID-19 continues to disrupt essential health services in almost all countries across the globe. The magnitude and extent of disruptions within countries has not significantly changed since Q1 2021, though all countries have intensified efforts to respond to health systems challenges, bottlenecks, and barriers to care brought on by the COVID-19 pandemic. The survey also highlights the impact of pre-existing health systems issues that have been exacerbated by the pandemic' (WHO et al., 2022).

8.5. COVID-19 effects on healthcare systems

At the time the writing of this book was being completed, the results of the third Global Pulse Survey (WHO et al., 2022) represented the most updated and comprehensive set of evidence on the role the COVID-19 pandemic on the supply of healthcare services across settings and platforms. The survey evaluated a total of 66 services, including primary care, emergency, critical and surgical care; rehabilitative and palliative care; and community care. In general, the results were dramatic, presenting a situation in which in 95 countries all service delivery settings and platforms were disrupted, including primary care (53% of 80 countries), emergency and critical care (38% of 76 countries), rehabilitation and palliative care (48% of 66 countries), and community care (54% of 69 countries). The results also confirm the situation recorded in the previous two surveys, with the notable exception of emergency care, which was disrupted more frequently in round 3.

Looking at individual settings and services, Fig. 8.13 shows that primary care setting services, such as routinely scheduled visits, unscheduled primary care visits, and prescription renewals for chronic medications, were disrupted in more than half of the countries covered by the survey. Such significant disruptions in primary care services should serve as a cautionary reminder for healthcare managers. They have a critical role in shaping population health outcomes, which calls for careful consideration and decision-making. As the WHO has repeatedly stated, primary care is 'at the foundation of achieving universal health care coverage (UHC), and any disruption in this setting can have a major impact across the health system on service delivery and general health and well-being of patients'. Life-saving emergency, critical, and surgical care interventions have also been disrupted, raising great concern among professionals. In particular, the data show that emergency service disruptions increased from 29% of 67 countries in Q1 2021 to 36% of 58 countries in Q4 2021. The postponement of elective surgeries was also widespread, with 59% of 71 countries reporting that they had done so. This is the main reason why backlogs have increased as the pandemic has dragged on, with 40% (21 out of 52) of countries reporting an increase in backlogs for elective surgery and procedures in the past six months. Rehabilitation services (52% in 71 countries) and palliative care services (44% in 61 countries) were also similarly affected. Some of these results, which are, on the whole, in line with those obtained in several more detailed and country-specific studies, are summarized below.

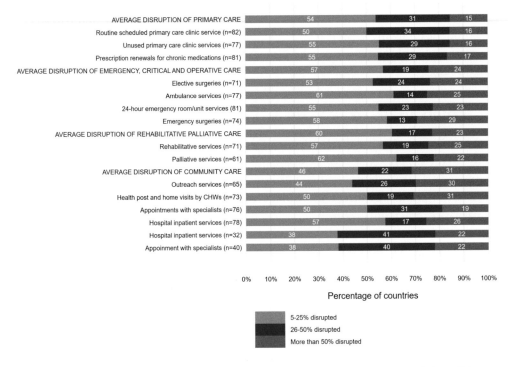

Figure 8.13 Service disruptions across service delivery settings. *Source*: WHO (2022b).

Heart disease. According to the meta-analysis done by Nadarajah et al. (2022), using data covering 158 studies in 49 countries, the negative of the pandemic on heart disease and care has been especially strong. In particular, the authors find that, in all types of heart disease and all countries studied, there were fewer hospitalizations, treatments, and healthcare appointments than before the pandemic. The impact was the most severe in low- and middle-income countries, which recorded significant increases in heart disease deaths in hospitals.

Diabetes. According to Valabhji et al. (2022), diabetes care and services were widely disrupted throughout the pandemic, from new diagnoses to critical screening and treatment programs. In England, death rates (excluding deaths caused by COVID-19) were higher among people with diabetes in 2021 compared to previous years, and this can be attributed to disruptions in routine care caused by the pandemic. This effect was unequally distributed across the population, and patients in the most deprived groups suffered harsher consequences compared to those in the more advantaged groups. Sometimes diabetic patients saw their health worsen simply because they could not have access to insulin. Two separate studies, one encompassing 163 countries and the other 47, discovered that the ongoing treatment of chronic diseases like hypertension and diabetes was the most disrupted or impacted during the COVID-19 pandemic (Chudasama et al., 2020; WHO, 2021). In Portugal, for example, the number of foot tests for diabetes treatment decreased by 24% between 2019 and 2020, while in a nationally representative sample in the United States, two-fifths of adults living with at least one chronic health condition reported delayed or forgotten treatment during the pandemic (Gonzalez et al., 2021; Serviço Nacional de Saúde, 2021).

Immunization services. This is a key sector, which could be responsible for medium- and long-term problems, as it deals with communicable diseases (not just COVID-19). According to a study by Shet et al. (2022), using data from 170 countries and territories, compared to the situation before the pandemic, the administration of vaccines for common childhood illnesses underwent a marked decline. This was mainly due to the interruption in supply and demand flows and the availability of fewer healthcare professionals to deliver vaccines. People may also have been reluctant to go to get their vaccinations due to fears of contracting COVID-19. Most of the problems were found in lower and middle income regions, which are those territories where communicable disease outbreaks could occur very easily and a low vaccination rate can cause a rapid spread of the infection (this is particularly true for future vaccine-preventable disease outbreaks, as we saw during the Ebola epidemic in Africa).

Cancer prevention and care. According to a 2022 report by Cancer Research UK, cancer care provision decreased in all areas. In the first year of the pandemic, one million fewer screening invitations were sent, 380,000 fewer people saw a specialist after an urgent suspected cancer referral, ten times more people were waiting six weeks or more for cancer tests, and almost 45,000 fewer people started cancer treatment. The same report also shows that by November 2021, cancer waiting-time standards in the UK were missed by wider margins than ever before. Riera et al. (2021) found similar results across the world. In particular, when looking at the impact of the COVID-19 pandemic on health services for cancer treatment, they found that 'the most frequent determinants of disruptions were provider or system-related, mainly due to the reduction in service availability.' The studies identified 38 different categories of delays and disruptions with impact on treatment, diagnosis, or general health services. The delays or interruptions most investigated included reduction in routine activity of cancer services and the number of cancer surgeries; delay in radiotherapy; and delay, reschedule, or cancelation of outpatient visits. The interruptions and disruptions largely affected the facilities (up to 77.5%), the supply chain (up to 79%) and the availability of personnel (up to 60%)' (Riera et al., 2021).

In terms of prevention, cancer screening, including mammography and colonoscopy, is an important component of prevention programs, with early detection of cancer strongly associated with higher survival rates. The available data indicate that all screening activities were canceled or significantly delayed. In seven OECD countries with comparable annual data, the percentage of women screened for breast cancer in the past two years decreased on average by 5 percentage points in 2020, compared to 2019. The decline was particularly significant in the early part of the pandemic. According to OECD (2020a) in Italy, the screening rates for breast (−54%) and cervical (−55%) cancers decreased significantly between January and May 2020 compared to the same period in 2019 and remained at lower levels throughout the year than in 2019 (OECD/European Observatory on Health Systems and Policies, 2021b).

Similar values were recorded in all OECD countries where data are available. For example, in France, breast cancer screening decreased significantly in the second quarter of 2020 (−44% compared to the second quarter of 2019). However, from September onward, screening activity exceeded the levels observed in previous years, with weekly screenings in January and May 2021 13% higher than the corresponding numbers in 2019 (OECD/European Observatory on Health Systems and Policies, 2021a).

These delays and reductions had an obvious negative impact on mortality due to associated delays in cancer diagnosis. For example, it is estimated that a delay in cancer surgical treatment by four weeks increases the risk of death by approximately 7%, while a delay in therapies (such as chemotherapy) or radiation therapy by four weeks can increase the risk of death by up to 13% (Hanna et al., 2020). Delays have been reported in many OECD countries, including Australia, Belgium, Canada (Ontario), Denmark, Finland, France, Ireland, Italy, Korea, the Netherlands, Slovenia and Sweden. According to OECD (2020a) in Belgium, due to the interruption of cancer treatment during the pandemic, the number of new cancer diagnoses between March and September 2020 was 5000 less than expected. In England, diagnostic delays are expected to increase mortality to five years for four types of cancer from about 5% (lung cancer) to 16% (colorectal cancer) (Maringe et al., 2020).

Reduction in access to primary care and tele-medicine. According to OECD (2020a), the tightening of restrictions in health and other areas meant that in May 2020, referrals to the General Practitioner (GP) had decreased significantly, with the number of visits to the GP falling by 66% in Portugal, around 40% in Australia, 18% in Austria and 7% in Norway, compared to the same month in 2019 (see Fig. 8.14). However, annual data between 2019 and 2020 indicate that the number of medical consultations (both GPs and specialists) per capita had not changed significantly in some countries. In these cases (e.g., Australia, Israel, and Norway), teleconsultations and tele-medicine may have helped compensate for the drop-in visits in person.

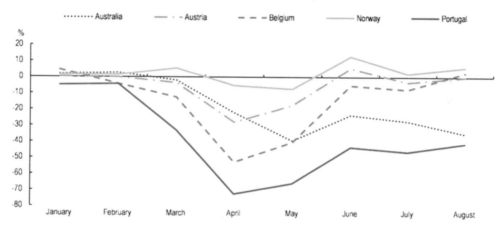

Figure 8.14 Monthly change in the total number of medical examinations in attendance (2020 vs. 2019), in selected OECD countries. *Source*: Source: Australian Institute of Health and Welfare (2020), 'Impacts of COVID-19 on Medicare Benefits Scheme and Pharmaceutical Benefits Scheme service use', https://www.aihw.gov.au/reports/health-care-quality-performance/covid-impacts-on-mbs-and-pbs/data; Helsedirektoratet (2020), 'Konsultasjoner hos fastleger', https://www.helsedirektoratet.no/statistikk/statistikk-om-allmennlegetjenester/konsultasjoner-hos-fastleger; INAMI (2020), 'Monitoring COVID-19: L'impact de la COVID-19 sur le remboursement des soins de santé', https://www.inami.fgov.be/fr/publications/Pages/rapport-impact-covid19-remboursement-soins-sante.aspx; Leitner (2021), 'Number of e-Card consultations: Analysis of eCard consultations during the pandemic/during the lockdown in 2020', Serviço Nacional de Saúde (2021) 'Consultas Médicas nos Cuidados de Saúde Primários', https://transparencia.sns.gov.pt/explore/dataset/evolucao-das-consultas-medicas-nos-csp/export/?sort=tempo. Note: Total number of monthly consultations of the GP in person in 2020 compared to the same month in 2019. The data exclude tele-medicine services and refer only to face-to-face consultations and home visits.

In general, tele-consultation services expanded in all countries. However, while the pandemic has clearly driven the adoption of tele-medicine services, it is not yet clear to what extent tele-consultations were able to compensate for the decrease in in-person visits for a wider set of countries (Fig. 8.15). As with the adoption of other digital tools, the use of digital health technologies in medicine was not evenly distributed among the population, with some groups - including the elderly, those on lower incomes, and those with lower levels of education - less likely to seek health information online.

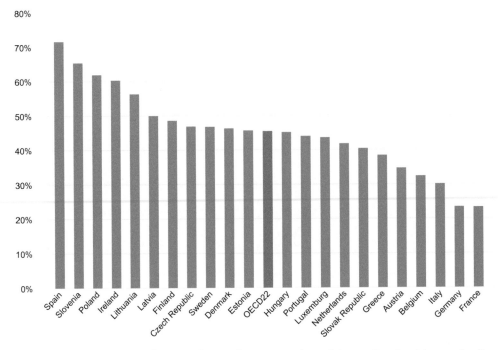

Figure 8.15 Percentage of respondents who reported receiving online health care (online medical advice or telephone) from a doctor since the beginning of the pandemic. *Source*: Ahrendt et al. (2020), available at the web address: https://www.eurofound.europa.eu/sites/default/files/ef_publication/field_ef_document/ef21064en.pdf. Note: Results based on an online survey may not be representative of the entire population. The data for Luxembourg are of low reliability.

Nevertheless, the COVID-19 pandemic has highlighted the importance and useful-ness of tele-medicine to provide a way to connect patients and healthcare professionals when a consultation in person is not possible. In the hope of curbing COVID-19, the authorities took several measures that significantly expanded access to tele-medicine during the pandemic. Tele-consultations are a safe and effective way to assess suspected cases of COVID-19 and to guide the diagnosis and treatment of the patient, mini-mizing the risk of transmission of the disease. Tele-medicine also allows many of the major clinical services to continue to operate regularly and without interruption during a public health emergency. This is even more important if access to medical care is to be provided in health systems weakened by the pandemic.

As a consequence of these developments, tele-medicine has become a useful and available tool in many countries to ensure patient care and reduce the risk of exposure to COVID-19 for patients and healthcare professionals. There is significant evidence of tele-medicine-managed patients' satisfaction with the services they received. This evidence also suggests that tele-medicine has helped to evaluate, diagnose, select, and treat patients with COVID-19, avoiding the potential complications of an emergency room or outpatient visit.

Some challenges have been pinpointed though: while telemedicine offers great potential, it cannot entirely replace in-person consultations, and there are substantial issues related to privacy, regulation, and insurance that need to be tackled by policymakers. Moreover, there are particular implications for the vulnerable population, namely people living with disabilities, migrants, the homeless, etc. These groups of people already face challenging circumstances, aggravated by the pandemic, and may not necessarily have access to tele-medicine. Proactive vulnerability-based strategies must be developed to address their specific needs. In essence, while tele-medicine can help overcome some barriers to entry, such as for people living in remote communities, it is possible that the adoption of digital services during the pandemic may have also exacerbated some of the inequalities that preceded the pandemic.

On a broader front, and in spite of the expansion of tele-medicine services, several negative effects of the pandemic on general health care were observed. Non-urgent elective surgery was postponed and waiting times increased. Treatment backlogs - of people waiting to receive treatment existed before the pandemic, but the pandemic made them much worse. To increase the capacity of health systems and address the COVID-19 wave, many countries postponed non-urgent elective surgery. As a result, the amount of time patients spent on waiting lists for many surgeries underwent significant increases. In seven OECD countries with available data, waiting times for three elective surgeries - cataract, hip replacement, and knee replacement - increased in 2020 compared to 2019. For patients waiting for surgery, the average number of days on the waiting list before undergoing the procedure increased in 2020 by 88 days for knee replacement, 58 days for hip replacement, and 30 days for cataract surgery, compared to 2019.

The number of elective surgeries that require hospital stays, such as hip or knee implants, also decreased in many countries in 2020, with a reduction of more than 25% in the number of knee implants in the Czech Republic and Italy (Fig. 8.16). Similar reductions were observed for hip replacement and cataract surgery.

Although the first months of the pandemic had the greatest impact on increasing waiting times and reducing completed treatment pathways, subsequent spikes in hospitalization for COVID-19 interrupted treatment even further. In the UK, for example, treatment activity decreased dramatically between March and May 2020, before again declining between November 2020 and January 2021, although at much lower rates

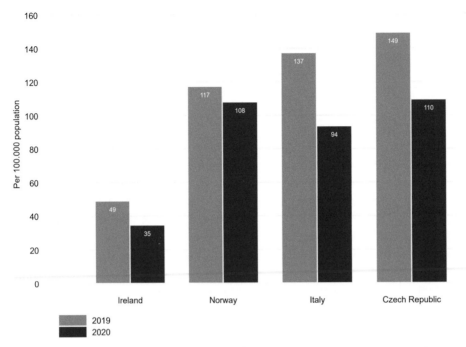

Figure 8.16 Knee replacement surgery in selected OECD countries, 2019-20. *Source*: OECD (2021a), available at the web address: https://doi.org/10.1787/health-data-en.

than during the initial decline period (Suleman et al., 2021). A report from the NHS (NHS-UK, 2022) showed that in England about 6 million people were on the waiting list for elective care (care planned in advance, instead of emergency care), compared to 4.4 million before the pandemic. For instance, in Finland, the waiting period for elective surgeries escalated by one third compared to the pre-pandemic era. However, the frequency of such surgeries rose by one fifth after the easing of lockdown measures.

Addressing the backlog of patients who need elective intervention will be challenging, particularly in countries with more limited hospital capacity and may require additional resources. Treatment delays, including surgeries, can increase preventable deaths and harm health. Delays in receiving medical care are also associated with anxiety, depression, and poor quality of life among patients and caregivers.

In anticipation and response to COVID-19 patients needing hospital care, many countries tried to increase the number of hospital beds available by redesigning hospital discharge policies and postponing scheduled hospitalizations for non-urgent care. Consequently, in five OECD countries where data is available, there was a decline in overall hospital admissions between 2019 and 2020, with reductions ranging from approximately 7% in Denmark to about 30% or more in Lithuania, Italy, and Chile (Fig. 8.17).

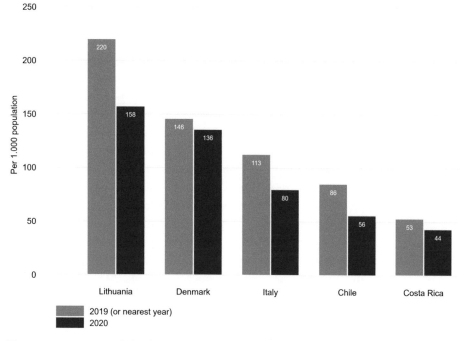

Figure 8.17 Hospital discharge rates, 2019 vs. 2020. *Source*: OECD (2021a), available at the web address: https://doi.org/10.1787/health-data-en. Note: Excludes the resignation of healthy children born in hospital (3-10% of all resignations.)

Many OECD countries have also observed a decrease in emergency visits and hospitalizations. In general, the presence of emergency declined in 2020 by more than 20% in Canada (24%), Portugal (28%), and the United Kingdom (England) (21%) compared to 2019. Reductions in activity were particularly pronounced in March and April 2020.

Another significant facet of COVID-19's effect on healthcare is the reduction in patient visits related to heart and brain vascular incidents. Data from the early months of the health crisis indicate that hospitalizations for cardiovascular events, including acute myocardial infarction and strokes, decreased by 40% or more in many countries, including Austria, Brazil, France, Germany, Greece, Spain, the United Kingdom, and the United States. Although hospitalizations for cardiovascular events declined at the beginning of the pandemic, mortality rates and complications of myocardial infarction appear to have increased dramatically (De Rosa et al., 2020; Primessnig et al., 2021). These changes are likely to be associated with fewer hospital visits among patients with milder cardiovascular events. Hospitalized patients had more severe cases than in the same period in 2019, with a higher risk of complications and worse short-term outcomes and mortality (Primessnig et al., 2021). At least some of the drivers of this increased mortality are likely to be associated with interruptions in care pathways due to health system constraints and restrictions, including ambulance response times and times required to implement critical interventions (Scquizzato et al., 2020).

Increase in health inequalities. The impact of COVID-19 on people has been highly uneven, with those most at risk and unable to access primary care resources being more severely affected. Social inequalities, exacerbated by the pandemic, have played an important role in this regard and have been particularly pronounced among some communities and minorities, historically marginalized, that were already receiving less quality care (McGowan and Bambra, 2022). Two appear to be the main causes of the reduction and/or lack of adequate healthcare care for people. First, many resources and staff had to be diverted to activities related to COVID-19 monitoring and treatment, substantially reducing the supply of care for all other conditions. Second, fears of exposure to the virus led to a significant drop in patient demand for treatment. Several earlier studies have paid attention to the impacts of the pandemic on the management of non-COVID-19 diseases and appear to be useful in the development of policies for the design of a post-pandemic healthcare system (Kendzerska et al., 2021). In the design of these policies, the most important factors are those related to healthcare delays, especially in urgent care, which are linked to higher rates of mortality and morbidity. These factors may be related to the healthcare system or long waiting times in the emergency department or may be due to patient-related factors with individuals declining care due to COVID-19 anxiety.

8.5.1 The reasons for services disruption

According to the Global Pulse Survey (WHO et al., 2022), disruption of healthcare services was caused by a combination of demand and supply side factors. This includes factors such as resource scarcity, deliberate alteration in service provision, and a reduction in patients seeking care through major healthcare platforms. Fig. 8.18 shows that the 'predominant reasons for the disruptions were intentional service delivery modifications (in 40% of countries) - such as temporary closures or postponements of services - and the lack of health care resources (in 36% of countries) - such as challenges related to health worker availability and capacities, availability of essential medicines, diagnostics, vaccines or other health products, facility infrastructure and space capacities. Decreased care-seeking due to community fear, mistrust, financial difficulties during lockdowns or other barriers to care was also commonly reported, most frequently for primary care services (in 36% of countries)' (WHO et al., 2022).

Functional supply chain systems are critical to ensure that the necessary health products are available in the right quantities to provide essential health services. Disruptions in supply chain systems can limit capacity across the continuum of care. Such issues were reported by 46% countries (38 of 83). Countries in the African region and the Americas were more likely to report supply chain problems than countries in other regions.

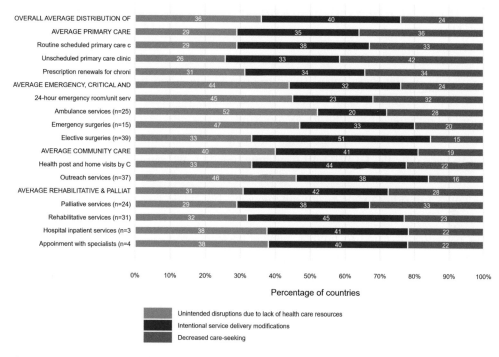

Figure 8.18 Percentage of countries reporting reasons for service disruptions. *Source*: WHO et al. (2022).

In the African region, 59% of countries (20 of 34) reported disruptions, while in the Americas, 67% of countries (12 of 33) reported similar issues. Comparing the responses over time (all countries responding to any round) shows that more countries reported disruptions in the supply chain system in Q4 2021 than in Q1 2021.

Problems in the healthcare supply chain as a cause of health service disruption (the first reported cause) deserve further review. A variety of factors led to the slow response of healthcare supply to the COVID-19 emergency. Most of them were related to export bans put in place by countries where protective garments, medical equipment, and pharmaceuticals were manufactured, port choke points and trucking bottlenecks, and the dependence on a few manufacturers of essential products. These events in turn caused additional disruptive phenomena on production and transport, through shortages of workers who were out sick or who did not show up to work. Organizational failures contributed to poor alignment and coordination between state, regional, and local agencies, as well as health care institutions. These shortcomings impacted clinical care, leading to insufficient testing capacity, lack of care coordination, and rationing of supplies.

8.6. The COVID-19 pandemic: a boost toward the European Health Union?

On 30 March 2021, the WHO released a statement signed by several top government officials and by the President of the European Council and the Director-General of the WHO, recognizing how COVID-19 had shown that united action is needed for a more robust international health architecture. In this important statement, the signatories solemnly declare the following: '[...] today, we have the same hope that as we fight to overcome the COVID-19 pandemic together, we can build a more robust international health architecture that will protect future generations. There will be other pandemics and other major health emergencies. No single government or multilateral agency can address this threat alone. The question is not if, but when. Together, we must be better prepared to predict, prevent, detect, assess, and respond effectively to pandemics in a highly coordinated way. The COVID-19 pandemic has been a stark and painful reminder that no one is safe until everyone is safe. Therefore, we are committed to ensuring universal and equitable access to safe, effective, and affordable vaccines, medicines, and diagnostics for this and future pandemics. Immunization is a global public good and we will need to be able to develop, manufacture, and deploy vaccines as quickly as possible'.

More generally, the pandemic has taught us that multilateralism and international collaboration are critical to the success of health care. The lesson has been learned in Europe, where, in November 2020, the European Commission produced a set of proposals released under the heading 'European Health Union' (EHU), limited to preparedness and resilience. It is the first time such an agreement is signed, and, we hope, could be considered the first step toward a potential more harmonized European health care system. The purpose of the proposal is to ensure the health security and safety of healthcare and to develop a stronger and more equitable European Health Union with a more harmonious collaboration between member states and stronger institutions. More importantly, the proposal implies a strong global responsibility, where multilateralism is crucial.

The proposal addresses several significant short- and long-term topics. The first group highlights issues that the pandemic has emphasized. Each of them is tackled in a manner that fosters collaboration and integration. Among these, there is a set of regulatory suggestions to broaden the scope of the European Medicines Agency and the European Center for Disease Control. Additionally, the proposal includes adopting a new regulation that enables the EU to respond to and coordinate efforts during severe disease outbreaks. Further recommendations concern the definition of the new pharmaceutical strategy, the institution of HERA[8] and the Europe Beating Cancer plan,[9] and a regulation on cross-border health threats that gives the Commission a broader right to declare a public health emergency and formalizes the role of the Health Security Committee. The last important change deals with the joint procurement of medical goods, where the Commission proposes to exclude the possibility for Member States to hold parallel negotiations with (vaccine) manufacturers as long as they want to be part of the joint efforts organized by the Commission.

On the long-term side, the EHU will address challenges related to antimicrobial resistance, the health impacts of climate change, the aging population and pressures on health systems, and evolving disease patterns. As such, it incorporates elements of the EU4Health program and its related initiatives. In specific areas of concern – namely the development and procurement of medical countermeasures (vaccines) and wider pharmaceutical supply chains – additional initiatives have been launched. The vaccines strategy aims to accelerate the development, manufacture, and distribution of COVID-19 vaccines, including through joint procurement by the European Commission, while the Pharmaceutical Strategy seeks to address structural issues within the pharmaceutical sector, primarily by reviewing the regulatory framework.

Nonetheless, creating more union in public health policy and transforming these proposals into efficient and effective institutions and processes will not be an easy task. Four issues can be briefly mentioned to better explain the challenges that must be faced. First, the EU has supplementary and coordinating competence in public health, which means that it supports and guides member states in their efforts to maintain public health. Second, the EU has the power to shape the single market for healthcare products, through the EMA, which regulates the scientific evaluation and supervision of medicinal products within the EU. Third, the EU can intervene in emergency situations to combat major cross-border health threats and stimulate cross-border cooperation (based on Article 168(5) of the TFEU[10]). The EU used this mechanism to make joint advance purchases of vaccines on behalf of member states on an emergency support mechanism designed for humanitarian assistance in the event of natural disasters (Art. 122 of the TFEU).

---

[8] HERA, or the European Health Emergency Preparedness and Response Authority, is a new agency that was established by the European Union in 2021. The agency's mission is to strengthen the EU's preparedness and response to health emergencies, including pandemics.

[9] The Europe Beating Cancer plan is a set of proposals that were put forward by the European Commission in 2020. The plan aims to reduce cancer deaths by 40% by 2040. The plan includes a number of measures, such as improving early detection, providing better access to treatment, and investing in research.

[10] The Treaty on the Functioning of the European Union (TFEU) is one of the two main treaties that form the basis of the European Union (EU). The TFEU was signed in 1957 as part of the Treaty of Rome, and it was revised in 2007 as part of the Treaty of Lisbon.

Fourth, there is still a high sensitivity to health privacy and the General Data Protection Regulation[11] (GDPR) at both the EU and the national level. Several problems encountered by the ECDC[12] in collecting accurate and up-to-date data on disease contagion will persist even after the EU has proposed to improve and expand its capabilities. The same problems will be faced by HERA. This highlights the need for a discussion with all EU member states about the definition of a set of comparable data that should be collected and shared in a timely manner by national systems. Similarly, the lack of a single capital market and delays in a truly single patent law or a single EU legal framework pose major obstacles for the EMA,[13] even though the biomedical industry is a core part of Europe's economy. These issues highlight the ongoing challenges facing the EU in its efforts to effectively regulate the healthcare sector and promote public health.

In conclusion, while the EHU is undeniably a significant step forward in terms of healthcare and safety for EU citizens, it is primarily comprised of a collection of proposals. As highlighted by Nabbe and Brand (2021), advancements in the EU's health competence have historically been driven by crises, particularly in the realm of cross-border threats and crisis management. However, given that the main objective of the EHU is to be concerned with health for all, the range of initiatives can go far beyond that. How this will evolve will strongly depend on what the EU wants, how the competences will be shared with member states, and how the EHU will be able to improve all the deficiencies revealed during this crisis. More important, a key problem is to obtain the data needed to formulate a better view on where the EU stands in public health policy and in health research, and the solution depends on sensitive issues of personal health data. 'In the coming years, a treaty change does not seem realistic but the development of EHU is possible inside the current treaties, depending heavily on political climate and choices. The main issue is to find common ground on what is wanted regarding health at the EU level. In this sense, debates on the topic and exchange on the willingness of stakeholders, EU institutions, Member States, and European citizens for the future should be encouraged to move forward on the EU health competence' (Nabbe and Brand, 2021).

Finally, as suggested by Andriukaitis (2021), the following three potential scenarios may advance and enhance the health and well-being of Europeans: *i)* doing business-as-

---

[11] The General Data Protection Regulation (GDPR) is a regulation in EU law on data protection and privacy for all individuals within the European Union (EU) and the European Economic Area (EEA). The GDPR aims primarily to give control back to citizens and residents over their personal data and to simplify the regulatory environment for international business by unifying the regulation within the EU.

[12] The European Centre for Disease Prevention and Control (ECDC) is an agency of the European Union (EU) that was established in 2005. The ECDC's mission is to support the EU member states in preventing and controlling communicable diseases.

[13] The European Medicines Agency (EMA) is the agency of the European Union (EU) responsible for the scientific evaluation of medicines. The EMA's mission is to ensure that medicines are safe, effective, and of high quality for human use.

usual (using existing legal, financial, and managerial instruments, strengthening institutions and improving the implementation of new and existing policies), *ii)* supplementing existing instruments by means of secondary legislation and the creation of new institutions, *iii)* amending the TFEU to provide the EU with the explicit legal competence in health policy to construct a real European Health Union, while preserving subsidiarity where functional. Although not fully supported by all member states, this last scenario is probably the best for the EU, as it can allow citizens to enjoy the many benefits that come from greater cooperation in the governance, planning, and management of health.

CHAPTER 8 - Take-home messages

- COVID-19 pandemic has led to dramatic loss of life worldwide and represents an unprecedented challenge to public health. Almost 6 billion people have had a significant impact on health due to the COVID-19 pandemic. Three years after the start of the pandemic, about 675 million cases of infection and approximately 6.9 million deaths from the virus were reported around the world. Europe has contributed with nearly 247 million cases and about 2 million deaths to this performance.

- Preliminary estimates also state that all-cause mortality in 2020 and 2021 increased by 13% compared to the 2015-2019 average. Life expectancy decreased by 1.2 years during the pandemic, from 83.6 years in 2019 to 82.4 years in 2020.

- The pandemic has significantly and negatively impacted mental health and led to a 38% drop in breast cancer screening and a sharp increase in healthcare spending as a percentage of GDP.

- The upheaval caused by the pandemic was devastating: Tens of millions of people worldwide have been and are still at risk of falling into extreme poverty, while the number of undernourished people, currently estimated at almost 690 million, is estimated to have increased by another 132 million by the end of 2022. All this will put great pressure on all health care systems, mainly those already strained before the pandemic start.

- Fragile people and workers in the informal economy are particularly vulnerable because most lack social protection and access to quality healthcare and have lost access to productive resources.

- Global solidarity and support are needed to overcome the health, social, and economic problems imposed by the pandemic and prevent the escalation of long-term humanitarian and food security catastrophes.

- International collaboration is essential for enhancing the resilience of the global health system. It also serves as a vital element in fostering a more cohesive and effective European Union.

PART 4

The policy analysis

CHAPTER 9

What did we learn after more than 6 million deaths?

Contents

The main objective of the previous chapters has been to explain the origin and development of a complex phenomenon that has affected humanity in different areas of daily life. The story returned so far has highlighted the many facets of a problem that has invested the whole sphere of human relations (not only those medical or epidemiological), impacting people's behaviors and expectations. If looking at the relentless dynamics of infections and deaths is one aspect that can deeply touch and make resurface the suffering of so many people, other changes, less obvious and currently difficult to quantify, could be even more worrying in the medium to long term.

While in the previous chapters we discussed the effects we should expect in the coming years, here we aim to distill the lessons learned during this challenging period to make it a treasure for the future. Therefore, we will produce a reasoned synthesis of the many phenomena and themes analyzed, trying to highlight the valuable lessons that can be learned. Given that we are still not sure whether the pandemic has ended, or is

The COVID-19 Disruption and the Global Health Challenge
https://doi.org/10.1016/B978-0-44-318576-2.00023-8

crossing an unknown transition phase, more lessons can also be learned in the coming months. Over time, any piece of information and any new evidence that will be added will help reconstruct a comprehensive framework that will allow us to understand the mechanisms of virus action, the effects on health, the strategies of intervention, and the solutions to be proposed. In what follows, to help present the chapter content, the lessons learned are divided into two categories: those related to managing the events that occurred between January 2020 and September 2022 and those associated with managing events that may happen in the future. Although in the first case it is only a matter of preserving the past, the lessons learned in the second case may help better manage future events.

In general, we aim to elaborate further on the topic of market and government failures, which seem to plague our current economic system. These failures have manifested themselves in a spectacular way in the case of the pandemic, determining disruptions that have caused great human suffering and economic losses. On the one hand, governments have not responded effectively to the outbreak of infections and have been shown to be unprepared on a broad front, ranging from health management to logistic. On the other hand, the market has been shown to be highly vulnerable, with too long and insufficiently diversified value chains, uncoordinated trade models, and fragmented reactions. The general public has long appeared to be at the mercy of failures, disorientation, and panic, under a flood of often contradictory instructions and unclear or misleading information.

At the same time, some positive and extraordinary responses have been obtained from the pandemic disruption, resulting, for example, in the great acceleration of scientific work, as well as institutional coordination. We thus must also ask whether the nature of market failures, as unplanned externalities, may not be, in the end, at least partly benevolent. In the long run, short-term disruptions generated by the pandemic can have a positive impact if they contribute to the advancement of knowledge and the improvement of state and market institutions that can improve the future capacity of the system to prevent and control similar outbreaks. As in the case of the creative destruction process described by Schumpeter but reversing the order of causality, disruptions may be necessary from time to time to induce sufficiently large innovations.

9.1. More than 6 million deaths… and we are still learning!

Ever since SARS-CoV-2 was recognized as a global threat, several studies and scientific collaborations have been initiated in each disciplinary area to help combat and prevent this pandemic. Since then, scientific publications on the subject have grown steadily. According to https://covid19primer.com/dashboard, from March 1 to December 31, 2020, the number of documents collected increased from less than 700 to more than 95.000. From January to June 2020, the Journal of the American Medical Association

(JAMA) alone received more than 11.000 submission requests, a value almost triple that of the previous year (Bauchner et al., 2020). The science of COVID-19 has been produced and circulated (and still is) at an incredible rate. During the initial phases of COVID-19, the median time from submission to acceptance of an article was only six days, compared to approximately 100 days during the pre-pandemic (Palayew et al., 2020). The data in Fig. 9.1 show that the knowledge about this topic is still fluid and slowly consolidating over time. Much of what was considered a clinical best practice in March 2020 could now be recognized as ineffective or, in some cases, as serious clinical errors.

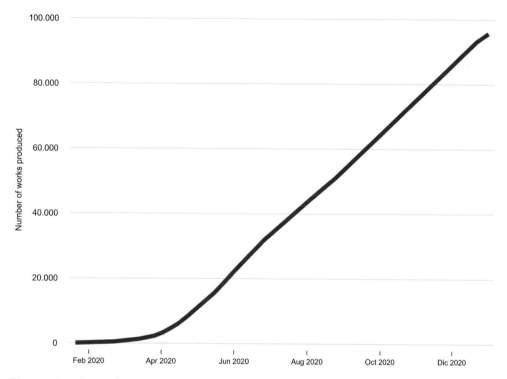

Figure 9.1 Scientific progress in the fight against COVID-19: number of works produced (January 15 - December 31, 2020). *Source*: COVID-19 Primer available from: https://covid19primer.com/dashboard.

Pandemics and health crises typically present remarkable chances for swift generation of scientific information, but the speed and extent to which this opportunity has been taken with COVID-19 was rather unexpected. This can only be partially explained by the extraordinary amount of research funding made available by governments as well as targeted funding opportunities and collaboration with industry partners. However, this 'super-production' yields at least two important side-effects. The first concerns the difficulty of taking into account what is being published and by whom (to avoid duplication of research activities). The second has to do with the quality of research and the results' reliability, which may not always be at an acceptable level.

As Rochwerg et al. (2020) document, several international and government organizations have worked to collect and organize the vast amount of scientific material that has been produced about COVID-19. For example, the WHO maintains an updated real-time COVID-19 website that contains reliable information on the epidemic (http://www.who.int/health-topics/coronavirus). The website of the US CDC includes updates on the status of viruses in the United States, travel restrictions, and a world map highlighting areas with COVID-19 cases (http://www.cdc.gov/coronavirus). The Johns Hopkins University operates a website (http://www.gisanddata.maps.arcigis.com) that provides up-to-date and reliable data describing the number of people infected by gravity and by country, as well as the number of deaths' (Rochwerg et al., 2020).[1]

The second side effect is even more critical, as unreliable data increase the level of disinformation (especially in the early stages of knowledge development) and could thus contribute to create misconceptions or even public hysteria. In periods of uncertainty, it can be challenging to understand how reliable the information made available by the various media is. Usually, in the early stages of an epidemic crisis, journalists only have access to partially reliable information. Still, this limit is not enough to stop their work of diffusion and commenting on the bits available, which paradoxically may give rise to controversies and more entertaining communication. In fact, it is not uncommon for media to rely on information that is not fully controlled, much of which ends up proving wrong once the entire story becomes clear.

The coronavirus pandemic has clearly shown that a large amount of information can be produced quickly and easily shared. For example, Santos et al. (2020) performed a bibliometric analysis of global academic research activity (not only medical) on COVID-19 topics by collecting information on more than 40.000 articles whose publication period was between January 2019 and July 2020, indexed or made available until July 2, 2020 (when data were collected). All articles are available in Scopus databases (https://www.scopus.com/), PubMed (https://pubmed.ncbi.nlm.nih.gov/), arXiv (https://arxiv.org/covid19search), and bioXiv/medRxiv (https://connect.biorxiv.org/relate/content/181) and are easily accessible to all researchers. However, the high-speed and accelerated information throughput process highlighted

---

[1] Another important WHO site is the one available at the web address: https://www.who.int/emergencies/diseases/novel-coronavirus-2019/global-research-on-novel-coronavirus-2019-ncov. This site aims to collect in a single access point all the results obtained by scientists and global health professionals from all over the world to accelerate the research and development process and produce new knowledge to contain the spread of the coronavirus pandemic and help treat infected people. The project was activated to accelerate the development of diagnostics, vaccines, and therapies to combat the coronavirus.

in this survey is likely to bring both costs and benefits, with no clear tendency for one of the two variables to prevail over the other.

Costs are mainly associated with the asymmetrical nature of the information available, which consists primarily of specialized knowledge and is thus challenging to convey to those who are not experts in the field. This asymmetry of information also leaves room for disinformation, which can result in incorrect health-related decisions such as isolation orders, travel bans, population quarantines, and even discriminatory actions against individuals from specific countries or of particular ethnic origins. Furthermore, costs may be incurred for using non-approved therapeutic or prophylactic interventions or for using them outside the context of an approved clinical trial (a classic example was hydroxychloroquine, endorsed by former US President Trump). According to Santos et al. (2020), in the best case, 'these effects tend to increase the amount of noise, thus limiting our collective ability to discover new ways of treating patients'.

The benefits of multiple sites are the availability of access to an enormous amount of information in near real-time. The significance of this is well illustrated by the occurrences during the H1N1 influenza (swine flu) epidemic. According to PubMed, although more than 20.000 citations related to H1N1 influenza were published since 2009, the vast majority (about 14.000) were published after 2011, i.e., more than two years after the main phase of the pandemic. The main determinants of research delays include the competing interests of the investigators, regulatory barriers, the time spent developing the protocols, ethical approval, peer review, and the uncertainties related to the publication process. This classical research model does not fit well with pandemic research, where rapid information is needed to fill the gaps and address public concerns. For COVID-19, some of these traditional delays have been by resorting to unprecedented procedures. For example, many major general medicine journals, including the Journal of the American Medical Association (JAMA), the New England Journal of Medicine (NEJM), and The Lancet, have prioritized COVID-19-related publications. This has been facilitated through invitation-based content and accelerated peer review processes. JAMA, The Lancet, and NEJM also run a Coronavirus Resource Center that includes research and multimedia content (http://www.jamanetwork.com/journals/jama/pages/coronavirus-alert), where most online content is free. Providing peer-reviewed and easily accessible content has also helped overcome some of the widespread disinformation in non-specialized media (Santos et al., 2020).

We can expect that many more studies will become known in the coming months. They will partially confirm what we already know, but are also likely to somewhat change our knowledge and improve the process of managing infected patients. Like all scientific processes, this will be a long and troublesome process. The extent of difficulties experienced will depend greatly on how researchers, doctors, and the media will be able to cooperate.

9.2. The scientific studies' validity problem

Editorial boards of scientific journals (which have accelerated peer review times for coronavirus articles) and managers of preprint servers (which have become an alternative consolidated publication system) have helped to rapidly disseminate knowledge among researchers (avoiding, for example, duplicating efforts) and useful responses among public health experts, political officials, doctors, and patients (all seeking a guide among several choices and behaviors).[2]

This new approach to fast information sharing has the undisputed merit of quickly circulating content on a research topic where the time factor is crucial. Being able to quickly answer questions like 'can a healed patient get the disease again' or 'what techniques will provide faster and more accurate tests?' or finally, 'how effective can social displacement be in slowing the spread of the disease?' high-speed information sharing may appear essential to issue 'now or never' recommendations to avoid contagion and often death. However, this process is not without problems. In recent months, researchers have been divided between the need to release fully controlled information, which can take weeks or months, and the urgent public need for information during a devastating pandemic. As Yassine Khalfalli pointed out in an article in the Undark digital magazine, publishing more or less quickly is not simply an academic issue for the research community; more and more policymakers have turned to the scientific process to guide their decisions, not to mention doctors who trust it to find ways to treat their patients and save lives.[3]

The trade-off between fast sharing and controlled information arises when the pre-print process cannot ensure an adequate quality of the information that is made public. With pre-prints, researchers send their research work to a server that posts it online quickly (in a few days). At that point, without going through a formal peer review process, the work is available to anyone, conveying the positive (solutions) and negative (errors or inaccuracies) aspects. The most worrying aspect is that information is read by non-experts, including journalists, policymakers, and researchers in other fields, who may not be able to discern the good from the bad, contributing to the dissemination of misinformation. According to an editorial in The Lancet, preprints allow researchers access to much more information, but without peer review and editorial control, some results may be completely unreliable and even dangerous to public health. The editorial

[2] The first platform, arXiv, was created by an American physicist in 1991 to allow a relatively small community of physics and mathematicians to share their scientific results. Subsequently, in 1994, a similar service was developed for the social sciences and humanities. Biomedical sciences arrived only in 2013 with the *preprint* platform for biological sciences, bioRxiv, followed in 2019 by one dedicated to medical disciplines, medRxiv. Since 2014 there has been an explosion of similar initiatives with the birth of at least 18 new servers in various scientific disciplines.
[3] The full article is available from the following web address: https://undark.org/2020/04/01/scientific-publishing-covid-19/.

board of the Lancet believes that, in general, preprints are an important part of the spread of science and argues that they will continue their pre-press offering with the journal, emphasizing the importance of explaining the preliminary nature of that kind of contribution (Lancet, 2020).

This aspect has been particularly emphasized in the context of the COVID-19 crisis. But this can lead to critical communicative consequences that affect the lives of millions of people. During the first months of spread of the contagion, there have been several examples of this type. One of the most cited examples of this was when a WHO expert stated on June 10, 2020 that the spread of COVID-19 by people without symptoms was 'very rare.' However, the WHO retracted this statement the next day, admitting that it was based on a 'misunderstanding' of a few studies. Another critical example occurred in early June 2020 when researchers from leading research institutions withdrew two papers related to anti-COVID-19 therapies, one on hydroxychloroquine and another investigating blood pressure drugs. In both cases, the results were alleged to be based on data from studies involving thousands of patients in hundreds of hospitals around the world. However, the company that managed the data collection had refused to provide the referees with the entire data set.

Two other interesting cases have received considerable attention. In the first, a group of Chinese researchers claimed to have found evidence of a higher risk of SARS-CoV-2 infection associated with patients with the type 'A' blood group. Subsequently, the study was rejected but gained extensive media coverage (from 126 newspapers and 13 blogs and was mentioned in 3.958 Twitter accounts). The second case concerns research published in January 2020 on a preprint platform by a group of scientists from the Indian Institute of Technology in Delhi. The blog propagated conspiracy theories that the virus was a man-made biological weapon. It claimed evidence of a 'strange similarity' between the new coronavirus and HIV. Although the research was retracted merely two days after publication, the articles remained in circulation online and have continued to spread misinformation about the virus's origins.[4]

Many additional examples of disinformation from insufficient control of research studies can be cited. According to the Retraction Watch website, the number of articles once published and later withdrawn as containing wrong information (negligent or malicious) is much higher than one might think. Founded in 2010, during its first year of operation, the site initially reported about 200 retractions, which had increased to 21.792 in January 2020.[5] Compared to the COVID-19 theme, at the end of 2020,

---

[4] This case was further relaunched when Luc Montagnier, 2008 Nobel laureate for HIV studies, repeatedly cited the study to support a conspiracy theory suggesting that Chinese scientists created the virus while they were developing an HIV vaccine.

[5] Retraction Watch is a blog that monitors retractions of scientific articles and related topics. It was launched in August 2010 by Science journalists Ivan Oransky and Adam Marcus, respectively, vice president of the

more than 40 papers were withdrawn.[6] The phenomenon of retraction of scientific articles involved prestigious journals such as The Annals of Internal Medicine, Cellular & Molecular Immunology, The Lancet, and The New England Journal of Medicine.

Of course, this evidence has led many people (often non-specialists) to question the role of science and, especially, the role of scientists, criticized for publishing hasty and potentially poor articles during the pandemic. This has also led to a parallel pandemic of rumors, unverified claims, and harmful falsehoods. Many have argued that the blame is to be placed not solely on politicians and social media, and even proposed the broader question of whether it is still possible to trust science. However, from the beginning, most scientists working in the field were very clear in their warnings about how little was known about the 1918 influenza pandemic, SARS and MERS, and how their initial views on coronavirus transmission were anchored to their understanding of SARS and MERS, which turned out to be a totally misleading reference for COVID-19. If this is an usual way for scientists to proceed (an update made possible by new information that has arrived), this experience led many non-specialists to give up the idea that specific competencies should always be heard with priority (Van Doren, 2020).

9.3. Lessons learned during the epidemic management

As mentioned in Chapter 3, the initial stages of the pandemic and its management were characterized by confusion and lack of information. Despite the existence of previous events and alarm bells by experts and organizations, all countries have been unprepared to manage a pandemic, some even more culpable given the examples that preceded them (see, for example, what happened in the United Kingdom and the United States).

In many countries, decisions were made without considering all the necessary information. One major issue was the lack of an infrastructure capable of consistently gathering and analyzing data, not just related to the spread of the virus. This meant that contact tracing was not possible during the initial outbreaks. Decision-makers struggled to regain control of the situation, especially since the majority of infected individuals showed little to no symptoms. Additionally, there were delays in making decisions, and the measures implemented varied across different areas, despite the need for uniform action. These missteps generated a cascade of related effects, preventing the ability to understand the severity of the pandemic and plan appropriate economic policies. After almost three years since the outbreak of COVID-19, several lessons appeared to have been learned from these early mistakes. Here, we discuss the most relevant ones.

Medscape website and editor of the Medical Journal Gastroenterology & Endoscopy News. The link to the website is https://retractionwatch.com/.

[6] The list of removed works is available at the following web address: https://retractionwatch.com/retracted-coronavirus-covid-19-papers/.

First, it is clear that the asymmetric nature of the information affected all prominent political leaders in the world and their scientific advisers, who were unwilling or unable to engage in effective communication with scientists and health experts and systematically underestimated the danger of SARS-CoV-2. The initial underestimate of the problem came from no other than China. Chinese authorities minimized the threat of the epidemic when it first started in December 2019 by not communicating crucial information and allowing millions of people, including potentially infected individuals, to travel around the world. The reports of the new coronavirus first emerged in Wuhan in December 2019, but many reputable research institutions have raised serious concerns that the virus may have been circulating as early as November 2019. On December 31, 2019, the Chinese government issued a statement to the WHO stating that the disease was 'preventable and controllable'. Only at the end of January 2020 Chinese authorities identified the disease as a new strain of coronavirus and recognized the possibility of spreading by contact among humans. But the government allowed millions of people to travel before starting the blockades. The mayor of Wuhan stated that one million people had left the city before the lockdown began, while the New York Times said that at least 7 million people had left the city.

Second, as mentioned in Chapter 3, the WHO also appears to share responsibility for the mismanagement of information at the beginning of the pandemic. Despite its crucial multilateral role, the WHO delayed the release of information from Beijing and declared the emergency almost a month after the situation in China had reached a point similar to the SARS crisis. Internal communication documents also suggest that the WHO did not pressure China and only raised complaints internally without taking any official position.

Third, as the second most affected country in the world, Italy was itself responsible for many information and action delays, extensively discussed in Chapter 3. For example, on February 23, 2020, Council President Giuseppe Conte downplayed the spread of the virus, attributing an increase in positive cases to the increase in testing. A few days later, Italian Foreign Minister Luigi Di Maio accused the media of inflating the gravity of the virus, stating that only the '0,089%' of Italians were in quarantine and defining the subject media coverage as a case of 'infodemia'. We need to wait until March 10, 2020, as cases of contagion continued to rise, to see the Government taking a more serious approach, implementing a national blockade, and warning the rest of the world to 'be ready'.

In the US, former President Trump had consistently minimized the outbreak from January to March 2020, making statements such as 'we have (the virus) completely under control', 'it will disappear', and 'America will be open to business again and soon'. In an election rally at the end of February, Trump called the coronavirus a 'new hoax' created by the Democratic Party. In addition, he downplayed the threat of the epidemic in early March by comparing it with seasonal influence. Also, the country did not establish a national lockdown: each State established its own guidelines, and eight States hesitated for a long time before issuing lockdowns or even masks and social distance orders.

In Britain, Prime Minister Boris Johnson and his Cabinet considered the coronavirus a 'moderate risk' until late February 2020, seeking at first to apply a 'herd immunity' strategy rather than implementing containment measures. In early March, Johnson said that he had shaken hands with coronaviruses in a hospital setting. A few days after the announcement, Johnson was infected and had to be transferred to intensive hospital care. By insisting on underestimating the problem, in mid-March, when bars and restaurants began closing across Europe, Johnson merely urged citizens to avoid eating out of their homes. The United Kingdom soon became one of the countries with the highest mortality rates from the pandemic. In Spain, Prime Minister Pedro Sánchez allowed large gatherings of people in stadiums and sports events just as the COVID-19 cases soared dramatically. On International Women's Day, he agreed that a rally should be held in Madrid with 120.000 people. He did not embark on a national blockade until March 2020, which many observers criticized as out of time (although there were fewer infections in Spain than when Italy, the UK, and France declared their lockdowns).

Many countries adopted a wait-and-see approach to the COVID-19 pandemic, even though other countries, such as Brazil, had already shown the negative consequences of this strategy. Underestimating the severity of the situation led to many mistakes, as leaders failed to follow the necessary protocols for dealing with an epidemic, including the following 'golden rules':

- isolate and trace not only symptomatic cases but also their contacts to immediately stop them on the ground to avoid spreading the virus;
- not underestimate the number and role of asymptomatic or pathological symptoms, which may favor the spread of the virus. Testing should be done to all, not only symptomatic patients, practice initially adopted in many countries (including Italy);
- inform and engage citizens who have a history of being responsive to risk (such as avian alarms and SARS) and who have demonstrated a high degree of responsibility in previous crises;
- provide sufficient stocks of diagnostic tools and protective equipment for health professionals;
- have an efficient system of tracking and monitoring;
- activate pathways of access to alternative care and dedicated health areas;
- isolate hospitals and nursing homes for the elderly from the rest of the world.

Catastrophic results were predictable outcomes of the failure to observe the basic and well-known rules to contrast the pandemic. This was overwhelmingly caused by the chronic asymmetry of information held and generated by experts, policy makers, and the general public. Information on the disease was difficult to communicate, but also withheld and manipulated in a context of low cooperation with the Chinese scientific community, which was suspected of the dubious quality of many studies that initially examined the virus. Many deaths caused by the pandemic could have been avoided, especially among healthcare professionals. They were caused by ignorance and/or failure to apply basic protocols against contagion within hospital settings, lack of connection between hospitals and territory, problems of encroachment of intensive therapies, lack of materials to ensure the medical staff's protection, and many other organizational failures.

The effects of these errors were later exacerbated by a series of biological and economic factors that led to the opacity and unpredictable development of the disease. The main biological factor was the ability of the SARS-CoV-2 virus to spread quickly and the presence of so-called 'super spreaders', that is, infected persons who, for reasons not well understood, infect a large number of people. A second factor was the novelty of the virus, so that the lack of immunity in the population made it easier for the virus to spread. Furthermore, many infected individuals did not show symptoms, making it difficult to identify and control the spread.

Globalization also facilitated the rapid spread of the virus. However, the fact that the virus originated in a city of 11 million inhabitants, which is hyperconnected by land, sea, and air, and in a country, China, which is considered the 'world factory', made the global spread of the infection especially large and dramatic. The high degree of connectivity between regions and countries facilitated the rapid spread of the virus across borders, making it difficult to control and contain. In sum, the combination, in time and space, of many errors and unprecedented conditions produced a 'perfect storm', which, by the end of 2020, caused more than 83 million infections and about 1,850,000 deaths. And this was just the start!

9.3.1 Maintain maximum transparency and clarity in the communication of decisions

In a global pandemic context, it is essential to have clear and accurate public health information to allow the population to make informed decisions about health protection behavior (Garfin et al., 2020). In addition to creating confusion and poor decision-making, the unclear information reinforced by the media could also undermine the trust of people in science, a key long-term factor in population behavior in cases of epidemics (Brzezinski et al., 2020; Plohl and Musil, 2020). Unfortunately, as we've previously noted, significant imbalances marked the available information, which made it difficult to share scientific knowledge. This resulted in the potential for misinterpretation and manipulation by media and policy makers. Therefore, clarity and precision in communication were often difficult to achieve and were not pursued with sufficient determination. As a consequence, incorrect information and disinformation were often spread, putting lives at risk.

In addition to the possible asymmetry of the underlying information, however, it should be stressed that the exercise of communication on public utility issues and public policies is not always straightforward: the articulation of the arguments, the difficulty of formulating a synthetic and clear message, and the specificity of the language make the task of the media professionals particularly difficult, and their behavior prone to errors and misunderstandings. A special condition of exceptional circumstances and unprecedented events can partially justify why so many improvisations, misleading messages, retractions, and announcements have changed previous statements in public communications since the beginning of the COVID-19 crisis. However, both governments and public health agencies, as well as journalists and media, share the responsibility of failing to ensure that their messages were clear and compelling.

With the COVID-19 situation constantly evolving, it was difficult to properly communicate the government's actions and the preventive measures imposed on the population. There was a shortfall on the part of communicators, who leaned on traditional information processes and communication languages in the absence of a reliable reference model. The last major global pandemic occurred more than 100 years ago, making it difficult to develop effective communication strategies. On the other hand, government leaders were forced to address the COVID-19 crisis in real time, often responding more to people's feelings than to the suggestions of experts (who often did not have adequate information on which to base their judgments). This meant that most leaders tried to walk a fine line between their commitment to protecting public health and their attempt to spread some optimism on scenarios for the recovery of normality, almost always not knowing which was the best choice.

Communicating proved to be more difficult as time passed because political bodies tended to be progressively less accommodating to the government than they had been in the past. In Italy, for example, a conflict between parliament and the government had begun to mount during the pandemic well before the latest national elections and may have been partly responsible for the election results. Several months after the outbreak began, many people were able to compare situations between different jurisdictions and realities, questioning local and national political choices.

To better understand how challenging it is to communicate about the issues raised by the pandemic, another interesting aspect to consider is the presentation of the data. According to a study by Romano et al. (2020), the way data are presented 'has important consequences on how people understand and react to the information transmitted'. The media usually report on the progress of the epidemic using charts showing the evolution of the number of COVID-19-related infections and deaths in a given period and in an area using a linear scale (The Washington Post; Vox) or logarithmic (Financial Times; The Guardian; New York Times). Experts in the field of epidemiology and statistics recognize that the type of scale used for visual representation can significantly influence our understanding of the phenomenon at hand. A logarithmic scale, for example, is often useful in highlighting the rapid, compound growth often seen in epidemics, showcasing how cases can double or triple over a specific period. On the other hand, a linear scale tends to provide a clearer picture of the time-based progression of the epidemic, helping us understand how the situation evolves day by day or week by week. However, non-experts may lack the capacity to discriminate between these two types of information, so that the message that they take from the mere visual inspection of the diagram may result in highly distorted impressions.

For example, through a series of experiments, Romano et al. (2020) demonstrate that 'when people are exposed to a logarithmic scale they have a less accurate understanding of how the pandemic has happened so far, they make less accurate predictions about its future and have different political preferences than when exposed to a linear scale'. Ryan and Evers (2020) also obtained similar results, confirming that the scale of the graph affects policy preferences and that people have trouble understanding the implications of logarithmic scales or even logarithms altogether.[7]

Another important issue of transparency and communication is the role plaid by disinformation activities. Disinformation depends, at least in part, on the asymmetric nature of information since this prevents clear communication and allows various types of manipulation on the part of those who hold or claim access to 'inside knowledge'. In the words of OECD (2020b), 'misinformation is influencing countries' responses to the global pandemic by undermining confidence, increasing fears and sometimes leading to harmful behaviors. When public trust and compliance with measures (from blockades to hygiene guidelines) are of the utmost importance, a wave of disinformation is undermining governments' responses to the COVID-19 pandemic and putting people's health at risk. Unproven medical treatments, prevention techniques, and other information have been flooding the Internet and disseminated by users whose concerns were reinforced by the vast volume of conflicting information. Therefore, the fight against the so-called info-demics has been one of the priority lines in the management of the coronavirus pandemic' (see also WHO, 2020b). The problematic information that circulates about the virus has also become increasingly complex. Unlike the first phase of the pandemic, a smaller share of the content is completely invented in the current widespread disinformation. According to a Reuters Institute sample analysis, up to 59% of false COVID-19 content relies to some extent on accurate information that has been manipulated, while 38% is completely invented (Brennen et al., 2020).

The activist group Avaaz has published a report that highlights the importance and widespreadness of health disinformation on Facebook during the pandemic. The report also emphasizes how society's failure to keep people safe and informed during the pandemic has had devastating consequences.[8] At the same time, anti-vax communities,

---

[7] It remains true that sometimes, even among experts, there are problems in understanding and interpreting graphs that use the logarithmic scale (Heckler et al., 2013; Menge et al., 2018).

[8] Avaaz is a global web movement to bring 'people-based political decision making' everywhere. The Avaaz community campaigns in 15 languages and comprises a *core team* available on 6 continents and supported by thousands of volunteers. They act by signing petitions, funding media campaigns and direct actions, sending emails, calling, and lobbying governments, and organizing 'off-line' protests and events. All this

conspiracy theories, and fake healthcare were rampant on websites and social networks. In the period examined, for example, false or misleading health articles were seen about 3.8 billion times on Facebook and peaked during the pandemic. Avaaz' research focused on 82 websites known for spreading false or misleading health stories and had millions of views by April 2020, amid one of the worst months of the pandemic. The gravity of the situation can be better understood by looking at the graph in Fig. 9.2, which shows that in 2020, the content of the first ten websites that spread health disinformation on the Facebook platform received almost four times more views than the equivalent content of the websites of the 10 most important health centers. These include WHO, ECDC, and the leading health institutions in France, Germany, Italy, the United Kingdom, and the United States. Avaaz also stated that the 82 websites examined represented a limited sample and that the extent of the problem highlighted by the report is likely a conservative estimate.

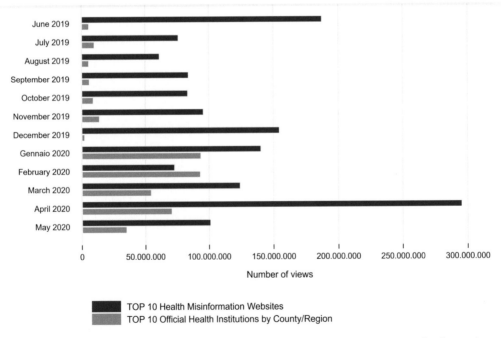

Figure 9.2 Overall content views of the top 10 disinformation *vs.* top 10 sites of information (January-May 2020). *Source*: Avaaz (2020), available at *web*: https://avaazimages.avaaz.org/facebook_threat_health.pdf.

aims to ensure that people's opinions and values around the world inform decisions that affect everyone. The report quoted here is available at the following website: https://avaazimages.avaaz.org/facebook_threat_health.pdf.

Disinformation threatens the effectiveness of emergency measures from a behavioral and cognitive point of view. It also contributes to an information overload that can obscure or hide important content (Bawden and Robinson, 2020). The problem of infodemics requires more effort from the public, which often shows a limited capacity for attention. Polarization and mistrust resulting from disinformation have long-lasting negative implications for government action, democracy, and inclusive growth. In the words of OECD, 'the implication for public policy is that increasing the volume of official and truthful information will not necessarily be more effective unless this content is made more convincing and provided to various audiences through their preferred channels and with an understanding of behavioral and psychological prejudices' (OECD, 2020b). This is particularly important for the young audience, who tend to access news primarily through social networks.

Lastly, it's worth noting that during crises, it's vital for policymakers to convey the inherent uncertainties surrounding issues and impending decisions, ensuring the public fully understands the context. It's equally crucial to closely align messaging with security and intelligence sectors combating foreign influence efforts, as well as with civil defense agencies spearheading crisis management.

9.3.2 Do not waste time (even) if you have time

One crucial lesson learned is that time plays a vital role in effectively managing a pandemic. It is imperative to invest early in preventive resources that can help mitigate the risk of virus transmission. This includes expanding and enhancing detection capacities and screening programs, and leveraging technology-driven tracing opportunities to ensure real-time contact research capabilities. The window of opportunity for full containment, especially in cases of asymptomatic transmission, is limited to a very low number of infected persons, which may occur in a reduced period. Therefore, it is necessary to develop early warning systems with rapid detection capabilities to reduce the time required to intervene. Such systems should be based on response times of hours instead of days: timing is crucial during a crisis, since, for example, a delay in introducing a blockade can add thousands of deaths.[9]

Unfortunately, in the first weeks of the epidemic, the decision-making process did not respond to these urgent requirements and became instead lengthy and cumbersome. Governments around the world gradually and slowly moved from bland suggestions to recommendations. Only when the situation had fallen out of control did governments and other empowered institutions take effective action. The process of public risk perception and the move toward compliance with the new rules also took time.

---

[9] In terms of option theory, in this scenario, the cost of inaction is higher than the cost of action, even though we may not have all the information we would like. Delaying action could lead to permanent harm, even though we are making the decision based on incomplete information.

According to Hoffman and Silverberg (2018), in global epidemics, it may be physiological to observe significant delays between the onset of the infection and the response through collective action. However, studying the experience of the three public health emergencies of international interest (PHEIC) of the last 20 years (H1N1, Ebola, and Zika), Hoffman and Silverberg (2018) conclude that deferred global mobilization (due to politics) is a greater source of delays than poor surveillance technology. Additionally, response times are reduced when the event occurs not in the midst of the holiday season and when the population is not affected by other epidemic events (e.g., influenza). In contrast, it appears that the severity of the pandemic (in terms of mortality) or the number of people involved (in terms of contagion) can reduce response times. It is clear that in the case of COVID-19, delays have accumulated in terms of joint efforts and the ability to monitor infection. Furthermore, the fact that the epidemic was partially coincident with the Chinese New Year holiday period in China and that Europe was under seasonal influence further contributed to the delay.

Several studies, e.g., Anderson et al. (2020); Berger et al. (2020); Chater (2020); Emanuel et al. (2020); Hansen (2014); Lazzerini and Putoto (2020) analyzed how policymakers intervened in the management of the COVID-19 pandemic, considering the mindset and the psychological approach of these policymakers in responding to the crisis. A first consideration emerging from these studies concerns the brain trying to organize a large and often confusing mass of partial information in a coherent interpretation (Chater, 2020). Efforts to distill a decision from an overwhelming and muddled influx of news may lead to choices based on a single interpretation of the available information, without giving credit to further alternatives. Sometimes, such behavior can help and lead to the right decision; some other times, it may be the beginning of a series of wrong and even disastrous actions.

According to Berger et al. (2020), policy makers around the world received different information and views on potential epidemic scenarios, resulting from divergent expert assessments or different modeling forecasts. Two basic behaviors can be distinguished: *i)* balance the various alternatives and wait, or *ii)* choose one and act. According to Chater (2020), three common and dangerous ways of interpreting complex and uncertain situations using a stylized narrative can be identified: 'a storm in a teacup', 'the house on fire', and 'holding the tide':

> *'The first interpretation is the mind's natural default: most alarms are false alarms; most panics are overblown—so probably this one is too. China's now-notorious early attempts to suppress news of the outbreak make sense only in the 'teacup' interpretation; so too does the US's initial downplaying of the crisis with the US President's comment on 24 February that the virus is 'very much under control in the USA.' This interpretation aims to avoid the danger of unnecessary panic in the belief that the problem will resolve itself on its own (for example, with the arrival of warmer weather). The 'house on fire' interpretation responds to an opposite urge, and it has driven unprecedented lockdowns in China, South Korea, and Japan, followed by Europe and the US. According to this viewpoint, tackling the virus is such an overwhelming priority that all other problems become irrel-*

evant. The economic and social impacts of closing or drastically reducing sports, restaurants, pubs, flights, and many more will be great. Nevertheless, they would have to endure rather like collateral water damage, however severe, caused by the firefighter's hose. This interpretation also implies taking the strongest action as soon as possible. The right time to start fighting a fire is immediately! Until the Prime Minister's press conference on Monday, 16 March of 2020, the UK government appeared to be working within the third narrative. The 'holding the tide' viewpoint sees beating down the virus as workable only as a temporary stop-gap: we can build temporary defenses against a rising tide, but eventually inundation is unavoidable. If this is right, ultimately, containment of the virus is not possible. Therefore, our objective should be to minimize the impact of its spread. This can be achieved, for example, by actively 'flattening the peak' to manage the burden on healthcare services and to reduce the possibility of flare-ups during winter, when those services are under the most strain. The end-game would be herd immunity, which it would be hoped to begin to establish when perhaps 60-80% of the population had been infected and recovered from the virus. From this point of view, immediate aggressive countermeasures may not be appropriate - what is required is a staged approach to manage the smoothest possible progress of the virus through the population'.

The three narratives described above appear to characterize the modes of intervention that have guided all policymakers around the world. However, according to psychologist Philip Johnson-Laird, the most common mistake in human reasoning is precisely the tendency to cling to a single narrative and making decisions on that basis. Instead, the optimal strategy should have been to recognize 'when we do not know what interpretation is right and make decisions that are as strong as possible, whatever interpretation is correct' (Chater, 2020). The information we gained later makes it possible to conclude that the 'teacup' response was wrong. Of the three strategies, the burning house was the most prudent (especially since it would allow one to act quickly), and China appeared to be clinging to it more than any other country. However, in China, as elsewhere, this strategy should have been put in place as the only rational political response before any contagion developed. Of course, even in August 2020, when the outbreaks started to increase, it would have been helpful to use more prudence by investing more resources to extinguish the fire immediately. In essence, recognizing that the cost of inaction would likely be higher than the cost of taking immediate and necessary measures. Unlike in February, policymakers were in a situation of greater availability of information and it was entirely irrational not to recognize it.

9.3.3 You can't manage an epidemic hoping for natural herd immunity

Since the Wuhan pandemic, policy makers have struggled to contain infections around the world. They learned three measures to stop contagions based on current and past experiences with infectious diseases. The first consists of imposing substantial restrictions on people's freedom of movement and assembly and aggressive testing. During the first wave in 2020, without other instruments, this has been the strategy most used. The second measure is the administration of a vaccine that could protect everyone. This measure was implemented thanks to a vaccine developed at an unprecedented speed.

However, it only became available ten months after the initial outbreak, leaving us vulnerable for a considerable amount of time. The third measure, widely advocated especially during the early stages of the pandemic, is the pursuit of herd immunity. This type of measure is basically a 'hand off' policy that lets people be contaminated until a condition of mass immunity is established, since so many people have been infected and have become immune, that the virus is unable to find a sufficiently large number of non-immune individuals to spread further.

If there is no chance that an effective vaccine will be ready a few weeks after the first outbreak, policymakers usually only have two options: keep people away from each other or let natural herd immunity take over. Since the first weeks of the expansion of contagion, the most debated topic worldwide was which of these two strategies to use. As discussed in more detail in Chapter 5, the costs and benefits of these two options are different and not easy to account for. People who support the natural herd immunity strategy claim that the costs to the economy will be lower than the health costs. People who support social distance and the lockdown strategy have the opposite point of view.

According to the view held by most scientists and by international bodies (in particular, the WHO), attempts to reach herd immunity by exposing people to a virus are both scientifically problematic and unethical. Letting COVID-19 spread through populations of any age or health status will lead to unnecessary infections, suffering, and deaths. Opponents of this view argue that a natural herd immunity approach should have been implemented to stop the spread of SARS-CoV-2. This would have involved allowing low-risk groups to get sick, while keeping vulnerable groups separate. But the evidence collected so far suggests that this plan was unfeasible (see, for example, Section 5.1.2 in Chapter 5) and ridden with many risks, since many things (some of which are known and many unknown) can go wrong.

First, even though the number of deaths from infection might be low due to a low lethality rate, a new pathogen could cause a significant number of fatalities as most, if not all, of the population will not be immune to it. Sequencing high-risk groups is not a good idea because diseases that first spread in low-mortality groups can also spread to high-mortality groups. Also, there has not yet been a large-scale example of a successful herd immunity strategy based on a planned infection. In the case of COVID-19, available estimates suggest that the herd immunity threshold would have been reached for a range of infected individuals composing between 60 and 75% of the population. To fix ideas, if we take the United States as an example, of the 330 million people who live in the United States, based on early estimates from the WHO that 0.5% of people who get an infection from COVID-19 would die from it, about 198 million people in the United States would have needed to be infected and survive to reach a herd immunity threshold of approximately 60%. Without this, several hundred thousand additional lives would have been lost. Achieving herd immunity is a time-consuming process, which is a scarce resource during a pandemic. Most importantly, relying on natural herd immunity is a risky gamble that can easily fail. For example, assume that less than 30% of the population were infected until vaccines were rolled out and that infection-induced immunity lasted between 2 and 3 years (the length of time is still unknown). In this case, it is not likely that infection-induced herd immunity would be able to stop the pandemic.

We are still learning about immunity to COVID-19. Most people infected with COVID-19 develop an immune response within the first few weeks, but we don't know how strong or lasting that immune response is or how it differs for different people. People infected with COVID-19 for the second time are relatively common. Some people may, in fact, contract the same strain as a common cold coronavirus more than once a year, as it is fairly well known that other viruses in the same family (including SARS, MERS, and some cold viruses) do not produce a lasting immune response, as, for example, measles does. Until we better understand COVID-19 immunity, it will not be possible to know how much of a population is immune and how long that immunity lasts, let alone make future predictions. Few examples of an infection leading to long-lasting herd immunity are known. Although the Zika virus in Salvador and Brazil is one such example, according to ECDC, we still have much to learn about the possibility of acquiring immunity after an infection.

Early on in the COVID-19 pandemic, in late February and early March 2020, when other countries in Europe were locking down, Sweden chose not to. At first, some local officials and journalists called this the 'herd immunity strategy'. The promise was that Sweden would do its best to protect the most vulnerable, but it would also try to get enough people infected so that there would be true herd immunity based on infection. Counting on citizens to use common sense while doing their daily routines without limits, the Swedish authorities hoped that by May 2020 a large enough percentage of people – 40% in their original estimate – would become ill. They hoped that this would help to protect the rest of the population.

At the end of March 2020, Sweden had abandoned this strategy and instead opted for active interventions. Most universities and high schools were closed to students, travel was restricted, working from home was encouraged, and gatherings of more than 50 people were prohibited. In April 2020, COVID seroprevalence[10] in Stockholm was less than 8%, which was about the same as in several other cities (i.e., Geneva and Barcelona). Therefore, the strategy had not worked as expected. According to a study released in August 2020, only 15% of the Swedish population had become sick enough to build the antibodies necessary to be immune by May, leaving the remainder of the population without immunity and unable to help prevent the spread of the virus. As a result, Sweden experienced higher rates of viral infection, hospitalization, and mortality than neighboring countries that had implemented strict lockdowns by March 2020.

[10] COVID-19 seroprevalence is defined as the percentage of people who have antibodies to the SARS-CoV-2 virus.

UK presents a similar story, with the proponents of the herd immunity philosophy having influenced the UK government in the first phase of policy response, allegedly delaying the 'circuit breaker' advocated by the government's own Scientific Advisory Group for Emergencies (SAGE). Estimates suggest that an additional 1.3 million cases of infection occurred in the UK since the government's inaction to the circuit breaker in September 2020. Resistance to lockdown measures stemming from a desire to maintain economic activity undoubtedly influenced the policies of many European nations. Undoubtedly, it resulted in massive increases in cases and deaths in the second wave of the pandemic.

In conclusion, a strategy of herd immunity remains a 'hypothetical possibility' not borne out by available evidence. However, what appears to be a dangerous illusion is the belief that the economy will benefit if the pandemic goes unchecked. The entire experience of the pandemic points to a basic lack of trade-off between the economy and health, at least for a wide range of containment measures. Except for the East Asian and Australasian economies, which successfully dealt with the earlier phases of the pandemic, there is no evidence that economic performance can be restored if the pandemic is not first dealt with. The challenges imposed by natural herd immunity should preclude any plans to increase immunity within a population by allowing people to become infected. Furthermore, although older people and those with underlying conditions are at increased risk of severe disease and death, they are not the only ones threatened by infection and its short- and long-term consequences. Finally, while most people infected have mild or moderate forms of COVID-19 and some do not experience disease, many become seriously ill and must be admitted to the hospital. We are only beginning to understand the long-term health impacts among people who have had COVID-19, including what is described as 'long COVID'.

9.3.4 Is it possible to develop an immune response? The role of T-cells

A major policy problem in the design and evaluation of infection control is the role of the immune response in the short and long term. The immune system is how the body recognizes and defends itself against foreign invaders, such as bacteria, viruses, and toxins. It does this by recognizing and responding to antigens, which are molecules that are unique to foreign invaders.[11] A debate on the natural capacity to generate immunity was primarily started by the decision to implement a 'herd immunity' strategy in the UK in the first weeks of the pandemic. Apart from all the ethical problems involved (see Section 9.3.3), this debate was ignited by the lack of information on the strength as well as the distribution of immunity across the population and the human body's capacity to

---

[11] Antigens are substances (often proteins) placed on the surface of cells, viruses, fungi, or bacteria. Non-living substances, such as toxins, chemicals, drugs, and foreign particles (such as a splinter), can also be antigens. The immune system recognizes and destroys or tries to destroy substances that contain antigens.

organize an immune and lasting response once infected. Much work has been done in recent months and many studies have been conducted to improve our understanding of activated mechanisms capable of providing a durable immune response. In this section, we describe how these issues relate to policy interventions and the level of knowledge available today.

9.3.4.1 *The human immune system and economic epidemiology*

Immune mechanisms constitute a test case for economic epidemiology (EE) for various reasons. These reasons include the use of economic concepts to construct plausible narratives of the interaction between microorganisms, human evolution, and bioresilience, as well as certain forms of 'economic imaginary'. This is a concept introduced by Brown and Nettleton (2018) defined as '. . . ways of projecting and materially restructuring economic and political orders through motifs, metaphors, images, and practices'. It echoes the more traditional notion of a 'collective imaginary', studied in different ways by social scientists (Maffesoli, 1996; Taylor, 2009; Cabanes et al., 2014), psychoanalysts (Castoriadis, 1987), and philosophers (Jung, 1991; Bachelard and Jones, 2002; Durand, 1993).

As part of commonly confused narratives depicting layman interpretations of scientific progress and futuristic concepts (Scandizzo, 2009), the immune system is also the subject of increasingly popular storylines that drive expectations and play a role in mobilizing resources for technological advancements (Van Lente, 1993). Narratives and storylines have a particularly strong influence in promoting biosciences as drivers of what is known as the 'bioeconomy'. This is defined as an economic system based on the sustainable use of biological resources to produce goods and services that meet human needs while also preserving the environment. A paradigm of a sustainable bioeconomy in turn rests on the vision a new generation of public and private investments aimed at creating bio-values through fast expanding bio-science markets and industrial activities. These investments include, for example regenerative medicine (Waldby and Mitchell, 2006), transplantation (Beck, 2011), and global infrastructural investment in bio-banking (Brown, 2018; Brown et al., 2011).

Narratives are also important to understand the problems associated with the social experience of a pandemic and the cultural models associated with the acceptance or rejection of recommended preventive and health care practices. Policy measures are generally justified by claiming that public interventions are aimed at strengthening or replacing natural mechanisms. In turn, the net benefits for society of preventive measures and of vaccines themselves plausibly depend both on the degree of social compliance and on the economic costs that the measures may impose on individuals and businesses. The narratives try to integrate these two often opposite tendencies by describing mechanisms and processes of bio-interaction, reinforcement or replacement of natural immune defense processes. These processes have developed in the course of evolution and therefore reflect refined mechanisms of control of the relationships of different organisms. The immune system is thus often explained in the context of intuitive narrations of the body protecting itself against attackers, through co-evolution and dynamic interactions with potential predators and offensive agents over time.

The search for a balance between costs and benefits generated by these processes is the consequence of two general economic principles. On the one hand, immune defenses require resources, such as, for example, energy, which could be used to improve longevity or fertility (van Boven and Weissing, 2004). These costs tend to grow with the level of protection and with the risks due to 'friendly fire', i.e. the danger that defense strategies can be harmful to the very organism that generates them. This may happen, for example because they give rise to excessive inflammatory reactions or because they do not discriminate sufficiently between prey and predator, especially when the latter lurks within the organism that seeks to protect itself. On the other hand, the benefits of a strong immune system are obvious, both in preventive protection and, so to speak, in the fight with the aggressor. The presence of costs and benefits determines a situation for which it is never convenient for the prey to put in place sufficient costs to completely eliminate the predator, and this obviously also applies to the predator, which for its very survival needs the prey's population not to perish completely. An elegant and effective mathematical model describing the equilibrium between prey and predator was formulated independently by Lotke and Volterra (L-V) in 1920. It relies on a system of nonlinear differential equations that converges to an equilibrium solution for a defined range of parameter values but which becomes chaotic outside this range.

Economic epidemiology (EE) is a branch of economics that studies the interaction between economic factors and the spread of infectious diseases. EE uses mathematical models that in many ways can be considered to originate from L-V as systems of differential equations that represent stylized behavior on the part of decentralized agents. However, in these models, the course and outcome of infection are predicted as the result of endogenous decisions on contacts and preventive measures of rational individuals, who may differ from each other by income, health status, and other characteristics (Fenichel et al., 2013). Therefore, the discourse on the mechanisms of 'checks and balances' created by the endogenous co-evolution of the infectors and the infected is extended and, at the same time, drastically changed by EE models through the concept of incentives (Perrings et al., 2014; Perrings and Espinoza, 2021). Incentives are naturally provided by the health risks and the costs of preventing the pandemic. They may also be essential to align individual motivations and social goals, serving as reinforcements of command-and-control measures such as lockdowns and vaccinations.

EE models show that acknowledging the presence of policy measures to contrast infection may contradict conventional epidemiological predictions. For example, classical epidemiology assumes that the degree of prevalence, defined as the percentage of the population infected, is positively related to the spread of the disease. However, this does not take into account that rational behavior may change under pandemic pressure. The demand for prevention may be 'prevalence elastic', so a higher share of infected will create an incentive to seek more prevention, resulting in a reduction in prevalence. On the contrary, when prevalence is reduced, the demand for prevention also decreases and so does the willingness to undergo vaccination, while prevalence may increase again. Therefore, the growth of the disease will be self-limiting because its spread will encourage prevention. At the same time, eradication will also be unlikely because an equilibrium point will be reached between demand and supply of prevention before the disease is eliminated (Philipson, 2000).

Similar considerations apply to EE analysis of policy interventions, such as, for instance, testing to discover asymptomatic carriers (people who do not know that they have been infected). For example, in a study of targeted lockdowns in a multi-group model where infection, hospitalization, and fatality rates vary between groups, Acemoglu et al. (2020) observe: 'Testing enables isolation of infected individuals, slowing down infection. But increased testing also reduces voluntary social distancing or increases social activity, exacerbating the spread of the virus. We show that the effect of testing on infections is non-monotone. This non-monotonicity also implies that the optimal testing policy may leave some of the testing capacity of society unused'.

Social distancing will also tend to be non-monotonically related to prevalence, since people will not only be willing to incur into costs from reducing contacts with other people, but will also invest resources to detect other infected individuals. They will value contacts with the general public and with valuable business or social partners differently, thus becoming more discriminatory in their 'mixing' decisions as the infection grows or other conditions change (Fenichel et al., 2010). This effect can be particularly high in the case of vaccination, since the benefits of vaccinated people tend to spread over those not vaccinated, but at the same time, unvaccinated people may account for a disproportionate proportion of new infections (Fisman et al., 2022).

On the empirical side, EE studies report evidence that risk-averse individuals tend to engage in behavior that reduces their risk of exposure to disease. This behavior is influenced by the individual's perception of the cost of the disease and its trajectory over time and is therefore endogenous (Mummert et al., 2013). People tend also to reduce the precautionary behavior carried out in the increasing stages of infection when risks appear to decrease, and their responses are motivated by subjective rather than objective perceptions of risk (Fenichel et al., 2013).

The economic consequences of the disease and its prevention activities through the modification of individual incentives also go beyond the prediction of anything similar to a market equilibrium. On the one hand, it can be argued that there will always be a tendency for chronic underinvestment in the prevention of infectious diseases. This is because, as these diseases become less common due to successful prevention measures, the perceived need for services like vaccines and medications might also decline. Additionally, the inability of future generations to afford preventive investments will lead to market failure, resulting in undesired allocation and distribution outcomes. On the other hand, the damage caused by infection will also generally be understated by the usual measures based on health damage and direct economic losses. Preventive behavior, in fact, acts as a random tax whose excess burden is given by the value of the activities it helps discourage or avoid (Philipson, 2000).

Ultimately, by insisting on the endogenous nature of agents' behavior, EE studies tend to discredit the prey-predator model and, in general, the metaphor of a war between two opposite organisms, which has long been accused to be both naive and culturally loaded. As Ludwick Fleck wrote in 1935 (Fleck, 2012) 'The concept of infectious disease is based on the notion of the organism as a closed unit and of hostile causative agents invading it. The causative agent produces a bad effect (attack). The organism responds with a reaction (defense). This results in a conflict, which is considered to be the essence of disease. The whole of immunology is penetrated with such primitive images of war... It is very doubtful whether an invasion in the old sense is possible, involving, as it does, interference by completely foreign organisms in natural conditions. A complete foreign organism could not find receptors capable of reaction, and thus could not generate a biological process'.

Somewhat in contrast to the stylized features of EE models, popular explanations of the immune system provided by media or professional sites tend to combine intuitive descriptions with technical details, even though at best their accounts aim to be only a gentle introduction to very complex topics that only experts may hope to master. For example, most descriptions of 'how it works' present the intuitive notion of the immune system as a plethora of cellular agents consisting of an 'innate' and an 'adaptive' component. 'The immune system has two arms of defense, nonspecific and specific. As with the immune system in general, there are many elements of the nonspecific arm of the immune response. These have in common their lack of strict recognition of foreign material and lack of memory. The nonspecific arm is also known as innate immunity, as it is present at birth, unlike specific acquired immunity, which develops after birth and is also known as adaptive immunity. The innate response is the first line of defense against infectious disease and, if effective, can completely eliminate the agent before the specific adaptive immune response is called upon. It also interacts with the adaptive immune response, aiding its activation, and modulating the response' (https://www.dvm360.com/view/immunology-dummies-painless-review-basic-concepts-proceedings). These explanations also typically include a summary classification of adaptive components: antibodies, B-cells, and T-cells, describing B-cells as generating antibodies, whereas T-cells have a more complex mechanism of activation and defense.[12]

Beyond its intuitive appeal to the general public and the fact that it hides considerable complexities, the popular description of the immune system tends to generate

[12] CAR-T cells, in particular are T-cells that have been genetically engineered for immunoterapy.

different 'imaginaries'. Although more complex motivations may also be involved, a simple reason for people's different reactions is that antibodies (and, by implication, associated B-cells) are perceived as akin to more traditional medical treatments, while T-cells appear to be more related to genetic engineering and disruptive technologies. For example, Kruining et al. (2021) distinguishes two 'sociotechnical' visions for the future: *i)* an imaginary 'Pharmaceutical Inclusion', based on the expectation of a highly inclusive community-based health system, characterized by cooperation between government and pharmaceutical companies, and *ii)* an 'academic driven' future dominated by personalized medicine, academic CAR-T development, and less influence from big pharma.

In the COVID-19 case, people's expectations have been influenced in different ways by the explanations offered by experts through the media and official communications. Since the beginning of the infection, acute respiratory distress syndrome (ARDS) and multi-organ failure appeared to be distinctive features of the disease, suggesting a hyper-inflammatory immune response (Yazdanpanah et al., 2020). Although this indicated a robust immune response in most people, it appeared to also to be the main cause of death, making recovery difficult. It also did not square with the current communication paradigm on infection, the prey-predator model, and with the 'herd immunity' theory.[13,14]

While hyper-inflammation was soon recognized as the first enemy to fight in the outbreak of infection, for several months, public discourse was primarily framed by a 'war against the virus' metaphor, so that much of the conversation focused on antibodies (Wicke and Bolognesi, 2020). This focus increased with the advent of vaccines because it was imagined that the persistence of immunity induced by them was critically dependent on the durability of the antibodies produced as a consequence of vaccination. The war metaphor was reinforced by expert explanations that antibodies represent only a part of the complex and coordinated set of tools (a 'true army of combatants') available to the human body, each with its unique attack modalities. Among immune instruments, T-cells appeared to be a specially skilled set of defendants, as they were described as capable of forcing infected cells to self-destruct, while T_h cells, nicknamed 'helpers', were causing B-cells to transform into machines that produce antibodies.

---

[13] However, inflammation response is only one of the possible biological responses of the immune system and occurs when tissues are damaged by bacteria, trauma, toxins, heat, or any other cause. Damaged cells release chemical substances, including histamine, bradykinin, and prostaglandins. These chemical substances cause fluid leakage from blood vessels into tissues, causing swelling. This process helps isolate the external substance from further contact with body tissue.

[14] These same chemicals also attract white blood cells called phagocytes that 'eat' dead or damaged germs and cells. This process is called phagocytosis. The phagocytes eventually die. The waste material of this process is pus, which is formed from a collection of dead tissue, dead bacteria, and live and dead phagocytes.

The narrative on the immune system hinged on its power as a natural warlike machine, but at the same time, it was laced with doubts about its robustness in the face of increasingly threatening developments of the pandemic. The first danger, variously described by experts and most media sources, was the decay of immunity. Unlike T-Cells, antibodies were characterized as having an expiration date, disappearing from blood a few weeks or months after their production, while B-cells and T-cells could survive for much longer periods and be detectable years, sometimes decades, after the end of the infection (Qi et al., 2022; Le Bert et al., 2020). Another more insidious danger appeared to be the virus's mutating capacity, which looked like a dynamic threat that challenged both natural and vaccine-engineered immunity. The narrative developing around the Omicron variant is, in this respect, emblematic, since it combines an account of its reported mildness with a structural description of its capacity to defy most forms of immunity acquired through infection or vaccination. 'Omicron has, so far, produced mild symptoms, its genome contains 60 mutations including 37 in the spike protein and 15 in the receptor-binding domain. Thirteen sites conserved in previous SARS-CoV-2 variants carry mutations in Omicron. Many mutations have shown evolution under positive selection. The giant mutational leap of the Omicron has raised concerns as there are signs of a higher rate of virus infectivity, pathogenesis, reinfection, and immune evasion' (Rajpal et al., 2022).

The narrative on the role played by innate and acquired immunity was also affected by the gyrations of public mood and policy makers' actions due to changing expert advise and the disorienting messages from several epidemiological models, whose publicized results proved both contradictory and then wildly off in terms of both explanations and predictions. Examples of widely publicized estimates that largely missed their mark, although in different ways, are a study released on March 16, 2020 by Imperial College researchers, and state-level estimates by the Institute for Health Metrics and Evaluation (IHME) of the University of Washington (Avery et al., 2020b).

The performance of epidemiologists was defended with several arguments, ranging from the necessity to take prompt action to the 'split loyalties' required by scientists' involvement in policy making. For example, Eleanor Murray (2020) draws a distinction between applied and academic epidemiology, where the former would aim to guide immediate decision making under pressure and the second would be to obtain more reliable models and parameter estimates over a longer period of time and, presumably, without the urgency requested by an ongoing pandemic. As an example of the second line of defense, Warren Pearce (2020) argues that expert advice can be fatally affected by the conflation of the role of scientists as knowledge producers, in modeling and estimating the impact of the pandemic, and knowledge users, in translating estimates into specific recommendations. According to Pearce, in providing advise to policy makers, scientists are affected by the uncertainty of their estimates and are thus liable to downplay their model predictions, especially if, as in the case of the COVID-19 pandemic, these estimates may lead to recommend lockdowns or other drastic command and control measures.

9.3.5 Do not underestimate the unknown COVID-19 effects

Since the start of the pandemic, more and more people have been infected. However, in addition to the acute phase of the disease, several signs pointed to a potentially worrying chronic evolution of its consequences. Because of this, medical attention is shifting from treating acute infections to learning about the long-term consequences of the pandemic ('long-term COVID'). This topic appears also important for policy making to adjust health and social care response, in ways that not only ensure preparedness to new infections, but effectively deal with their long-term impact on patients and health structures.

In the early days of the COVID-19 pandemic, clinicians knew that the disease primarily produced a respiratory infection similar to flu that affected the lungs. After looking at thousands of cases, they realized that the virus could affect all body organs. Gupta et al. (2020) present an interesting and extensive review of all studies published until June 2020 on the effects of COVID-19 on human body organs outside the respiratory system.[15] According to the review, in addition to respiratory problems, the virus can have many other effects. First, the virus can cause significant difficulties to the cardio-circulatory system, increasing the risk of heart attack and coagulation disorders. Patients with severe cases of COVID-19 are also at risk of stroke caused by blood clots and delirium. The neurological system is also highly involved, with symptoms that include, among others, headaches, dizziness, fatigue, and loss of smell and taste. Finally, significant damage can also be caused to the kidneys with severe kidney failure problems.

A few weeks after the work of Gupta et al. (2020) had been published, a series of new studies highlighted additional effects that had previously been overlooked. According to Muller et al. (2020) and Lania et al. (2020), the virus appears capable of disrupting thyroid hormones with a significant share of COVID-19 patients suffering from thyrotoxicosis, a condition of a damaged thyroid, characterized by an excess of thyroid hormone in the blood. These studies indicate that patients with COVID-19, especially those with severe disease, should be monitored for symptoms of thyroid-related problems.

Clinicians believe that these complications may be caused by a systemic inflammatory response, which occurs when the immune system overreacts to the virus's attack on the body, especially the cells that line blood vessels. When the virus attacks the cells of the blood vessels, an increasing inflammatory reaction develops, and the blood begins to form clots, large and small. These blood clots can travel throughout the body, causing

---

[15] The same paper also reviews the mechanisms underlying these systemic effects and provides an interesting clinical guide for physicians to help them intervene better.

extensive damages to the organs, and perpetuating a vicious cycle. Furthermore, modulated signals downstream of the immune system can go out of control in more severe cases, contributing to these widespread effects.

In the most recent survey on the problem, Davis et al. (2023) conclude that 'Long COVID' is a complex and multi-faceted condition that affects multiple organ systems, leading to various symptoms, such as ME/CFS, dysautonomia, and vascular and clotting abnormalities.[16] The authors also note that there is evidence that Long COVID has impacted millions of people globally, and the number of cases continues to rise, and yet '...diagnostic and treatment options are currently insufficient, and many clinical trials are urgently needed to rigorously test treatments that address hypothesized underlying biological mechanisms, including viral persistence, neuro-inflammation, excessive blood clotting, and autoimmunity' (Davis et al., 2023).

Although the long-term effects of COVID-19 remain uncertain, insights have been gained by examining the disease progression, which can be categorized into three distinct phases:

- The viral phase: the virus replicates very quickly in the respiratory system. The symptoms are similar to the common flu and spontaneously disappear after V–1Pdays (approx.).[17] This is a case that involves approximately 80% of patients;
- the pulmonary phase: the remaining 20% of patients may develop pneumonia. It is a type of pneumonia that attacks both lungs and thus causes respiratory failure. Many of them spontaneously recover;
- the serious phase: about 10% of the total hospitalized sufferers develop a 'cytokine storm', an inflammatory response of the immune system that can be uncontrolled, responsible for most critical conditions, until death.[18,19]

---

[16] Dysautonomia is a medical condition that occurs when there is dysfunction in the autonomic nervous system (ANS). The ANS is responsible for regulating various automatic bodily functions, such as heart rate, blood pressure, digestion, and breathing. Dysautonomia can cause a wide range of symptoms, including lightheadedness, fainting, rapid or slow heartbeat, fatigue, nausea, vomiting, constipation, diarrhea, bladder dysfunction, and sweating abnormalities. The severity of symptoms can vary from mild to debilitating and can affect an individual's daily life.

[17] V–1P days is a shorthand way of referring to the number of days from the onset of symptoms to the peak of viral replication in the respiratory system. The V–1P value is typically between 3 and 5 days, but it can vary depending on the individual and the virus.

[18] Cytokines are small secreted proteins released by cells, with a proven specific effect on the initiation and persistence of chronic pain through the activation of sensory neurons.

[19] It is worth noting that this latter percentage is still subject to high uncertainty. In an interview with Forbes, Dr. Randy Cron, of the University of Alabama, one of the leading researchers on cytokine storm syndrome, argues that of the 20% who are hospitalized, it is completely unknown what percentage has a cytokine storm. For this reason, it is difficult to assess the mortality of COVID-19-related cytokine storms'. The interview is available at the following web address: https://www.forbes.com/sites/claryestes/2020/04/16/what-is-the-cytokine-storm-and-why-is-it-so-deadly-for-covid-19-patients/.

Since the earlier phases of the pandemic, many studies have analyzed cytokine storms, clarifying the mechanism that determines them and the effects they cause (Ragab et al., 2020; Simon et al., 2020; Sinha et al., 2020). These studies have been followed by a series of more specific investigations (see, for example, Jiang et al., 2022 and Thampy et al., 2023). Cytokine storms occur in intensive therapies.[20] Often they are caused by other infections and even some medications. Unfortunately, there is no treatment other than 'support' to protect the patient's vital organs (ventilators or drugs to control blood pressure or corticoids to reduce inflammation). According to Sinha et al. (2020), although there is no consensus on its definition, a cytokine storm can be described as a hyperactive immune response characterized by the release of chemical mediators such as interferons, interleukins, tumor necrosis factors, chemokines, and many others. These mediators are part of a well-preserved innate immune response necessary to effectively eliminate infectious agents. In the case of ordinary pneumonia, germs damage the lung tissue, prompting the immune system to initiate an inflammatory response. This response releases chemicals that cause blood vessels to leak fluid into the tissues, resulting in swelling and creating a protective barrier for body tissues. In the case of COVID-19, the immune system appears to overreact by releasing large amounts of chemicals. This becomes the motor of an uncontrolled inflammatory response, not only in the lungs but also in other parts of the body, which is harmful to host cells. The over-reaction also explains why kidney, intestinal, coronary, and brain damage (encephalitis) can occur, with very rapid degradation in people of any age.

Therefore, for COVID-19 infection, clinical evidence suggests that the occurrence of a cytokine storm is closely associated with the rapid deterioration and high mortality of severe cases. This is because in severe cases of COVID-19, the immune system response to the virus can become overactive, leading to a surge in cytokine production and subsequent inflammation. This can lead to acute respiratory distress syndrome (ARDS), a severe and potentially life-threatening lung disease that can cause breathing difficulties, low oxygen levels, and can require mechanical ventilation. The cytokine storm is also associated with other severe complications of COVID-19, such as multiple organ failure, sepsis, and thrombosis.

In conclusion, physicians are increasingly led to think of COVID-19 as a multisystemic disease, with possible and largely unknown long-term effects. Although clotting problems are relevant, it is also essential to understand that many patients suffer damage to the kidneys, the heart, and the brain. Doctors must treat these diseases together with respiratory diseases. Understanding multi-systemic involvement is undoubtedly a step forward in providing better care and developing a comprehensive follow-up plan once the patient is dismissed.

---

[20] Situations of patients suffering from cytokine storms often occur during the flu season. In these cases, hospitalizations are limited and there is time to focus on the individual patient, which was impossible during the COVID-19 emergency.

The lack of knowledge about the causes and predisposition to hyperinflammation in Long Covid is compounded by a lack of understanding of why a significant number of patients are infected with symptoms that persist for months, even longer than those who are hospitalized for respiratory and non-respiratory complications of COVID-19. According to Prof. Sally Singh of the University of Leicester, who is leading the development of a COVID-19 rehabilitation program for the UK Health Service, 'the share of COVID-19 patients with persistent problems is much higher than that observed for other viral diseases such as influenza. Problems are also more varied and often include pulmonary, cardiac, and psychological symptoms'. These patients were called long haulers, meaning 'long-term' or 'long-range'.

With more than 675 million confirmed cases worldwide at the end of February 2023, strong evidence begins to emerge of people struggling to recover from the effects of the infection. A work by Greenhalgh et al. (2020) concluded that up to 60.000 people in Britain have long-term symptoms, although only about 5% of the UK population (about 4 million people) seemed to have been infected with the virus. In Italy, Carfi et al. (2020) have verified that 87% of hospitalized patients still had symptoms after two months; a study by Arnold et al. (2020) in the UK has found similar trends. In a German study, which included many healed patients at home, Puntmann et al. (2020) found that 78% had heart abnormalities after two to three months. Finally, a CDC working group in the US found that a third of the 270 non-hospitalized analyzed had not returned to their normal health status after two weeks (about 90% of people who contract the flu heal within that money-saving time) (Tenforde et al., 2020).

A meta-analysis of 63 studies (Domingo et al., 2021), with sample sizes ranging from 58 to 1733 people, for a total of 16,336 patients, looked at the frequency of symptoms after a lab-confirmed SARS-CoV-2 infection. Overall, 53% of people with an infection still had one or more symptoms after 12 weeks. More frequent symptoms were tiredness, pain or discomfort, shortness of breath, problems with thinking, and mental health problems.[21] In a later meta-analysis of 18 studies with at least one year of follow-up (Han et al., 2022), study samples ranged from 51 to 2433 (8591 in total). Results show that 28% of the participants were tired or weak, 22% were anxious, 18% could not catch their breath, 19% lost their memories, 18% could not focus, and 12% could not sleep. However, the authors mention several problems with the study, such as a small sample size, a lack of representativeness, and a low response rate.

To address some of these challenges, Hastie et al. (2022) more recently published a new analysis that examines the frequency, nature, causes, and consequences of long-

---

[21] It must be noted that in 18 studies, the authors found a high risk of bias, and in the other 45, a moderate risk of bias. This was due to factors such as convenience sampling and study populations that were not representative. Half of the studies looked at groups of people who were in hospitals. The only two studies with more than 1000 participants were a cohort of people in hospitals and a cohort of healthcare workers with mild infections.

term COVID in the general population of Scotland. They conducted this study using a large-scale nationwide sample that included people with severe, mild, and asymptomatic infections, as well as a group of people who had never been infected. They also assessed both self-reported outcomes and outcomes that were linked to routine health records. A total of 33,281 people with laboratory confirmed SARS-CoV-2 infections and 62,957 people who had never been infected were tracked using questionnaires sent at 6, 12, and 18 months and links to hospitalization and death records. Of the 31,486 infections with symptoms, 1856 (6%) had not disappeared and 13,350 (42%) had only disappeared in part. Hospitalized infection, age, female sex, deprivation, respiratory disease, depression, and having more than one illness were all related to no recovery. Previous infection with symptoms was associated with a lower quality of life, problems with all daily activities, and 24 persistent symptoms, such as shortness of breath, palpitations, chest pain, and confusion. An infection without symptoms was not associated with poor outcomes. When people were vaccinated, they were less likely to have severe symptoms.

The presence of many symptoms and effects and the lack of information about their determinants puts physicians in an environment of great uncertainty. While medicine is science-based, it is also largely dependent on field work and quick decision-making. The great uncertainty that practicing physicians and other medical professionals are now experiencing is creating significant problems and many others will create in the future. Especially concerning appears to be the impact that long-term COVID-19 problems may have on patients who, for a variety of reasons, have not been tested (initially did not fall within the protocols necessary to be tested) or resulted in false negatives (which will potentially increase the number of patients who may be suffering from these problems in the future).

Unfortunately, long COVID has received some attention from the public, but not nearly as much as acute COVID-19 infection, despite the fact that its effects on health and the economy are likely to be comparable to those of acute illness. Health care costs are also likely to be significantly affected, although the population prevalence of long COVID is not fully understood. This is because many estimates are based on convenience samples from people who are in COVID-19 support groups or who have had a severe acute disease. For example, data from the British population suggest that 22% to 38% of people with the infection will have at least one COVID-19 symptom 12 weeks after the first symptoms appeared, and 12% to 17% will have three or more symptoms (Groff et al., 2021; Whitaker et al., 2022). If the order of magnitude of these prevalence rates were confirmed by further studies, the number of potential patients would be huge, as would the economic burden. In a paper analyzing the economic cost of Long Covid, Cutler (2022) reports: 'as of May 5, 2022, the US Centers for Disease Control and Prevention think that there have been approximately 81 million cases of COVID-19 and 994,187 deaths from COVID. Even the lowest estimate, that 12% of people with 3 or more symptoms of long COVID have it, means that 9.6 million people in the US may have it. This is about 10 times the number of people who were estimated to have died from COVID-19. No one knows how long people with long COVID will have symptoms, but recovery may be very slow for those who have it in the first year'.

In some countries, such as the UK and Belgium, specialized COVID-19 rehabilitation programs have been established for those recovering from the disease. Unfortunately, the number of people enrolled is reportedly always much higher than the places available. This is partly due to the slow response by governments and health authorities and their delayed recognition of the problem. More than two years since the pandemic started, in many countries, the issue of Long Covid has not yet been brought to the attention of decision makers and the media, and the list of symptoms does not fully reflect the range of medical problems that should be addressed.

9.3.6 Invest in training and more effective drug treatments

Once again, the global coronavirus pandemic has shown how unprepared the world is for infectious disease outbreaks. Although they lead the world in medicine and have well-established systems for public health and preventing epidemics, many developed countries were not ready. One of the main reasons appears to be the lack of knowledge on what it takes to predict, prepare for, and treat an epidemic.

The COVID-19 pandemic has unleashed an unprecedented wave of medical improvisation. While it is true that at the beginning of March 2020, most doctors in western countries had never seen a sick person with COVID-19, it must be noted that infrastructure for public health, training in the medical profession, research on infectious diseases and the development of new treatments have been neglected for decades. Before the pandemic, our longer lives and the possibility of preventing and treating some infections simply made us less concerned about infectious diseases than we used to be.

A few months after the outbreaks, almost all emergency and intensive care physicians developed some familiarity with the disease. During that period, health workers learned how to provide the best treatment available to patients. However, they followed the same approach adopted during the initial phase for a long time, despite the accumulation of new evidence on both symptoms and cure. The same applies to general practitioners who learned to recognize symptoms and could identify an infected patient before seeing the results of a specific diagnostic test. Medical improvisation was a common practice in both outpatient and inpatient settings. In March 2020, every patient management process was essentially improvised or experimented with little guidance provided by institutions and recognized national and international medical centers. Little was known about the management of patients outside the hospital, trying not to make them worse to avoid hospitalization. Those who worked to intervene outside the hospital before the health conditions of the patients worsened did much better. But even in those cases, general practitioner activities were mainly driven by improvisation, with little aid from the institutions.

Three years ago, what is today a routine in the hospital setting was also mostly improvisation. In the early months, patient management recommendations changed quickly. According to many medical professionals, it was challenging to keep up with the pace of the changes in the protocol, which was very disorienting to both doctors and nurses. The best form of learning was word of mouth among colleagues and information learned through blogs, medical training podcasts, and social networks. Following Florence Nightingale's teachings, any new protocol was immediately implemented if suggested by a reputable source in a Facebook group and appeared 'low risk' for the patient and generally helpful.[22]

Recent examination of health systems has also revealed deficiencies in the ability of health professionals to detect and treat infectious diseases. According to a report from the Committee on Science and Technology of the House of Lords in the UK (House of Lords, 2003), the lack of qualified professionals in the field of infectious diseases was already recognized 20 years ago. The report stated that not enough physicians knew how to find and treat complex or rare infections. This means that patients may not receive the best possible care and that important events that could be signs of an outbreak could be missed until the start of a major epidemic. In addition, the report also voiced concern about shortage of people with the necessary skills to operate laboratories and provide patient care. Medical microbiology jobs were found difficult to fill, even with the increase in academic microbiology fellowships. Additionally, the report identified a shortage of people with epidemiological skills to track and predict disease outbreaks.

Since the early days of the pandemic, intensive care units have been established in various locations, and hospital COVID-19 tents were set up in numerous parking lots. Medical professionals from all fields, including urologists and orthopedists, were enlisted to help. The same happened for nurses who were thrown into the fray regardless of their career paths and previous experience. Finally, medical students were hastily awarded their first degrees to complete ranks (even before they had satisfied completion requirements such as, for example, thesis discussion). However, the availability of intensive care and trained professionals was not the only tool that saved many lives.

As clearly described by Mulligan et al. (2020), pharmacological treatments have been an essential part of the fight against COVID-19 for two reasons. First, they were an indispensable tool to manage the pandemic before vaccines were available. Second, even after vaccine availability, drug treatments were still necessary to handle cases where vaccines did not work or to treat those who did not get vaccinated.

---

[22] According to Megan Ranney, an emergency physician and associate professor at Brown University's Department of Emergency Medicine, the practice of asking COVID-19 patients to put themselves in a prone position (stomach down) was born through word of mouth and on social media. When someone is supine (stomach up), her organs squeeze her lungs, making it more difficult for the airways to expand fully. When a person is on her stomach, her lungs have more space to fill with air. The advice began to circulate in the medical community before there was an official, published study of the practice.

In mid-April 2020, a review published in JAMA reported that 'no therapy has been effective to date' (Sanders et al., 2020). However, pharmacological therapies changed enormously with time, and several drugs were used and experimented with different degrees of success and failure. In addition to hydroxychloroquine, which was initially administered, but proved ineffective, only two drugs demonstrated a therapeutic effect against COVID-19 in randomized clinical trials: the antiviral Remdesivir and the standard steroid Dexamethasone. Remdesivir has been shown to reduce recovery time in infected patients, and dexamethasone has been shown to reduce mortality in the most severely ill patients. Changing protocols for ventilator use also improved hospital survival rates.

In the early months, trials were conducted mainly for the most severe patients, whose life expectancy was very short with current treatment. For everyone else, there was little to do. Today, things have changed and doctors are much more attentive to a wide range of possible further events (e.g., blood clots). However, it remains true that active interventions for patients with less severe symptoms are still more or less the same as they were in March 2020. Furthermore, changes to protocols for non-severely ill patients have been more challenging to obtain, partly because it is riskier to try something new for these people. If someone is not in a life-threatening condition, there is nothing to gain from an experimental treatment that could potentially cause harm.

The search for more effective drugs has lagged behind for a long time. At the end of July 2020, scientists were studying almost 300 potential treatments. The US National Institute of Health has launched extensive clinical trials on some of the most promising approaches, including studies on antiviral monoclonal antibodies. These medications help regulate the immune system and thin the blood to prevent complications caused by blood clots. Throughout the pandemic, the pharmacological management of COVID-19 has evolved from anecdotal evidence and case studies to robust randomized clinical trials. The role of regulatory agencies has been critical in ensuring the proper oversight and rapid approval process of new diagnostics and therapeutics. The most exciting data on new COVID-19 therapeutics have recently emerged with the development of new oral antivirals, which, if administered early in patients at high risk of progression to severe diseases, can reduce hospitalization and death. However, the potential of the treatment intervention to limit the spread of the disease is still somewhat limited.[23]

Finally, it must be recognized that if a new virus were to rage in the next few years with little knowledge of its characteristics and how it interacts with the human body,

---

[23] A recent statement by BMJ, the Journal of the British Medical Association, notes that: '...systemic corticosteroids (particularly dexamethasone), interleukin-6 receptor antagonists (such as tocilizumab), and Janus kinase inhibitors (such as baricitinib) reduce mortality and have other benefits in patients with severe COVID-19, such as reducing the length of hospital stay and the time needed on a ventilator'. It also notes that the antivirals molnupiravir (Lagevrio), nirmatrelvir/ritonavir (Paxlovid), and remdesivir (Veklury) have been shown to be effective against non-severe COVID-19.

it would be like starting from scratch. The knowledge from dealing with SARS-CoV-2 might offer limited insights, just as our learnings from SARS, MERS, and Ebola provided little assistance during the COVID-19 outbreak. A possible lesson learned is to avoid hospitalization of patients when they are in desperate conditions and, where hospitalization is necessary, to avoid any further contact with infected patients.[24] New pharmaceutical treatments are thus important as it will likely be necessary to handle situations similar to COVID-19 determined by other viruses in the future. Having more tools to reduce the virus's impacts and allow patients to avoid cascading effects is a goal to be set. This entails creating robust health systems, bolstering the economy, and ensuring a prolonged life. In addition, this could reduce the urgent need to vaccinate millions of healthy people and allow a better balance between public health and full reopening of the economy.

9.3.7 Gaining citizens' confidence

One critical aspect of the management of the pandemic has been the quality of leadership provided by governments and institutions. Government officials and other persons in position of authority and prestige have found themselves suddenly invested at first with high responsibilities and trust, and after a short period of time, subject to close scrutiny and possibly devastating criticism. A famous English proverb says: 'Cometh the hour, cometh the man'. The reference is to the idea that the right leaders will come to the fore during times of crisis. A good leader is someone who, in addition to showing the right way to overcome a difficulty, is also able to reassure citizens and convince them to support the government in implementing the decisions taken. This is especially important when measures such as social distancing, with its knock-on effect on employment, come at high personal cost. Leaders who have been honest in declaring the limitations they have faced in fighting the virus have been more likely to gain credibility among the population and be more effective in fighting the disease than those who have been unwilling (or able) to admit the difficulties.

Where the trust in the leaders is low, it may be necessary to intervene with sanctions and other direct and indirect means to enforce the rules. These interventions can trigger undesirable side effects, with non-compliance spreading on the wave of open defiance of governments' command and control measures. To a large extent, the measures imposed

---

[24] This is what did not happen in LTC residencies worldwide. High rates of morbidity and mortality in residents were observed, as well as high rates of absence from staff due to SARS-CoV-2 infections. According to ECDC, in several EU and EEA countries, deaths among residents accounted for more than half of all COVID-19-related deaths. As discussed in Section 3.6.2 in Chapter 3, in the earlier phases of the pandemic, up to a half of the deaths from COVID-19 occurred in long-term care facilities, particularly nursing homes for the elderly. Surprisingly, this happened even though residents of these institutions were subjected to drastic isolation measures before the general population (OECD Policy Responses to Coronavirus (COVID-19), 2021).

by governments during an epidemic concern modes of behavior that can be considered trust goods, that is, goods whose value depend on the trust that buyers accord to sellers if their information is asymmetric or incomplete. In economics, this trust relationship is identified as a principal agent problem, where the principal (in this case the citizens) depends on the agent to provide the right information and agents in turn depend on the principals to trust such information. In the case of the COVID-19 outbreak, the success of anti-epidemic measures depended on public trust in not only the government, but also health and scientific institutions and the media. Such trust would have resulted in individuals being more inclined to follow the guidelines and instructions provided by these organizations.

Asymmetric information and principal agent relations carry two possible dangers. The first is due to the non-communicable nature of some of the information, which may be too technical and cumbersome to be quickly transferable from the informed agent to the uninformed principal. This is the case, for example, of medical information possessed by medical professionals that cannot be readily transferred to their patients. The second danger arises from the fact that the informed agents may use some of their privileged information to acquire gains at the expense of the uninformed principals. These two dangers are always haunting the principal agent relationships, which depend on trust and can be easily compromised by signs of untrustworthy behavior. In this sense, public trust depends on communication between leaders and the public, in an often complex relationship based on beliefs, past performance, and the capacity to convey the right messages convincingly. Even a solid relationship between political leaders and the public can become fragile under the pressure of extraordinary events such as a pandemic, where leaders are perceived as untested and possibly unreliable under the current circumstances. For example, according to Boin et al. (2005), a common mistake by leaders is to create a message that lacks coherence and can quickly be reported and emphasized by the media. A prime example is the early messages, later withdrawn, from the British Government, stating that the epidemic could be solved by achieving herd immunity. Overall, both the initial message and its retraction have contributed to increased confusion and distrust in the government among the populace.

To promote trust, Boin et al. (2005) recommend that good leadership should keep in mind that it is always better to avoid proposing itself with a 'paternalistic sense to children who need to be protected from bad news' and instead face the population 'as adults who will make a long-term effort'. Without such an approach, the public can quickly perceive deception, which would reduce the credibility and trust of the government in the policies implemented. In general, these authors suggest that an optimal message is a message that 'offers a credible explanation of what happened, offers guidance, instills hope, shows empathy, and suggests that leaders have control'. Failure in only one of these points may lead to a loss of trust in citizens.

Communication in a crisis should take into account the new normal under which citizens receive information from many sources, and thus government statements can be easily suspected to be partial, untruthful, or misleading. The recommendation to rely on official information is more likely to be followed if the authorities establish and maintain their credibility as reliable sources of information. However, during the COVID-19 epidemic, we have witnessed numerous examples of social media effectively spreading misinformation and fostering skepticism about official communications, particularly when individuals are confined to their homes. The experience of the fight against Ebola also shows that doubts about official information can undermine the response to the crisis.

Mutual trust embedded in collective behavior and civic sense is also a critical component of a good relationship between institutions and citizens, as shown by research on civic culture and its role in the pandemic (Doganoglu and Ozdenoren, 2020; Durante et al., 2020). A typical example of how to test this role is the imposition of measures of social distancing and lockdowns, adopted by almost all countries around the world without previous experience. As Doganoglu and Ozdenoren (2020) point out, 'a large proportion of the population must respect the social distance to be effective', and paradoxically, compliance with these measures is greater in countries where citizens have little confidence in their peers' behavior. From several studies, trust in others has emerged as one of the driving forces shaping the measures recommended and the citizens' response. For example, in a thoughtful article, the Economist reviews the past experience with non-pharmaceutical measures and concludes: 'Countries that trust have generally implemented looser blocs. Rather than strictly enforcing the rules of social distancing, their governments rely on citizens to observe the guidelines voluntarily'.[25] However, the same article highlights how these situations can create side effects: in countries with a high level of trust among citizens, it is possible to teach a false sense of security that tends to lower the guard, as the case of Sweden shows.

Within Europe, polls consistently note that residents of northern countries have a lot of confidence in their governments and citizens, while those in southern and eastern countries do not. When the World Values Survey (WVS) asks Swedes if they think most people are trustworthy, more than 60% say yes. In Italy, this percentage drops to about 30% and in Romania to 7%. Another study, the European Social Survey (ESS), asks respondents to rate their confidence in politicians on a ten-point scale. In 2018 the Dutch had expressed an average of 5.4, the Poles 3.1, and the French and Germans were placed in between. In Bulgaria, the average was below 1 (one in eight Bulgarians gave their politicians a zero). Of course, these results reflect profound socio-cultural differences. Greater trust is correlated with greater wealth, fewer crimes, and other welfare metrics. It also appears to influence population responses to COVID-19.

According to Durante et al. (2020), as a combination of human and social capital, 'civic capital' can help mitigate the consequences of a pandemic in a context of a

---

[25] The full text of the article is available from https://www.economist.com/europe/2020/05/02/do-low-trust-societies-do-better-in-a-pandemic.

standard public good (the absence of contagion) and positive and negative externalities (compliance or non-compliance with the rules imposed), since 'it is likely that internal-ization of externalities produced by personal mobility depends on the ability of citizens to contribute to the public good'. In the context of a pandemic, reducing contagion generates a public good that benefits everyone and any individual behavior restricting contagion has a positive externality since helps slow the spread of the virus, thus low-ering the probability of infection for all community members. Thus, compliance with rules such as social distancing and mask-wearing can have positive externalities, while non-compliance can have negative externalities. As documented by several studies (e.g., Putnam, 1993 and Herrmann et al., 2008), communities with higher levels of civic capital appear to perform better in acting collectively and providing public goods. 'The higher civic capital, the slower the social expulsion rate, the slower the spread of the epidemic. A high level of civic capital can already encourage social distancing in the epidemic's early stages, even in the absence of government intervention and persuasion campaigns' (Durante et al., 2020). Using provincial data on individual mobility in Italy, these authors show that complying with social distance directives was higher (mobility decreased more) in areas with more civic capital.

However, somewhat in contrast to this result, Doganoglu and Ozdenoren (2020) find that when people believe that others will work hard to protect themselves from the disease, they feel safer going out without the risk of contracting it. The study suggests that in countries where social norms on protective disease behavior are more demanding, people are led to go out more. This leads to the paradox that trusting others and respecting rules can be a double-edged sword. When social standards become more demanding, people are more likely to adopt precautions against disease, but this can also lead to a sense of complacency and an increase in social interaction.[26]

Identifying the best strategy is not an easy task. For example, a policymaker aiming to minimize the number of people infected with the disease might prefer to confront weaker social norms in the presence of government interventions and persuasive cam-paigns. Weak social norms can mean that people are less likely to comply with measures such as wearing masks and practicing social distancing, which can lead to higher rates of infection. In this scenario, the policymaker may need to focus on strengthening social norms through education campaigns and community engagement programs, in addi-tion to implementing government interventions. On the other hand, if the policymaker aims to maximize social welfare, it may be more helpful to have stronger measures that promote compliance through sanctions and more rigid social norms, as these can help reduce the need for lockdowns and mitigate their negative consequences on the econ-omy. Overall, the best strategy for dealing with a pandemic depends on the specific

---

[26] For example, when wearing masks becomes the norm, people find it less costly to wear them. Similarly, when others do not expect a handshake as a greeting, it becomes easier to avoid shaking hands. In both cases, social interaction may increase.

goals and values of policy makers and society. Minimizing the number of people infected with the disease may be the priority in some situations, while maximizing overall welfare may be the priority in others. In practice, policy makers must consider a range of factors, including social norms, government interventions, and public health campaigns, to identify the most effective strategy for their particular context.

To better understand to what extent governments have gained public confidence during the COVID-19 crisis, it is useful to report the results of research promoted by the PEW Research Center.[27] Through a survey conducted from June 10 to August 3, 2020, among 14.276 adults in 14 countries (United States, Canada, Belgium, Denmark, France, Germany, Italy, the Netherlands, Spain, Sweden, the United Kingdom, Australia, Japan, and South Korea), PEW analysts investigated people's attitudes towards their government's pandemic response (see Fig. 9.3). Public opinion was generally positive, with 73% median approval. However, there was a diversity in responses between the various countries. The population was very divided in countries such as the United Kingdom and the United States. In the UK, 54% of respondents disapproved their government's response, while 46% gave the green light to the efforts of the authorities. In the United States, which recorded the highest number of infections worldwide at that time, 52% of the population believed that the national response was poorly managed compared to 47% who believed that the government responded well. In contrast, moving a little further north to the Canada border, the survey finds that only 20% of the population in this country were happy with the government's response during the crisis.

Overall, 58% of the respondents indicated that their lives had changed much or enough due to COVID-19. Women appeared to have been more affected than men. In nine of the countries involved, this gender gap reached double digits, and in Sweden, the United States, and France, the differential was 15 percentage points. Somewhat surprisingly, in a quarantine period characterized by much controversy over the role of the WHO, an average of 59% believed that increased international cooperation would have been able to reduce the number of coronavirus cases facing their country. Missed collaboration opportunities to reduce coronavirus cases were particularly felt in Europe, where the lack of coordination of initial responses led to sudden and severe outbreaks in northern Italy and Spain. More than half of the respondents in seven of the nine European countries considered in the survey expressed the opinion that increased cooperation would have reduced coronavirus cases.

---

[27] The PEW Research Center is an independent (non-partisan) organization that aims to gather opinions to inform the public about the problems, attitudes, and trends that are shaping the world. It conducts public opinion polls, demographic research, media content analysis, and other empirical research in the social sciences. The reported results are from nationally representative surveys of 14,276 adults from June 10 to August 3, 2020, in 14 advanced economies. All surveys were conducted over the phone with adults in the US, Canada, Belgium, Denmark, France, Germany, Italy, the Netherlands, Spain, Sweden, the UK, Australia, Japan, and South Korea.

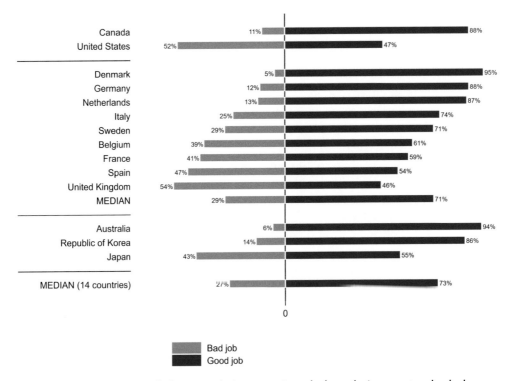

	Bad job	Good job
Canada	11%	88%
United States	52%	47%
Denmark	5%	95%
Germany	12%	88%
Netherlands	13%	87%
Italy	25%	74%
Sweden	29%	71%
Belgium	39%	61%
France	41%	59%
Spain	47%	54%
United Kingdom	54%	46%
MEDIAN	29%	71%
Australia	6%	94%
Republic of Korea	14%	86%
Japan	43%	55%
MEDIAN (14 countries)	27%	73%

Figure 9.3 Percentage of the population convinced that their country had done a good/bad job of countering the outbreak of COVID-19. *Source*: Pew Research Center, August, 2020, 'Most Approve of National Response to COVID-19 in 14 Advanced Economies', available from: https://www.pewresearch.org/global/2020/08/27/most-approve-of-national-response-to-covid-19-in-14-advanced-economies/. *Note*: Period June 10 to August 3, 2020. Those who did not respond do not appear. In Australia and Canada, the question was asked on 'COVID-19'. In Japan, 'new coronavirus' has been asked, and in South Korea, 'Corona19' has been asked.

In all geographical areas examined in the research, those who thought their government had done a poor job with the coronavirus epidemic were more likely to argue that their country was more divided than before the pandemic. This suggests that sentiments of national division are tied to mutual distrust. In many countries, those who expressed lower trust in others were also more likely to find their country more divided than those who argue that most people can be trusted to be trusted. In 11 of the 14 countries examined, this gap between trusting and not trusting people exceeded 10 percentage points.

9.4. The lessons for the future: are we prepared for the next 'black swan'?

In a celebrated study, Taleb (2007) came up with the modern name for a 'black swan' event, as is characterized by three distinctive features: *1)* It is an unusual event that goes against normal expectations; *2)* It has a considerable effect; *3)* After the event, it is possible to explain its causes. Outbreaks of infectious diseases fit perfectly the above description representing rare events that may have a significant impact. They also fit Scholthof's (2007) thesis that most disease outbreaks are caused by three factors: the pathogen, the environment, and the host. These factors create an epidemiological triangle which influences the severity and transmission of the disease. As new pathogens appear, the environment changes and hosts become immune; these factors constantly fight to keep up with each other. Emerging pathogens occur when a pathogen copies itself or moves from one animal to another and makes a new serotype or strain. Changes in migration patterns, the growth of cities and the lack of resources all lead to natural and man-made crises, which are made worse by the environment.

Does this mean that pandemics will become more common in the future? There is no doubt that this is one of the risks of our current conditions. Despite the knowledge gained since the first age of pandemics, the world still does not effectively manage the risk of outbreak infections. In fact, we could reduce this risk even further by improving human and animal sanitation. For example, nearly a third of the world's people go to the bathroom in open pits or in the fields. This makes the spread of diseases from the toilet to the mouth even more dangerous. Too often, animals in factory farms have to stand on piles of their waste. This infectious threat is worsened by the practice of giving antibiotics to livestock regularly. This practice in turn increases the chance that bacteria will become resistant to antibiotics, which are one of our best ways to fight disease (Deshpande et al., 2022). Finally, with regard to host susceptibility, vaccination is the best way to increase host immunity. But problems with vaccine production, distribution, and effectiveness prevent their full effectiveness and thus make hosts more likely to get sick.

9.4.1 A pandemic, certainly not the last

COVID-19 will not be the last pandemic in this deeply interconnected world. Unfortunately, it also will likely not be the worst. We will probably be exposed to other pandemics in the next few decades. These events will almost certainly occur, as suggested by the recent increase in the frequency of major epidemics (such as SARS, MERS, and Ebola) and due to social and environmental changes by humans that appear to have contributed to the emergence of COVID-19. As already mentioned in Chapter 2, scientists have long warned that factors such as climate change, deforestation, population growth, industrial agriculture, and globalization have produced many of the conditions that lead viruses, such as SARS-CoV-2, to emerge from the dark recesses of nature to infect the environment. They are the so-called 'zoonoses' or 'species leaps', and scientists' warnings imply that the world must be prepared to see more. Although it appears to be a bizarre biological phenomenon (the so-called 'black swan event'), the COVID-19 pandemic cannot be considered a casual accident.

As discussed in Chapter 2, warnings of a pandemic strangely similar to SARS-CoV-2 appear in published research and reports over several decades. SARS-CoV-2 could be the beginning of an era when a natural world besieged and attacked by so many human activities can introduce more and more infectious diseases into increasingly globalized humanity. These are not new events: humanity has always been susceptible to epidemics, but today disease outbreaks are spreading much more easily and quickly worldwide as trade and global travel increase. As a side effect of globalization, the hyper-connected world has changed the virus 'travel conditions'. While in the not so distant past people used to travel on foot and live in low-density settings, today we travel by plane and live in higher-density areas (e.g., attending supermarkets, shopping malls, stations, theaters, etc.), which makes virus life much easier. Human activities have significantly altered three-quarters of the earth and two-thirds of the oceans, changing the planet to such an extent that a new era has emerged: the 'Anthropocene'. Changes in soil use that bring wildlife, livestock, and humans closer to each other facilitate the spread of diseases, including new strains of bacteria and viruses (Kilpatrick and Randolph, 2012).

Where the next viral challenge will come from might be impossible to know, as the viral lot displays impressive numbers and varieties of life forms. More viruses than any other living organism reside on the planet, with higher numbers than bacteria, animals, and plants. Viruses have been everywhere and in circulation since life started on the planet, probably playing a vital role in the early stages of evolution. According to Jean-Michel Claverie, Professor of Genomics and Bioinformatics at the Faculty of Medicine of the University of Aix-Marseille in France, 'there are so many viruses in the environment that we don't even know. Many regularly infect humans, but unlike the new coronavirus, they are not so infectious, dangerous, and certainly not deadly and, therefore, go unnoticed'.[28]

We must prepare for a new pandemic at any time, and the phrase 'at any time' could also overlap with the current one. We only have to hope that the next virus will arrive as late as possible. Indeed, the threat from a new virus would be even harder to detect now, given that all resources (physical equipment, health professionals, public health experts, financial resources) are strained and often about to run out because of the past and current pandemic needs.

But there is an opportunity. While politicians and other key decision makers are fully focused on a global problem, this may be an excellent time to rethink how countries

[28] https://www.courthousenews.com/scientists-expect-more-pandemics-in-age-of-globalization-climate-change-assault-on-nature/.

work together and how they organize their value chains. As Nathan Wolfe said in an article on Time, 'we have two profoundly different possible futures: one in which we put our heads in the sand as we have constantly done, and one in which mankind takes the decisive and necessary steps to protect itself'.

Some steps forward have been taken, but this is just the beginning. For example, a series of reports produced around the world by various institutions have begun to spur policymakers and researchers to learn from this pandemic and past epidemics by switching to the holistic approach, termed 'One Health', which brings together expertise in human, animal, and environmental health. Examples in this direction are the report by The Lancet One Health Commission (Amuasi et al., 2020), which called for more trans-disciplinary collaboration to solve complex health challenges, and the World Wide Fund Italy report (World Wide Fund Italy, 2020), which highlighted the intimate connection of human, animal, and environmental health.

9.4.2 Preventing the next pandemic will cost only 2% of the COVID-19 expenditure

In general, the saying 'prevention is better than cure' is a valid statement under a variety of medical conditions, but it is dramatically appropriate in the case of pandemics. Yet, as the world struggles with its coronavirus pandemic budget, those concerned with countering the emergence of new viruses warn that the funds needed to prevent the next zoonotic disease outbreak are severely lacking. Although it may be irrational and all have a specific interest in preventing a further pandemic from emerging, there are few initiatives to improve and develop prevention activities.

Evidence suggests that new epidemics such as SARS, MERS, Ebola, and HIV now emerge every four or five years (see Fig. 2.1 in Chapter 2). We could say that the COVID-19 epidemic is more similar to the 1918 Spanish flu than to SARS and MERS in recent years. Yet, it is also true that, compared to 1918, the world has changed. A series of intensified activities in recent decades have made the appearance of new epidemics more likely than in the past. Dobson et al. (2020) have estimated that an expenditure of 260 billion dollars in the next decade would considerably reduce the chances of shortly facing another pandemic with effects similar to COVID-19. To be precise, the research estimates that the amount needed may vary between 22.2 and 30.7 billion dollars per year, an amount of resources useful mainly for protecting and monitoring pristine forests and the trade in wild animals. This seemingly extravagant amount represents only 2% of the estimated 11.5 trillion dollars (with a minimum of 8.1 and a maximum of 15.8 trillion) that the COVID-19 pandemic is projected to cost the global economy. This cost would amount to around 500 billion dollars, which is nearly twice the investment needed for proposed preventive measures. The estimate takes into account the loss of GDP and the years of productivity lost due to the deaths of hundreds of thousands of workers worldwide, assuming that only two zoonotic viruses emerge each year. The current risks are much higher than those in the past, mainly due to the ongoing destruction of natural environments.

Most of the 22-30 billion dollars that should be devoted annually to prevention should be used primarily for forest protection, since disease transmission from wild animals to humans frequently occurs near the edge of tropical forests. Reducing all forms of deforestation by at least 40%, breeding and trading livestock, and selling exotic animals implies reducing the likelihood of human contact with wild animals. Furthermore, the work of Dobson et al. (2020) points out that contact with wildlife is more likely when 25% of original forest coverage is lost and that 'the clear link between deforestation and the emergence of the virus suggests that a major effort to keep forest coverage intact would have a great return on investment even if its only benefit was to reduce the virus's emergency events'. Reducing deforestation would have an additional benefit: lowering the CO_2 levels that currently fuel the climate crisis.

In addition to the preservation of forests (tropical forests in particular), Dobson et al. (2020) recommend some other measures, such as i) more effective early diagnosis and control, ii) end wild meat consumption in China, and iii) reduce the spread of diseases through livestock. The consumption and breeding of wild animals are among the main contamination problems, especially in China, where the sector is worth around 20 billion dollars and is estimated to employ around 15 million people. Furthermore, this is an area where health and safety regulations are seriously lacking (just remember the hygienic conditions of the Wuhan market from where COVID-19 seems to have started).[29] Finally, researchers also propose increasing funding to create an open source library of known virus genetic codes, which could help quickly identify the source of emerging diseases before they can spread.

Further work has been done by Smith et al. (2009) on behalf of McKinsey Global Institute to measure the costs of the pandemic compared to the investments in prevention. According to the authors, the economic distress caused by the COVID-19 pandemic could cost between 9 and 33 trillion dollars, many times more than the expected cost to prevent future pandemics. The authors have estimated that spending from 70 to 120 billion over the next two years and 20 to 40 billion a year may substantially reduce the probability of future pandemics. A comprehensive program to strengthen the health system at all levels would cost significantly more. The study concludes that infectious diseases will continue to emerge and a capacity development program should prepare the world to respond better than it has done to date to the COVID-19 pandemic. Although Smith et al. (2009) estimate higher costs and benefits than those proposed by Dobson et al. (2020), the relations remain essentially the same and always in favor of the choice to intensify investment in prevention in the coming years.

---

[29] In March 2020, China introduced a law banning the consumption of wildlife (primates, bats, pangolins, civets, and rodents) for food or related trade.

The economic evaluation of investment in prevention can also be seen as the equivalent of a premium paid to buy insurance that protects against an adverse event. An article by Evan Ratliffe published in June 2020 on Wired addresses the issue of pandemic risk insurance and how this type of insurance has been first conceived and subsequently developed and proposed to the public over the years (Ratliff, 2020). According to the article, insurance against a COVID-19 pandemic had already been proposed by two giants of reinsurance (Munich Re and Marsh), which designed packages both for individual customers and businesses, but with not much success at selling any of them.

In the same article, Ratliff interviews Nathan Wolfe, a professor of epidemiology at UCLA and the founder of Metabiota, a company that makes epidemiological studies and predictions using a worldwide network of disease surveillance, and had tried to push the insurance program as a business partner with Munich and Marsh. In his interview, Wolfe argues that the lack of willingness to pay to insure for damages from rare events is easy to explain, since our brain 'is not particularly suited to solving these kinds of risks, particularly rare ones' and '...Companies are led by humans who suffer from the same failures of imagination that they all suffer, unable to truly internalize the one-year disaster on 100 until it reaches its doorstep'.

However, there are signs that the private sector is gearing up on this front, and insurance companies are organizing programs to take into account the economic consequences of future pandemics. Some insurance experts also speculate that one day banks will grant loans to companies, for example, in industries such as travel and hospitality, only if they have epidemic insurance, perhaps imposed or promoted by government policies. Government intervention would be needed also to counter the danger that demand for infectious disease insurance could outstrip the ability of reinsurers and other investors to cover losses.

9.4.3 We need to understand better the limit between personal freedoms and the right to health

Implementing early and comprehensive tests has been shown to be effective in controlling the virus and has even allowed certain countries to avoid shutdowns and the block of the economy. In these countries, tracking and re-tracking people who tested positive for COVID-19 to warn those who had been in contact with them also helped limit the spread of the disease. This practice has been considered controversial, given that some governments have requested direct access to personal data from mobile telephone networks to monitor people's movement. At the same time, digital surveillance may have helped uncover new outbreaks that would otherwise have gone unnoticed, even though it came at the expense of individual privacy rights.

Privacy advocacy groups express significant concern when intelligence agencies spearhead the development of coronavirus tracking programs, as seen in Pakistan and Israel, or when governments involve private companies by outsourcing these initiatives.

The United Nations has warned that the potential for abuse is high. Although an emergency can justify access to personal data, there is a danger that it will become standard after the crisis. Automated contact tracking applications that protect people's data and store it only on the person's phone can be a viable alternative. These apps do not use a central database to store personal data on movements that could be used to re-identify people and reveal with whom they spent time.

9.4.4 Past mistakes are always paid with interests in times of crisis

Chapter 2 discussed how the current COVID-19 pandemic had been vastly underestimated. In this regard, it suffices to consider that in 1993 Robin Marantz Henig published a book in which, for the first time, he mentioned the concept of 'emerging viruses', a term recently coined by the virologist Stephen Morse (Marantz Henig, 1993). The book spoke about how experts were identifying conditions that could lead to the introduction of new potentially devastating pathogens. Such conditions were then and now represented by climate change, massive urbanization, and the proximity of humans to farms or wild animals. This situation was compounded by other problems, including events such as war, globalization of the economy, and international air travel that could have facilitated the worldwide spread of these pathogens. Several other books that had also treated similar topics with a significant impact on the public and the media were, for example: 'The Hot Zone' by Richard Preston (1995), 'The Coming Plague' by Laurie Garrett (1995), and more recently 'Spillover' by David Quammen (2012). Each of these studies describes similar catastrophic scenarios and calls for alarm about humanity being woefully unprepared for global infection outbreaks.

The global unpreparedness for the pandemic can be partially attributed to the concept of 'availability heuristic,' which was introduced by two behavioral economists, Tversky and Kahneman, in 1983. According to this concept, an availability heuristic problem arises whenever we tend to estimate the probability of an event based on the vividness and emotional impact of a memory rather than on its objective probability. This depends on how events are recorded: we tend to remember more of something more important than other events or factors we don't care about. This tendency also gives more weight to the most recent information and the most glaring news. These behavioral features explain, for example, why traveling by plane seems more dangerous than traveling by car. In fact, plane crashes tend to receive more media coverage than traffic accidents, even though the number of annual road fatalities is significantly higher than that of aviation-related deaths.

The same phenomenon also works in reverse. If an event occurs very rarely, its expected impact is underestimated. Pandemics are a classic case. The prolonged absence of a pandemic such as COVID-19 may have led people to underestimate its probability and (especially) its impact. To this, it must be added that investing in essential public health activities to prepare for diseases never seen before is a recent policy notion that dates only to about 20 years ago.

Andrew Lakoff, a sociologist at the USC Dornsife College of Letters, Arts, and Sciences, states that since the 1990s, 'it has been a challenge to convince the political authorities, and even some public health authorities, to take pandemic preparedness seriously'. It is about asking people to invest resources to face a potential threat whose probability is impossible to calculate' (https://dornsife.usc.edu/news/stories/why-u-s-wasnt-better-prepared-for-the-coronavirus/). This may have led political and industrial leaders to focus on what they perceived as the most urgent priorities in recent years. One of the most striking results was that of the United States, where in 2018, the Trump administration cut funding for National Health Programs.

Although this may seem to be a plausible explanation for why we were not prepared to deal with the pandemic, it is nonetheless important to acknowledge that crucial mistakes famously were made by those responsible for national security issues. President George W. Bush said: 'If we wait for a pandemic to occur, it will be too late to prepare. And one day many lives could be lost without need because we have not acted today' (https://abcnews.go.com/Politics/george-bush-2005-wait-pandemic-late-prepare/story?id=69979013). From this perspective, the grief and outcries of those who have today highlighted the collapse of health care systems should be contrasted with the silence of those who have, in recent years, often considered health expenditure a burden on public budgets, starting with costs incurred for prevention investments. The inadequate response to the COVID-19 pandemic is also linked to longer-term trends of neglect and underfunding of healthcare and social security systems, the failure to innovate and adapt to changing societal needs, and the magnified costs of past inaction and lack of investment. As discussed in this chapter, the errors of not investing or acting in the past are magnified today in a ratio of 1 to 500 in terms of costs (Dobson et al., 2020).

The essence of these observations suggests that a significant number of challenges we face today stem from past errors. Consequently, we are bearing the burden of these mistakes through a worsened situation characterized by inadequate infrastructure and insufficient emergency management personnel. Once this crisis subsides, its impact will be comparable to an expensive war in terms of economic toll and human lives lost, much of which could have been anticipated or alleviated with proper consideration of recent history.

9.4.5 We need to invest more in Global Health

Although the COVID-19 pandemic has been declared over by the World Health Organization (WHO), its health and economic costs continue to rise. The pandemic has been ongoing for three years or more, and its long-term effects are still unknown. The official estimates of COVID-19 deaths are over 6 million, but reliable sources believe the actual number to be between 16 and 20 million, which is similar to the number of deaths in World War I. Despite these staggering numbers, there are signs that we are better equipped to handle the pandemic now than we were in the early days. We have learned more about how to prevent the spread of the virus, and we have developed vaccines that are effective in protecting people from serious illness. We are also better prepared to treat people who do get sick.

However, we should be able to do a lot better. Over the past 200 years, the world has made progress against disease by making connections less dangerous, such as through improved sanitation, hygiene, and medical treatments. Due to economic growth and new technologies, we now have more power than ever to deal with pandemic threats. In this situation, hiding behind national walls is not only selfish, but also a costly and risky way to act. Instead, we need global cooperation to ensure that everyone has access to sanitation and basic health services, including prevention, vaccinations, and treatments. We also need a much stronger surveillance system that covers the whole world and better ways to research, develop, and make tests, treatments, and vaccines on a global scale. The costs of these steps could be hundreds of billions, but the costs of a pandemic are trillions.

If we do not pay enough attention or make enough mistakes, we could let communicable diseases take over as the most common way to die. From what we know about history, a change like this would change the next century more than almost any other event, even more than climate change. The threat is about the same as a small-scale nuclear war. Even if this complete threat does not materialize, we could let global progress be slowed down by how we handle new diseases like COVID-19. Yet, we have the tools and knowledge to end the plague cycle for good and limit damage to people and the economy. By taking a collaborative and proactive approach to public health, with economic growth and new technologies, we have more tools than ever before to prevent and manage outbreaks of infectious diseases, as well as to stop a new age of pandemics.

According to the 2022 IMF's World Economic Outlook (IMF, 2022), the pandemic will cause about $13.8 trillion in lost production by 2024. The global economic recovery is slowed down by the fact that not everyone has the same access to tools to prevent and treat COVID-19 and that policy support to deal with the effects of the pandemic is not the same everywhere. The fact that the world is in the economic grip of the pandemic, even though this may look like the end of its medical phase, is also a cause of some of the problems the global economy is facing: supply problems, inflation, and persistent uncertainty. COVID-19 is also expected to have a long-lasting effect on the economic potential of many countries.

Although everyone agreed that ending the acute phase of the pandemic was a major global priority, neither the ACT Accelerator (ACT-A)[30] nor the IMF Pandemic

---

[30] The American Rescue Plan ACT of 2021 is a $1.9 trillion economic stimulus bill that was passed by the 117th United States Congress and signed into law by President Joe Biden on March 11, 2021. The ARP

Proposal for the delivery and use of available countermeasures were met (Agarwal and Gopinath, 2021a,b; WHO et al., 2021).[31] In terms of vaccines, 86 countries did not reach the goal of getting 40% of their people vaccinated by the end of 2021, and there are still significant differences in how people can get vaccines and how often they get them. With how things are going, more than 100 countries are falling short of their vaccination goals by mid-2023, and some may never get it. On tests, the daily goal was set at 1 per 1000 people, and although the Omicron wave occurred, about two out of three developing economies still test below this goal. During this time, countries with high incomes have testing records 80 times higher than countries with low incomes. Similar differences are observed for oxygen availability, treatments, and protective equipment. And monitoring diseases through testing and genomic sequencing, which is a crucial way to find new virus strains and make sure that vaccines and treatments work, is done in different ways in different countries. This means that we look for variants wherever we find them, which is not always where they are happening.

9.4.5.1 How much do we devote to DAH?

According to the Development Assistance for Health (DAH) database, among the many effects of COVID-19, the health development aid has increased significantly. Institute for Health Metrics and Evaluation (IHME) (2021) estimates that $13.7 billion in DAH went to the health response for COVID-19 activities, such as spending on treatment and logistics. Due to this help, the total DAH increased by 35.7% between 2019 and 2020, which had never occurred before. However, so far, health development aid for COVID-19 has not always gone to the places where the virus has caused the most damage. For example, the Latin America and Caribbean region has only received 5.2% of DAH for COVID-19, against reports (as of August 15, 2021) of 28.7% of all deaths from COVID-19.

In 2020, an estimated amount of $81.1 million was spent in development assistance for health or about 1.5% of the total DAH in the same year. Unfortunately, this money is insufficient to prepare for pandemics that can have terrible and long-lasting effects. One of the main lessons from COVID-19 is how important it is to have strong health systems, of which preparation for pandemics is just one part. Government and DAH spending on Global Health can help policymakers figure out where health systems can be improved. The Institute for Health Metrics and Evaluation (IHME) (2021) document the increase in the total amount spent on health worldwide, reaching $8.5 trillion in 2018, the most

---

was designed to help the United States recover from the economic and health impacts of the COVID-19 pandemic.

[31] Access to COVID-19 Tools Accelerator (ACT-A) is a group of international health organizations working together to speed up the development, production, and fair access to COVID-19 tests, treatments, and vaccines. It started in April 2020 after the G20 leaders asked for it.

recent year for which data on total health spending are available. This is an increase of 3.0% over the total for 2017. Here is how 2018 spending is broken down by type:
- 59.4% of government health spending, or $5 trillion, goes to Medicare;
- $1.9 trillion, or 22.1%, of private spending were paid in advance;
- $1.5 trillion, or 18%, was spent out of pocket;
- $39.2 billion, or 0.5%, donor financing.

In 2018, about 7.7 billion people lived on the planet, of which approximately 3.6 billion (46.7%) in one of the 79 countries with low or low middle income. Although DAH only made up 0.46% of global health spending in 2018, it is still important for the countries that rely on it. DAH can be a catalyst by focusing investments on activities or populations that otherwise might not receive attention or resources. It can also be a lifeline for some countries.

As noted, the total DAH in 2020 was 35.7% higher than the $40.4 billion estimate for 2019. Although it has grown in the last 30 years – in 1990, DAH was $8.1 billion, which is a rise of 574.0% – it has never grown as much as it did between 2019 and 2020. In fact, between 2010 and 2019, the aid for health development remained mainly the same, ranging from $35.3 billion to $40.6 billion per year. Even though the growth in 2019 and 2020 is high by historical standards, much more money is needed to ensure that countries with lower incomes do not fall behind. Without the cost of COVID-19, the DAH for 2020 was $41.2 billion, which was about 1.8% more than the DAH for 2019.

Classifying countries by World Bank income criteria, DAH is still used the most in low-income countries (25.0% of health spending in 2018), while out-of-pocket spending is used the most in lower and middle-income countries (55.9% of health spending in 2018). Most health spending in high-income countries comes from government and pre-paid private spending (86.5% of all health spending in 2018). Between 2019 and 2020, the DAH for HIV / AIDS, malaria, and tuberculosis decreased by 3.4%, 2.2%, and 5.5%, respectively, but increased for COVID-19. The DAH for NCDs increased by about 4.7%, from $846.9 million in 2019 to $887.0 million in 2020. However, the DAH for newborn and child health and reproductive and maternal health decreased between 2019 and 2020, with the funding for newborn and child health decreasing by 2.6% and the DAH for reproductive and maternal health decreasing by 6.8%. Although between 2019 and 2020, COVID-19 spending did not cause significant drops in other health areas focused on assistance, DAH did not grow for many health areas, and the pandemic has made it harder to figure out how to spend money on development and get care to people who need it most.

9.4.5.2 Who will be able to afford investing more in Global Health

According to Kurowski et al. (2021), in the coming months, the world economy would have been likely to grow faster than it had after any recession in the last 80 years. Unfortunately, this prediction was denied by the outbreak of the Ukraine war, and countries must make tough decisions to keep government health spending from decreasing. Before the war, the per-capita government spending in 126 countries was expected to increase until 2026, even though it was predicted to go down in some years, especially in 2021 and 2022. However, in 52 countries, government spending per person was expected to decrease and remain below what it was before COVID-19. If bold decisions aren't made to prioritize health, per-capita government health spending will remain below 2019 and drop in many of these 52 countries.

To keep their public health growing at the same rate of the pandemic, the governments of these 52 low-income countries would have to increase their spending on health by an average of 100%, from 10% before COVID-19 to 20% in 2026. Those in lower-middle-income countries will have to increase their spending from 8% before COVID to 13.50% in 2026. In the coming years, these and other differences in the macro-financial outlook could increase inequality in health spending across countries even more. This could hurt the COVID-19 recovery in a large way. Most low-income countries would not be able to pay for their share of a COVID-19 vaccine rollout, let alone invest in better preparedness and response capabilities. As the health divide increases, countries in cash shortages will find it increasingly difficult to spend their money on health care.

Even though it will not be easy to increase health development aid at a time when some wealthy donor countries are also having trouble, high-income countries have a fundamental reason to help the world get better. The development of human capital and a total return to inclusive growth everywhere depend on making progress toward universal health coverage. Together, countries can solve health care affordability problems so that everyone can have a healthier, safer, and more prosperous future.

For numerous countries, the capacity to invest in healthcare is further jeopardized by the increased interest payments on their public debt. In an effort to curb inflation, interest rates have risen, making it challenging for many governments to manage their public debt. In 2027, interest payments are estimated to limit the amount that low-income and lower-middle-income countries can spend on health by an average of 7% and 10%, respectively. Interest payments have very different effects on each of these countries. Some of them are expected to limit their ability to spend on health care between 15 and 30% in 2027. If nothing is done right away, many low-income and lower-middle-income countries will fall behind on the road to health and economic recovery.

Many of these countries are expected to cut government spending, making it harder for them to prepare for pandemics and slowing their progress toward universal health coverage. This will make the differences between countries more prominent, threatening global stability and prosperity. To contrast these undesirable developments, governments could bring in more money, put health at the top of their budgets, and make health spending more efficient and fair. Investing in low- and lower-middle-income countries can help to prevent future threats to peace and prosperity. By providing financial assistance, high-income countries can help to promote sustainable development, reduce poverty, and improve health and well-being. This will make the world a better place for everyone.

Although there are hopeful signs of renewed awareness, the international community has not yet been able to gather the level and breadth of political support needed to fully fund ongoing international measures against COVID-19 and longer-term measures to prepare for and stop future pandemics. In March 2022, the Coalition for Epidemic Preparedness Innovations (CEPI) replenishment summit was a good example of how hard it is to deal with this problem. While a positive result was achieved in that the sponsors agreed to pay $1.5 billion for CEPI's plan to encourage the development of new vaccines within 100 days of a future pandemic being identified (and make sure they are distributed fairly around the world), that amount was still far less than the $3.5 billion that had been requested.

Since then, international efforts have seen the further development of several initiatives. These include the International Monetary Fund (IMF) Resilience and Sustainability Trust, the World Bank-hosted Financial Intermediary Fund for Pandemic Preparedness and Response, and Pfizer's new initiative to supply patented drugs, including vaccines, at cost price in low-income countries.

Despite these initiatives, it is still not clear whether the current lack of global governance will be fixed. Much will depend on how long the war in Ukraine will last. However, efforts to bring together national initiatives and multilateral coordination are not very promising, especially with regard to long-term viability and concern for global equity. Meanwhile, the invasion of Ukraine has put much extra financial pressure on western governments, and aid budgets may need to be adjusted to deal with a possible global food crisis. At the moment, it is increasingly difficult to find money to support international health projects.

9.5. Appendix: antibodies and T-cells

While it was soon recognized that hyper-inflammation was the first enemy to fight in the midst of infection, for several months, when discussing the responses of the immune system to the coronavirus, much of the conversation focused on antibodies. This focus even increased with the advent of vaccines, because the persistence of immunity was imagined to be critically dependent on the durability of the antibodies produced as a consequence of the vaccination. However, antibodies represent only a part of the complex and coordinated set of tools available to the human body, each with its unique attack modalities. Among the immune instruments, T-cells force infected cells to self-destruct, while T_h cells, nicknamed 'helpers', can cause B-cells to transform into machines that produce antibodies.

Unlike T-Cells, antibodies have an expiration date. They are inanimate proteins and not living cells, cannot reintegrate and therefore disappear from blood a few weeks or months after their production. Furthermore, they tend not to be detectable in all patients, especially those with less severe forms of COVID-19. In general, antibodies appear shortly after a virus has broken physical barriers and then vanish when the threat dissolves. Most B-cells that produce these first antibodies also die. But even when not under siege, the body retains longer-lived B-cells that can churn out antibodies if needed. Some of these antibodies are detectable after years, sometimes decades, from the end of infection.

This principle has enabled the presence of infection to be verified through the analysis of antibodies (IgG, IgA, and IgM). However, as explained by Le Bert et al. (2020), one of the main features of corona-viruses is to trigger long-lived anticorpal T-cell responses in infected patients. While SARS-CoV-specific antibody levels have fallen below the detection limit in 2-3 years, SARS-CoV-specific memory T-cells have been detected 11 years after SARS.

Knowledge in this field has improved enormously in recent times, but we still do not have a final answer to many complex questions involving immune responses and cell roles. This is partly due to a standard speculative process that allows new questions to arise only once the original questions are answered. For example, in the early days of the pandemic, an important question was to understand if, in the absence of a vaccine, it was crucial to determine whether exposed or infected people, particularly those with asymptomatic or very mild forms of the disease, which probably act inadvertently as the main transmitters, develop robust immune responses against SARS-CoV-2. Many researchers have launched studies in response to this doubt, with early work focusing on antibodies because these are relatively easy to measure with a blood test. At first, the results appeared to suggest that the number of antibodies in those recovering from the virus decreased within a few months of infection (Long et al., 2020; Seow et al., 2020).[32] These results sparked much skepticism about the possibility of obtaining long-lasting virus immunity, also shedding uncertainties about the potential of having a vaccine that could protect for an extended period.

As seen previously, the adaptive immune system also involves the T-lymphocytes. According to Grifoni et al. (2020) and Mateus et al. (2020) SARS-CoV-2 virus generates a strong T-cells response, in particular to the spike protein used by the virus to enter the cells (see Chapter 1). A study conducted by researchers at the Karolinska Institute in Sweden found T-cells responses in people with mild or asymptomatic COVID-19,

---

[32] Long et al. (2020) conducted a study with a small group of patients in China showing that in asymptomatic and symptomatic individuals with COVID-19, antibody levels decreased significantly during recovery and that levels became undetectable in the 40% of the asymptomatic group. Similarly, for example, Seow et al. (2020) in a study of British patients showed that antibody levels dropped substantially within months of infection and that people with less severe diseases had fewer antibodies.

even when antibodies were not detectable (Cell Press, 2020). Therefore, even without antibodies, T-lymphocytes may track the infection. But the strength of the trace may depend on the severity of the infection. In an interview on Scientific American, Zania Stamataki, a professor at the Institute of Immunology and Immunotherapy at the University of Birmingham in England, 'Memory is proportional to the perturbation of insult: how frightened the immune system is [...]. And it is possible that some people can eliminate SARS-CoV-2 by using the inborn immune system without developing any memories of it. If they re-encounter the virus, they may potentially contract COVID-19 for a second time'.

However, according to Sekine et al. (2020), mild cases of COVID-19 can trigger robust T-cell responses, even without detectable virus-specific antibody responses. Such responses can be a significant immune component in preventing recurrent episodes of serious illness. Therefore, it is possible that SARS-CoV-2 specific memory T-cells will prove crucial for long-term immune protection against COVID-19.[33] In this regard, numerous anecdotal reports of people reinfected with the novel coronavirus, but no substantial evidence has been established. However, according to Stamataki, there may be other explanations. People with weak immune systems may not eliminate the virus, or tests may detect its remains that are not infectious. While true reinfection is not impossible, it would probably only occur in a small minority of people.[34]

A recent review on the role of T-cell immune response against SARS-CoV-2 by Moss (2022) shows that we can learn from infections from previous coronaviruses. A subset of T-cells primed against seasonal coronaviruses cross-reacts with SARS-CoV-2 and may contribute to clinical protection, particularly in early life. More than 90% of adults have been exposed to one of the four seasonal HCoVs that cause the 'common cold' (Beta and Alpha-coronaviruses). If infected by these viruses, antibodies to HCoVs do not last long, and it is common to get infected again within 12 months. There are T-cell responses to HCoVs, but they are not very strong, and it is not clear how long they last, since they do not occur frequently in older people. The case of the SARS-CoV-1 virus suggests a more hopeful picture. Even though antibody and B-cell responses don't last long and are often no longer visible after 4 years (Tang et al., 2011; Wu et al., 2007), T-cell responses can still be triggered after 17 years (Le Bert et al., 2020). The responses of T-cells to the Middle Eastern respiratory virus (MERS) are also interesting and seem to be stronger and last longer than humoral immunity. Recent evidence from abattoir workers in Nigeria on their cells response to MERS supports the idea of cellular sensitization without seroconversion12. Together, these facts show that cellular immunity is the most important part of stopping HCoV infections.

---

[33] It should be added here that a debate started in recent months on the problem of reinfections.

[34] Dr. Stamataki statement is available in a interview on Scientific American available at the following web address https://www.scientificamerican.com/article/concerns-about-waning-covid-19-immunity-are-likely-overblown/.

Since HCoV-specific antibodies and cellular responses did not provide sterile immunity, it is believed that immunity against SARS-CoV-2 will also not last long. The information we now have gives us a mixed picture. Reinfection with SARS-CoV-2 can occur, but a previous infection protects about 87% of people from 6 to 13 months and keeps them safe for at least 10 months (O Murchu et al., 2022). The emergence of the highly contagious Omicron viral variant led to a significant increase in the number of breakthrough infections. However, the fact that the vast majority of T-cell immune responses against Omicron are still present may help explain why the clinical severity is not as bad. Vaccinated people were also found to harbor sustained T-cell response up to 6 months after vaccination (Guerrera et al., 2021).

Although there has been great progress, many important questions about T-cell immunity to SARS-CoV-2 still need to be answered. It is not clear what the best way for cells to work together is when they are first infected. To control the pandemic, it is also important to know if different vaccine schedules can obtain the best cellular responses and how they will help protect against the spread of viral variants such as Omicron.

CHAPTER 9 - Take-home messages

- Pandemics and health crises provide scientists with rare opportunities to gather a wealth of information quickly. COVID-19 is no exception: research activities have increased tenfold. Unfortunately, the outcomes produced have often not been adequately reviewed, and, due to the enormous possibilities of mass media dissemination, have often generated disinformation and confusion.
- One of the lessons learned in the early stages of the pandemic is that underestimation has been the most common error, causing tens of thousands of extra deaths.
- Public health communication during the pandemic needs to be more transparent, clear, and informative. A lack of these qualities has contributed to the spread of misinformation and disinformation, also known as the infodemic.
- Policymakers and health authorities understand that they have a precious and scarce resource: reaction time. Contagion mitigation, monitoring, and tracing practices must occur quickly, under penalty of significant loss of effectiveness of the strategies adopted.
- Herd immunity is not a viable strategy for COVID-19 because we do not yet know the long-term behavior of the virus. If we were to wait for herd immunity to develop naturally, it is likely that many people would die before we reached that point.
- COVID-19 will be neither the last nor the most severe pandemic to occur in humanity. To avoid the next global health crisis, it is necessary to spend only a tiny fraction of the costs incurred during the current epidemic.

CHAPTER 10

Bioeconomy, biodiversity, and the human footprint
A new 'One-Health' perspective

Contents

The evidence presented in Chapter 2 raises significant concerns about whether we are entering a new era of pandemics. It is worth noting that we have long been aware of the rising trend in outbreaks and epidemics of infectious diseases. Smith et al. (2014) conducted a study on more than 12,000 reported disease outbreaks worldwide since 1980, revealing a consistent increase in both the number of diseases and the number of outbreaks over time. They further concluded that due to the rapid mutations of these newly evolved microbial threats, humanity will continue to face an ever-increasing rate of infection. Moreover, in a world where billions of people are linked in globally connected cities, the potential for rapid spread is always present. In this respect, hard quantitative evidence is difficult to obtain, but a recent study by Metabiota has estimated that the probability of a future zoonotic spillover event that results in a pandemic of magnitude of COVID-19, or greater, can be as high as 3.3% each year.[1] This annual probability, which at first sight may appear low, adds up to about 90% (of having a pandemic at least in one year) in 100 years.[2] If we combine this information with the increase in the aging global population, the risk of serious illness or death from the threats that emerge could be higher than what we have seen even during the COVID-19 pandemic.

---

[1] Metabiota is a company describing itself as having the mission to estimate, mitigate, and manage epidemic risk, supporting global health security and sustainable development (https://metabiota.com/).

[2] This calculation has been performed by Max Rosen of OurWorldinData and presented in the following twitter https://twitter.com/MaxCRoser/status/1499652503423627264?ref_src=twsrc%5Etfw%7Ctwcamp%5Etweetembed%7Ctwterm%5E1499652503423627264%7Ctwgr%5E%7Ctwcon%5Es1_&ref_url=https%3A%2F%2Fthreadreaderapp.com%2Fthread%2F1499750222305583105.html.

The (re)emerging infectious diseases (e.g. SARS, COVID-19, Zika, and Avian flu) developed during the early 21st century, together with the problems raised by climate change and environmental sustainability, represent important and urgent threats to be addressed. Their prevention and management raise complex issues that include both the way humans interact with themselves and with nature and should be adequately addressed by a trans-disciplinary and integrated approach, such as 'One Health' (OH). This is a health model based on the integration of different disciplines and institutions, and a joint management and surveillance of health-impacting processes.[3] OH originates from the recognition that human, animal, and ecosystem health are inextricably linked and aims to apply a comprehensive strategy and trans-disciplinary collaboration on all aspects and dimensions of health.

In recent decades, the OH concept has expanded to include not only medical and veterinary sciences but also gradually an increasingly wide range of synergistic disciplines, including food safety, public health, health economics, ecosystem science, social sciences, and animal health (Xie et al., 2017). The OH approach is also based on the recognition that environmental factors, including chemical contaminants in animals and products of animal origin, residues of veterinary drugs, and plant protection products, play a significant role in human, animal, and plant health. All this has allowed the OH concept as the extended notion of 'One Health – One Environment' to be widely accepted among industry experts.

The holistic and multidisciplinary view of OH is also a consequence of scientific progress, as in the past two decades, several studies have shown a close connection between natural capital, as a bioeconomic machinery that provides ecosystem services, and health for all living things (Fisher et al., 2021). The connection is not a simple cause-effect relationship but may include synergies and co-determined states of the world that signal deeper and perhaps more fundamental ties between nature and the determinants of development and well-being. The World Health Organization's Ottawa Charter on Health Promotion (Organization et al., 1986), for example, states: 'The fundamental

---

[3] Many organizations internationally recommend and support the OH approach. Among the many, we should mention the WHO ('One Health is an approach to the design and implementation of programs, policies, legislation, and research in which multiple sectors communicate and work together to achieve better results in public health' (WHO, 2017), the EcoHealth Alliance ('One Health as an interdisciplinary approach to strengthen systems globally and locally by recognizing the shared health of humans, animals, and the environment' (Consortium, 2016), FAO ('The One Health approach is an integrated way of preventing and mitigating health threats through the Animal-Human-Plant-Environment interface'). The Sustainable Development Goals (SDGs) adopted by the United Nations General Assembly in 2015 embody a One-Health strategy: healthy people living on a habitable planet (Cf, 2015). The Green Deal of the European Commission is an integral part of the implementation strategy of the United Nations 2030 Agenda and the SDGs. The experience gained with the COVID-19 pandemic should also lead to drawing important conclusions about the close links that connect the environment, health, and the economy.

conditions and resources for health are peace, shelter, education, food, income, a stable ecosystem, sustainable resources, social justice, and equity. Improvement in health requires a secure foundation in these basic prerequisites'. Although nature was not explicitly mentioned in this and similar earlier statements, the links between nature and human health were formally introduced in the WHO contribution to the Millennium Ecosystem Assessment (Assessment, 2005) and the State of Knowledge Review jointly conducted with the Convention on Biological Diversity (CBD) and WHO (Romanelli et al., 2015).[4]

The notion of OH is related to natural capital by the key concept of biological equilibrium. This concept, in turn, refers to two distinct functions of natural capital: *i)* the provision of services that ensure a dynamic balance between preys and predators (including humans and pathogens) and *ii)* a sufficient diversity of lifeforms to support evolutionary processes. Several effects on health are invoked in the literature to explain these functions. The first effect, which is mainly dependent on biodiversity, is dilution. It is based on the hypothesis that high vertebrate species richness reduces infectious diseases among humans by diluting pathogens through a high number of animal reservoir species (Schmidt and Ostfeld, 2001), and a lower prevalence of infection in vectors (Johnson and Noguera, 2017; Ostfeld and Keesing, 2017). Although higher species richness can also imply higher pathogen richness (Dunn et al., 2010), the net effect will still be a reduction in the risk of infectious diseases among high vertebrates (and humans).

A second effect is amplification and may result from a change in the environment that increases the number of vectors and/or their level of infection and infectivity (Faust et al., 2017). Habitat fragmentation, as the process of dividing large continuous natural habitats into smaller isolated patches due to human activities, tends to increase the risk of infectious diseases by reducing biodiversity and, at the same time, increasing the probability of encounter of higher vertebrate species with pathogens. Furthermore, in a fragmented habitat, the sum of fragments may not be able to sustain the same diversity and prevalence of species (pathogens, reservoirs, and vectors) as in the original habitat (Murray and Daszak, 2013). Fragmentation also leads to a longer boundary between the habitat(s) and those of other communities. This in turn increases the chance of encounters between communities of hosts and vectors.

Therefore, the loss of biodiversity appears to be a rather subtle source of risk if the abundance of species and the richness of biomass cover are seen as an ecological buffer that protects us from direct contact with potentially destructive pathogens. The breeding and consumption of wildlife for food may be one direct cause of the so-called 'leap' of viruses or other pathogens from animals to humans, due to the easier contacts between

---

[4] All official sources played also an important role in pointing to such negative aspects as pollution and environmental degradation, even though they failed to discuss the positive and negative effects of ecosystems different characteristics, biodiversity, demographic and economic drivers.

humans and wild animals in natural landscapes destroyed by chaotic anthropogenic interaction. However, threats go beyond this simple cause-effect relationship, since the loss of bio-complexity of natural ecosystems facilitates the spread of infectious diseases, by providing contiguity along the entire chain of parasites and potential hosts that would normally be separated in diversified landscapes.

In an evolutionary context, the OH approach is closely related to biodiversity. By ignoring the interconnected issues related to human, animal, and environmental health, we risk overlooking the fundamental importance of biodiversity for human well-being. Biodiversity's role in nutrition is particularly significant, as it supports diverse and sustainable agricultural production, leading to nutrient-rich diets. Additionally, biodiversity serves as a valuable resource for medical research and traditional medicine. Its decline could impact reservoirs of infectious agents and potentially contribute to the spread of diseases. An often overlooked link between biodiversity and human health is at the level of microorganisms, the most abundant form of biodiversity on Earth. The importance of human microflora to the immune system, its role in noncommunicable diseases, and its links to the biodiversity of ecosystems on a larger scale are all areas of the biodiversity-health nexus that need to be further explored. The same is true of the links between urban biodiversity and physiological and mental health, which are sometimes described as the spiritual co-benefits of biodiversity.

Crucial for the protection of biodiversity is the maintenance of intact landscapes that are not degraded by agriculture, mining, and infrastructure, such as roads, railways, and power lines. Degradation of nature destroys habitat, releases sequestered carbon, and exposes people to viruses with which they had no previous contact and for which they have no natural immunity. COVID-19 offers an unfortunate but essential opportunity to demonstrate the fundamental importance of the One-Health approach. The world has noticed all this significantly.

In summary, a renewed vision of OH is today not only relevant, but necessary for a set of anthropogenic factors. The pressure at the interface between humans, animals, and their environment is increasing and detrimental to health. Demographic pressure, growing urbanization, and increasing migrations and movements are all elements of globalization that break down barriers and increase contacts between humans, animals, and natural areas. Furthermore, pollution and some production methods contribute to the loss of biodiversity. Climate change and natural disasters are changing the face of the planet and changing man-nature interfaces. All these elements have an impact on ecosystems at various levels and, together with these, on the general well-being of the planet.

10.1. The role of biodiversity

According to one of the most recent estimates (Mora et al., 2011), the natural world contains about 8.7 million species, but the vast majority have not been identified, and cataloging them all could take more than 1000 years. How can so many species coexist? The current interpretation among ecologists is that each species has a unique set of traits and corresponding limiting factors, so that each of them thrives in a sort of specific environmental niche, which partitions the environment and, at the same time, makes it uniquely suited to the bio-diverse populations living in it. The universality of niche assembly theory and potential alternative explanations for biodiversity are central questions in Hubbell's 'Unified Neutral Theory of Biodiversity and Biogeography', Hubbell (2011). Hubbell's 'Neutral Theory' unifies the twin concepts of biological resources and diversity by making the controversial 'neutrality assumption' that all individuals within a particular trophic level have the same chances of reproduction and death regardless of their species identity. The theory master equation describes the dynamics of abundance of an arbitrary species in the community as a function of the transition probabilities between different abundances of the species. Because all individuals are demographically identical, these dynamics also characterize the distribution of abundances of all species and, ultimately, the evolution of biomass in the community.

Competition for resources and/or different sensitivities to disturbances is of central importance to explain the evolutionary trends of biomass. Much of the literature on ecological evolution has dealt with this issue to explain the dynamics of vegetation in semi-arid areas, where various types of continuum ranges are hypothesized, such as, for example, the continuum between 'tree only' and 'grass only' vegetable cover. However, the more apparent phenomenon today is the fragmentation of highly diverse natural landscapes, resulting in a mosaic of agricultural patches that invade and deteriorate the natural environment. The ensuing loss of biodiversity is not only the abrupt destruction of species variety, but a sort of slow and silent agony of the biological basis of the ecosystem, with a continuous degradation that may degenerate in a sudden collapse when a tipping point is reached. This is, for example, the case of the peatlands of Indonesia, where deforestation proceeds through selective land clearing and slash-and-burn agriculture, with the ensuing loss of organic matter and soil fertility, increased forest fires, inundation and subsidence.

Zoonotic potential is at the root of the main threats to human health through the increase in the probability of spillovers of natural pathogens to anthropogenic settlements. Fragmented natural landscapes are instrumental in creating interfaces between former ecosystems and patches of anthropogenic activities, the 'ecotones', where boundaries between adjacent ecological systems are crossed and biological and ecological processes are concentrated and intensified. Ecotones are transition zones where spillovers may become epidemics through new forms of hyper-connectedness: the pervasive interdependence of global networks. This interdependence is the consequence of long trade value chains and fast and easy international travel that act as positive feedbacks on infectious diseases and are able to transform in record time local episodes into global events.

The increasing dangers of environmental degradation are epitomized by bridge species that mediate the passage of pathogenic agents from wildlife to humans and emerge from their niches as a consequence and a cause of loss of biodiversity. Bats are one of the main examples of these species among the most numerous highly adaptive wildlife mammals in anthropized rural and urban environments. As the only winged mammals originating 64 million years ago, they uniquely span space and time in a way that might have allowed the development of co-evolutionary processes between them and different viral species. As such, they appear to offer a unique combination of hospitality and biological resistance to viruses, especially those of the coronavirus variety. In anthropized environments, bats can accommodate their roosting and hunting needs, in conditions of low biodiversity, population density, and high interspecies contact rates, with increased risks of transmission of viruses favored by environmental and weather stressors.

In a broad sense, modern ecological theories of sustainable development suggest that biodiversity is the necessary basis of a robust, flexible, and resilient bio-economy, where this name is meant to characterize the intimate relationship between biology and socio-economic systems. The ability of organisms to respond and adapt to changes in the environment depends on their genetic makeup, which essentially consists of resources that cannot be easily changed due to past events. As a result, both biodiversity and society rely on a combination of different landscapes and organisms with diverse competencies and capabilities to emerge and regenerate. Generalist organisms possess versatile competencies that can be applied to a wide range of future situations, providing them with more options for action. Examples of generalist organisms include humans and bats, which have built up resistance to a diverse range of pathogens and can therefore act as both biological reservoirs and potential spreaders of diseases.

In contrast, specialist organisms have highly specific competencies that only give them a competitive advantage in a few specific situations. They include viruses that fill niches of time and space and compensate with their reproductive activity and genetic variations their dependence on suitable hosts. Generalists thus operate with confidence over rugged territory, and their contingent wealth is given by the sum of the value of many alternative options that they can exercise according to the circumstances. Specialists are confined to ecological niches, where they can prosper by exploiting sequential deepening options. Their success is thus the result of few high-value options that can be exercised only under the narrow set of conditions and states of the world that define their ecological niches.

The differentiation between generalist and specialist organisms is very important for the bio-economy because it implies that a shift towards increasing specialization and seamless continuity of global growth could pose a significant systemic risk and could become a major threat to sustainable development. In the field of agriculture, for example, modern farming practices have heavily emphasized specialization and simplification, concentrating production on only a few crops with similar characteristics. Technologies that use specialized inputs to save labor and land have also been developed to improve sector performance. However, this increased specialization has led to a decrease in biodiversity and genetic variation, which increases the vulnerability of crops to diseases and other stressors, putting the entire bio-economy at risk.

The promotion of commodity-based agriculture has discouraged functional diversification across different types of farmers, ignoring traditional differences in the personal roles and characteristics of rural society. This loss of diversification has negative impacts on nutrition and the creation of long and standardized food value chains, which have contributed to the growing concentration of sugar and fats in diets worldwide. These value chains are vulnerable to sudden disruptions and contaminations, leading to health and nutrition problems such as an increasing incidence of obesity, cardiovascular disease, and a greater vulnerability of human populations to infections and pandemics.

In the history of life on Earth, there have been five mass extinctions – episodes in which large numbers of species became extinct in a short time (Jablonski, 2001). All these mass extinctions were caused by natural catastrophes, such as volcanic eruptions, oxygen depletion in the oceans, and meteorite impact. However, a sixth different mass extinction seems to be ongoing, caused by human activity (Ceballos et al., 2015). According to a natural physiological extinction rate, one would expect to lose two species for every 10,000 species present every 100 years. Current figures show that 477 for every 10,000 species have become extinct in the last 100 years. Under a natural extinction rate, we would have expected to have only nine extinctions; in other words, there were 468 more extinctions than would be expected in the last century! This evidence is dramatic and tragic. At this rate, we may lose a large proportion of vertebrates, including mammals, birds, reptiles, amphibians, and fishes, over the next two to three decades. The loss of them has many consequences, including the loss of part of the 'dilution' effect exercised on pathogens. Vertebrates are essential to maintain ecosystem services, which are all the benefits we receive free from the proper function of nature. The combination of atmospheric gases, the quality and quantity of water, soil fertilization, and pollination are ecosystem services. By losing species, we are eroding the conditions of the Earth that are essential for human well-being. Habitat loss, due to agriculture, forestry, mining, urbanization, over-fishing and over-hunting, pollution, greenhouse gas emissions, and human overpopulation are the leading causes of this ongoing mass extinction. The One Health-One Environment approach promises to be a key instrument to avoid the most tragic consequences of the sixth mass extinction: saving endangered animals is the only way humanity can save itself.

10.2. The bio-economy

The COVID-19 pandemic has exposed the anthropogenic risks for the environment in many ways, ranging from threats of collapse of the health systems of countries to unexpected benefits in reducing traffic and pollution from lockdowns. In fact, the experience of most countries during the pandemic has been suggestive of how a combination of population pressures, conflicts, and mass tourism may be at the root of the condition of the planet as one of both exposure and vulnerability to anthropogenic risks. A bio-economy strategy of conservation appears to be a compelling need, with development aimed at transforming the present dynamics, by drastically changing the pattern of resource uses, service provisions, and resource management. This would involve reversing the abandonment of forest and water management and developing services for mitigation and sustainable adaptation to climate change, involving innovative blue growth trajectories: bio-technologies, food, and deep sea resources. Combining these strategies with new patterns of sustainable tourism and green energy usage and the OH model should also be part of a new approach to a global blue economy where circularity and bio-materials secure pollution-free water, biodiversity, and progressive independence from fossil fuel energy.

The trend towards regional poly-centrism is closely linked with the OH paradigm, which emphasizes the inter-connectedness of human, animal, and environmental health. This paradigm is expected to drive the reorganization of production networks and the separation of living and working spaces, which in turn will promote ecological sustainability and support the development of a bio-economy. As production and consumption activities are more widely distributed across larger areas, the concept of territorial capital will be redefined to include a wider range of ecological and economic resources. This will create new opportunities to enhance regional heritage and promote sustainable forms of tourism, while also shaping a new bio-economic model that supports both human and environmental health.

In summary, the pandemic may have introduced or accelerated a series of transformative innovations that bring previous developments and certainties back into play, creating new risks and new opportunities. On the natural and human capital side, these innovations concern the complementarity between private and public activities and between cultural heritage and environmental heritage, both of which are affected by the changes in work organization, production, and lifestyles that are taking place. These include the promotion of regenerative agriculture and the development of circular food systems that minimize waste and promote sustainable use of resources. They may also involve the transformation of traditional forms of mass tourism into new 'sustainable' or 'regenerative' forms of tour operations and land travel made possible by the expansion of distance working and new patterns of territorial settlement. In general, these transformations point to a potential major shift toward more sustainable and holistic models of economic development. By promoting decentralization, diversity, and resilience, these changes can help create more sustainable and equitable regional economies, while reducing the negative impacts of economic development on the environment and society.

10.3. COVID-19 and the bio-economy

The 'OH' concept is linked to the emerging interest in natural capital as a key factor of economic development and reflects the unifying paradigm of nature as a primary source of well-being. Although health is a general notion of a form of well-being that transcends market values, it can be regarded as a universal public good that needs to be produced and maintained, whose value depends both on its accumulated stock and the services it provides to those who enjoy it. In this sense, health is consistent with the Knightian (Knight, 1944) paradigm of capital as the ultimate source of value. Knight's thesis can be expressed in the following four propositions (Buechner, 1976): *1)* capital is the only factor of production; *2)* capital is a fund of value; *3)* production and consumption are simultaneous; and *4)* capital is permanent.

Therefore, Knight's position has two important elements that are relevant to current research on capital and the growing importance of the notion of natural capital. First, all factors of production are services of some type of capital, including, in particular, the three basic forms of physical, natural, and human capital. Second, capital is a sort of permanent fund of value that, once established in a steady state, is independent of time. This second characteristic is more in line with the modern concept of natural capital and, more specifically, of renewable natural resources. In Knight's view, this characteristic is well exemplified by a forest or, in a somewhat mythical form, by the 'crusonia' plant, a gift of nature that is constantly subject to harvest for consumption and growth. Being in the steady state can also be taken as equivalent to health as an optimal condition, characterized by homeostasis and stability, as well as by a form of intrinsic harmony within and between itself and its surroundings.

The concept of OH, which emerges from the reconsideration of nature as a primary value source, has been around for a long time and was implicitly invoked by many environmental theories based on the interdependence of natural and human capital, such as, for example, those based on the Gaia hypothesis (Lovelock, 1995), which assumes that organisms co-evolve with their environment. However, a more specific paradigm of OH emerged at the beginning of the new millennium, mainly as the confluence of two different currents of thought: holistic medicine ('one medicine') and the theory of a systemic interdependence of all living things and their environment. Although holistic medicine invokes a comprehensive notion of health and care based on balance between living things, OH embeds this notion in a wider model of interdependence between humans, animals, plants, and the ecosystem.

However, the notion that an ideal combination of natural, human, and non-human capital can achieve a time-independent 'steady state,' as suggested by the natural capital theory, is neither a widely accepted theoretical concept nor supported by substantial empirical evidence. Most ecologists and environmental scientists tend to reject the very notion of OH as defining an objective property of an ecosystem consisting of an equilibrium state of a super-organism of some sort. A difference between an ecosystem as a connected set of different systems (a Prudential OH Approach) or an interdependent system of different components (a Radical OH Approach) also appears to be relevant in the current debate (Sironi et al., 2022), bringing to two contrasting sets of prescriptions for both proactive and adaptive actions. An alternative to the equilibrium view is based on the concept of resilience, defined in a seminal contribution by Holling (1973) and further developed in the theory of social-ecological systems, as a property linked to 'the persistence of systems and their ability to absorb change and disturbance and still maintain the same relationships between populations or state variables'. Furthermore, different variants of a systemic approach to health have been proposed, although they tend to converge at least in some of their basic characteristics, such as the importance of preventive measures at the individual and systemic levels (de Garine-Wichatitsky et al., 2021).

Vulnerability has been defined as '... the capacity to be wounded' (Kates et al., 1985), while resilience can be conceived as the ability to make wounds less serious and more likely to be cured immediately. Vulnerability to external shocks is a condition that combines exposure and limited ability to adapt to undesirable changes in climatic conditions, market functioning, and other factors. Exposure depends on external conditions, especially location in places and at times where shocks are more likely to occur, as well as on specific capabilities to avoid danger. Resilience depends on the ability to adapt to external shocks, for example, by reducing harmful consequences or by exploiting beneficial opportunities. By and large, these capabilities depend on some form of human and/or non-human capital, such as savings, education, institutions, and access to technology. They also depend critically on the capacity of to make decisions and revise them in an inter-temporal, uncertain context, acting on variables that manifest themselves across time periods, such as savings, storage, borrowing and lending, and various types of investment.

The increasing interdependence brought about by globalization in the value chains of virtually all sectors, but especially critical for food supply, nutrition, and health, is a key feature of the bio-economy. In this sense, the sustainability of the present growth model appears particularly risky and is exposed to both ordinary environmental threats and the more important challenges posed by climate changes, mass extinction, and pandemics. Natural resources directly linked to the cyclicality of life-sustaining processes, such as water and biodiversity, appear to be the basis for a growing systemic risk for the global economy. Biodiversity is considered the main asset of the current economic capabilities of the world and, at the same time, a source of potentially catastrophic liabilities under unfavorable scenarios of degraded landscapes, biomass degradation, and climate change.

The risk of biodiversity has many facets. On the one hand, biomass growth and diversity depend on water supply (both annual runoff and soil moisture), which is likely to be significantly reduced in the long run, as a consequence of the unfavorable evolution of rainfall and of the long-term consequences of glacier melting. On the other hand, extreme events such as flash floods, droughts, heat waves, and wildfires are increasingly expected to occur with progressive global warming. Intense tropical and mid-latitude storms with heavier precipitations and winter speeds are also projected to become more frequent and devastating. Systemic risks from the combined effect of ordinary climatic variations and extraordinary climate change are exacerbated by population growth and increased urbanization as well as several other ominous characteristics: *i)* the association of catastrophic events with relatively high probabilities, *ii)* large areas of ambiguities and lack of information on biomass evolution, landscape dependence and anthropogenic influence, *iii)* dependence of outcomes on abrupt changes, discontinuities, and tipping points.

Because biodiversity is an intrinsic aspect of natural capital, which is a significant source of various ecosystem services, its concept revolves around a fundamental ambiguity between existence and differentiation. If natural capital is considered the result of a stochastic evolutionary process, biodiversity can be seen as cross-sectional trace, at any point in time, of the volatility of the evolutionary process. In this respect, biodiversity is like the variability of the geological record or of any collection of objects whose plurality and specificity provide an accessible documentation that, in a broad sense, recapitulates a process, albeit in a non-exhaustive way. As the variability of the biological record of the history of the earth, biodiversity can therefore be considered as an indicator of the spread of biomass distribution throughout the world.

The concept of biodiversity as the observed variability induced by a stochastic process regulating biomass evolution suggests that it may be more important to value natural capital than to quantify its diversity dimension. Due to the nature of the evolution of natural capital as a collection of risky assets, subject to exogenous and unpredictable shocks, biodiversity can be seen as a portfolio of resources (species, genes, and habitats) that provide similar benefits but suffer from different sensitivities, so that their performances are not perfectly correlated. Solow et al. (1993) were the first to develop this idea and derived a measure of the diversity of a collection of species that reflects the probability that at least one species in the collection provides a given ecological service. Therefore, the potential value of natural capital as a generator of future opportunities depends on the important role of biodiversity as a 'library' (Goeschl and Swanson, 2007; Weitzman, 1995), which contains information that can be valuable in the future (see also Farnsworth et al., 2015), as a repertoire of possible ecosystems and ecosystem configurations.

10.4. The exposome

In 2005 Christopher Wild coined the term 'exposome', defining it as the result of cumulative environmental exposures throughout life from the prenatal period onward

(Wild, 2005). Since then, this definition has been extended and refined by enumerating the constituent components of the exposome, as well as suggesting metrics and measurement methodologies. In general terms, the exposome is an attempt to define in more meaningful and documentary detail the environment variable in the equation phenotype = genotype + environment. It is semantically characterized by a parallelism with the genome and has been interpreted as an index of 'nurture', as opposed to the genome as an index of 'nature'. As Miller and Jones put it: '... the exposome is even more expansive than what Wild described 9 years ago. The exposome captures the essence of nurture; it is the summation and integration of external forces acting upon our genome throughout our lifespan' (Miller and Jones, 2014). The exposome seems also to recall the notion of human capital (Becker et al., 1964), itself a somewhat ambiguous term, which can also be interpreted as embodying the genome and the exposome, both interacting to produce the unique quality of a lifetime.

The exposome promises to be a bio-statistical revolution because it substitutes the usual model based on data collection with the notion of cumulative value of multivariate exposure factors. This concept is occasionally utilized to gauge stock variables like capital and wealth, but it generally fails to incorporate the diverse interactions that characterize the accumulation process. Because the exposome is defined as the cumulative measure of external exposures on an organism (external exposome), and the associated biological responses (internal exposome), its empirical measurement needs a time-dependent set (ideally, a lifetime dependent) of exposures and reactions over long periods or, alternatively, in selected windows of time. Therefore, the paradigm of exposome research is based on a radically different approach from traditional data analysis to causal inference due to three distinctive features: multiple exposure domains; integrated exposure and response data across multiple scales of variation, such as populations, time, and space; and data-driven discovery from the investigation of multiple exposure-response relationships (Stingone et al., 2017).

This implies that cohort investigations rather than random control trials will be the dominant observational studies of subjects identified for their exposure to complex risk factors. It also implies that these studies will need large cohorts and sophisticated statistical methodologies to be able to disentangle the cause-effect relations among the myriad of exposures and the high level of association across different risk factors that characterize the different individuals. More generally, bio-statistical studies will require a new type of research infrastructure to support the expanded exposure assessment activities required by a meaningful analysis of the exposome. These infrastructures will have to combine the capacity to collect, store, and make available data on the genome and the exposome on an extended basis for large population samples and long periods of time. At the same time, they should also be the main depository of analytical, bioinformatic and statistical tools to process, integrate, and analyze high-dimensional data on individual genetic and exposure characteristics over time.

Exposome observational studies include a new generation of mega-cohort research projects, such as the UK Biobank (Julkunen et al., 2023) or similar US (e.g., (Rozmus, 2023), and Asian cohorts (Shimoyama et al., 2023)). In these studies, large samples of individuals over time are used to investigate the impact of genetic variations, exposures to different environments and environmental changes, lifestyles, general health conditions, diseases, and deaths. The sizes of these samples require a complex and yet untested mixture of methodologies which are largely still being developed and, at the moment, are no more than a mixture of traditional methodologies. These include, for example, different variants of multiple regression, 'retrofitted' to adapt to large longitudinal samples with a high degree of association across most potentially explanatory variables. At the same time, the mega-cohort studies appear to require substantial investment, not only for direct data collection but also for the need to store and make available detailed exposome data in bio-banks beyond their present capacity of collecting, preserving, and providing access to bio samples. For example, UK Bio-bank has recruited half a million people at a cost of around £60 million in the initial phase. The proposal to establish a 'Last cohort' of 1 million people in the United States or a similar-sized Asian cohort (Wild, 2005) would presumably exceed this sum. These costs have also to be considered in the context of a new wave of investment in sharing and open data research infrastructures, such as bio-banks and biological resource centers (Ozerskaya, 2008).

Traditional methods of assessing the environmental impact on human health are generally based on multiple regression techniques, by which the effects of a few critical variables on health indicators are explored with multivariate models of statistical correlation. Estimates seek to attribute specific health effects (from increased morbidity to anticipated mortality) to exposure to one or more pollutants, often under the assumption that these exposures are not correlated over time and that no joint effect is determined by simultaneous exposures of the same subject. Recent studies have tried to apply the exposome research paradigm by investigating human health response to multiple exposures and their interactions with a variety of methods. These are selected on ad hoc basis to manage the extensive dimensionality of these data and the intricate statistical inference routes conceivable for interpreting them. They include for example mixture analysis, integrating the selection, shrinkage, and grouping of correlated variables (e.g. LASSO, elastic-net, adaptive elastic-net), dimension reduction techniques (e.g. principal component, partial least square analysis) or Bayesian model averaging (BMA), Bayesian kernel machine regression (BKMR), etc.) (Stafoggia et al., 2017; Lazarevic et al., 2019).

Because of the still exploratory nature of the exposome studies, these methods combine statistical techniques with new software capabilities and typically are not framed by a theoretical model. They are only used to reduce the dimensionality of the phenomenon (the exposome as a set of multiple exposures). Consequently, they lack model selection stability (shrinkage methods), lack interpretability of the latent variables (dimension reduction), and computational inefficiency (Bayesian models). In addition, they are rarely applied in the context of large (>100 variables) and heterogeneous exposome data (omics, categorical/continuous variables).[5]

A further challenge of the exposome-based approaches concerns data integration. This includes matching exposome and genome data from various sources and integrating different data flows over space and time to account for multiple exposures. While the ideal way of data collection would be through longitudinal studies (prospective or retrospective) of lifetime exposures combined with dynamic genomic data, in practice, this type of collection still lacks strong theoretical foundations and is being developed only slowly, since it is very difficult and costly to implement.

However, while the data produced with traditional techniques are not ideally suited to study multiple exposures, primary data from cohort studies can be combined with secondary data from environmental monitoring systems. The key integrating principle is that exposure to a single environmental insult (e.g., a high level of PM 2.5) is part of a multidimensional event that includes multiple exposures at a particular moment and their accumulated effects over time. An individual may thus appear to exhibit a response to a single exposure, but this response can only be understood in the context of a combination of concurrent simultaneous exposures and is conditioned by the accumulation of all her exposures (external exposome) and responses (internal exposome) in the past. Moreover, the concept of accumulating and concurring exposures over time is readily applicable to the environment itself. For example, an eco-exposome can be defined as the cumulative results of all exposures of an ecosystem to both nature and human induced environmental changes over time. Similarly, an urban exposome can be conceived as the qualitative and quantitative assessment of environmental and health indicators that describe the framing and evolution of urban health and its interactions with urban infrastructure, climate, and small area (neighborhood) features (Andrianou and Makris, 2018).

The exposome concept also suggests a different way to consider statistical evidence and organize routine data collection on the part of national and international institutions. At the moment, statistical data is gathered through a variety of traditional means, such as surveys based on questionnaires and baseline or periodic samples, administrative sources (including medical and biographical information), and increasingly with direct observations through remote sensing machinery. The great heterogeneity resulting from all these different methods is not significantly reduced by standardized classifications, because of the lack of integrating methodologies and procedures. Unlike the traditional

---

[5] For example, shrinkage methods are a statistical approach used in health cohort studies to estimate the association between a set of predictor variables and an outcome of interest, while controlling for other confounding variables. These methods aim to improve model performance by reducing the variance of the estimates and controlling for overfitting. However, they lack model stability because they are sensitive to tuning parameters that they often use to balance the bias-variance tradeoff. This instability in the model can make it difficult to interpret the results and limit the generalizability of the findings.

environmental epidemiology approach, which focuses on final outcomes such as mortality and morbidity, the concept of exposome suggests that data collection should pay special attention to intermediate and more subtle effects related to environmental exposures. This includes internal biomarkers of exposure and response, using information increasingly available from biobanks and the application of 'omic' technologies.[6] The main idea is that not only data collection should aim to document the accumulation of exposures through lifetime (as expressed by the motto from biography to biology), but also that it should attempt to capture the sequence of intermediate molecular changes that characterize the process of emerging health outcomes under the pressure of the environment (Vineis et al., 2020).

The denomination of exposome suggests a natural association with 'exposomics', a term evoking the recent development of 'omics' technologies (http://omics.org/), which are aimed at developing high-throughput analysis of a set of molecules. Omic technologies (OT), of which the first to appear was 'genomics' (as opposed to genetics), allow us to combine two very important properties of modern data analysis: interrogation and intervention. OT are able to increasingly exploit these two properties through: i) data recovery and processing from a variety of sources, and through a plurality of computing facilities and ii) automated analysis and selection of production and application options available through modeling. The two main distinctions of competitive options in the data collection space are their automated preparation via search and processing algorithms, and the use of Artificial Intelligence and Deep Learning to mine and manage available information. A common data model makes analytics more agile because it is designed without being constrained to the individual data models and business definitions of a particular tool. Valid unified OT platforms for life sciences are also intimately linked to their advantages as part of the industry integration process through horizontal infrastructure, interdisciplinary communication and analysis, and their agility and self-service capabilities. OT technologies are themselves natural vehicles of high throughput and instant communication, as shown by the unprecedentedly fast development of mRNA vaccines in the current pandemic. Their main basis is data integration through array technologies, producing gridded data, complemented by a wealth of associated metadata, such as information on space and time, as well as on anomalies and other distinctive features. Combined with deep learning and other automated information algorithms, array technologies have been able to spread to a variety of applications, including disease analysis. For example, 'volatolomics', a high-throughput, array-based technique that studies the relationship between volatile

---

[6] Omics technologies are a group of high-throughput molecular analysis techniques developed in the early 2000s. The first omics technology to emerge was genomics, which focuses on the study of the entire genome of an organism. Genomics has been followed by other omics technologies, such as transcriptomics (the study of gene expression), proteomics (the study of proteins), metabolomics (the study of small molecules involved in metabolism), and epigenomics (the study of epigenetic modifications).

compounds and molecular patterns, has been successfully used to discriminate patients with COVID-19 through breath analysis (Mougang et al., 2021).

Causal inference models may be significantly influenced by the exposome paradigm's novel approach to causal attribution. To enhance causal attribution, it is essential to document the intermediate steps involved in this process, even within the traditional inference model. For instance, a combination of external variables (such as various sources of pollution) and omic technologies could be utilized. A new generation of epidemiological and economic models that consider simultaneous exposures and responses is required to unravel the exposome through a more comprehensive causal analysis pattern than that of typical observational studies, including those based on randomized control trials. These models should not be designed to predict outcomes of endogenous variables in response to exogenous changes, but '... to systematically explore possible counterfactual scenarios, grounded in thought experiments – what might happen if determinants of outcomes are changed' (Cunha and Heckman, 2007). New models of economic and social epidemiology should also be used to address the truly most challenging task of policy analysts: 'Forecasting the impacts of interventions (constructing counterfactual states associated with interventions) never previously implemented to various environments, including their impacts in terms of well-being' (Heckman and Pinto, 2022).

10.5. OH and the present health system

The concepts of One Health (OH) and exposome, as well as the rapidly evolving technologies described, have shown great promise for improving public health. At the same time, the recent COVID-19 pandemic has highlighted the limitations of existing health systems. These systems are often reactive and focused on intensive disease treatment and hospital care, rather than being proactive and preventive. This narrow scope has left many people vulnerable to preventable diseases and health problems. In addition to its direct effects, as we have argued in a section of this chapter, COVID-19 has also unveiled the dramatic side of a more general threat to health as natural capital, which is rooted in the destructive anthropogenic impact on the environment and the encroachment of wildlife. It has also manifested an unsuspected power to neutralize adaptive responses and attempts of mitigation on the part of existing institutions. Some of the induced changes, such as the flight from contact jobs in the health system, may well be temporary, while others, such as building back better institutions, are, we hope, part of a structural and long-term legacy.

10.6. Health and nutrition

The world has changed by the COVID-19 experience, but we do not know how persistent and widespread these effects will be in the near and distant future. The forces at work are related to technology and behavior, and in both cases their long-term consequences are unclear. Lockdowns and the associated alterations of human contacts have altered traditional social patterns, reduced the probability of encounters with unfamiliar counterparts, although they have intensified meeting and communicating with distant and diverse interlocutors through digital connectivity. Distance working and confined interaction have been spectacularly effective in multiplying long-range communication and diverse and remote partnerships, with growing global links and small-world effects. By reducing population pressures, lockdowns, and social distancing, these developments have also provided temporary, but suggestive, glimpses of new ways of sharing actions for a sustainable collective life in densely populated urban areas. Lockdown consumption has been polarized between food and delivery and communication services, with most other items receding in a background of suspended animation. Most traditional patterns of consumption are likely to rebound, although the experience of the pandemic has helped to focus attention on the quantity, quality, and healthy characteristics of the local food supply. Some dietary patterns appear to also have changed. They may be part of a new set of strategic priorities to improve the safety and security of food value chains through the support of small-holder food producers and ecosystem services and the maintenance of local biodiversity.

The pandemic has also exposed the critical vulnerability of the present patterns of international specialization and trade embedded in food value chains. Although it is already part of the emerging bio-economy based on great bio-digital convergence and advanced technology and science, food production around the world is still organized in a wasteful and haphazard way, with an enormous amount being squandered every year by rich countries (about US $400 billion of food annually, an amount that could feed around 1.26 billion people a year). Food value chains are not only long and unsafe, but also socially unsustainable, with most farm operators being destitute or disenfranchised small cultivators and migrant workers.

The case of food consumption is relevant to understand the impact of pandemics and its future developments for several reasons. First, providing food to everyone appears again as a planetary challenge of ever-growing Malthusian proportions and drama. Although almost 850 million people are still hungry, the world's population is expected to reach nearly 10 billion by 2050, a figure that requires global food production to increase by at least 50% to meet demand. Pandemics have shown that even these orders of magnitude may be optimistic, as they do not take into account the possible impacts of widespread diseases, climate changes, and other global disruptions. Additionally, increasing pressures on arable land, energy, and water threaten the very foundations of a sustainable world economy.

Second, food and shelter are the heart of personal security and have been the main drivers of lockdowns and other responses during the most acute phases of pandemics. However, since the beginning of urbanization, food has been channeled to consumers by a series of intermediaries that include producers, processors of various degrees, commercial agents, and distribution centers. Logistic is a key element of these food channels and so is the cold chain, an uninterrupted sequence of low-temperature vectors and intermediate storage components that ensure end-to-end delivery with minimal risk of contamination and degradation. Efficient logistics and extensive supply chains have facilitated prompt food delivery globally, but minor disruptions in production or transportation of few critical nodes can dramatically affect system continuity and overall performance.

Third, long supply chains are associated with a high degree of specialization, so key commodities are concentrated in a few producing countries. Progressive adoption of agri-food standards has led to loss of diversity in dietary habits and material culture. In the bustling cities of most Mediterranean countries, for example, the renowned 'Mediterranean diet' based on fresh fruits and vegetables risks being overridden by northern European dietary habits led by meat and rich in fats and sugar. Obesity and cardiovascular diseases are increasing and catching up faster than incomes and other development indicators. The agricultural revolution expected from biotechnology has not yet arrived and optimistic expectations are shattered not only by insufficient technological progress, but also by shifting perceptions and lack of producer confidence. Consumer trust in processed foods is also declining, with the growth of new, healthier, and more broadly bio-based agriculture emerging as the only hope for sustainable development.

What should be the characteristics of this new 'bio-economic' agriculture? While a nutritious, healthy, diversified, and universally affordable food supply is certainly a priority, a bio-based agriculture should be redesigned to go beyond the food system. This may seem too ambitious given the challenge of feeding 10 billion people in the next 25 years. We know, however, from past experience and from science, that the only way to escape the Malthusian trap is through innovation and technological progress. On the supply side, a new bio-based agriculture has to go beyond the simple provision of food; it should take advantage of Engel's law, which states that as people's income increases, they tend to spend a smaller percentage of their income on food and a larger percentage on other goods and services. It should gain resources from consumers' willingness to pay for more diversified and sophisticated goods and services, including energy, fibers, and other materials of various kinds, and active principles for pharmaceutical, physical, and chemical products. As a vast and renewable reservoir of potential products of increasing attractiveness for direct consumption and effectiveness for industrial processing, the bio-economy can become part of a new impetus for development and, at the same time, the resolution of the apparent trade-off between economy and ecology.

Rebuilding disrupted value chains after COVID-19 appears to offer a chance to construct new bio-based mechanisms of production with built-in tendencies to recycle, treating the environment as a source of renewal rather than as a waste reservoir. In these mechanisms, the economy would become part of ecology by using and 'recovering' natural resources to build materials and products that minimize waste and negative impacts on the environment. This outcome cannot be obtained by mere goodwill. It requires visionary planning and innovative investment at a global scale and reach, with substantial resources, and the public and the private sector joining in a whole series of actions aimed at constructing a new bio-based circular flow of production, distribution, and consumption. Its success would depend not only on mobilizing financial resources, but also a critical amount of international cooperation and consensus on the pursuit of economic development through a new circular relationship between the economy and the environment.

The 2030 Agenda of the United Nations, the Paris Agreement, and the 2017 G7 have dictated fundamental policy guidelines to pursue sustainable development and improve resilience to climate change (greenhouse gas emissions). However, due to the exceptional amount of resources to channel into rebuilding the economy from the present emergency, it is possible to turn production processes into a bio-based circular system, through product design, manufactures, services, and a whole new set of transformative projects. Given the prevailing economic model of open and linear input output type 'take, make, and dispose', the new bio-economy model should aim to a global sustainable alternative based on 'closed-loop' processes that minimize waste and maximize the duration and re-use of materials as well as their natural renewal on the basis of biological processes. In turn, these algorithms will be made possible by the design of innovative, bio-based products that, in terms of materials, energy requirements, components, engineering, and chemistry, maximize the options of maintenance, repair, multiple use, recycling, and renewal. In this respect, it is important to underline the difference between a mere circular economy, based on closed recycling loop processes, and a circular bio-economy. This ultimately relies on the autonomous capacity of the natural environment to provide the basis for sustainable use and management of biological resources and involves activities such as the production of bio-based materials and products, and the use of biomass as a source of energy and feedstock for industry (Sharma and Malaviya, 2023).

10.7. Bio-economy and energy

COVID-19 is the latest and most spectacular occurrence of an increasing series of zoonotic diseases that has shown how closely intertwined are men and nature. Many have speculated that this pandemic was long overdue. It had to be expected, as a sort of impersonal, but powerful, organismic response of a threatened environment to man as its main damaging agent. The bio-economic impacts of pandemics may be misunderstood, as some environmental pressures decrease, while new forms of global contamination arise from substantial biomedical waste. A clear indication appears to be the need to develop a global biomass governance model. Unlike other chemicals and inputs for production, biomass has an intrinsic unity as both the basis of life on the planet and the ultimate renewable material. Today, biomass is used in the ranking order for the production of feed, food, energy, fuels, and chemical feedstock and represents around 18% of global final energy consumption. In energy production, biomass is of critical importance in the process of decarbonization, a technology transition that aims to drastically reduce the emission of carbon dioxide and the other greenhouse gases that contribute to global warming. Scientists have set a goal of 450 ppm of CO_2 in the atmosphere (or lower). This would give us a 70% chance of keeping global warming below 2 degrees Celsius (C) and a 50% chance of keeping it below 3 degrees C. This objective has been generally accepted as a 'doable' reduction of CO_2 in the atmosphere – an increase of approximately 65 ppm from current levels, but is far from obvious. To reach a peak at this level of CO_2 in the atmosphere requires emission reductions of 50–80% from current levels. To grasp the enormity of the task, consider the fact that current emissions in the US are around 20 tons of CO_2 per capita, and this amount would have to drop to 4 tons per capita to achieve the goal of a 80% reduction in emissions.

These quantifications give an idea of the technological challenge that climate change offers as well as the opportunities to rebuild the world after the trauma of the pandemics. The COVID-19 pandemic has caused an increase in awareness at the international level, in a way that has not been seen in the last 20 years, where a sense of complacency and inertia has prevailed in the international community. There are signs that the pandemic has been a wake-up call, reminding people of the importance of working together to address global issues. The pandemic has highlighted the interconnectedness of the world and the need for coordinated international action in the face of a global crisis. As a result, there a higher level of international awareness and has perhaps begun to emerge.

In addition to climate change, the pandemic has also shed light on other global challenges, such as economic inequality and social injustice. The global response to the pandemic has created an opportunity to address these issues with renewed vigor and a greater sense of urgency. The pandemic has also shown, without a shadow of doubt, that only through international cooperation and the massive application of multiple technologies and transformative innovations can we take effective action and reverse the current scenario. As the successful experience of vaccine development has shown, the world needs technological solutions, but innovations must also be tested quickly. They depend not only on the experimentation of great solutions, but also on specific applications and their likely spread, with the bio-economy playing an essential role, for example, in the field of biotechnology, energy efficiency, technology of sequestration and storage of CO_2. To achieve these goals, we need a collective effort and a new and more ambitious round of international cooperation. At the same time, we must also move the frontier of business innovation and direct it in the direction of a new and sustainable bio-economy, through a better functioning of the economic and financial sector and the construction of a framework of appropriate incentives.

In some respects, the COVID-19 crisis has led to an unprecedented situation in energy-producing countries rich in natural resources. Before the Ukrainian crisis, the fall in global demand and in prices had led to lower gains in the production and export of fossil fuels and related products. On the other hand, the same prices and the negative development of the macroeconomic situation had rendered the consumption of renewable energy less attractive and possible investments more burdensome than the local use of oil, gas, and coal itself. Falling oil prices had also been seen as an opportunity to eliminate the ubiquitous and expensive fuel subsidies, thus reducing one of the biggest and most difficult distortions to remove in the domestic economy.

Prior to the Ukrainian war, investing in natural gas appeared more viable than before, both as a step towards diversifying from coal and expanding the production base for increased electrification access by a growing number of businesses and consumers. The fall in the prices of fossil fuel exports, which mainly affected oil and its derivatives and had less weight on local and regional markets, appeared to reinforce the prospect of natural gas development as the main component of a gradual energy transition. The abundance of resources of this fuel, detected by the most recent discoveries, therefore appeared to be a real strategic reserve and a crucial tool for the resilience of developing country economies affected at their most crucial point by the supply and demand shocks of the pandemic.

The invasion of Ukraine by the forces of the Russian federation drastically altered these expectations and may have created a completely new scenario of energy transition. The dramatic surge in fossil fuel prices was followed by a reduction, intensifying the need for transition, but simultaneously dampening the hopes for swift progress. From this new point of view, the role of the energy industry and the entire international community appears particularly delicate. In terms of energy resources, producers' countries were suddenly rescued from experiencing great economic difficulties, with falling incomes and investments, and partially irreversible losses in productive capacity and human capital. The pandemic has also highlighted critical issues in the organization of the extractive industry, particularly offshore investments. There is a need to redesign production processes, taking into consideration the remote locations and limited working spaces of many mining operations. Additionally, increased focus should be placed on environmental protection and worker health and safety.

The need to diversify their energy sources and reduce dependence on Russian imports has created new incentives to invest in the traditional energy value chain. However, due to the large investment needed to meet the new demand for health security and safety, a 'build back better' phase of the pandemic may also offer the international community an opportunity to actively accelerate the energy transition in developing countries. New energy investment should include innovation, through locally tailored technologies that can expand rural communities' energy capacity and use through mini-grids and renewable sources, and allow technology leaps and diffusion processes to take place through scalable projects, fast developing within and beyond the local community levels. The substantial investment required may also exercise an important stimulus to alleviate the tendency of the current downturn of the economy to become a worldwide recession, reducing the dangers of a depressive spiral that could also affect the advanced countries.

10.8. The geopolitics of a new OH system

COVID-19 has exacerbated world inequality in many ways, including exposure to existing fragilities in income distribution and patterns, as well as the enforcement of egotistic and rival behavior in consumption habits, irrational fears, and willingness to cooperate. Lack of coordination has been a major problem at both the national and international levels, and the rich nations of the world have exhibited a more recent nationalistic policy for vaccine procurement and usage. According to the UN Secretary, the 10 richest countries in the world have *de facto* monopolized both vaccine consumption and supply. However, the apparent early end of their acute pandemic crisis, achieved through successful mass vaccination, may be a temporary illusion, as the pandemic appears as a persistent threat, especially, where the lack of financial resources and international coordination makes vaccine supply short, ill-distributing, and insecure. This condition is yet another manifestation of the unsustainable nature of the current pattern of development. Unsustainability in this case hinges not only on the emergence of another dimension of inequality and global distress, but also on the reality that the pandemic cannot be truly contained unless a sufficient level of immunization is achieved globally. This must be accomplished in an impressively short amount of time due to the virus's proven ability to mutate into increasingly infectious variants.

Against this background of unsustainable social consequences and the urgency of international actions, the decision of the G7 to endorse a minimum global corporate tax of 15% on the profits of multinational companies is good news. In October 2021, the OECD estimated that as much $81bn in additional tax revenues each year could be raised under the reform, but this amount would probably be much larger if the tax were enacted immediately and could collect some of the vast extra profits created by the energy crisis. While the tax has been hailed as an assist to the coffers of the rich countries strained by the economic crisis, in the short run its main use would have to be the international financing of a sustainable mechanism of mass vaccination for developing countries, as well as the provision of basic health services and monitoring.

Would this be enough to reverse the current pattern of unsustainable drift of the pandemic and its dangerous health and social consequences in the poorest part of the world? If the tax is raised in a timely and effective manner, it would. At the time the tax was approved, the International Monetary Fund estimated that the overall cost of a far-reaching campaign aimed at vaccinating 60% of the world's population in 2022 would be 50 billion dollars, that is, an amount that even a lower bound application of the tax would be able to collect only in one year. Against this limited cost, the IMF estimates of cumulative benefits, including vaccinations, diagnostics, and therapeutics, were around 9 trillion dollars by 2025, with more than 40% of this gain going to advanced economies. The benefits would be widespread and could reverse some of the unsustainable aspects of the pattern of spreading of the economic and social consequences of the virus.

However, while these interventions meet short-term demands, they do not address the structural issues that underlie the current biomedical divide between rich and poor countries. This divide is not only another manifestation of global inequality and unsustainable social disparities. To a significant degree, the existence of the pandemic and its dramatization highlight a crucial structural issue in delivering health as a global public good. Indeed, the nature of global public goods in healthcare stems from the distinct situation where their supply and demand rely on direct relationships between consumers, communities, and ultimately, countries.

In this regard, the pandemic has shown that, unlike ordinary consumption, inequality in primary health services is intolerable. While ordinarily a given person may consume to satisfy uneven states of need compatible with those of other subjects, in fact, in the case of health, consumption gives rise to direct and reciprocal perception of an indivisible state of need common to the world community of all subjects affected. A moral order requires that the state of need of the entire international community is perceived and satisfied with the common help of all subjects involved, in conditions of reciprocity, where each subject feels his or her need together with the states of need of the others. The interdependence of the states of need of all the components of the community gives rise to a communion of states of need which finds its foundation in direct relations of reciprocity across all consumers.

This makes it possible to better define the nature of health services and the role of the WHO and other international organizations. As with all pure public goods, the consumption of health services, along with a merit character, includes a socially compulsory element. This is especially true for vaccines, but it should be extended to all prevention practices, including hygiene, lifestyle, and nutrition. However, despite its merit nature, the fact that health is a global public good cannot completely obliterate the autonomy of people to judge the quantity and quality of the services consumed. In pursuit of this objective, it becomes essential to produce and allocate health protection services through a 'quasi-market' framework. This approach should be defined by the regulatory involvement of public entities, including the State and multilateral organizations. Within this context, a dynamic industry and flourishing innovation can coexist, while simultaneously ensuring equitable access to both the quantity and quality of services for everyone.

Individual autonomy, which the organization of the health area must respect, must be ensured through the fulfillment of the conditions necessary for its full satisfaction. An overriding principle of fairness dictates that compliance with the public rules is counterbalanced by the acceptance of a sustainable and fair tax burden on the part of the international community and of the tax payers. In this respect, it appears desirable that specific international tax proceeds be earmarked to the provision of international health services, as may occur with the provision of vaccines and the global corporate tax. The fair distribution of services implies, furthermore, that they should be evenly distributed from the point of view of access to their use, so as to mitigate inequalities and eliminate any social discrimination caused by differential access to health services.

Although the experience of COVID-19 has been poor and painful, its social consequences and its learning aspects present an inescapable opportunity to reorganize health services in a manner that is both efficient and sustainable for our planet's well-being. Public health institutions at the country level, including the Ministries of Health and regional and state government agencies, have re-claimed a leading role during their fight against COVID-19 against the bustling private health sector. Despite the stress on national health systems, in developed countries, this has given new legitimacy to government actions and public good provision, while in developing countries, performance has been mixed, due to lack of resources and governance problems. At the international level, after a phase of intense critique and pressure on the WHO, multilateral organizations appear to have gained new grounds for legitimacy from the international community. The consensus and positive expectations appear to have increased, in spite of current funding difficulties, also for NGOs and civil society initiatives.

More broadly, the pandemic has revealed significant vulnerabilities in our current health systems. These systems operate under a demand-driven paradigm where patients passively consume pharmaceutical products and medical services. COVID-19 has shown that these systems were not prepared for the pandemic, lacked emergency plans everywhere, and were neither robust nor resilient in relation to the stress caused by the crowding of patients and the frequently desperate conditions. The scientific community and the pharmaceutical industry responded quickly and effectively to the COVID-19 pandemic by developing new vaccines. However, the industry was not prepared for the scale of the pandemic, and the response was inadequate. This was evident in the large gap between vaccination rates in developed and developing countries.

Several market and government failures appear to be at the root of these systemic under-performances. On the one hand, the health industry is torn between the search for profit and the difficulty of maintaining a feasible business model in the face of the long-term payoffs and the high risks of pharmaceutical and biomedical research. On the other hand, the remarkable success of biomedical practices in extending patients' lifespans has led to a higher percentage of older individuals with disabilities or chronic diseases. This fact has been a key ingredient in the high infection and mortality rates of COVID-19, as the virus was especially infectious and lethal to older people with health fragilities. The fact that most patients in developed countries have chronic diseases also implies that willingness and capacity to pay are concentrated in older consumers and richer countries and that vaccines and other preventive medicines are undervalued and under-financed.

This brings us again to the problem of medicine as a crucial case of under-provision of public investments. As a global public good, health services are under the direct responsibility of the state and of inter-governmental bodies, with NGOs and civil society assisting on the sidelines. However, while health products and medical services are easy to identify, health infrastructure is much more difficult to plan, administer, and assign. In addition to public hospitals and public campaigns for specific diseases, public investment in the health sector has been confined to financing selective medical schools and high-level research facilities. Both the overall size, the scale, and the scope of public health investment worldwide have been lacking. Public investment has been falling despite the fact that health expenditure in all countries has grown larger, with the USA reported by the WHO as the top of the list of developed countries with more than 18% of GDP, with 46% coming from the government. In contrast, 11 OECD countries spend less than 12% of GDP, but their government contributes more than 80% of total health spending. A whole series of actions that became a priority during COVID-19 had not received earlier support from the public sector. As a consequence, in most countries, there had been virtually no public investment in information sharing, public health prevention measures, protection of health workers, promotion of healthier behavior, ensuring the continuity of essential health services, and establishing reliable supply chains. Public health laboratories were established successfully during the pandemic, but were often improvised and could not count on accumulated knowledge, information, and an established and reliable supply of chemicals and biomedical equipment.

The lack of public investment in health has been caused by a mix of circumstances that include peculiar forms of market and government failures. On the one hand, there is a robust association between the demand for health and income, and consumers increasingly invest in the quality of life, of which health appears to be a key ingredient. However, lack of foresight and excessive discounting affect private investment in health as a form of human capital in ways similar to private under-investment in retirement and perceived under-remuneration of pension plans. As a consequence, consumers spend too little on health when they are young and healthy, and their later and increasing health expenditure is typically of a remedial nature, because it acts as a late and poor substitute for investment in preventive medicine. Government failure depends on the fact that public investment tends to second private behavior by accepting earlier patterns of under-saving in human capital and by concentrating on the provision of medical services needed by older consumers with pathological conditions. These tendencies are accompanied by private and public over-investing in health services for chronic diseases at the expense of medical and research services for other illnesses, including vaccines and preventive and curative anti-infectious medicine. Paradoxically and contrary to the panicky experience caused by the pandemic, before COVID-19, the concept of precision medicine seemed to be paving the way for a future where the medical field was gaining more control, and patients could rely on personalized and progressively more efficient medical care.

Although the impact on consumer behavior and medical community's attitude is hard to predict, COVID-19 may have significantly altered the perception of infections as minor issues in a world increasingly dominated by precision medicine. The very notion of the priority of increasing life spans and chronic conditions in private and public medicine has been challenged. The pandemic has accelerated the emergence of a new paradigm, where the central focus of medical services shifts from individual treatment to collective health. In this paradigm, prevention and effective public policies play a pivotal role as the primary instruments for achieving public well-being.

Under these conditions, vaccination does not appear to be a simple option, but a solution that embeds a whole series of enduring characteristics to rebuild the entire health system. These include far-reaching research capabilities at both national and international levels, potential and permanent production capacity on a global scale, and a continuous program with universal access, zero social or country discrimination, to increasingly effective immunization biomedical technologies. The global tax that the G7 have agreed to raise and whose income could range in the hundreds of billions of dollars would seem to be an adequate point of departure for the huge public investment needed to implement this new paradigm.

CHAPTER 10 - Take-home messages

- The increasing trend in outbreaks of infections and zoonic diseases is changing the paradigm at the basis of health systems around the world.
- These changes follow the outcomes of scientific progress which have led to a more holistic and multidisciplinary understanding of health as the basis of human development.
- The new paradigm, grounded in the concept of One Health and a sustainable bio-economy, has been driven by research showing the close connection between natural capital, ecosystem services, and the health of all living things.
- The new medical paradigm prioritizes collective health over individual treatment, with prevention and effective public policy as essential tools for achieving public well-being.

PART 5

The solutions

CHAPTER 11

How to manage the risk of new pandemics

Contents

As discussed in Chapter 2, there is widespread agreement among experts that facing new pandemics in the next 10 to 20 years will become a question of 'when' rather than 'if'. Considering this, it is crucial for health systems to take immediate action and become more proactive and effective than they were during the SARS-CoV-2 outbreak. Learning from past mistakes and valuing the experience and knowledge gained is essential.

In Chapter 9, we observed that numerous lessons from the current pandemic could be helpfully applied to managing future pandemics. However, these learnings alone are insufficient for effectively addressing pandemic risk management. In this chapter, we will try to illustrate the directions that we think should be followed. In this discussion, we will first emphasize the complexity involved in the emergence and spread of an epidemic or pandemic. Then, we will outline the primary interventions that can be implemented today to minimize the risk of future pandemics or mitigate their impact if they do arise.

11.1. A future with the double burden of infectious and chronic diseases

As already discussed in Chapter 2, one of the most important theories that has influenced health systems around the world in the twentieth century is the so-called 'epidemiolog-

The COVID-19 Disruption and the Global Health Challenge
https://doi.org/10.1016/B978-0-44-318576-2.00026-3

ical transition' (ET). The theory, originally formulated by Omram (1971), describes a phenomenon by which, at least in richer countries, the totality of communicable (and preventable) diseases was considered to have substantially disappeared, replaced by the emergence of so-called non-communicable diseases (NCDs). More generally, the ET described a complex process of change involving age population patterns, mortality, fertility, life expectancy, and causes of death.

According to ET, these patterns tended to produce a population structure characterized, together with a higher life expectancy at all ages, by lower fertility rates and progressive aging, with the consequent prevalence of chronic and degenerative diseases and a gradual disappearance of threats of infection. The theory suggested a tendency that could easily be supported by aggregate data, but was never satisfactorily tested against reliable population and health statistics. In the course of the years, several critiques (see, for example, Santosa et al. (2014)) have also revealed several limitations of the theory. These include its simplistic narrative features, its exclusive basis on mortality trends, the neglect of the decline in fatality rates, and the parallel increase in morbidity. Despite several corrective attempts by additional transition models, the role of poverty in determining disease risks and mortality was also undervalued or misunderstood, with insufficient account given to the concurring contributions of diseases and lifestyles to the various causes of death and overall mortality.

Frenk et al. (1991b) formulated a 'protracted and polarized' transition theory for Latin America, as a non–Western variant of the transition model, in which the sequence of epidemiological changes varies between population sub-groups with overlapping and persistent stages. In a revised formulation of the theory, Omran (1998) suggested that the transition in non–Western populations mimicked the Western transition model, albeit with a range of complex variations, including stage overlap and different transition speeds. For example, Japan was characterized by a rapid transition, with virtually no overlapping of stages, while other countries showed intermediate- and slow-transition models in accomplishing delayed mortality and fertility reduction. On more general grounds, Kunitz (1990) warned against the implicit historicity assumption that epidemiological stages would follow the same pattern and sequence everywhere. In this sense, recent epidemiological studies point to the need to consider the impact of shifts in age distribution through a life course perspective and the interaction with the environment as part of a continuous process of disruption of biological equilibria and the relationship between vectors and immediate and longer-term consequences of infectious diseases. The experience of several pandemics has also shown that any long-term evolution of the pattern of diseases cannot be limited to one country or one group of countries and that the only convincing transition theory would necessarily have to be a global one.

Because of the appealing narrative and the superficially persuasive evidence of epidemiological transition theory, at the end of the 1970s, there were many who believed that the chapter of infectious diseases could be considered closed. However, economic growth since the 1960s, which was partly due to healthier populations, had some unexpected side effects, which led to a resurgence of infectious diseases, that became more pronounced in the late 1990s. According to Bedford et al. (2019), one of these side effects was the demographic transition from rural to urban, which has seen, for the first time in the history of mankind, the number of people living in urban realities often densely populated surpassing that of the inhabitants of rural areas (see also Section 1.2 in Chapter 1). To this important effect, many others have been added. Among these, agricultural techniques have changed, becoming more and more intensive, modifying the way in which people and animals interact in their respective habitats. The world has also become increasingly interconnected, both in terms of tourist and commercial flows. Travel is more accessible worldwide and the travel costs of goods are decreasing thanks to new technologies that have improved logistics. All this has led to greater migrations, trade, and tourist flows that connect more people and therefore extend and accelerate the transmission of diseases. Climate change is increasingly affecting many ecosystems and environments, with an impact on the change in habitats and migratory habits of many of the vectors of infectious diseases.

Due to these significant changes, the population's health needs have increased as both acute infections and chronic diseases have risen. Additionally, there is an emerging long-term link between earlier infections and subsequent chronic conditions. The negative consequences of unhealthy lifestyles have also exacerbated the demand for medical care and support, as well as the need for effective prevention. The increasing demand for healthcare services has not been met with a corresponding increase in the supply of these services, especially in countries with poor healthcare systems. This is reflected in an increase in inequalities, which in turn are at the root of a growing distrust in national structures and institutions, exacerbating vulnerabilities of persons and fragilities of institutions and social arrangements. Many conflicts in recent years have also played a major role in regions of the world with high levels of poverty and vulnerability, feeding migration flows and increasing demands for health services.

One of the end results of this complex series of phenomena, is the re-emergence of communicable diseases, generating epidemics that are becoming more frequent and more difficult to prevent and contain. The emerging problems are thus complex and affect all aspects of daily life. The challenges posed by the epidemics of the 21st century are real and pressing: future epidemics will be fueled by conflict, poverty, climate change, urbanization, and the wider demographic transition.

To be able to respond to these challenges, it will be necessary to consider epidemics no longer as 'discrete events', but as 'connected cycles' for which it is possible to prepare, while recognizing that it will be difficult to predict specific outbreaks. With at least 150 pathogens harmful to humans identified since the 1980s (Smith et al., 2014), and recognized as emerging, re-emerging or evolving, we will need to get used to a 'new normal'. While living at a time when humanity possesses more advanced science and technology, all over the planet a virus has massively disrupted economies, healthcare and education systems, going so far as, in some cases, to jeopardize some fundamental rights.

As stated in Morens and Fauci (2012), the threats arising from the so-called Emerging Infectious Diseases (EID) will represent some of the most important challenges for the health of populations and the growth of global economies. However, given the emerging exploitation by pathogens of a hyper-connected world, if we continue to act according to business-as-usual policies, the damages and economic costs caused by EID will increase exponentially. Therefore, it will be necessary to think differently by integrating multiple disciplines. The management of pandemics will have to move towards interdisciplinary science, with an integrated approach of medical science and public health with social sciences, diplomacy, biomedical sciences, big data, information technology, artificial intelligence, statistics, meteorology, biotechnology, economics, and ecology. All of these areas will need to be combined to provide an integrated cycle of prevention, preparedness, response, and recovery.

11.2. Epidemiological or health transition?

Epidemiological theories suggest that as health services improve, mortality rates decline. However, morbidity rates, which measure the prevalence of disease, may not decline as quickly. This is because morbidity is often caused by chronic diseases that are not easily treated. Since these trends depend in large part on the success of modern medicine and the expansion of health services, it has been proposed to replace the concept of the 'epidemiological transition' by that of a 'health transition' (Frenk et al., 1991a; Mackenbach, 2022), even at the cost of introducing a confounding variable (health services) that is both a cause and an effect of higher morbidity from lower mortality. The dynamics of supply and demand of health services is important to explain the evolution of both diseases and cures. This is because health services are a crucial element in the operation and governance of the common good in modern societies.

The crucial, often overlooked, feature of health services, which distinguishes them from other public goods, is that their consumption takes place through agency relationships. Consumers of health services are, in fact, forced to rely on health experts for most of their choices: doctors, nurses, and other specialized operators. Before acquiring the final health service, consumers-patients often buy an intermediate service, such as a medical examination. Only later, on the basis of the diagnosis and recommendations of experienced operators, can they access additional medical care services. The agency relationship in the health system is also characterized by the fact that the reliance on experienced operators is a long-term relationship: the patient invests in experienced operators and they invest in their patients, according to a model that, for intensity and duration, is difficult to reflect in contractual relationships of another nature. The patient-doctor relationship is therefore an agency contract in which the patient is the principal and the doctor the agent. This implicit contract is characterized by trust and uncertainty, since patients must trust that their doctor is acting in their best interest, under conditions of inherent uncertainty in the diagnosis and treatment of medical conditions. The relationship between doctor and patient is also the basis for a series of other agency relationships with other physicians, doctors, nurses, and health facilities. The dual characteristic of trust and uncertainty embedded in these relationships means that health services are fiduciary public goods, that is, goods whose choice by the patient is conditioned and ultimately determined by the fiduciary relationship with experienced health professionals.

This principal-agent relationship, which is especially flourishing for chronic diseases and is continually cultivated throughout the patient's interaction with his health operators in ordinary times, suffers sudden disruption in the case of epidemics. As was demonstrated by the experience of COVID-19, the massive characteristics of the infections and the health measures needed in the pandemic emergency tended to radically deny any personal relationship between doctors and patients and to transform the search for personal care and cure into an impersonal mass response. In fact, not only do epidemics tend to overshadow long-term care and prevention measures, but they also make the care of chronic diseases and disability conditions very difficult and often impossible to pursue in an orderly fashion. At the same time, infections tend to be more likely to spread to people who are fragile because they are elderly, immuno-deficient or chronically ill.

Because infections can spread quickly and lead to widespread outbreaks, ensuring public health becomes crucial. It is important to eliminate the option of free-riding or of deciding whether to consume or not essential health services. This can be achieved by enforcing a minimum threshold of consumption to safeguard the well-being of everyone. This is also in consideration of the position of consumer weakness with respect to the market for health services. The phenomenon of forced riding, or the requirement that people consume a minimum amount of health services, is similar to the case of basic education services and nutrition (Scandizzo and Knudsen, 1980; Knudsen and Scandizzo, 2005). Obligatory nature characterizes the consumption of so-called pure public goods of merit, understood as those goods (for example, vaccines) whose consumption, when it is lower than the total produced and offered for reasons attributable to imperfect knowledge or opportunistic behavior, can cause damage to the entire community.

The social implication of the obligatory consumption of such goods is that every single consumer of a given collectivity cannot be the sole 'judge' of what is 'good' or 'bad' for himself. Therefore, the intervention of a public authority is necessary to correct the outcome of uninformed or opportunistic actions of individual consumers in their condition or final satisfaction. In this case, the option of use is invoked and exercised by the State. In developed countries, the increasing focus of the healthcare system on the treatment of chronic diseases has weakened the connection between overall health of patients and the importance of collective prevention and health education. This, in turn, has led to less emphasis on promoting healthy habits and preventive measures to avoid chronic diseases in the first place. Furthermore, the focus of the healthcare system on reducing morbidity that arises as people live longer has come at the expense of preventive measures.

To a large extent, the 'forced riding' of health services that has been seen as imposed on all citizens (and resented by many) by recent vaccination campaigns is something new in the usual experience of health services as a matter of individual choice. Somewhat paradoxically, it has developed as a mirror image of the epidemic as a potential condition of communal sharing of a social plague. An epidemic could infect the whole population and thus determine a situation where all its members could be affected simultaneously by the disease, either because they or their relatives, friends, and associates fall ill with it or because they would be likely to contract it in the future. In this sense, the transition of the health service has seen a gradual shift from lifesaving cures to life-supporting care, with an increasing use of remedial and palliative drugs to prolong survival and ease pain, in the face of pathologies that could have been avoided by investing in health education and prevention.

11.3. What can be done?

Any intervention that can be envisaged to reduce the risk of a new pandemic should be able to respond to the challenges described. According to Pike et al. (2014), imagining interventions to prevent a pandemic or mitigate its effects is not so different from finding measures to solve the problems of climate change. First of all, it must be borne in mind that the factors underlying pandemics, in a way similar to those of climate change (e.g. the increase in levels of CO_2), are growing in a non-linear and possibly explosive manner. Second, both climate change and the emergence of a pandemic are two 'global' problems, which require policies that must be coordinated worldwide to be effective. Finally, in both cases, important constraints limit the scope of possible interventions to be carried out.

The most important constraint is geopolitical in nature, because policies for climate change and health challenges can be influenced by political and economic factors that differ across countries and regions. Therefore, solving these problems is never a win–win operation; on the contrary, it imposes costs and benefits that are not evenly distributed. A further constraint has to do with the debate between interventions that favor adaptation, such as technological solutions, and interventions that favor mitigation or reduction of underlying causes. In the case of the pandemic, this means that ex post adaptation policies seek to reduce the problems caused by the massive spread of the disease, while ex ante mitigation policies seek to reduce the factors underlying the disease emergency and the frequency with which new EIDs emerge. In theory, mitigation actions should be preferred to adaptation actions, but in practice, policy makers are faced with finding an acceptable balance between the two measures. (Pike et al., 2014).

Until now, the prevailing trend has been towards the implementation of adaptive measures, through business-as-usual pandemic management programs that increase the capacity and speed of investigation and outbreak reporting, implement emergency control measures such as social expulsion and travel restrictions, and ensure the availability of medicines and vaccines. In contrast, in the case of interventions to mitigate pandemic threats, efforts are made to implement programs that, in most cases, act on socio–economic and demographic factors. The goal of these programs is to alter the evolution and spread of viruses, starting with those that originate in animals and then spread to humans. Programs may include measures such as improving sanitation and hygiene, increasing access to healthcare, and implementing policies that reduce contact between people to slow the spread of the disease.

11.4. Mitigating actions

Global efforts to combat pandemics, like those for climate change, are mostly adaptive. This means that they focus on reducing the impact of a pathogen after it has emerged. As explained in a well documented study by Pike et al. (2014): 'Pandemic mitigation programs include those that foster multisectoral collaboration among governmental or intergovernmental agencies for health, environment, and agriculture (the 'one health' approach); that conduct targeted pathogen discovery in wildlife, coupled with international development programs to address underlying socioeconomic drivers and promote behavioral change in at-risk populations; and that increase farm biosecurity to reduce the risk of novel zoonoses originating in wildlife or livestock, particularly in EID hotspot countries'. In this context, this section discusses the role of international cooperation in preventing and responding to pandemics. It examines the current status of the World Health Organization (WHO) as a supranational body capable of ensuring such cooperation. It then discusses the One Health approach as a holistic solution to pandemic problems.

11.4.1 A firm point: increased international cooperation is needed

In order to understand the role that international cooperation can play in helping solve global problems such as the emergence of a pandemic and its successful management, it is first necessary to recall why international institutions exist, how they function, and in what relationship they are with the sovereign states that created them.

International organizations have been established with the support or involvement of one or more powerful national governments, and are thus often considered as manifestations of power rather than instruments of cooperation. However, they are also the effect of shared interests and common goals among multiple countries. For example, the UN was established in 1945, as a result of the collective efforts of 50 states to promote international cooperation and prevent future wars. Similarly, the EU was formed as a voluntary association of European countries that sought to promote economic, social, and political integration and cooperation. International organizations have also various membership and decision-making processes that allow participation and input from a wide range of countries. Although a superpower such as the United States may have more influence on certain organizations, decisions are typically made through a consensus-based process involving multiple member countries.

As explained in Hooghe et al. (2019), three main approaches to the theory of international organization go under the name of realism, functionalism, and constructivism. Realism is a political theory that views international organizations as instruments of power politics, with national states acting in their own self-interest to maximize their power and influence. Functionalism views international organizations as a means of achieving common goals and solving shared problems through functional cooperation, rather than power politics. Proponents of functionalism believe that international organizations can provide a forum for national governments to work together to address global challenges and that the development of international organizations can lead to increased cooperation and peace. Constructivism considers international organizations as social constructions shaped by the norms, values, and ideas of member countries. Constructivists argue that international organizations are not just neutral forums for cooperation, but are also shaped by the standards and goals of an ethical community. In this view, the actions of member countries within the organizations are motivated by a desire to promote and reinforce their own norms and values, rather than simply to maximize their power and influence.

Once established, international organizations can function effectively only if member countries recognize their significance and permit them to address issues that would otherwise remain unresolved due to their transnational nature. To achieve this without friction and full effectiveness, the following basic conditions must be met in principle: *i)* a clear mandate, *ii)* adequate resources, *iii)* broad membership, *iv)* effective decision-making, *v)* accountability, *vi)* flexibility and adaptability. These conditions require, in turn, an effective legal system with checks and balances and fully implemented treaties, conventions, and agreements, as well as transparency and trust between the various members. In this context, the organization's objective is to take joint action 'with a sense of co-ownership and shared responsibility. Joint positive decisions will create a virtuous circle that will generate greater trust between states and more productive cooperation for the benefit of the entire international community' (Montella, 2020). Of course, just as an organization that respects the above conditions can be the solution to many problems, one in which these principles are lacking could become the cause of further difficulties, especially in cases where limits are placed rather than solutions, as has recently happened in Europe.

The principles outlined above have been the basis for the success of the many international institutions created after World War II, which have often managed to resolve important crises and ensure stability in the world. A collective response 'to unite in hardship' was first created the League of Nations after World War I, the UN after World War II, and the OECD from the Marshall Plan to promote cooperation between countries and for the reconstruction of Europe. The results of these initiatives were generally positive and especially so in the field of health. Smallpox eradication, HIV control, containment of SARS, MERS, and Ebola infections are just some of the examples in which international collaboration has been the main driver for achieving unprecedented success.

A positive climate for international cooperation appeared to persist at least until the terrorist attacks of September 11, 2001. Since then, the relationship between the United States and three major international institutions has been deteriorating. The first relationship to worsen was that with the UN, because the US could not persuade other member states to fight against countries that allegedly sponsored terrorism, such as Afghanistan, and could not convince NATO to punish Iraq as an alleged producer of weapons of mass destruction. On that occasion, countries such as China and Russia blocked the UN Security Council from passing any resolution supporting US policies, while many US allies did not support its foreign policy in Iraq. From that point on, the multilateral institutions created to maintain a status quo in line with the American strategic vision seemed to fail to respond to the task for which they had been born. These conflicts have subsequently expanded from security to economic policies, opening up to the influence of other global powers. According to Allen and West (2020) one of the most important changes in world foreign policy occurred when the most prominent 'rising' power, China, challenged the US for leadership in international institutions, a field in which the US had long dominated.

These problems were further exacerbated by the election of former U.S. President Donald Trump, who began to pursue nationalist security objectives and a protectionist economic policy. Trump's 'America First' and 'America Only' slogans reflected a unilateral and nationalist foreign policy orientation that damaged relations with not only potential adversaries and neutral states, but also with many of US allies. The simultaneous rise of populism greatly weakened the essence of international collaboration and the effectiveness of multilateral organizations. The United States also withdrew from some important international multilateral platforms, putting an end to its political and financial support. As a consequence of these actions, international organizations experienced a dramatic loss of their effectiveness and responsibility. With a domino effect, this has also led other countries to question the role of such organizations.

The COVID-19 pandemic has underscored the difficulties in international collaboration and revealed the constraints of international entities when tackling global health emergencies. The WHO, responsible for the promotion of health care for all peoples and their protection from public health emergencies, has been heavily criticized for its inaction towards China over the failure to provide accurate and timely information on the progress of the epidemic, thus at one point becoming perhaps the main culprit in the poor management of the global spread of the virus (see Box 5 in Chapter 3). This situation has generated heterogeneous responses according to the different pre-existing ideological positions of individual countries. For those countries that were and remain convinced that they are in an increasingly globalized world, the need to maintain strong international cooperation remains valid and, in the case of health, this can only be done through strengthening the WHO (see also Section 11.4.3 in this chapter). In contrast, for those countries that consider China's growing influence on international affairs to be the main problem to be addressed, this episode is proof of Beijing's influence and China's ability to manipulate international institutions. For those who tend to mistrust globalization itself, the spread of COVID-19 is a 'textbook case' for initiating the process of reversing globalization and thus gradually weakening international organizations.

As UN Secretary General António Guterres pointed out: 'The COVID-19 pandemic is a tragic reminder of how deeply connected we are. To combat it, it is necessary to work together as one large family. Viruses know no boundaries, spread everywhere in the same way and affect everyone. But if this seems to be an obvious statement, it remains difficult to understand why it should not seem so obvious that a pandemic is a global problem requiring a global response. And global responses (both short-term and long-term strategies) must be coordinated among national states. Unfortunately, international organizations can at times become ideal scapegoats, offering national and local governments an easy outlet to channel the discontent of their constituents elsewhere.

In conclusion, given the constraints faced by international organizations, enhancing collaboration and reinforcing its global standards seem both beneficial and essential. In this sense, it is worth recalling that after a first ineffective phase, the WHO has played a vital operational role in responding to the pandemic. This has been particularly true of those countries, which are the vast majority, that do not have an equivalent center for disease control and prevention as in the United States (CDC) or Europe (ECDC) and rely on the World Health Organization for information and analysis on the disease itself. As rightly pointed out by Jenkins and Jones (2020): 'The WHO Secretariat has helped many governments to form rapid response teams to manage contact research; helped governments to reorganize their hospitals and emergency assistance centers to address the specific characteristics of the COVID-19 outbreak; and has delivered test kits and equipment to more than 120 countries around the world. WHO is also playing a key role in coordinating rapid scientific work to generate progress towards the treatment of the disease. Its most important effort in this field is the so-called solidarity process, in which scientific institutions from 100 countries have come together in an effort to quickly test four different sets of drugs for their potential in treating the disease[...] And in March 2020, the WHO published the first road map of coordinated efforts to develop a vaccine, a key activity that will allow more efficient allocation tests and trial efforts to accelerate vaccine development. The WHO, however, is only a part of a broader set of national and international institutions involved in responding to the crisis. Several other actors have mobilized to respond'.

Among the different international bodies, the international governmental organizations (IGOs), such as the United Nations, the International Monetary Fund, and the World Health Organization, stand out because of the importance of their mandate and of their limitations. This is because the pursuit of their global and wide ranging role depends critically on the collective will of governments, especially those of the most powerful countries. When governments disagree, IGOs cannot act. It is easier to blame an IGO for its inefficiency than individual members, just as at national level, one can blame Parliament as an institution for lack of action, rather than individual members making up the legislative body.

11.4.2 WHO as an example of the limitations and weaknesses of current international cooperation

To provide a clearer picture of how and how limited IGOs can be today, it is useful to analyze some mechanisms that regulate the scope of WHO action and the level of funding it receives in the context of the current COVID-19 pandemic. According to Mathews et al. (2020), everything that the WHO is able to do is defined by the International Health Regulation (IHR), which establishes its duties and responsibilities, as well as very clear limits on what it can do. In particular, the IHR rules do not empower the WHO with sufficient investigative authorities and resources to guide and coordinate adequate international responses to pandemics, largely because Member States are reluctant to expand those authorities and their funding. These limits reduce the capacity of the WHO to promote pandemic prevention, detection, and response. Furthermore, they do not allow the WHO to promote the full compliance of individual member states' responses to IHR rules. The WHO plays a coordinating role, but cannot operate in countries without permission from national governments. Furthermore, it does not have an independent capacity to collect intelligence information and, above all, cannot impose the application of the IHR requirements on information sharing and transparency. Although binding on the Member States, the IHR does not give the WHO the authority to impose sanctions against countries for non-compliance: at most it can publicly recall recalcitrant governments.

Finally, it should be noted that, in the face of a large mandate, the WHO had a very limited annual budget of only $5 billion in the period 2018-19, which is far lower than, for example, the budget of some of the leading US hospitals, such as, for example, the New York Presbiterian Hospital (The Kaiser Family Foundation, 2020). For the two-year period 2018-19, the WHO was able to devote $554 million, less than $300 million annually, to implement its core activities in health emergency management and lacked sufficient resources and a large-scale emergency response capacity. For example, according to Mathews et al. (2020), in addition to the COVID-19 pandemic, the 2020 WHO emergency program was also managing the international response to the Ebola outbreak in the Democratic Republic of Congo, health emergencies in Syria and Yemen and the Rohingya crisis in Bangladesh. The program also responded to hundreds of 'acute' global health events.

An excellent example of how international collaboration and international organizations can play a key role in similar emergencies is the efforts being made to develop the COVID-19 vaccine. Since February 2020, enormous efforts have been made to accelerate as much as possible both the discovery of the vaccine and its subsequent production and dissemination. The availability of a safe and effective vaccine for COVID-19 was recognized as a key tool for contributing to pandemic control. The challenges and efforts needed to develop, evaluate, and rapidly produce such a vaccine on a large scale were enormous, and the chances of success required a great deal of collaboration between all countries.

However, the first phase of the pandemic seemed to be characterized by the absence of a collective response to develop and distribute a vaccine for all, with rich countries implementing selfish vaccine procurement strategies with individual contracts to meet the therapeutic needs of their own citizens. According to Oxfam, a group of rich countries, which represents 13% of the world's population, reserved about half of the supply of vaccines that were expected to be available in the first year of production. The United States, the United Kingdom, the European Union, Australia, Hong Kong and Macau, Japan, Switzerland, and Israel signed agreements with manufacturers to get their hands on 51% of doses when production begins. This type of 'vaccine nationalism', which follows the desire of countries to develop a vaccine on their own, is a tendency that can contribute rather than solve this or other health crises, and can perpetuate a pandemic rather than stop it.

In the case of COVID-19 vaccines, which will likely be needed in the coming years, an international plan for equitable distribution is crucial. Without such a plan, unnecessary price fluctuations and uneven allocation may arise, leading to potential shortages in some regions while causing excess supply in others. This situation is not in anyone's interest and, in any case, represents a short-sighted strategic choice. In an interconnected world, no country is safe until every country is safe. Vaccine nationalism could also condemn many countries to prolonged suffering, which means slower economic recovery around the world. Therefore, global coordination is essential. This will ensure that vaccines are distributed wherever they are most needed. Cooperation minimizes the risks of failure in domestic production, ensuring that each country has access to a solution. In these cases, a collaborative approach is both ethical and prudent. It is also more productive and facilitates a speedier return to normalcy.

However, cooperation requires effective and proactive international institutions. For example, in February 2020, the WHO brought together 400 of the world's leading vaccine researchers to identify research priorities. Then, to help find an effective treatment quickly, it launched a 'Solidarity Trial', an international clinical trial involving 90 countries. Along these lines, the WHO subsequently developed research protocols that are used in more than 40 countries in a coordinated way, and around 130 scientists, funders, and producers worldwide have signed a declaration committing to work with the WHO to accelerate the development of a COVID-19 vaccine.

This massive effort led by the WHO was also instrumental in the launch of the worldwide vaccine-enabling initiative named 'COVAX'. This is the acronym for 'COVID-19 Vaccines Global Access' and indicates a Global Vaccines Facility that aims to ensure that vaccines are available to people all around the world, regardless of their income levels or where they live. By combining the financial efforts of rich countries, COVAX aimed to achieve greater efficiency, accelerate the development and manufacture of COVID-19 vaccines, and ensure that they are distributed fairly to countries regardless of their income levels. The initiative brings together governments, international organizations, manufacturers, scientists, and civil society organizations to achieve this goal. In return for their participation, rich countries would be guaranteed supplies of vaccines to cover between 10 and 50% of their population, while poor countries will be guaranteed a share of vaccines that can cover up to 20% of the population. The initiative is co-led by two non-profit groups, GAVI, the Vaccine Alliance[1] and the Coalition for Epidemic Preparedness Innovations, together with WHO. 170 countries have joined it, thereby resisting the false fascination of vaccine nationalism. On June 27, 2023, one of the COVAX leaders, Gavi, made a series of decisions that will expand the portfolio of vaccine programs available to lower-income countries in the coming years. It also committed to providing COVID-19 vaccines to high-risk groups until 2025 and approved an innovative plan to enable pilot investment in reserves of investigational candidate vaccines for Marburg and Ebola Sudan viruses. This will help to respond more quickly to outbreaks of these viruses.

Technically, the initiative works in such a way that COVAX jointly funds research and purchases billions of doses of vaccines from many companies, anticipating many of the funds so that companies can immediately invest in production. This increases production capacity, ensuring that all countries in the consortium have simultaneous access to successful vaccines. In the first phase of distribution, doses have been allocated based on population size to protect those at the highest risk of infection and serious illness, including health and welfare workers, the elderly, and individuals at high risk. The winning principle of the COVAX initiative is to provide a form of 'insurance' to individual countries due to its large portfolio of vaccines and the high probability

---

[1] Gavi, the Vaccine Alliance, is an international organization founded in 2000. It brings together public and private sectors to increase equitable and sustainable use of vaccines, saving lives and protecting people's health.

of having at least one successful product. It also means that all countries can plan the timing and size of the vaccine distribution for optimal global recovery. Furthermore, having more than 170 countries in the program ensures a large-scale offer and the best collective price possible, which could be only one-tenth of what countries that have embraced the 'nationalist vaccine' strategy may have to pay. In the second phase of distribution, as more doses are produced, the vaccine will go to groups with less likelihood of being infected or at risk. At the time of its launch, the program aimed to deliver at least 2 billion doses by the end of 2021. Although this was estimated to be not enough for everyone, it provided a critical contribution to end the acute phase of the crisis and put the world on the road to recovery. In the words of WHO Director-General, Tedros Adhanom Ghebreyesus: 'Ending the pandemic and restarting our economies are in our collective hands. Over the next few days, the decision that countries will take whether to share or maintain an excess supply of vaccines will not only determine the speed with which the world comes out of the crisis, but it may also be a defining moment of this new decade and establish a new norm for international cooperation'.

Has COVAX been a success? It certainly has, at least in part. Although it fell short of its earlier, more ambitious targets, between April 2020 and March 2023, COVAX had shipped a total of 1.9 billion COVID-19 vaccine doses to 146 countries. It had also delivered 15.5 million syringes to 19 countries, 305,175 safety boxes to 22 countries, and cold chain ultra-low freezers with a storage capacity of 8,490 liters. As a result of these efforts, AMC92 countries[2] had achieved a primary series coverage of 55%, up from 28% in January 2022. Moreover, as discussed in Budish et al. (2022), they had achieved these results, by experimenting and enacting a series of often effective public distribution and market strategies. However, what COVAX had not anticipated was that a major challenge for vaccine distribution would come from lack of demand, with many developing countries turning away from mass-strategy vaccination, as a consequence of pandemic fatigue and a general feeling that the pandemic had subsided. As countries move from managing COVID-19 as an acute phase to an endemic phase, and as a vertical approach to vaccination becomes less effective, countries are starting to integrate COVID-19 vaccination into their primary healthcare systems. The COVID-19 Vaccine Delivery Partnership (CoVDP) is transitioning out of its temporary structure and back into the partner agencies, incorporating the new ways of working that were developed in 2022 and 2023.

However, it is important to note that during the initial stages of the pandemic, some affluent countries chose to act independently in securing their own supplies, instead of participating in a collaborative effort. The United States signed a $1.5 billion deal

---

[2] AMC92 countries are 92 low- and middle-income countries selected on the basis of their income level and their ability to finance their own vaccine programs.

with the biotechnology company Moderna to buy 100 million doses of its COVID-19 vaccine. It also invested billions of dollars to develop its own vaccine, distributing the financing among eight potential candidates, including AstraZeneca, Moderna, and Pfizer. AstraZeneca signed agreements with France, Germany, Italy, and the Netherlands. The UK government secured 60 million doses from Glaxosmithkline and Sanofi, as well as millions of doses of vaccines developed by Pfizer and BioNTech. Similar behavior was also practiced by China and Russia. The Institute of Microbiology of the Chinese Academy of Sciences and the Research Institute on Vaccines and Serum Mechnikov in Russia established a joint laboratory to test vaccines and conduct tests.

In spite of the WHO efforts, and partly because of its limitations, COVID-19 has exacerbated world inequality in many ways, that include the exposure of existing fragilities in income distribution and patterns, as well as the enforcement of egotistic and rival behavior in consumption habits, irrational fears, and willingness to cooperate. The WHO's limited response has resulted in a significant coordination challenge both domestically and globally. Furthermore, the world's affluent countries have exhibited a nationalistic approach in acquiring and administering vaccines. According to the Secretary of the UN, as vaccines became available, the 10 richest countries in the world *de facto* monopolized both their consumption and supply. However, the apparent end to their acute pandemic crisis, achieved through the successful mass vaccination, may be a dangerous illusion. The pandemic in fact is still threatening all countries, and specially those whose lack of financial resources and international coordination render vaccine supply short, poorly distributed and insecure, and also vaccine demand dwindling in the face of more pressing economic and social priorities. This condition is another manifestation of the unsustainable nature of the present pattern of development. Unsustainability in this case depends not only on the opening of yet another dimension of inequality and global misery, but also on the fact that neither this pandemic nor the ones that may come in the future can be prevented unless a sufficient amount of immunization can be realized at the global level, and this can be done in sufficiently short times because of the proven capacity of viruses to mutate into increasingly infectious variants.

More generally, vaccine equity transcends just the coordination efforts of the WHO or other entities and extends beyond merely guaranteeing access to vaccines. As highlighted by Jensen et al. (2023), while ensuring access to vaccines is important, this is not enough. Vaccines are not simple pharmaceutical objects but key elements of the world bio-infrastructure. Both equity and efficiency cannot be pursued through single, however enlightened, multilateral initiatives. Vaccines' crucial role in shaping the global bio-infrastructure of health prevention and care requires a broader approach to the political, legal, and logistical problems that bound and regulate their production and distribution and contribute to their chronic tendency to be under-provided and unevenly distributed.

Although the production of effective vaccines in record time appears in general as a success story, the COVID-19 pandemic has also exposed the vulnerability of the global system of vaccine development, which is largely dominated by a few powerful countries and companies. The concentration of R&D in a few places has perhaps accelerated vaccine discoveries and deployment (Ndwandwe and Wiysonge, 2021), but has also resulted in significant inequities in their availability and affordability. In addition, the logistical constraints of vaccine distribution have also contributed to vaccine inequity. The COVID-19 pandemic has highlighted the importance of a robust and equitable global distribution system for vaccines, as well as the need for efficient and effective transportation and storage systems that can ensure that vaccines reach those who need them. However, these properties are unlikely to be possessed by a fragmented bio-infrastructure that follows the national lines of the richest countries. As Mbembe and Shread (2021) argue, 'for lack of a common infrastructure, a vicious partitioning of the globe will intensify, and the dividing lines become more intense'. Fragmentation and inequities also depend on the legal constraints that govern vaccine development, production, and distribution. Intellectual property rights and other legal barriers limit the production and distribution of vaccines in low-income countries and contribute to high prices and limited access. The ghost of decoupling (see Chapter 6) is always around the corner.

In conclusion, vaccine equity is not simply a matter of ensuring access to vaccines, but requires a greater understanding of the political, legal, and logistical constraints that contribute to the uneven distribution of vaccines. The COVID-19 pandemic has highlighted the importance of addressing these structural inequalities to ensure that all people, regardless of their level of income or location, have access to life-saving vaccines. Achieving vaccine equity requires global collaboration, political will, and a commitment to building a more just and equitable world.

In summary, beyond its poor character and the widespread suffering caused, the COVID-19 pandemic has important social implications and learning aspects. Public health institutions at country level, including Ministries of Health and regional and state government agencies have re-claimed a leading role during their fight against COVID-19 in the face of the bustling private health sector. In spite of the stress on national health systems, in developed countries this has given new legitimacy to government actions and public good provision, while in developing countries the performance has been mixed, because of lack of resources and governance problems. At the international level, after a phase of intense critique and pressure on the WHO, multilateral organizations appear to have gained new grounds for legitimacy and positive expectations from the international community. The consensus and positive expectations appear to have increased, in spite of current funding difficulties, also for NGOs and civil society initiatives.

11.4.3 A wider adoption of the One Health approach

The evidence available clearly demonstrates that the spread of zoonotic diseases is not a new phenomenon. In fact, the number of outbreaks of infectious diseases from animal reservoirs has increased over the past three decades. These diseases can jump from animals to humans, which can have serious consequences for human health. Epidemics caused by viruses such as Ebola, avian influenza, and the Nipah virus remind us that 'human, animal, and environmental health are interconnected and that an adequate and timely response to emerging zoonotic pathogens requires a coordinated, interdisciplinary and cross-sectoral approach. As our world becomes more connected, emerging diseases are a greater threat and require coordination at the local, regional, and global levels' (EClinicalMedicine, 2020). The size of the problem is significant, since the data from WHO (2018b) show that more than 70% of emerging infectious diseases are zoonotic and originate from wildlife.

Extensive prevention measures are thus needed, but difficult to carry out because the events that cause the emergence or re-emergence of zoonoses are complex and influenced by many factors. Whenever a zoonotic outbreak occurs, the public surveillance system must be able to identify its signs from the earliest possible moment and be able to react promptly. The COVID-19 outbreak was a striking example of the failure of this process, as its emergence was initially observed almost by accident, but was perceived only much later as an urgent threat by decision-makers.

The lessons learned from the COVID 19 oversights, together with other recent experiences of overlooked first signs of infection and/or inadequate reactions, highlight the need to move to a more integrated, holistic, and proactive model for the organization of a global health system, to prevent the emergence of new diseases and reduce the impact of future epidemics. The system should be capable of identifying premonitory signs of infections before new diseases arise, thereby preventing potential epidemics or significantly mitigating their effects. Over the years a series of proposals to achieve these goals have gathered around the so-called 'One Health approach'. According to this approach, first proposed in 2000 and increasingly adopted by international institutions (Zinsstag et al., 2023), infectious disease problems should be analyzed and assessed in light of all the links between human, animal and plant health and their shared environment. 'As such, the approach allows for a deeper understanding and ability to address the complex ecosocial determinants of health and to address threats more effectively and efficiently through coordination between disciplines and sectors. One-Health approaches are increasingly recognized for their value in addressing the threats of emerging infectious diseases, as most EIDs are derived from wild animal reservoirs in biodiversity landscapes that suffer strong human pressures, including growth in human population, change in land use, and extraction of natural resources' (Kelly et al., 2020). 'One Health' designates a multisector, transdisciplinary, and collaborative approach to address these complex health threats with a comprehensive economic, social, and environmental strategy.

As discussed in Chapter 10, over the years, the One Health approach has increasingly been recognized as valid and capable of addressing problems that appeared to have no solution under the traditional segmented model of policy intervention. This has led to the development of several national and international initiatives. These include the creation of dedicated divisions within US federal agencies (for example, in the United States the National Park Service One Health Initiative, for Disease Control and Prevention One Health Office, and the U.S. Department of Agriculture One Health Coordination Center), interagency working groups, national multisector coordination agencies and mechanisms (such as the One Health Secretariat of Bangladesh and the One Health Coordination Platform of Liberia), and international One Health networks and consortia (e.g., the FAO/OIE/WHO tripartite collaboration, One Health Workforce, One Health Alliance of South Asia, Southeast Asia One Health University Network, One Health Central and Eastern Africa). In addition, nearly 50 countries have signed up to the Global Health Security Agenda (GHSA), launched in 2014 to promote One Health approaches in different areas of the world and strengthen the capabilities to prevent, detect, and respond to disease threats.

Despite significant advancements, numerous challenges persist and may potentially give rise to future events like COVID-19. Many countries lack formal mechanisms to coordinate and integrate efforts across the domains of public health, agriculture, and the environment. Traditionally, these sectors have been handled by separate ministries or agencies with differing mandates pertaining to activities and expenditures. A classic example is that of many African countries where global livestock markets and rapidly changing socio-economic conditions have made these regions increasingly vulnerable to public health problems. As pointed out by Vandersmissen and Welburn (2014), '...Traditional farming practices continue along with innovative methods to increase livestock productivity, but weak regional regulatory systems and national disease control responses often imply that rapidly evolving systems have the potential not only to cause the emergence and resurgence of zoonotic infections, but also, more importantly, to further alienate populations of already marginalized small farmers, as we have seen recently in outbreaks of disease Bird flu in Asia. Humans who live in close proximity and/or have frequent contact with wild animals and livestock and share the same ecosystem contribute to the emergence of zoonotic disease. The lack of community awareness, the lack of effective surveillance in humans and animals, and the limited access to human health care and veterinary services only exacerbate risk'. Similar situations are also seen in the regions of South and Southeast Asia, where it is estimated that in the coming years the absolute highest growth in livestock production and consumption will be recorded over the next 40 years.

There are numerous opportunities to apply One Health approaches in the situations described, going beyond merely managing epidemic situations retroactively. One Health approaches can address the management of zoonotic diseases (both emerging and endemic) as practical and economic solutions to alleviate poverty, addressing at the same time ecosystem management, animal and human health surveillance, and community participation in the mitigation of disease risk. To construct a system that effectively prevents disease outbreaks, it is important to understand the links between human and animal health, the environment, people's livelihoods, and political processes. This understanding should inform policy recommendations and be supported by inter-disciplinary approaches combining epidemiological, socio-economic, and socio-cultural research methodologies.

11.5. Adaptive measures

In determining which adaptive interventions can be most effective in countering a pandemic, it will first be necessary to understand that some of these may generate important trade-offs between health and economic objectives. The implementation of such interventions may also be particularly difficult as it takes place in the full pandemic phase, a situation very often governed by chaos and, for this very reason, potentially supported by small and unreliable information. Pandemic plans can be very helpful, but their practical implementation can be challenging. This is because there may be political, social, or economic inconveniences for different groups of people, the proposed interventions may not be communicated effectively, and there may be practical difficulties in implementing them on the ground. As a result, it is not easy to determine the most effective measures to combat a pandemic.

Another essential aspect to consider is that these interventions are 'second-best' solutions compared to the previously mentioned mitigating interventions. In fact, once the outbreak has occurred, they impose significant costs (monetary and non-monetary) on the system, considerably less than those that would otherwise have been done to combat the pandemic, but certainly more than those necessary to prevent the pandemic (see also Section 11.5.1).

11.5.1 Improving and updating prevention plans

In Section 9.4.2 of Chapter 9, we highlighted that the pandemic revealed worrying shortcomings in multilateral agreements for global health security, including a lack of coordination between nations and a breach of the rules established in the International Health Regulations, the main international agreement that governs the events of dangerous diseases. The primary responsibility for these weaknesses lies with the national governments, which remain torn between the desire for effective global health governance and the reluctance to expand authorities, funding, and capacities of the WHO and other international agencies.

However, the problem of lack of international coordination is only one of the reasons of the failure to manage the pandemic globally. A key factor was the lack of preparation to address infection as a public emergency. In countries where plans existed and were up-to-date and operational, the results in pandemic control were excellent. A good example is South Korea, which was better prepared because it had an infectious disease surveillance system that could provide guidelines for the investigation and management of outbreaks for different types of infections. Thus, early and widespread testing, traceability, and case isolation, together with government advice based on evidence of distancing, were critical to achieving early control of the disease and limit human and economic losses. South Korea was not alone in this success. Several countries such as Canada, Germany, New Zealand, Norway, Rwanda, Taiwan, and Vietnam were more prepared and thus more able to respond effectively to the pandemic.

According to the WHO, pandemic preparedness means 'putting in place national response plans, resources, and capacity to support prevention, detection, and containment measures, as well as programs that respond to and mitigate the problems arising from the spread of pandemics, such as IPR deficiency, limited hospital capacity, and vaccine acquisition and other countermeasures'. However, once emergency management plans have been developed and agreed upon, they are often archived and rarely revised (see the case of Italy). In contrast, these plans should be implemented more frequently and become flexible documents that should be constantly updated in line with developments in new experiences and technologies. This lack of ongoing review and re-evaluation meant that, as COVID-19 began to spread, many healthcare systems did not have adequate plans to define roles or adjust supply and command chains.

The pandemic has also shown that supervision is essential, but that the current system is not up to the task. Vigilance requires continuous emergency preparedness, but in much of the world the tendency has been to shut down the supervisory systems once the danger has passed. After a crisis, people often become complacent as memories fade and processes return to a pre-crisis level of alertness. Numerous experiences worldwide have demonstrated that support for pandemic preparedness plans is subject to the cycle of crisis and complacency. Health threats have caused far more victims than terrorism has, but the funds allocated to counteract them are nowhere near comparable to the global funding allocated to the fight against terrorism each year (Our World in Data, 2017; Pike et al., 2020; Zucchi, 2018). As can be seen in Fig. 11.1, despite the adoption of the International Health Regulations (IHR), multiple pandemic threats, and numerous reports calling for increased investment, international pandemic preparedness assistance has never exceeded 1% of the funds available to the WHO for global health care.

Given the weaknesses exposed by COVID-19, one key issue is to ensure that pandemic plans clearly show their ability to meet the needs of a new pandemic. Specifically, it would be helpful to clarify and define the following issues that have been flagged as problematic in several recent studies (see, for example, Ravaghi et al. (2023) and Sundararajan et al. (2022)):

- Pro-active resource management. Over the years, health systems, and hospitals in particular, have learned to improve management efficiency by importing management models from other industrial sectors. In particular, hospitals have adopted just in time techniques (JTT) to manage inventories. However, as in other sectors

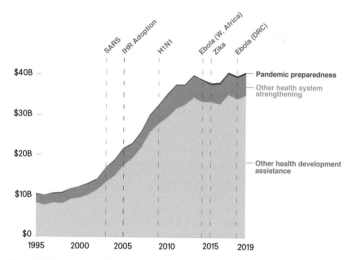

Figure 11.1 Global Financing for Health Development. *Source*: adapted by Institute for Health Metrics and Evaluation. *Note*: The dotted lines indicate the year of outbreaks and the entry into force of the International Health Regulations (IHR), a binding agreement with rules on sharing critical information on epidemic threats and pandemic preparedness capabilities.

of the economy, JTTs have caused quite a few problems since critical bottlenecks have emerged in the overly long supply chains characterizing several internationally traded health products. Therefore, health organizations are rethinking their strategies to identify and activate the supply flows of various products, always checking the levels of stocks and turning them around to avoid expiration.

- Personal Protective Equipment (PPE) management. The previous point refers to the management of PPE and technological installations in resuscitation rooms. In the hope that future pandemics do not occur with high frequency, technological progress could lead to improvements in these products, making those already in stock obsolete. This is an important risk that health care systems and hospitals will have to learn to manage with dedicated staff, possibly centrally. Additionally, the arrival on the market of new and more effective products also implies the need to organize training and updating on the ground.

- Involving stakeholders in planning. One of the most significant issues with pandemic plans was that the individuals responsible for executing them were often unaware of their existence or lacked adequate knowledge about them. Rather than remaining a document restricted to the few experts who drafted it, a pandemic plan must involve all departments and decision-makers in planning emergency preparedness. In particular, participation in the planning of front-line physicians (whether hospitals or territory) improves understanding of their experience during an emergency, helps with participation and adoption of new processes, and gives doctors a vested interest in emergency preparedness rather than simply giving them orders to execute.

- Enhancing the technological infrastructure. The plan should provide for the improvement of technological infrastructure, such as the development of strong telemedicine and remote work capabilities. This means that the plan should include specific investments and initiatives to improve technological infrastructure, rather than simply assuming that these improvements will happen on their own. These investments ensure business and operational continuity during a crisis, as well as dramatically reduce the chances of contagion by healthcare personnel.

11.5.2 Strengthening the territorial and hospital health system

Healthcare systems in affluent countries are primarily designed to address the needs of patients with non-communicable diseases (NCDs). Until now, these systems have had limited experience in managing communicable diseases like seasonal influenza, which can be controlled through vaccination. As a result of their structure, current health systems had limited opportunities to develop automatic response mechanisms among health professionals, patients, or the broader community. Consequently, the emergence of a disruptive epidemic can present challenges, despite the existence of pandemic plans that have documented and codified many of the necessary responses in advance. Therefore, health systems are structurally unprepared to face infectious diseases and need to be rethought and strengthened in many ways.

The systems' structural faults, and their fundamental unpreparedness to the challenges that arose between February and April 2020, were evidenced by several major shortcomings. These included the lack of separate pathways for contagious and non-contagious patients, inadequate protective equipment, and insufficient measures to protect vulnerable individuals, as well as limited integration between hospitals and community care. These shortcomings were the consequences of mistaken choices made mainly in the first phase of the outbreak but only partially resolved in the second phase when between May and June 2020, the talk of 'reopening' and 'recovery' started. The lessons learned from the COVID-19 pandemic have also highlighted the lack of preparedness in hospital and community health settings. These settings require significant investment in infrastructure, vocational training, and resources at all levels, including doctors, nurses, administrative, and technical staff. The need for these investments is critical and underscores the importance of developing a robust and adaptable healthcare system that can better respond to future emergencies.

In the case of hospitals, strengthening preparedness will require identifying which facilities will be used to treat infected patients by providing separate pathways to prevent contagion. Additionally, it will be necessary to build resilient hospitals that can adapt to the unique needs of each situation, thereby avoiding the demand for a sudden increase in the provision of intensive care beds for infectious diseases during a full-blown epidemic. These measures will be important to better respond to future emergencies and ensure the safety of patients and healthcare workers. Future hospitals must be designed and restructured to better respond to pandemic events. This includes interventions to isolate different areas of the hospital with separate air treatment units and filtration systems to mitigate cross-contamination. But hospitals will also have to be structured in such a way as to avoid long queues and waiting rooms packed with patients about to be visited. Patients should remain in the hospital for the shortest duration necessary while still receiving adequate care. Hospital designs must adapt to include telemedicine infrastructure, as a physical presence is not always essential for effective patient care. This could also be an opportunity to improve patient care services by developing applications that allow for accurate booking of appointments. While social distancing rules may create problems, they could make contact with the hospital easier and safer.

Additionally, front-line health workers in hospitals deserve special attention. During the first phase of the pandemic, they experienced periods of stress due to a lack of mental health support. This has undermined the resilience of the skilled workforce needed to deal with future waves of the pandemic. This suggests that future hospitals should be designed to offer emergency medical professionals psychological support and stress management training for critical events. In many parts of the world, general practitioners (GPs) were left to cope on their own during the pandemic's most challenging moments, especially in hot spots. One lesson to draw appears to be that their mode of operation will need to change significantly, which will require them to be reorganized and provided with better training on the appropriate procedures to follow when interacting with patients. Additionally, it will be important to ensure that GPs have adequate equipment to prevent infections and protect themselves and their patients from the spread of disease. These measures will be essential to ensure that they can respond effectively to future emergencies and provide high-quality care to their patients.

The pandemic has probably accelerated, but also somewhat changed the direction of an already existing movement toward a new form of health care and assistance. Patients, physicians, and the general public will gradually become accustomed to more frequent and intensive virtual interactions, online transactions, and home operations. After experiencing the convenience of virtual consultations for minor injuries and routine illnesses, patients and doctors may prefer this option over in-person visits to the doctor's office, emergency room, or clinic. According to some experts, up to 50% of family doctor appointments could occur virtually. General practitioners (GPs) will need to adopt procedures that are similar to hospital care, including separating patients who are infected from those who are not.

A further aspect to be considered concerns the technologies with which GPs will have to operate. In very little time, technology has revolutionized the way they intervene around the world, without the support of adequate information and training processes.

Virtual visits, often viewed with suspicion in the past due to possible confidentiality and security problems, have become the focus of GPs action plans. As a British doctor tellingly reported in an interview with The New York Times, 'we are seeing 10 years of change in a week. In the past, 95% of all contacts with patients were face-to-face: go to your doctor, as it was for decades, centuries. But that changed completely, while the rapid dissemination of these technologies has been beneficial, they can also pose initial challenges due to the presence of numerous alternatives, which can limit the ability to find collaborative working solutions or ensure interoperability'. To ensure that general medicine is effective and efficient, it seems thus important for its ecosystem to be based on open standards and to have transparent evaluation and assurance frameworks. This will allow for the sharing of data and best practices, and will help to ensure that patients receive the best possible care.

The third aspect to consider is the training of the personnel to support change. Staff will need to improve their skills in new ways of working. These skills will no doubt include digital health and information aspects. Communication will take a completely new form, including virtual support for patients and family members among usual practices and the need to engage effectively with colleagues who manage social welfare and local government aspects. It will also be important to equip doctors and nurses with all the skills that will enable them to better understand current technological trends and solutions, to determine which are the most effective and which may be useful for their clinical practice. This training should begin at the university level.

This transformation process will need a regulatory framework that can in some way help the process of technological and digital transition. The legislation will have to be revised in such a way that it does not hinder the adoption of new digital health technologies and should offer a certain degree of flexibility for the future. However, adequate procedures must be implemented to ensure appropriate levels of risk management and patient safety.

Finally, the whole process of strengthening medicine on the ground cannot fail to take account of a review of incentives and funding flows, which, where possible, must be restructured to take account of the nature of digital solutions (e.g. recognition of performance in telemedicine). Other steps to strengthen general medicine include overcoming the current concept of 'silos budget' at the national and regional levels, and the widespread adoption of a result-based payment model (to avoid the adoption of new technologies simply because they are the latest models on the market).[3] Additional resources and expertise, where appropriate, should be allocated to support the implementation of all these solutions.

---

[3] A silos budget is a budgeting system in which different levels of care, such as hospitals, clinics, and long-term care facilities, have separate budgets. This can make it difficult to coordinate care and ensure that patients receive the services they need.

11.5.3 Establish real-time contagion monitoring

The early months of the COVID-19 pandemic made it clear that one of the most critical challenges was the lack of effective surveillance and tracing systems to track the spread of the virus. As already seen in several chapters of this book, the lack of information to trace outbreaks often prevented the implementation of adequate policies to combat the epidemic.

The Tech-Health world has made great strides in this regard, releasing in the first months of the pandemic a series of new tools aimed at monitoring the spread of the disease and facilitating its management: tools for digital epidemiology, electronic health record governance, and rapid response test kits. Facebook, Amazon, and Google have been in contact with the WHO to discuss their role in combating the spread of disease and disinformation by making available the data and tracking technologies in their possession. In particular, Apple and Google launched the Application Programming Interface (API) that will allow interoperability between iOS and Android products through official public health apps (see Chapter 4).

The development of these technologies is crucial to epidemic management, given the importance of Track-Test-Treat strategies. This approach, which was advocated by many at the beginning of the epidemic, is now the only available way to effectively monitor and manage outbreaks in the first months of spread, ensuring they don't spiral out of control. The importance of these tools is crucial because there is a strong need to develop a real-time monitoring system of contagion trends on the ground. As stated by Dewatripont et al. (2020), 'The real problem in the coming months will not be fear of trading with an insolvent counterpart, but fear of working with someone who is contagious. [...] Restarting production in the economy requires reliable identification of those who will not contract the virus or pass it on to others, regardless of whether they have previously shown the associated symptoms'.

Another crucial reason for implementing a real-time monitoring system is the possibility of enduring the pandemic for a prolonged period, resulting in the need for intermittent solutions, as proposed by one of the studies conducted by researchers at Imperial College London (Ferguson et al., 2020). This approach would allow the economy to return to normal, but if a significant number of new cases of acute COVID-19 requiring intensive care were reported, emergency measures could be promptly be taken again.

11.5.4 Economic interventions

Economic interventions are generally designed to address the short- to medium-term context and are related to the duration of the epidemic. Their primary objective is to help economic operators, such as businesses and households, navigate the challenging economic conditions created by the epidemic without exposing them to critical situations that could threaten their survival or their ability to maintain their customary lifestyle. These interventions typically involve a range of actions, such as limiting difficulties, maintaining economic capacity, and stimulating economic recovery once containment and mitigation measures are relaxed. These measures are designed to support the economic well-being of businesses and households during the epidemic, while also positioning them for a return to growth once the crisis has passed. By taking these actions, governments can help ensure that the economic impact of the epidemic is minimized and that businesses and households are able to continue operating in the face of significant challenges.

It is useful here to emphasize that when actions of this kind are needed, it is because any other initiatives to mitigate the adverse effects of the epidemic, such as those analyzed previously, have failed and, therefore, the economic systems are fully affected. For these reasons, economic intervention is conceivable as a 'second-tier' intervention, which takes place only when first-tier (i.e., public health) measures have failed. Experience at the international level has shown that the lower the preparedness to contain the spread of contagion, the more significant the economic effort to mitigate the negative effects on the entire production base and on households.

At the level of individual countries, the consensus of the economists is that the best way to tackle the problem is to envisage measures to cushion the immediate impact of the sudden fall in economic activity on companies and households and, more generally, to preserve the productive capacity of the countries. This appears particularly appropriate in the case of low-income households and small and medium-sized enterprises, for which liquidity problems can be disruptive. Here, too, the effectiveness of intervention at the country level could be increased by multilateral cooperation and coordination at all stages of the recovery process and by strengthening the resilience of the global economy to future shocks (OECD, 2020b). To address the challenges of global health emergencies, international organizations such as the G20 should establish coordinated action plans that can be implemented across countries. This requires international coordination to ensure that the measures taken at the national level are effective and to support low-income and low-capacity countries in their response to the crisis.

After a pandemic, the progression toward recovery may not be smooth and even, and may result in uneven recovery across countries. To mitigate these types of problem, international coordination is essential. By coordinating efforts and resources, it is possible to reduce the time it takes for countries to recover from the pandemic and return to normal. In this regard, international financial institutions, such as the IMF and the World Bank, can play an important role by providing financial support to low-income and low-capacity countries, which will help ensure a more equitable and coordinated recovery for all. By working together, countries can better address the challenges of global health emergencies and minimize their impact on the global economy and society.

Identifying liquidity support policies for households and firms is another important outpost of economic intervention during the acute phases of a pandemic. This is especially the case when partial or total shutdowns of economic operations (lockdowns) are enacted to halt the spread of infections. In such cases, intervention plans must be available to provide support to families and businesses that may be running out of cash. The first form of aid must be to guarantee liquidity. Although a fiscal stimulus may be the wrong prescription at first, in the more acute phase, the only thing that can work is to provide liquidity to those who have lost their jobs, those who have contractual obligations to fulfill, and those who think they can invest in restructuring. Liquidity is a financial buffer to allow companies and people affected by an inevitable decline in economic activity to have enough money to survive the shock (Romano and Schivardi, 2020; Schivardi, 2020).

As reported by OECD (2020b), many measures were implemented throughout the world in the lockdown phase of the pandemic: 'The measures included the extension of time limits for tax returns, deferral of tax payments, provision of faster tax refunds, more generous loss compensation provisions, and some tax exemptions, including social security contributions, payroll taxes, or property taxes. Countries have also implemented wide-ranging measures to help companies retain their workers through short-time work programs or wage subsidies. There is evidence, from policies implemented in the wake of the global financial crisis, that keeping people at work through such schemes is an effective way to provide income support and limit job losses, while avoiding costly research and matching as recovery proceeds. Household income support has been extended in many countries, generally through targeted cash benefits rather than through tax cuts, given the need to provide support quickly. There are also cases in which access to sickness benefits has been facilitated and eligibility has been extended, with several countries having expanded the coverage of unemployment benefits in particular to the self-employed' (OECD, 2020b).

Of course, these measures come with significant administrative challenges. One of the most fundamental problem is how to effectively distribute liquidity to borrowers who need it. Here, too, speed of action is crucial. Borrowers cannot afford to wait for months for the necessary funds to be made available, whether through redundancy payments or business loans.

The COVID-19 pandemic forced businesses and employees to rethink how they work. Many workplaces had to implement new safety procedures, such as social distancing and mask-wearing. Some businesses had to suspend work altogether, while others had to completely rethink their way of working. The pandemic has shown that businesses need to be prepared for the unexpected and have plans in place to ensure the safety of their employees and customers.

While it is widely accepted that these measures are effective in reducing the spread of infections, it is also true that they may not always be possible to implement. In fact, for people whose job requires them to be physically present at work, it may not always be possible to maintain complete physical distance from others. In all these cases the safety and well-being of workers is one of those areas that must be well guarded during a pandemic in order to mitigate its negative economic effects. In addition to the health and safety risks of employees, a pandemic also increases stress in the workplace and decreases productivity, partially reducing workers' well-being and undermining the competitiveness of the business.

All workplace safety and health activities can provide practical support to minimize contagion during the pandemic and return to normal after relaxation of physical removal measures. It should also be borne in mind that during a pandemic, there are two distinct times when operating, which in turn require different actions: the emergency phase, focused on containment and mitigation and the recovery phase focused on restarting the economy and rebuilding society. The two phases may overlap, and they may not be linear.

For the emergency phase, detailed plans, sector by sector or even company by company, should be drawn in operational terms, going well beyond the 'general' protocols issued by governments. Some of these aspects were accurately highlighted by Lucifora (2020), who, in the COVID-19 emergency phase, recommends 'in addition to sanitizing the environment and providing personal protective equipment (who will guarantee it?), in addition to respecting the meter distance to be kept in public places, 40 square meters per customer in retail stores, the detection of temperature on the workplace and the need to inform healthcare providers of any positive cases, more needs to be done, and time is not wasted. When a worker is in the yard in front of the plant at the beginning of his shift or at the entrance of the business where he works, what should he do? How do I get to work? How to dress and how to properly wear devices and protect yourself? How do I alternate between work shifts? How can congestion be avoided in common environments, at the entrance, in the canteen or in the dressing rooms? All of this has to be defined in concrete, not in abstract terms. It should be recalled that, at the beginning of the infection, the lack or non-compliance with these procedures led to the uncontrolled spread of the virus in health facilities'.

In the restart phase, attention should be focused on other aspects. The COVID-19 experience has shown that, contrary to the rapid and immediate passage into the emergency phase, the restart phase may be slow and uncertain. In some ways, this is an advantage, as it allows for a more measured and deliberate approach to implementing the various activities. A crucial step for companies to limit exposure to contagion at the workplace will be to assess risks and thus implement the hierarchy of controls. This involves putting in place monitoring actions to primarily remove risks and, when feasible, lessen worker exposure. The approach should begin with collective efforts and, when needed, be complemented by individual protections, such as personal protective equipment (PPE). The health of the workforce should be a top priority for everyone, as it is a key issue in returning operations to a semblance of normality. This is, of course, a moral, ethical, and legal concern for all businesses. And from a business perspective, safeguarding employee well-being is essential because no plan to resume normal operations can be successful without them.

To do so, new protocols for deep cleaning and sanitation may be needed. It might also be useful to make changes to the layout of the workspace, such as increasing the distance between workstations or changing employee schedules to reduce the number of people in buildings at the same time. Companies may also consider setting guidelines for PPE use, such as masks and gloves, for the application of monitoring modalities such as temperature monitoring of employees and visitors before entering the workplace, and establish rules governing when employees can return to the workplace after recovering from contamination. Given the ubiquity of mobile devices, it may be useful to leverage technology (e.g. apps) to facilitate contact tracing and communicate with employees if they have been exposed to the virus and need self-quarantine. At the same time, whenever considering steps like these, it is important to ensure that protocols are in place to protect workers' personal data.

Ultimately, companies' ability to 'inform' employees will be crucial, a step that should be undertaken proactively before resuming operations, utilizing all feasible technologies at their disposal. Of course, safety in the workplace will not depend solely on the work of the supervisory bodies. Hence, establishing forms of cooperation between companies and workers will be essential.

In times of emergency, there is an opportunity to implement virtuous investment plans that can help businesses perform better. If well executed, these plans can create an environment that fosters innovation and supports the development of new business models. The current moment may be appropriate to determine what many economists refer to as the 'cleansing effect', which entails a process of shedding less efficient businesses and practices to pave the way for new, more innovative ones. Not all economists concur with the practical validity of this concept, but proponents contend that it exists and can benefit the economy in the long run. Companies enter and leave the market, plants are built and dismantled, and workers change jobs. New businesses can emerge that are better suited to the current economic and social context and can be more resilient, innovative, and capable of adapting to changing circumstances, which can ultimately contribute to long-term economic growth and stability. In this sense, temporary economic downturns would not necessarily be negative events, especially in a long-term perspective, because they would serve to 'clean up' the economy from inefficient production units. A related but distinct idea is the pit-stop view that recessions are times when productivity-enhancing activities are undertaken because of their temporarily low opportunity cost.

In sum, all these considerations lead to the conclusion that it may be desirable to envisage policies aimed at encouraging firms to invest, especially in sectors with high added value and a high innovation content (product, process, or organizational). Simultaneously, targeted measures appear suitable for both the most impacted sectors and those with the highest potential to promote recovery, effectively addressing the urgent need for resource allocation to tackle liquidity issues.

CHAPTER 11 - Take-home messages

- The challenged posed by epidemics are tangible and dynamic: Conflicts, poverty, climate change, and urbanization will drive future pandemics. We must be prepared, as the occurrence of emerging infectious diseases is anticipated to rise sharply.
- Interventions to contain the risk of new pandemics are of two types: adaptation, based on technological solutions, or mitigation, i.e., to reduce the underlying causes. In the past, the prevailing trend has been toward adaptation with control programs.
- Efforts to mitigate require global cooperation, which has diminished significantly since 2001. Instead, it should be fostered and strengthened. The WHO example is paradigmatic both because of the weaknesses demonstrated during the pandemic and because of the important role it can and must play in addressing the global health challenges.
- The One Health approach represents a multidisciplinary, integrated and proactive form of integrated health policy. However, for its application on an extended level, there are still open problems of political will, coordination, and integration between activities in different sectors related to human health.
- The idea that the regulatory route is the best way to combat the effects of a pandemic has been questioned in the light of behavioral science: nudging (a strategy of gentle encouragement), imagined as an alternative, can help, but in cases where people's behaviors generate massive negative externalities, the authorities should still opt for the normative way.
- Among adaptive interventions, pandemic prevention plans are a priority, with effective surveillance systems and strengthening of the territorial and hospital health system. The latter will have to evolve based on the critical issues highlighted by the pandemic (for example, by avoiding crowded waiting rooms) and provide for the implementation of real-time monitoring systems for infections.
- Among the interventions on the economic front, actions to guarantee liquidity to households and businesses and safety in the workplace are fundamental. Furthermore, the crisis can be perceived as an opportunity to take advantage of introducing necessary structural reforms.

CHAPTER 12

Epilog

While the world has experienced many pandemics in the past, COVID-19 has been the first truly global challenge for several reasons. First, it has spread rapidly and uncontrollably, demonstrating the illusory nature of national borders, material barriers, and geographic distances, and the true condition of a globalized 'small world' as a cluster of virtual communities densely connected by a myriad of links. Second, the likely zoonic origins of the virus and its intensive mutational activity have been a spectacular reminder of the possible catastrophic health consequences of the unsustainable pressure of human activities on the environment. Third, the disorderly reaction of most governments and institutions to the pandemic has exposed the fragility of national systems still based on local infrastructure and insular governments, with weak ties and lack of coordinating mechanisms for cooperation and joint action. Fourth, global infrastructure and long international value chains have also been shown to be inadequate and vulnerable to supply and demand shocks caused by the pandemic. Fifth, the management and impact of the disease have exposed a number of dramatic inequalities ranging from the well-known income and wealth disparities across countries to the more subtle, but not less important, digital divides and unequal accesses to science and technology.

Therefore, in several aspects, COVID-19 has been a test of the global economy. Today, the outcome of this test seems to suggest that the new global structures and remnants of the old economy represent an odd mix of unreconciled economic and cultural differences. This mix is evolving under the pressure of internal and external challenges and is affected by conflicts, intrinsic instability, and uncertainty. By shaking the global asset of the emerging world along its fault lines, the pandemic has challenged its main pillars, revealed vulnerabilities, and presented new opportunities for systemic reconstruction and improvements. But only time will tell whether it has been able to improve its robustness and resilience in the long run.

The pandemic serves as a test of globalization and a wake-up call, highlighting the growing unease in the relationship between humans and nature. This uneasy connection is evident through an increasing frequency of epidemics, as well as a rise in the number and intensity of other natural disasters, such as hurricanes and devastating floods. It reminds us of the close connection between climate change, pressure on natural resources, deforestation, and the spread of infectious agents and suggests the need to address these challenges in a unitary and global way. In this sense, COVID-19 has been a sobering experience for virtually everyone, showcasing the profound challenges natural disasters can pose to our individual lives, social ties, and mental well-being. To be sure, we have crossed many other calamitous events, like earthquakes and wars, but these had been

The COVID-19 Disruption and the Global Health Challenge
https://doi.org/10.1016/B978-0-44-318576-2.00027-5

experienced by most people as remote and confined in space and time. Unlike these cases, the pandemic has positioned itself as a major catastrophe with unknown impact and duration. It has generated a profound state of uncertainty, not experienced before by living generations, which persists today after more than three years since its inception.

As a stress test for the world's economic order, the combination of the last phase of the pandemic and the Ukrainian war has generated further uncertainties about the structural nature of many of the changes taking place and, among them, about the very continuation of globalization. In reality, while globalization can suffer slowdowns and setbacks as in the past, it does not appear to be the result of active policies, but rather an unstoppable form of cultural universalism, driven by a powerful and pervasive process of economic and financial integration. In this sense, the pandemic has been shown to be part of the same process, which does not coincide with specific phenomena of integration and internationalization. It rather appears to be a universal and ubiquitous tendency to change, in response to growing connections that extend beyond national boundaries, to individuals, social groups, institutions, and productive sectors under a variety of circumstances.

Therefore, rather than reversing globalization, the pandemic has revealed some of its crucial dimensions under stress. For example, both the spread of the disease and the vaccination campaign have been a glaring demonstration that globalization is neither uniform over time nor evenly distributed among possible participants. Developing countries, for example, are involved in very different ways, both in strength and scale, than developed countries. Indeed, one of the main problems of globalization is that it creates divisions based on different degrees and forms of achievement in interconnections and inter-dependencies and, through these, new bases for inequality, because not everyone can be equally globalized. The creation of inequalities is at the root of ongoing conflicts, both within and outside our democracies. However, it is also a process of creative destruction, as new connections tend to displace old ones in terms of density, number, and location. This process can completely overturn the way of life in preexisting societies. It is still too early to say whether the health emergency, the energy crisis, the higher inflation, and the slowing economic growth correspond to structural, or to temporary changes, but as with the energy crisis of the 1970s, the response of rational governance can only be through global institutions. Both evidence and rational thinking suggest that economic nationalism is not a good idea, whether it manifests as national insulation or vaccine nationalism. It does not improve well-being or help us address economic and environmental challenges. In fact, it only serves to feed political nationalism, which is based on the isolated defense of partisan interests.

Rather than reversing it, several signs suggest that COVID-19 had the opposite effect to enhance and accelerate globalization. The push for greater global integration was hastened by lockdowns and social distancing policies. Hyperconnectivity through the internet and digital communication played a crucial role in maintaining the resilience of the globalized world under pandemic stress. The internet was the crucial infrastructure providing immediate and powerful alternatives to daily communications and many social and economic activities. While no particular new technology appeared to emerge during the various phases of the pandemic, process innovation in internet and digital activities was huge. The internet became a part of everyday life in more pervasive and immersive ways. Moreover, the main drivers of this expanded role of digital communication were not limited to remote work and online shopping, but extended to all activities that could be performed and expanded online. Although technology was already available prior to the pandemic, the health crisis accelerated the development of numerous practical innovations that might have otherwise taken years to emerge. Suddenly, meetings were not limited by location, and people from distant places could be brought together for all kinds of gatherings, from scholarly seminars to working parties. Organizational and professional hierarchies were also modified in subtle ways as meetings became non-local and topic driven. Consequently, people could be chosen and literally summoned from all over the world, regardless of their position, professional standing, and seniority. Thus, new models of seamless communications were developed for individuals and institutions within a cyberspace, whose enlargement appeared to compensate for the parallel shrinkage of the possibilities to interact through ordinary means. These developments were effective in mitigating the social and economic impact of the pandemic and providing options to adapt to its effects. However, they were not without costs, especially in delicate areas such as education and government, and they took a heavy toll of material and immaterial damages on key aspects of social life. Their impact was also highly unequal: they generated benefits for those on the better side of the digital divide, while producing increasing isolation and social exclusion for the poor and the more vulnerable.

An unanticipated, yet temporary outcome of the pandemic was a renewed sense of trust in governments and public institutions among citizens. This effect was in part the realization of the nature of health as a unique public good, whose utility depends on the simultaneous and shared condition of all individuals in a community. However, viewing global health characteristics as a public good coincided with an increasing concern for one's local community members, ranging from family to national society. Contrary to the drive toward more global relations, spurred by lockdowns and the need to communicate, the desire for security tended to pull people back toward their local roots, while the ubiquity of the danger of infection pushed the concern all the way from families and primary communities to national borders. Governments thus came to be seen as the last line of defense against the virus, especially in the absence of any real action on the part of international bodies and little evidence of international cooperation. As institutions of last resort, national governments thus received unexpected consent and compliance during the height of the emergency. At the same time, this unprecedented condition of unity of public opinion was stifled by a number of failings that progressively reduced the credibility of institutions.

Several components of the health crisis conspired to determine the multiple failures of government handling of the COVID-19 pandemic. The health emergency highlighted the importance of scientists and, at the same time, exposed some of their intrinsic weaknesses as policy-making partners. While most governments immediately involved health scientists and other experts, they failed to secure an effective system of health governance of pandemic mitigation and adaptation processes. Scientific advice for policy making was sought under the assumption that 'technical' and 'political' choices can interact by maintaining the original distinction between expert knowledge and political empowerment. However, as most studies have shown, this presumption is unwarranted in the face of decisions that require value judgments, where scientific advisers are asked to define acceptable risks, available evidence, and uncertain trade-offs. Expert advice during the pandemic revolved around suggesting general policy measures, often based on common sense in the face of controversial scientific evidence, such as behavioral restrictions imposed on the population. Recommendations were also often reversed, showed different forms of inconsistency and seemed to be unable to control the behavioral rules and procedures of the health system. To a significant degree, the management of these matters was delegated to local governments and ordinary administrative structures. These followed protocols that varied across time and space, with no apparent adherence to standard criteria of efficiency and effectiveness, beyond mere rhetoric. Expert contributions also appeared to be lacking in the monitoring and evaluation of the response to the pandemic, from the deployment of resources to the analysis of the effects of the measures taken, which were limited to daily bulletins of infections and deaths. The latter were organized in such a way as to provide heterogeneous and statistically inaccurate information. The data published were typically incomplete, ambiguous, impossible to interpret correctly by the media and the general public, and often difficult to analyze even by experts.

The failure to combine science and policy making suggests that the pandemic may have been instrumental in exposing a broader problem of our society: the increasing failure of political institutions to keep pace with scientific progress and technological development. This may be the consequence of complexification, a phenomenon that creates a divide between increasingly complex fields of knowledge, such as science and its practical applications, and social and political processes, which are still operating under the relatively simple fundamentals of human behavior. The divide between science and everyday life is paradoxical in light of the dramatic changes that new knowledge and technological progress have made for everyone, especially in the area of public health, the increasing length of human life, and generally the comfort and security of many people. Yet, confronted with the products of the new technology and the somewhat impenetrable knowledge that they suggest, much of the present attitude of the man from the street resembles the reaction that his middle-age equivalent would have expressed towards magic or other forms of supernatural manifestation. As the high priests of a new form of religion, scientists are thus contemplated by an army of reluctant devotees, who are separated from them by a wall of asymmetric information and contingent mistrust.

Scientists, despite their genuine intentions to explain their concepts to the public, face the critical issue of asymmetric information. This issue not only leads to unequal access to shared knowledge but also hinders effective communication across the divide. As a consequence, scientists' recommendations seem often arbitrary or nonsensical, as they tend to oscillate between the sacred and the profane, i.e., the attempts to express the incomprehensible complexity of their knowledge and to reduce it to implausible and simplistic explanations.

Of course, all specialized knowledge entails some asymmetric information and is part of a principal agent problem. However, when individuals seek expert knowledge, they act as principals delegating powers and tasks to specialized and knowledgeable agents on the basis of trust and contractual agreements. But this relationship is not set in stone and can be terminated at any time if either party believes that the original conditions of trust and mutual obligation no longer exist. However, this is not the case for expert knowledge sought by policy makers, as the resulting agency relationship is imposed upon its ultimate principals (the people) by their political representatives. These are themselves delegates, and their recourse to experts creates a second-degree principal agent problem that requires much higher levels of trust, as not only politicians have to be trusted for their action and expert choices, but also the experts chosen have to be credible in their deliberations and recommended actions for the public. In fact, as the mixed performance of the scientific community with COVID-19 showed, the fiduciary links required in a medical emergency tend to be highly problematic. They may be continually challenged by lack of clarity, confusing statements, and expert disagreements. Experts themselves present an inevitably heterogeneous picture of specialization, knowledge, credibility, and communication capacity.

A more general question that the pandemic brought to the fore was the fact, pointed out by many philosophers and social scientists in the Marxist and postmodern traditions (see, for example, Flyvbjerg (1998)) that everyday politics inevitably falls short of the standards of communicative rationality (CR), i.e. on the idea that humans are rational beings who are capable of using language to reach agreement on common goals and values. As a process of rational and credible communication embraced on the basis of truth-seeking and trust, CR is perhaps a myth, even though it is constantly evoked in both modern democracies and autocracies, where the attempt to find rational patterns of behavior that reconcile individual and public interests is necessarily fraught with conflict and power struggles (Ryan, 1992). No matter how large the concession of trust to political authorities in an emergency may be, the contrast between government and civil society is omnipresent, even when, for temporary reasons, it may be momentarily suppressed by fear or dependence on government power. Moral outrage is continuously raised by authoritarian actions, simply because these actions can ultimately only be sustained by sanctions and a balance of power drastically pending against individual will (Flyvbjerg, 1998).

The information asymmetric nature of the relationship between science and public opinion was evident in the different phases of the pandemic. When, as in the case of total lockdowns, scientists' recommendations were simple and seemed to be validated by common sense, they were met with widespread compliance, at least for some time. As recommendations became more articulated, such as in the case of partial lockdowns, masks, and social distance, they also became more questionable. Compliance thus gradually relented, in part because health experts tended to disagree on technical details and on providing explanations for the recommended behavioral changes. But it was for the vaccines and the recommendations to vaccinate adults, older adults, and children that the most dramatic divergences occurred. The break of trust between scientists and the public in the case of vaccination was paradoxical since vaccines had been evoked since the beginning as the only truly effective solution to the pandemic. Furthermore, due to the fortuitous availability of powerful new technologies developed just before the explosion of the pandemic, highly effective vaccines for the new infection were obtained in record time and became available much before it had been believed possible. At the same time, several unexpected difficulties emerged that threatened to undermine public trust not only in vaccination, but in the whole management of the emergency.

First, the vaccine reports were confusing and incomplete. They were giving for granted common knowledge on key information on the nature and use of vaccines, which was instead ignored by much of the public and even policymakers. On the one hand, reports also celebrated the unprecedented speed of vaccine development while, at the same time, they tended to cast doubts on vaccine safety by alluding to possible side-effects and the fact that authorizations from health authorities had been granted without the usual lengthy testing and verification procedures. Figures on the incidence of negative side-effects and sometimes lethal dangers were often divulged without explaining their statistical nature and often seemed nonsensical, as, for example, when they reported percentages of lethal effects that seemed incompatible with the small number of trials performed. Reports on the effectiveness of different vaccines based on real world evidence were difficult to interpret. These reports were based on small trials and became obsolete due to virus mutations and the increasing capacity of its variants to defy natural and acquired immunity. Confusion also arose regarding the opportunity to vaccinate a sufficiently large portion of the population to prevent the spread of the contagion. On their part, vaccine deniers added to the confusion by citing uncertain evidence on the contagiousness of fully vaccinated individuals.

The vaccination campaign seemed also to have been put in place without adequate logistic reserve with respect to unforeseen events and inevitable obstacles arising from supplies, the difficulties of conservation and distribution, and the heterogeneity and reliability of sources. The timing of vaccinations appeared to follow a do-it-yourself logic in which local structures (governments, states, regional but also lower administrative bodies, such as districts or provinces) competed in execution speed, regardless of the distribution criteria among categories at risk, susceptibility to side effects, and other medical and socioeconomic parameters. Vaccine nationalism was also rampant, with developing countries being able to access a substantial supply of vaccines only very late in the development of the different waves of the pandemic. The quality distribution was unequal, with the most effective vaccines also concentrated in developed countries, in part due to the lack of infrastructure (e.g., the cold chain and reliable distribution channels needed by the highly effective vaccines). In the case of China, vaccines did not seem very effective, they were inexplicably withheld from the elderly, and did not reduce the lockdowns.

But vaccines also stroke a chord of disbelief toward official explanations that was surprising in its intensity and spread. Although in some countries vaccine mistrust appeared limited to a relatively small number of 'No-Vax' fundamentalists, in many US states they extended to Republican Party rulers and local crowds. The anti-vaccination posture of many politicians seemed to reflect an attempt to intercept a potential extreme fall of trust in the perceived social order on the part of a vocal minority. This was mainly traceable to an increasingly turbulent lower middle class whose apparent lack of rational thinking coincides with new forms of social exclusion, low-quality employment, and some propensity to violence. In other words, the resistance to vaccination, deemed irrational and even insane in modern society, seemed to expose the growing problems for democracy and social stability, posed by disenfranchised and protesting crowds of the people left behind.

More generally, the role of the government in the health emergency and the proliferation of command-and-control policies tended to collide with the libertarian instinct of much of the world population and to cause a gradual fall of trust between citizenship and public authorities. In part, this was due to a progressive divergence of interests of the parties involved, as prohibitions and mandatory measures were seldom accompanied by sufficient information, incentives, and compensations to align public and private goals. The compression of individual freedom, perhaps inevitable at the most crucial moments of the pandemic, also proved unsustainable if prolonged, as for example, in China. Rather than a trade-off between health and economics, the pandemic thus may have revealed a more basic trade-off between health and individual freedom, in the libertarian sense of the desire of the common man to be independent of external interference even when this may endanger health or welfare. Other causes and motivations were also at work, including the desire for sociality, educational problems, and a general feeling of helplessness in the face of unpredictable behavior from both the virus and the authorities. The heavy toll extracted by the chaotic decentralization of public sector response in most of the world's democracies was first negatively contrasted with the lower impact of the disease in China. In this country, where a continental strategy could be managed by a centralized authority, a policy of rigid lockdowns and social restrictions had undeniable success in terms of lives saved. However, in the longer term, this strategy appears to be unsustainable, given the accumulation of negative side-effects, including the economic consequences and the increasing fear of confinement and social isolation, combined with the failure to achieve effective mass vaccination. While extended lockdowns may not be sustainable, it is premature to conclude that the fragmented and makeshift responses of most countries are preferable to a more rigorous alternative, such as the strategy implemented in China, which serves as an imperfect yet informative example. On one hand, despite numerous indications of a return to normalcy, the virus remains undefeated. Moreover, current evidence points to a significant risk of a new surge in infections by the end of 2023. At the end of February 2023, three years after the start of the pandemic, the world still reported approximately 1000 daily deaths from COVID-19, mostly concentrated in high-income countries (900), of which 262 were in Europe and 276 in the United States. On the other hand, this concentration reflected a similar and higher incidence of all deaths from the start of the pandemic (2.03 million in Europe and 1.12 million in the United States). In comparison, the 5000 total deaths in China in the 2020-2022 period appear to be little short of a miracle, despite any alleged underestimation.

The Chinese case also points to another possible pattern of infection, whose diffusion recalls the well-known law of regression toward the mean. This term describes the statistical property of a large population, where successive measurements of a sample mean will tend toward the population mean when done in sufficient numbers, although there may be large variations in individual measurements due to random factors. While it may be too early to tell, the experience of COVID-19 suggests that a global tendency to regress toward a mean level of incidence of the pandemic may apply both across countries and over time, with the severity of the disease and the characteristics of the response converging toward a similar pattern over the three-year period of the pandemic, despite the very different paths of infection and means of controlling it exercised by different countries at any one time. In the later stages of the pandemic, although infection and death statistics still appeared to be concentrated in Europe and the United States, many countries that seemed to have escaped the brunt of the infection through luck or successful management showed a recrudescence of both the infectivity and lethality of the virus. Japan, for example, recorded COVID-19 deaths 70% higher in 2022 than in the combined 2020-2021 period; Iceland, Australia, and even New Zealand saw an unexpected surge of COVID-19 deaths from 5 to 10 times as high as in the same two years. As of December 2021, according to https://ourworldindata.org/, high-income countries recorded 2194 cumulative deaths per million people, upper middle incomes 991, with the US and Italy still the front runners with more than 3000 deaths per million inhabitants, but not too far ahead with the distance progressively diminishing over time.

In spite of country differences and common trends, in most countries, despite a burst of public activity in the emergency, the search for legitimacy on the part of the governments appeared to be shortsighted and not addressing the structural failings of the health system revealed by the pandemic. Although proposals for 'One Health' approaches and transformative reforms were praised around the world, recovery investment plans continue to focus mainly on infrastructure and energy and often appear to underestimate the structural aspects of the health emergency. In most cases, while the motto 'build back better' suggests a (moderately) transformative approach to post-pandemic reconstruction, both its implicit message and practical applications seem totally inadequate in scale and scope, given the size of the collective drama and the multiple failures of public health systems exposed by the experience of the pandemic. In many cases, recovery policies tend to echo generic proposals for reforms based on a territorial network of greater proximity to the sick and their families, while ignoring the total inadequacy of the current network of services in terms of human and social capital. The preliminary plans largely seemed to gloss over the pandemic as a harrowing experience, almost as if it could be sidelined or wasn't, to some degree, a consequence of the health system's shortcomings, the ineffectiveness of prevention measures, the dependence of chronically ill on hospital care and the lack of proximity between medical facilities and the sick.

Although the country patterns showed many quirks and idiosyncratic variations, with only a few exceptions, in most countries, the first response to the pandemic was based on restrictions on individual mobility and social isolation measures. These measures did not immediately affect the spread of the virus, but their impact on ordinary life was dramatic. Cities all over the world became suddenly empty landscapes, while dwellings were populated by inner crowds of forcibly reunited or alternatively separated family groups, trapped in small spaces for what seemed an indefinite period of time with uncertain perspectives. This first act of social drama was consumed almost immediately after the pandemic breakout, in a limbo of insecurity and lack of information, ranging from panic to denial. Deprived of their usual freedom to move freely, join social gatherings, and live an ordinary life, people tended to develop a reluctant form of compliance with the restrictions imposed by government decrees. The use of masks, which had been originally considered ineffective in the statement of some medical advisers, became commonplace, as the infection did not seem to stop and deaths and hospital emergencies continued unabated for a seemingly interminable time. After a while, the infection appeared to subside only to return with relentless virulence in several successive waves. Somewhat unexpectedly, while the second wave of the virus was at its full strength, vaccines became available. Their sudden appearance revealed the startling efficiency of the health industry and the unsuspected flexibility of regulators, which allowed vaccines to be immediately available with minimal delays from bureaucracies and standard procedures. At the same time, even though governments at first were unprepared to handle the mass vaccination campaigns needed to put the vaccines at work, in the end, vaccination proved to be a successful tool to curb the infection. And yet, this unexpected success was tainted by two concurrent failures: on the one hand, the vaccine's unequal distribution across countries due to vaccine-grabbing behavior encouraged by shortsighted nationalism and, on the other hand, the limited protection afforded by vaccination in the face of what seemed to be a frantic mutation activity on the part of an extremely resilient virus.

In the midst of a complex relationship between the government and the general population, the pandemic fueled increased social discontent due to its uneven impact on the most fragile members of society. The skewness of the impact reflected the underlying skewness of many major dimensions of well-being within and across countries, which generally disfavor the most disadvantaged, regardless of the plague or risk that society faces at any moment. The key variable in this perverse relation is not monetary income but access to goods and services, healthcare, and education, and all public goods whose distribution is overwhelmingly skewed between and within countries against the poor, the elderly, the women, and the other most fragile components of the population. In the case of COVID-19, the poorer people, among them the elderly and the women, were at higher risk due to their more frequent exposure to infectious encounters, lower capacity to understand and practice prevention measures, greater hesitancy to accept vaccines, and more precarious health conditions. At the same time, they were more likely to suffer dire economic consequences from the lockdowns, such as loss of employment from often precarious jobs and the backlash of the economic downturn. Once infected, they were more vulnerable and more likely to suffer severe health conditions, as well as long-term economic and health consequences of the disease. More than just a reflection of an underlying inequality condition, a new social gradient thus emerged from the pandemic from fundamental characteristics of poverty and social exclusion, such as low access to public services, low wealth and education, unhealthy lifestyles, and poor working conditions.

As part of its impact on globalization, the pandemic has revealed the inescapable connection between health, social justice, and the environment. This nexus is often ignored in national or international policies, although it will inevitably challenge policy choices in any recovery and reconstruction strategy with global ambitions. It is not only a question of considering social goals that go beyond the achievement of material growth as measured by GDP or other aggregate indicators of collective wealth. Rather, the pandemic has revealed the importance of reconciling private and social priorities in everyday life, by fostering new types of investment and drastic behavioral changes. The environmental challenge to health is highlighted by the global re-emergence of infectious diseases, more than 60% of which, as identified since 1940, are of zoonotic origin, including COVID-19, as shown by the genetic sequence of the virus analyzed by researchers from all over the world. The growing anthropogenic pressures on the environment are at the root of ecosystem degradation, including global warming, loss of biodiversity, and increased contamination of air and water. But challenges to health and well-being are only part of a more general problem: the ability of our societies to continue to function with an acceptable balance between opposing needs: growth and environmental conservation, short- and long-term economic rewards and social inclusion, and ultimately justice and freedom.

These concepts were called into question first by the pandemic and then by the war in Ukraine, two global events that dramatized the conflict between rights and needs. COP27 the United Nations conference on climate change held in Egypt in November 2022, has exposed the additional collective drama of the increasingly unsustainable and unjust environmental policies.[1] This drama is dramatically evident in the case of the African continent, which faces the greatest risks related to climate change, especially droughts (whose frequency has tripled in the last 30 years) and the severity of extreme events such as floods, heat waves, and devastating cyclones. These events have changed the ecology of the continent, dramatically increasing tropical diseases and infections in sub-Saharan Africa. The experience of COVID-19 has shown that Africa can effectively use its own resources and those provided by international solidarity by mitigating the first and, to some extent, the second wave of the pandemic. At the end of July 2023, approximately 13.1 million cases of COVID-19 and nearly 260,000 deaths were reported in Africa, in contrast to early epidemiological models, which predicted that up to 70 million Africans would be infected with COVID-19 by June 2020, with more than 3 million deaths. However, Africa is still far behind in the fight against the virus, with only 24% of the population having completed the primary COVID-19 vaccination series compared to 64% around the world. In addition, other emergencies are looming in terms of health and the environment, such as water quality and environmental hygiene-related diseases, worsening and rising sea levels, and increasingly extreme weather conditions. Environmental shocks and their negative effects on food safety also cause serious damage to mental health. In total, the climate crisis is estimated to have destroyed a fifth of the gross domestic product in the most vulnerable countries on the continent. Without an accelerated green transition in all countries, Africa is thus going to be hit twice, since its population's access to affordable energy will remain meager and, at the same time, most of its countries will experience the direst consequences of unchecked climate change.

Although it is primarily dependent on coal as an energy source, Africa is historically responsible for only 3% of carbon dioxide emissions into the atmosphere, while the industrial countries of the western bloc are responsible for more than 65%. The energy transition, which requires sacrifices of consumption and huge investments, cannot be accomplished without substantial financial contributions from developed countries (the Paris climate agreement provided for the aid of the order of at least 100 billion dollars a year, which was only partially provided and in the form of loans) from developed

---

[1] COP27 refers to the 27th annual meeting under the United Nations Framework Convention on Climate Change (UNFCCC). This event took place in Sharm El-Sheikh, Egypt, from November 6 to 20, 2022.

countries. At the same time, international public financial flows were at a record low of 10 trillion $ in 2019 (down 50% from their peak in 2017) and did not appear to have recovered. COP27 identified in an order of magnitude between 4 and 6 trillion dollars a year the investment needed for the mitigation of climate change and also reached a historic agreement to create a 'loss and damage fund' to help developing countries in their plight to adapt to climate changes. Although the mechanism of financing and the size of the resources that the fund could channel are just beginning to be negotiated, detailed simulations with state-of-the art economic models show that investments ranging between 500 and 700 billion $ a year would be needed to achieve reasonable objectives of adaptation and environmental sustainability.

The call to the international community, arising from the conditions revealed and partly determined by the pandemic, embraces the need to mitigate a growing threat to human survival and basic need provision and adapt to some of its inevitable consequences. The COVID-19 event has generated a crisis that, perhaps in different forms, was expected by many. Its sudden emergence from an indistinct border between nature and nurture signals that other crises may follow from a growing imbalance within human activities and from their impact on the natural environment. The crisis was expected and perhaps overdue. For a moment, lockdowns and sudden change in people's behavior have given us a glimpse of how environmental degradation could be mitigated by transforming housing and transportation habits and slowing explosive urbanization. At the same time, the likely zoonic origin of the virus has suggested that the start of the infection may be part of a more general drift towards 'black swan' events. These events, in turn, may follow from a 'regression toward the tail' process due to the crossing of critical thresholds between nature and nurture (Flyvbjerg, 2020).

Several lessons should be taken and much thinking should be done on the implications of these eye-opening messages and the whole experience of the pandemic. Of course, we should be better prepared next time, which probably will be soon than we might expect based on our previous experience of a more stable world. At the same time, the pattern of resilience displayed everywhere by governments and people, however different and perhaps conflicting, depends on a number of transformative changes, many of which should be encouraged because they favor better preparedness and, at the same time, indicate appropriate preventive measures. Prevention, above all, requires a new approach to health care based on the pursuit of a healthy ecosystem within a new global public health model. This is the One Health approach, still reposing on a general idea rather than a detailed plan, but clearly the only possible strategy in a globalized world where we are all connected by our social and natural linkages with society, the economy, and the environment.

APPENDIX A

The epidemiological models

A.1. Introduction

Beginning in mid–February 2020, the media introduced the public to various technical terms from statistical and epidemiological fields. As often happens, initial confusion led to growing curiosity about these previously unfamiliar concepts. Search engines witnessed a surge in queries as individuals sought to find clarity amidst the vast amount of information available. Concurrently, experts competed to provide detailed content that would help users understand these concepts while boosting their own media visibility. However, between growing demands for explanation and abundant content, confusion frequently reigned due to the inherent complexity of the subject matter. Surprisingly, TV anchors, journalists, commentators, and inexperienced 'experts' casually tossed around intricate concepts that typically require significant time for even statistics or epidemiology students to comprehend and assimilate.

During the early weeks of the pandemic, the world was flooded with information about infection spread estimates. Suddenly, after exposure to the news, the average citizen appeared to feel comfortable talking about the critical role played, for example, by the parameter R_0 and and the methods to estimate it. However, important factors were often overlooked such as data accuracy, basic assumptions, model specifications, behavioral responses, and numerous other technical aspects.

In contrast, the public debate about the epidemic requires precision, depth, and clarity. And math, which is necessary to produce statistical and epidemiological models helpful in predicting the epidemic's progress. Describing and providing basic information about the models enables us to comprehend why, during public debates with experts, they may refer to significantly different estimates. Indeed, models serve as essential tools for understanding contagion trends and the influence of policies on these trends. However, the approaches and assumptions underlying these models can differ substantially, leading to vastly diverse results. This appendix provides a simplified overview of how epidemiological models work, as a primer on the topic of modeling infectious diseases.

A.2. A short primer on epidemiological models

Models that predict the spread of the epidemic are fundamental in providing answers to the many uncertainties about the disease and suggesting better solutions to manage the crisis. Governments, organizations, and citizens all require accurate information on how to effectively combat the virus and predict the duration of the outbreak.

To provide a solution to the several questions posed by pandemics, researchers use models that estimate some key epidemiological characteristics of the disease, such as incubation period, transmissibility, asymptomaticity, and severity, as well as the likely impacts of various public health policy interventions, such as social distancing, travel restrictions, and contact tracking. While the existence of diverse objectives may be the primary reason for differing model estimates, it's important to remember that COVID-19 models are not infallible or undisputed oracles. Rather than predicting an exact future, these models aim to outline a range of possible scenarios based on the available information.

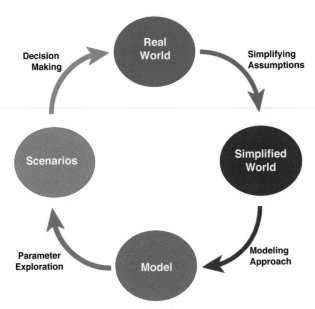

Figure A.1 The modeling process: from real world to results. *Source*: The graphic is adapted from Gonçalves (2020).

Grasping the basic assumptions that each model uses helps us better understand why some predictions differ from others. In general, the development of a model can be represented graphically as in Fig. A.1. Starting from the real world, researchers strive to create a useful representation of their observations, known as a 'model'. As such, a model is considered a simplified version of the world. Many of the differences between models arise in this phase: the simplifications may depend on various factors (lack of specific data, complexity of reality, mathematical intractability of the problem, methodological choices regarding the approach to the problem, availability of computational capacity, etc.). Once the primary hypotheses have been decided, modelers move on to implementing a framework (the 'model') that can be used to make predictions and evaluate the effects of different choices in terms of scenarios. Once available, the scenarios should help policymakers make 'informed' decisions, enabling them to design suitable policies for the real world.

In formulating her model, each researcher may make different assumptions about the coronavirus's properties. She will also quantify different properties such contagiousness, number of contacts between people, and the speed at which individuals succumb to the infection. Additionally, the models could be solved using various methodologies, leading to significantly different outcomes since they are generally based on multiple parameters. Finally, other hypotheses and/or model implications may concern what will unfold in the near future, such as people's behavior in response to a surge in infections.

In summary, it is worth remembering what British statistician George Box said: 'All models are wrong, but some are useful' (Box, 1976).[1] A model, by its very definition, is a representation of a system that highlights some components and ignores others and, as such, can never reflect all aspects of reality. Therefore, the question that needs to be asked is not whether 'the model faithfully represents reality' (it never will), but rather whether: 'the model is sufficiently adequate for a particular application'.

The main goals of an epidemiological model are the following:
- describe the characteristics and evolution of the virus or disease, even when some information is missing, by constructing plausible hypotheses;
- predict how many cases, deaths, hospitalizations, or other outcomes could occur in a given place and period;
- understand the potential effects of interventions and policies by observing the projections obtained starting from more or less plausible scenarios.

The mathematical modeling of vector-borne infectious diseases (viruses or bacteria) originated with a work by Sir Ronald Ross on the transmission of malaria in 1916 (Ross, 1916). Ross was the first to understand that vector-borne infections are governed by non-linear dynamics, making it difficult, if not impossible, to intuitively assess the natural trajectory of an epidemic and the effectiveness of public intervention without adequate mathematical modeling.

Mathematical models help clarify the dynamics of infectious diseases within hosts or populations, enabling comparisons between alternative control strategies and assisting policymakers in making informed decisions. In recent decades, the use of these models has intensified due to advancements in computational skills and capacities to solve them. They address a growing number of diseases and public health issues, exploring the significance of disease transmission (Fraser et al., 2009).

However, the origin can be traced back to two distinct families, which not only differ due to their unique methodological choices, but also because they focus on answering specific questions and addressing the uncertainty related to the results and parameters of the model in diverse ways. As evidenced by Holmdahl and Buckee (2020) epidemiological models can be divided into two main groups:

[1] Later in his life, it appears that George Box corrected the roll by stating: 'All models are approximations'.

- **statistical/forecasting models**. Their goal is to project future patterns by analyzing correlations. For example, starting from the observed relationship between the number of infected and the subsequent mortality rate and assuming that the relationship remains constant over time, it is possible to 'predict' how many deaths there will be tomorrow starting from today's infected. Therefore, under the assumption of parameter stability, this approach can be used to forecast the spread of the pandemic and its effects in the short term. However, if this assumption is violated, the results obtained could differ greatly from reality (Jewell et al., 2020).
- **mechanistic (process-based) models**. They are characterized by explicit mathematical modeling of the relations and laws that govern the development of the epidemic. Instead of solely relying on statistical correlations or empirical data, mechanistic models are grounded in the fundamental principles and laws (e.g., physical, chemical, biological laws) that govern the system in question. A typical example is the SIR model (or any of the several existing extensions), which uses a common epidemiological modeling technique to divide an estimated population into several groups ('compartments') – 'Susceptible' (S), 'Exposed' (E), 'Infected' (I) and 'Recovered/healed' (R). Mathematical rules are applied to the different groups to determine how people move from one compartment to another, adopting a series of basic assumptions about the evolutionary process of the disease, social mixing, the existence of public health policies, and other aspects. The SIR model captures population changes in each compartment with a system of ordinary differential equations (ODE) to model disease progression. This progression, or dynamics, is regulated by a series of parameters obtained through estimates or by observing the epidemic's progress.

In the following sections, we introduce the essential rules that govern mechanistic models, emphasizing the core parameters upon which the model relies and their significance in determining predictions. We begin by discussing the SIR model before exploring some of its extensions.

A.2.1 The SIR model

The SIR model (Susceptible-Infected-Recovered) is considered a 'compartmental' model that describes the dynamics of an infectious disease.[2] It is named the 'compartmental model' because it divides the population into compartments, where the compartments, which should have the same characteristics, are the following:

- The 'Susceptible' (they are 'healthy' individuals, who could be susceptible to infections);
- the 'Infected' (they are the infected individuals);

---

[2] The material reported in this section is partly taken and adapted from the following web address: https://towardsdatascience.com/infectious-disease-modelling-part-i-understanding-sir-28d60e29fdfc.

- the 'Recovered' (previously infected individuals, who cannot be infected again).

Specifically, in the 'Susceptible' group, we find people vulnerable to exposure to infected people, and they represent potential patients when infection occurs. The 'Infected' group represents people infected by the virus; they can transmit the disease to susceptible people but can be cured.[3] People 'cured' gain immunity so that they are no longer susceptible to the same disease. Therefore, the SIR model provides a logical *framework* within which it is possible to describe how the number of people in each compartment can change over time in the presence of communicable diseases.

The variables of interest in a SIR model are those listed below:
- N: the total population;
- $S(t)$: the number of susceptible persons on the day t;
- $I(t)$: the number of people infected on the day t;
- $R(t)$: the number of people healed on the day t;
- β: the expected amount of people an infected person infects per day;
- D: the number of days in which an infected person has infected and can spread the disease;
- γ: the percentage of infected people who recover per day ($\gamma = 1/D$).

Figure A.2 Graphical representation of the basic SIR model.

Graphically, Fig. A.2 clarifies how the three compartments are linked and also explains the conceptual simplicity of this model. The parameters of interest are represented by β and γ. In particular, β regulates the level of infection in the population, while γ regulates the number of healed.

A simple example can help one better understand how the model works. Imagine having a population of N individuals, with $N = 1000$, of which 400 people are known to be infected at the time t (e.g., $t = 7$ days after the outbreak), denoting it as $S(7) = 400$. The SIR model allows, just by entering some initial parameters, to obtain the values of $S(t)$, $I(t)$, $R(t)$ for all days t. Continuing the example, imagine having an infectious disease X. For this disease, the probability that an infected person can infect a healthy one is 20%. Also, let's imagine that the average number of people a person is in daily contact with (κ) is known to be 5.[4] So, every day, an infected individual encounters 5

---

[3] In more sophisticated versions of the model called SIR-D, a distinction is made between the 'cured' and the 'deceased' group.

[4] The parameter κ is independent of the population size.

people and infects each with a probability of 20%. Therefore, we expect this individual to infect 1 person (20% · 5 = 1) per day. This value is represented by the β parameter, which is the expected number of people an infected person infects daily.

Another important variable is represented by D, which is the number of days necessary for the course of the disease, during which an infected person can spread the disease.[5] If $D = 7$, this means that an infected person has 7 days to spread the disease, infecting 1 person per day (because $\beta = 1$). Therefore, an infected person is expected to infect 7 people (1 per day for 7 days). By combining the D variable with the β parameter, we can get a third variable, R, which represents the 'base reproduction number' ($R_0 = \beta \cdot D$), or the total number of people infected by an infected person.

A crucial piece of information required for estimating a SIR model is the relationship $\gamma = 1/D$. Since D represents the number of days a person is infected, γ is the percentage of individuals who recover each day from the disease. For example, if 30 people are infected and $D = 3$ (so they are infected for three days), a third will recover every day, so $\gamma = 1/3$. With $\gamma = 1/D$, then $D = 1/\gamma$ and $R_0 = \beta \cdot D$, so that $R_0 = \beta/\gamma$. Infections increase when this ratio is greater than 1, and an epidemic occurs. Conversely, when this ratio is ≤ 1, the epidemic tends to disappear.

More formally, a SIR model is a system of differential equations that allows one to describe the number of people in each compartment with a different ordinary differential equation, as reported below:

$$\frac{ds}{dt} = -\beta IS \tag{A.1}$$

$$\frac{dI}{dt} = -\beta IS - \gamma I \tag{A.2}$$

$$\frac{dR}{dt} = \gamma I \tag{A.3}$$

In general, in these models it is assumed that the population (N) remains constant (i.e., the effect of the death rate or birth rate is not considered) since SIR models assume that the development period of an infectious disease is much shorter than human life expectancy. A system like the one represented by Eqs. (A.1), (A.2), and (A.3) can be solved, upon estimation of the parameters β and γ and of the ensuing R_0 parameter.

The properties of the SIR model can be simply described by the following two theorems (Weiss, 2013):

Theorem 1. *If $R_0 \leq 1$, then I(t) monotonically decreases to 0 for $t \to +\infty$.*

Theorem 2. *If $R_0 > 1$, then I(t) starts to increase, reaches its maximum, and then decreases to 0 for $t \to +\infty$. A scenario with an increasing number of infected individuals is called an epidemic.*

---

[5] This period can include days of asymptomatic and symptomatic conditions. The longer the period of asymptomatic condition, the greater the risk of contagion.

As a result, an infection can invade and cause an epidemic in an entirely susceptible population if $R_0 > 1$ or $\beta > \gamma$.

Given a SIR model, the outbreak life cycle should appear as a S shaped curve when plotting the cumulative count of infection cases over time or as a bell-shaped curve when recording the daily cases over time (the new daily cases).

A.2.2 Extending the standard SIR model

Although the SIR model is reasonably reliable for understanding an epidemic's progression, it is far too simple to capture some of the key complexities of the real world. As a result, increasingly complex models have been proposed over time. These models offer a more realistic representation that allows for the identification of a greater number of steps in an epidemic's dynamics.[6]

The first modification to the SIR model involves accounting for the disease's incubation period. In the basic SIR model, infections are assumed to develop instantaneously, with no incubation period. However, in reality, an 'incubation' or 'latency' period affects the transmission dynamics of a contagion. How long should a potentially infected person be observed before they become contagious? This issue is addressed by the SEIR model, which adds a fourth compartment (exposed or latent), as shown in Fig. A.3.

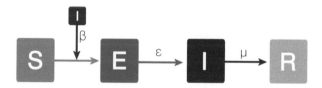

Figure A.3 Graphic representation of the SEIR model.

When a 'susceptible' person comes in contact with a contagious person (the beginning of the chain), they move towards being 'exposed' and move into the 'contagious' compartment at an assumed fixed rate ϵ. In the 'exposed' compartment, a person is 'incubating' the disease and perhaps even starting to develop symptoms but is not yet capable of infecting other individuals.

The improvement offered by this model's generalization lies not in estimating a more accurate number of infections (thereby altering the quantity), but rather in achieving a superior understanding of the infection curve's dynamics, which in turn influences the temporal progression of the epidemic. What usually happens is that the curve flattens

---

[6] Further information on the functioning of epidemiological models and the codes for building models presented below are available at the following web address: https://medium.com/data-for-science/epidemic-modeling-102-all-covid-19-models-are-wrong-but-some-are-useful-c81202cc6ee9.

and shifts to the right, significantly delaying and widening the peak of contagious cases. Such a change also makes it possible to estimate better the real need for intensive and sub-intensive care beds, allowing us to quantify the extent to which the health system may potentially face a crisis.

A second extension (the SEIRS model) can be obtained by abandoning the initial hypothesis of the SIR model that a person infected and cured is no longer susceptible to a new contagion an underlying concept in the calculations for the herd immunity threshold. Assuming that immunity can only be temporary, the model can be expanded simply by adding a spontaneous transition from the 'recovered' compartment to the 'susceptible' compartment, as shown in Fig. A.4.

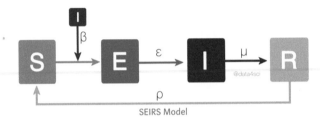

Figure A.4 Graphic representation of the SEIRS model.

This apparently harmless addition to the model has a significant effect, as it reconstitutes the group of people who can be infected again. The final result is that the epidemic never vanishes (it never runs out of fuel) and the disease becomes endemic, with a constant fraction of the population remaining infected. Fig. A.4 shows how the speed at which immunity is lost (ρ) has a decisive effect on the progress of the epidemic and the increase in endemicity. If ρ is small enough (immunity is more lasting), we may have several epidemic peaks before a steady state is reached for a fixed fraction of the population. If, on the other hand, ρ is sufficiently high, there is a process of endemization of the disease, as in the case of the common cold, which is also caused by a virus in the coronavirus family.

Another significant aspect of the model involves considering that a portion of infected patients exhibits symptoms, while the majority remains asymptomatic. For seasonal influenza, around 33% of cases are typically asymptomatic. For COVID-19, it is estimated that the proportion of asymptomatic cases could be up to 10 times larger than the number of symptomatic patients. This aspect can significantly distort the estimates obtained with a standard SIR model. Although it is debated whether asymptomatic individuals are less infectious than those with symptoms, it is nevertheless possible to assume that the level of infection is equal to a fraction of r_β. The previous model, depicted in Fig. A.4, can be generalized as shown in Fig. A.5. This extension is achieved by dividing the infected compartment into two parts: one symptomatic (I_s) and one asymptomatic (I_a). A fraction p_a of all those exposed becomes asymptomatic while the remainder $(1 - p_a)$ develops symptoms.

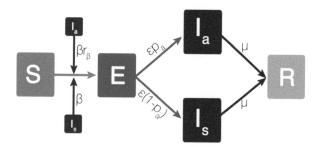

Figure A.5 Graphic representation of the SEIRS model with asymptomatics.

It is easily verified that the original SIR model is obtained if $r_\beta = 1$. On the contrary, recovering the original SIR value, if $r_\beta = 0$ (asymptomatic and completely non-infectious) the spreading parameter of the contagion R_0 is reduced by a factor $(1 - p_a)$ since the contagious population is much smaller.

A final interesting extension involves a scenario in which the infected population does not fully recover; instead, a portion could potentially succumb to the illness, as has sadly occurred worldwide. A simplification that can be made is to assume that only symptomatic cases can die from the disease or that all asymptomatic individuals who die from the disease are not counted as such. In this way, the model can be represented as in Fig. A.6.

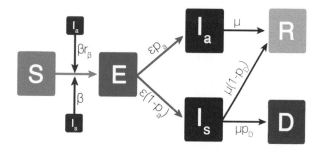

Figure A.6 Graphic representation of the SEIRS model with mortality.

The model now consists of 6 compartments and has 7 transitions and 6 parameters. At the start of an outbreak, most, if not all, of these parameters are partially or entirely unknown. As the outbreak progresses, more and more information is collected and more detailed models can be used. This constant refinement also helps improve the reliability of the scenarios that can be analyzed.

A.3. The theoretical foundation of public health interventions

The two theorems presented on the previous section provide a simple theoretical basis to understand what strategies public health experts must adopt to prevent an epidemic, with the aim of reducing R_0 below 1. For example, in the case of flu, it could be helpful to:

- reduce the duration of the D infection with antivirals;
- reduce the contact rate between people (κ) by by having susceptible people self-isolate (asking them to stay at home without going to school or work);
- reduce $S(0)$ by offering vaccines;
- reduce transmissibility by encouraging frequent hand washing and the distribution of protective masks.

A.3.1 Key parameters to understand epidemics

To understand the nature of an epidemic, it is essential to have a thorough grasp of the methods used to estimate the number of people infected.

The first parameter to consider is the rate at which susceptible (i.e., non-immune) individuals can become infected and infectious. This parameter is called 'infection strength' (it is usually denoted as λ): the higher it is, the greater the number of people who will be infected in the unit of time. The second relevant parameter is the 'recovery rate' relative to people's ability to heal (it can be denoted as γ).[7] Within the SIR/SEIR models, the parameter relating to the 'strength of the infection' is, in turn, a function of the ability of the virus to transmit itself and the frequency of contacts that an infected person may have. If the virus is difficult to transmit and people remain isolated, it is clear that the 'strength of the infection' will be very low and, consequently, the infection should spread very slowly. Therefore, the λ parameter is not a constant, but a function of the size of the 'infectious' compartment. The higher λ compared to γ, the higher the transmission speed of the infections that will result in an epidemic. The relationship between these two parameters results in the well-known indicator called R_0, which was discussed earlier, also known as the basic reproductive number. This indicator is most significant during the critical and intense weeks of the epidemic.

Other important parameters to consider are the parameters R_0 and R_e. Since the first outbreak occurs ('patient zero '), the spread of the epidemic depends mainly on the parameter R_0, or the 'base reproduction value which gives an idea of how contagious a disease is and is used to measure how quickly the virus is transmitted'.[8]

---

[7] In this context, we omit to mention the mortality coefficient, which instead defines the number of people who die from the infection.

[8] For a simple and detailed explanation of what it is and how can you estimate R_0 the interested reader can refer to the article available at the following link: https://www.iss.it/primo-piano/-/asset_publisher/o4oGR9qmvUz9/content/id/5268851.

Simply put, R_0, also known as the basic reproduction number, is the average number of secondary infections produced by a single infected individual in a completely susceptible population. It is a threshold parameter: if $R_0 > 1$, the number of cases will increase, leading to an outbreak or epidemic. If $R_0 < 1$, the disease will likely die out. R_0 is a constant for a particular disease and does not change over time unless there is a mutation or change in the underlying conditions.

Unlike R_0, the R_t parameter identifies the actual reproduction value, which defines the number of people in a population who can be infected by an individual at a specific time t. R_t changes as the population becomes more and more immunized, either by individual immunity after infection or by vaccination, or when people die. R_t is influenced by the number of infected people and the number of susceptible individuals infected people are in contact with. People's behavior (e.g., social distancing) can also influence R_t. The number of susceptible individuals decreases as people die or become immunized from exposure. It follows that the sooner people recover or die, the lower the value of R_t will be at any given time.

Fig. A.7 displays the contagion index, which is calculated by multiplying the basic reproduction number (R_0) by the proportion of the population P that is susceptible to the disease (S/P), for major infectious diseases that have occurred in recent years. It can be seen that MERS, more lethal than other coronaviruses, is characterized by a contagion index lower than 1 precisely because those who fall ill with MERS die very quickly and 'do not have time' to infect other people. Of course, the index value does not depend only on the lethality of the infection, but also on how contagious it is: for example, the Ebola virus, although lethal in many cases, is also very contagious and is therefore characterized by a higher contagion index. Finally, it is important to point out that the R_0 parameter measures the 'potential transmission' of a disease, not the speed at which the disease will actually spread.

However, the parameters R_0 and R_t do provide insight into the speed at which a disease spreads. They also enable us to address a crucial question: Can we prevent an epidemic by vaccinating only a fraction of the susceptible population (S)? We can answer because these parameters allow us to estimate that fraction, known as the 'herd immunity threshold' or HIT. Vaccination potentially prevents susceptible individuals from being infected, thus reducing the chain of contagion. The problem with vaccines is that even assuming full effectiveness (100%), it is costly to vaccinate the entire population and, above all, not everyone can take the vaccine given that for some individuals with compromised immune defenses or severe allergies. It may lead to more undesirable consequences than the illness itself.[9]

---

[9] This is a relevant problem well known to epidemiologists and experts of infectious diseases. For example, when dealing with the annual flu epidemic, even in the best years when a flu vaccine adapts well to circulating strains, its actual effectiveness remains around 60%.

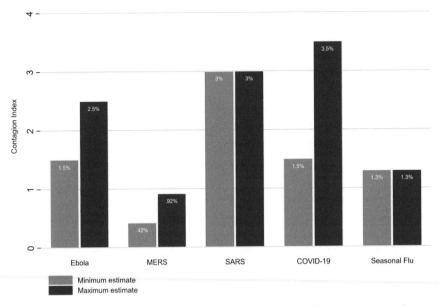

Figure A.7 Contagion index of the main infections currently in progress. *Source*: Calculations made by Statistica (2020) on data obtained from Asian Development Bank; Lancet; WHO; Journal of the American Medical Association. Available at the following address *web*: https://www.statista.com/statistics/1103196/worldwide-infection-rate-of-major-virus-outbreaks/.

In the simplest SIR model, the herd immunity threshold is only a function of the basic reproduction number R_0. Under the hypothesis that those who recovered from the infection gain immunity, the HIT value can be derived using the following simple logical reasoning. Given the basic reproduction number R_0 and defining p as the percentage of immune people and $(1-p)$ the percentage of non–immune people (susceptible), there will be an equilibrium situation as long as the product of these two variables is constant and equal to 1. More analytically, this result can be written as follows:

$$R_0(1-p) = 1 \rightarrow (1-p) = \frac{1}{R_0} \rightarrow p = 1 - \frac{1}{R_0} = \xi \qquad (A.4)$$

where ξ represents an estimate of the threshold value to obtain HIT. From Eq. (A.4), it becomes clear that if the reproducibility index $R_0 = 10$ (very high value), then the threshold value for obtaining herd immunity will be equal to 0.9, which implies that 90% of the population should not be susceptible (or must be vaccinated). If, on the other hand, $R_0 = 2$ then $\xi = 0.5$ and therefore only 50% of the population must not be susceptible. To prevent a smallpox epidemic with $R_0 = 5$, 80% of the population must be vaccinated. A highly communicable pathogen, such as measles, has a high R_0 (12-18) and thus a high proportion of the population (approximately 95%) must be immune to decrease sustained transmission. Finally, if $R_0 = 1$, then $\xi = 0$ and therefore the problem of herd immunity does not arise.

Since the beginning of the COVID-19 pandemic, most studies estimated that SARS-CoV-2 R_0 was in the range of 2 to 4. Assuming there is no existing population immunity and all individuals have equal susceptibility and infectiousness, the herd immunity threshold for SARS-CoV-2 is expected to range between 50% and 75% in the absence of interventions.

A.3.2 *Super-spreader*: a problem and an opportunity

A super-spreader (or *Super-diffuser*) is an infected individual who, in turn, infects an unexpectedly large number of people. Super-spread events have already occurred during the SARS and MERS outbreaks. Such events are not necessarily a bad sign because they may indicate that fewer people are generating an outbreak. Super-spreaders may also be easier to identify and contain, as their symptoms may be more severe.

This is why in addition to the R_0 parameter, scientists use a parameter called dispersion factor (k), which describes how the spread of a disease depends on a few individuals. The term 'dispersion factor' refers to the degree of variation in the number of secondary infections caused by a primary infected individual. In more technical terms, it measures the deviation from a perfectly homogenous transmission scenario. The lower the value of the k parameter, the more the transmission depends on a small number of people. In a 2005 article in Nature, Lloyd-Smith et al. (2005) estimated that SARS, in which *super-spreaders* played an important role, had a value of k equal to 0.16. The k estimated in 2012 for MERS was about 0.25. In the 1918 flu pandemic, on the other hand, the value of k was about 1, indicating that clusters played a minor role. Finally, Endo et al. (2020) estimated that the value of k for COVID-19 is less than 0.1, which implies that, potentially, only about 10% of infected patients infect approximately 80% of the susceptible population.

If the hypothesized data were correct, the theory of 'super-spreaders' in the COVID-19 epidemic would be very interesting because it would be able to clarify various doubts related to the 'timing' and the 'spatial' spread of infections worldwide. For example, it could explain some disturbing aspects of this pandemic, including why the virus did not spread around the world before its appearance in China and why the presence of some very early cases, such as the one in France at the end of December 2019 (reported only on May 3, 2020), apparently failed to ignite a larger outbreak. If k were indeed 0.1, then most of the infection chains would die on their own and the SARS-CoV-2 virus would have to be introduced to a new location (country) at least four times to have a uniform chance of spreading epidemically. Also, according to Endo et al. (2020) 'if the Chinese epidemic was like a great fire that sent sparks flying around the world, most of the sparks simply went out'.

A.4. Models of economic epidemiology

Economic epidemiological (EE) models differ from purely epidemiological (PM) models because they consider some of the infection parameters endogenous. This implies that both individual and public intervention strategies play a role in the pandemic, and contribute to determining the values of key parameters, especially the parameter R_0. As shown by Perrings (2014), EE studies were originally based on the simple concept of prevalence elasticity, or negative feedback between perception of the gravity of the infection and the preventive behavior of the population (Philipson, 2000). More recently, EE theories have been extended to modeling behavioral choices of social contacts and preventive strategies on the part of both individuals and policy makers (Gersovitz and Hammer, 2003; Barrett and Hoel, 2007). EE models tend to be more complex and less straightforward in their indications than PM models because they contain more variables and face the problem of agent heterogeneity. They must also consider the double feedback between behavior and spread of the disease: *i)* feedback from preventive measures taken individually, *ii)* feedback between individual behavior and policy measures implemented (Funk et al., 2010; Group, 2006).

In EE models, R_0 is the result of behavioral choices on the number of contacts (the degree of social activity), on the part of possibly heterogeneous agents, and on the conditional probability of transmission per contact. This probability, in turn, is a positive function of a number of contacts. The increase or decrease in R_0 is a measure of the prevalence response to individual behavior. When $R_0 > 1$ the pathogen may spread, when $R_0 < 1$ it will not. In analogy to natural resource management models, a basic reproductive ratio is established, along with several related metrics. These include the effective reproduction number, which quantifies infection growth in a partially susceptible population, and the control reproduction number, which measures infection growth in a susceptible population while implementing control strategies (see, for a discussion, Brauer and Castillo-Chavez, 2012). Both parameters are the consequence of behavioral decisions, as well as policy interventions and related incentives, on contacts between susceptible and infected individuals.

In most EE models, agents are assumed to maximize the expected lifetime utility as a positive function of social activity. The higher the degree of such activity, as a consequence of the agents' choice, the higher, *ceteris paribus*, social interaction, and the transmission of the disease. As Farboodi et al. (2021) explain: 'The assumptions that preferences depend on social activity, while disease transmission depends on social interactions are central to our view of social distancing. The former captures the idea that individuals value social activity (going to a restaurant, going for a walk, going to the office) and, absent health issues, are indifferent about whether other people are also engaging in social activity. On the other hand, if an individual goes for a walk and does not encounter anyone, they cannot become sick. Thus, interactions are critical for disease transmission'. Game-theoretic models of social distancing have emerged to account for this feedback between the state of the epidemic and people's behavior.

The feedback between people's behavior and the intensity of epidemics originates a series of modeling efforts of the game theoretical variety (see for example Kabir and Tanimoto, 2020; Reluga, 2010; Choi and Shim, 2020). In these games, agents modify their behavior in response to direct or indirect signals about the intensity and gravity of the infection. The infection, in turn, is influenced by mass behavior and subsides or surges based on whether people adhere to strict measures or act carelessly. This pattern of reciprocal control gives rise to what has been described as a condition of 'epidemic limbo', where agents' reactions and the epidemic continue for long time to coexist. As McAdams (2020) explains: 'People have an incentive to adopt precautionary measures once the epidemic has become sufficiently severe; so, the epidemic turns out to be not as bad as one would have predicted without accounting for behavioral response. However, as the epidemic wanes and there is less risk of being exposed, people eventually have an incentive to return to their usual behavior. Due to this self-limiting feedback, the epidemic can remain for an extended period of time in a limbo of intermediate severity: not so bad that all people take it seriously enough to distance themselves, but remaining enough of a threat that some people do so'.

To illustrate the economic version of an epidemiological model, we use the general formulation provided by Bloom et al. (2022). Consider a population N_t constituted at time t by susceptible (S_t), infected (I_t) and recovered (R_t) individuals such that: $N_t = (S_t + I_t + R_t)$. The evolution of infection can then be represented by a modified SIR model, describing the evolution of the three classes of individuals, but also taking into account their different mortality rates (μ_t^I for the infected and μ_t for everyone else) and of the newborns B_t.

$$S_{t+1} - S_t = B_t - \mu_t S_t - (1 - \mu_t)\alpha_t I_t \tag{A.5}$$

$$S_t = B_t - [\mu_t - (1 - \mu_t)\alpha_t I_t]S_t \tag{A.6}$$

$$I_{t+1} - I_t = (1 - \mu_t)\alpha_t I_t S_t - \mu_t^I I_t - (1 - \mu_t^I)\mu_t I_t - (1 - \mu_t^I)(1 - \mu_t) - (1 - \mu_t^I)\mu_t \gamma_t I_t \tag{A.7}$$

$$= (1 - \mu_t)\alpha_t S_t - \mu_t^I - (1 - \mu_t^I)[\mu_t + (1 - \mu_t)\gamma_t]I_t \tag{A.8}$$

$$R_{t+1} - R_t = (1 - \mu_t)(1 - \mu_t^I)\gamma_t I_t - \mu_t R_t = B_t - \mu_t N_t - (1 - \mu_t)\mu_t^I I_t \tag{A.9}$$

The rate of infection α_t, which can also be interpreted as a probability, is defined as a function of the number of people in the population and the level of three economic variables, respectively, for susceptible and infected individuals: consumption (c_t^S, c_t^I), employment (l_t^S, l_t^I), and prevention activities (v_t^S, v_t^I). The relationship is assumed to be positive with consumption and employment due to the correlation with social interaction, and negative with prevention.

$$\alpha_t = \alpha(N_t, c_t^S, l_t^S, v_t^S, c_t^I, l_t^I, v_t^I) \tag{A.10}$$

Individuals are assumed to maximize lifetime utility $u(c_t, l_t, v_t)$ according to the value function:

$$V_t^S = Max\{u(c_t, l_t, v_t) + \rho(1 - \mu_t)(\alpha_t I_t V_{t+1}^I + (1 - \alpha)_t I_t) V_t(t+1)^s)\} \tag{A.11}$$

$$V_t^I = Max\{u(c_t, l_t, v_t) + \rho(1 - \mu_t)(1 - \mu_t^I(\gamma_t V_{t+1}^R + ((1 - \gamma)_t) V_{t+1}^I) + \beta V_t^s\} \tag{A.12}$$

$$V_t^R = Max\{u(c_t, l_t, 0) + \rho(1 - \mu_t) V_{t+1}^R\} \tag{A.13}$$

The maximization process is governed by the dynamic budget constraint:

$$k_{t+1} - k_t = r_t k_t + \phi^i w_t h_t l_t - c_t - p_t v_t \tag{A.14}$$

where k_t is wealth, r_t the interest rate, w_t the wage rate per unit of human capital h_t, and p_t the unit cost of preventive activities.

The first-order conditions are as follows:

$$u_c - \lambda_t^S + \rho(1 - \mu_t)(V_{t+1}^I - V_{t+1}^S)\alpha_{cs} I_t = 0 \tag{A.15}$$

$$u_l - h_t(w_t \lambda_t^S + \rho(1 - \mu_t)(V_{t+1}^I - V_{t+1}^S)\alpha_{ls} I_t = 0 \tag{A.16}$$

$$u_v - p_t \lambda_t^S + \rho(1 - \mu_t)(V_{t+1}^I - V_{t+1}^S)\alpha_{vs} I_t \leq 0 \tag{A.17}$$

Under these conditions, λ_t^S is the shadow price of the wealth constraint in Eq. (A.15), while $\alpha_{cs} > 0$, $\alpha_{ls} > 0$, and $\alpha_{vs} < 0$ are the partial derivatives of the infection rate with respect to socially interactive goods (consumption and labor) and prevention.

The conditions outlined in Eqs. (A.11)-(A.13) have straightforward interpretations in terms of cost-benefit analysis. They indicate that consumption and labor supply will be reduced in response to the reduction of utility over time (the value function) expected from infection ($V_{t+1}^I - V_{t+1}^S < 0$). Preventive activities will instead be increased in response to its expected positive effects in reducing infection. The response will depend on the perceived stakes from infection, i.e., the expected balance between benefits and costs of each type of action and thus will tend to change in the course of the pandemic and to waiver as the gap between benefits and costs becomes smaller. This can happen because prevention reduces the probability of infection, but it can also happen because as individuals refrain from social interaction, their economic cost (the loss of value over time) becomes larger than the gain from prevention, or, at the limit, too large to be sustainable.

Infection depends not only on the contacts between infected and susceptible people but also on how these contacts come into place, how they are connected to each other, and how the virus circulates throughout the population. Links between two or more individuals constitute complex networks and may differ because they correspond to encounters of different frequency, intensity, and heterogeneity, which the virus may variously exploit to spread across and beyond a given community. Complex networks represent a wide range of systems found in nature and within society. Their structure is irregular and dynamically evolves over time, and the individual components can generally combine with each other in so many ways that it would take millions of years to try all possible combinations.

Interest in complex networks originated around the middle of the eighteenth century thanks to the work of the mathematician Leonhard Euler, who introduced graph theory, which is the basis of network analysis. In more recent times, graph theory was reused by Erdős et al. (1960), who developed the so-called 'random graph' theory. These are substantiated in a structure formed by nodes linked by connections following a random law. In the context of epidemiological models, random graphs correspond to the so-called random mix hypothesis, whereby infection spreads as a consequence of the random interaction between infected and susceptible individuals. A few years later, Milgram (1967) discovered an important property that was termed 'six degrees of separation' and characterizes random graphs as a 'small world' network. According to this property, in most networks, regardless of size, on average only six steps (sequential links) distance an individual from a stranger. This implies that two previously healthy people may discover to have been infected by a short chain of random links that involves only one common contact, as in the case where two people can have a short connecting path of acquaintances in a network, in spite of insularity or geographical and cultural barriers. For example, McAdams (2020) reports a case of a bacterium (Mycoplasma pneumonia), spreading through a psychiatric yard through caregiver workers, who acted as common acquaintances connecting otherwise isolated patients. As shown by Newman et al. (2002), three important types of networks can be distinguished: random, scale free and winner-takes-all networks.

Although random networks are characterized by probabilistic connections and tend to be formed by random growth, both scale-free and winner-takes-all networks appear to be the result of a clustering phenomenon based on 'preferential attachment' or the formation of links between nodes attracted to each other. Scale-free networks present a hierarchical structure, based on clusters where the probability that two nodes connected to a randomly chosen node are themselves connected is generally high and between 40 and 50%. The winner-takes-all category, on the other hand, is configured as a node with many interconnections and appears to be the result of an agglomeration based on a single node attracting all other nodes. For infections, the three possible structures of networks suggest three different paths of transmission: a random one, one through clusters of infection, and one through the so-called super-spreaders.

A.5. The limits of the models: why is it so difficult to estimate and predict the spread of infections?

For predictions to be reliable, a model must accurately reflect how the infection progresses in real life. Researchers typically utilize data from prior outbreaks of the same infection to develop their models and ensure that the predictions align with established knowledge. This method is effective for diseases like the flu, as there are decades of data available to help comprehend how outbreaks unfold in various communities. Influenza models are employed annually to inform decisions regarding vaccine formulations for flu seasons.

In contrast, modeling new outbreaks, such as COVID-19, is much more challenging because researchers know little about them. For example, what are the various ways in which the virus can be transmitted between individuals? How many hours can it survive on surfaces like doorknobs or Amazon packages? What is the duration between the virus entering a person's body and that person becoming contagious to others?

Besides these challenges, it is also important to consider the heterogeneity of the agents affected. This means that the many different factors that contribute to the spread of an epidemic may vary considerably from person to person and from place to place. Here are some of the most important types of heterogeneity that would be useful to incorporate into models:

- **Heterogeneity in the exposure**. The standard SIR model assumes a single transmission rate for the entire population. However, systematic differences in *routine* of daily life can result in different patterns of social interactions.
- **Heterogeneous response between people**. From observational data, correlations between age, sex, existing medical conditions, and the risk of infection from an epidemic can be estimated. For example, in the case of SARS-CoV-2, the observed mortality rate was at least 50% higher for men than for women in 12 of the 14 countries for which data are made available during the first wave. Furthermore, it is known that people of any age who suffer from previous chronic conditions may be at increased risk of developing complications from COVID-19.
- **Availability of health services on the territory**. The standard SIR model assumes a uniform medical response across space in terms of health services offered, regardless of the geographic location of the individuals.
- **Dosage and nature of exposure**. The standard SIR model does not adjust the risk of infection or mortality rates based on infection to consider the 'dose of exposure' or, more generally, the nature of social contact.

A.6. Conclusions

The analysis of epidemiological models, as highlighted in this appendix, plays a critical role in predicting the trajectory of infectious diseases, including the COVID-19 pandemic. Despite being simplified representations of reality, these models, employing various assumptions and methodologies, offer vital insights into potential disease progressions.

As previously discussed, there are two main categories of epidemiological models: statistical/forecasting and mechanistic. The SIR model serves as an example of the mechanistic type. This model, which segregates the population into specific groups and tracks disease evolution through defined parameters, has witnessed several enhancements to accommodate factors like temporary immunity and asymptomatic cases. The analysis presented underscores the significance of various parameters, including infection strength and recovery rate in deciphering epidemics, and delineates the complex roles of contagion indices, vaccination efforts, and super-spreaders in the diffusion of diseases. Economic epidemiology models introduce nuanced interaction between individual behaviors and public intervention strategies in determining disease parameters.

This appendix has briefly explored the role of endogenous variables within these models, which are shaped by individual actions and policy interventions, leading to a scenario termed as 'epidemic limbo' that results from the reciprocal relationship between human behavior and epidemic severity. In doing so it has recognized the challenges of forecasting disease propagation. These challenges arise from complex network structures and the virus's circulation patterns within them. Additionally, various factors increase the intricacy of modeling techniques, especially given the limited information available during new outbreaks.

References

Acemoglu, D., Chernozhukov, V., Werning, I., Whinston, M.D., et al., 2020. A Multi-Risk SIR Model with Optimally Targeted Lockdown, vol. 2020. National Bureau of Economic Research, Cambridge, MA. (p. 347).

Adams-Prassl, A., Boneva, T., Golin, M., Rauh, C., 2020. Inequality in the impact of the coronavirus shock: evidence from real time surveys. Journal of Public Economics 189, 1042–1045. http://www.sciencedirect.com/science/article/pii/S0047272720301092. (p. 277).

Adams-Prassl, A., Boneva, T., Golin, M., Rauh, C., 2022. The impact of the coronavirus lockdown on mental health: evidence from the United States. Economic Policy 37, 139–155. (pp. 181 and 277).

Addis, M., Bolici, F., Cannella, L., Cataudella, M.C., D'Orazio, A., David, P., De Simoni, M., Diana, G., Leonardi, S., Maiolo, S., et al., 2021. Riprogrammare la crescita territoriale. Turismo sostenibile, rigenerazione e valorizzazione del patrimonio culturale. Pàtron Editore. (p. 270).

Agarwal, R., Gopinath, G., 2021a. Pandemic economics. Finance & Development 58, 50–52. (p. 373).

Agarwal, R., Gopinath, M.G., 2021b. A Proposal to End the COVID-19 Pandemic. International Monetary Fund, Washington, DC. (p. 373).

Aguilera, X., Al-Abri, S.S., Anami, V., Annabi Attia, T., Aramburu, C., Blumberg, L., Chittaganpitch, M., LeDuc, J., Li, D., Maksyutov, R., et al., 2021. Report of the Review Committee on the Functioning of the International Health Regulations (2005) during the COVID-19 response. Technical Report. WHO. (p. 58).

Ahrendt, D., Mascherini, M., Nivakoski, S., Sándor, E., 2020. Living, working and COVID-19 e-survey data. http://www.eurofound.europa.eu. (p. 312).

Allain-Dupré, D., Chatry, I., Kornprobst, A., Michalun, M., Lafitte, C., Moisio, A., Phung, L., Power, K., Wu, Y., Zapata, I., 2020. The territorial impact of Covid-19: managing the crisis across levels of government. OECD policy responses to coronavirus (Covid-19). OECD Policy Responses to Coronavirus (COVID-19). OECD Report. Available from: https://read.oecd-ilibrary.org/view/?ref=128_128287-5agkkojaaa&title=The-territorial-impact-of-covid-19-managing-the-crisis-across-levels-of-government. (pp. 58 and 287).

Allen, D.W., Berg, C., Davidson, S., Potts, J., 2022. On Coase and Covid-19. European Journal of Law and Economics, 1–19. (p. 192).

Allen, J., West, D., 2020. Re-opening the world. How to save lives and livelihoods. The Brookings Institution Report. (p. 417).

Alsdurf, H., Belliveau, E., Bengio, Y., Deleu, T., Gupta, P., Ippolito, D., Janda, R., Jarvie, M., Kolody, T., Krastev, S., Maharaj, T., Obryk, R., Pilat, D., Pisano, V., Prud'homme, B., Qu, M., Rahaman, N., Rish, I., Rousseau, J.F., Sharma, A., Struck, B., Tang, J., Weiss, M., Yu, Y.W., 2020. Covi white paper. arXiv. Available from: https://arxiv.org/abs/2005.08502. arXiv:2005.08502. (pp. 109 and 111).

Amador, J., Di Mauro, F., 2015. The age of global value chains. Available from: https://cepr.org/voxeu/columns/age-global-value-chains. (p. 247).

Amendola, S., Spensieri, V., Hengartner, M.P., Cerutti, R., 2021. Mental health of Italian adults during Covid-19 pandemic. British Journal of Health Psychology 26, 644–656. (p. 301).

Amerio, A., Lugo, A., Stival, C., Fanucchi, T., Gorini, G., Pacifici, R., Odone, A., Serafini, G., Gallus, S., 2021. Covid-19 lockdown impact on mental health in a large representative sample of Italian adults. Journal of Affective Disorders 292, 398–404. (p. 301).

Amuasi, J.H., Lucas, T., Horton, R., Winkler, A., 2020. Reconnecting for our future: the lancet one health commission. The Lancet 395, 1469–1471. https://doi.org/10.1016/S0140-6736(20)31027-8. (p. 367).

Andersen, K., Rambaut, A., Lipkin, W., Holmes, E., Garry, R., 2020. The proximal origin of Sars-Cov-2. Nature Medicine 26, 450–452. https://www.nature.com/articles/s41591-020-0820-9. (pp. 16 and 197).

Anderson, R.M., Heesterbeek, H., Klinkenberg, D., Hollingsworth, T., 2020. How will country-based mitigation measures influence the course of the Covid-19 epidemic? The Lancet 395, 931–934. (pp. 235 and 340).

Andrianou, X.D., Makris, K.C., 2018. The framework of urban exposome: application of the exposome concept in urban health studies. Science of the Total Environment 636, 963–967. (p. 394).

Andriukaitis, V., 2021. A European health union as the way forward for the health of the continent table of contents former commissioner for health and food safety. FEPS Policy brief. (p. 320).

Antràs, P., 2016. Global Production: A Contracting Perspective. Princeton University Press. (p. 247).

Antras, P., Chor, D., 2021. Global value chains. (p. 247).

Arnold, D.T., Hamilton, F.W., Milne, A., Morley, A., Viner, J., Attwood, M., Noel, A., Gunning, S., Hatrick, J., Hamilton, S., Elvers, K.T., Hyams, C., Bibby, A., Moran, E., Adamali, H., Dodd, J., Maskell, N.A., Barratt, S., 2020. Patient outcomes after hospitalisation with Covid-19 and implications for follow-up; results from a prospective UK cohort. medRxiv. https://www.medrxiv.org/content/early/2020/08/14/2020.08.12.20173526. https://dx.doi.org/10.1101/2020.08.12.20173526. arXiv:https://www.medrxiv.org/content/early/2020/08/14/2020.08.12.20173526.full.pdf. (p. 354).

Askitas, N., Tatsiramos, K., Verheyden, B., 2021. Estimating worldwide effects of nonpharmaceutical interventions on COVID-19 incidence and population mobility patterns using a multiple-event study. Scientific Reports 11, 1972. (p. 125).

Assaad, F., 1983. Measles: summary of worldwide impact. Reviews of Infectious Diseases 5, 452–459. (p. 162).

Assessment, M.E., 2005. Ecosystems and Human Well-Being: Wetlands and Water. World Resources Institute. (p. 383).

Atella, V., Francisci, S., Vecchi, G., 2011. La salute degli italiani, 1861-2011. Politiche sanitarie 12, 165–189. (p. 30).

Authers, J., 2020. The coronavirus trade-off was always an illusion. Bloomberg Opinion. Available from: https://www.bloomberg.com/opinion/articles/2020-06-26/covid-lockdown-versus-economy-was-always-false-trade-off. (pp. 175 and 177).

Avaaz, 2020. Facebook's algorithm: a major threat to public health. Avaaz. (p. 338).

Avery, C., Bossert, W., Clark, A., Ellison, G., Ellison, S., 2020a. Policy implications of models of the spread of coronavirus: perspectives and opportunities for economists. https://ideas.repec.org/p/ces/ceswps/_8293.html. (p. 196).

Avery, C., Bossert, W., Clark, A., Ellison, G., Ellison, S.F., 2020b. An economist's guide to epidemiology models of infectious disease. The Journal of Economic Perspectives 34, 79–104. (p. 350).

Aylor, B., Datta, B., DeFauw, M., Gilbert, M., Knizek, C., McAdoo, M., 2020. Designing resilience into global supply chains. Boston Consulting Group. Available from: https://www.bcg.com/en-us/publications/2020/resilience-in-global-supply-chains. (p. 253).

Azariadis, C., 2018. Riddles and models: a review essay on Michel de Vroey's a history of macroeconomics from Keynes to Lucas and beyond. Journal of Economic Literature 56, 1538–1576. (p. 220).

Bachelard, G., Jones, M.M., 2002. The formation of the scientific mind a contribution to a psychoanalysis of objective knowledge. (p. 345).

Backer, J., Klinkenberg, D., Wallinga, J., 2020. Incubation period of 2019 novel coronavirus (2019-ncov) infections among travellers from Wuhan, China, 20-28 January 2020. Euro Surveillance: European Communicable Disease Bulletin 25. (p. 105).

Baker, R.E., Mahmud, A.S., Miller, I.F., Rajeev, M., Rasambainarivo, F., Rice, B.L., Takahashi, S., Tatem, A.J., Wagner, C.E., Wang, L.F., et al., 2022. Infectious disease in an era of global change. Nature Reviews. Microbiology 20, 193–205. (pp. 35 and 44).

Bakewell, O., 2000. Uncovering local perspectives on humanitarian assistance and its outcomes. Disasters 24, 103–116. https://doi.org/10.1111/1467-7717.00136. https://onlinelibrary.wiley.com/doi/abs/10.1111/1467-7717.00136. arXiv:https://onlinelibrary.wiley.com/doi/pdf/10.1111/1467-7717.00136. (p. 56).

Baldi, E., Sechi, G.M., Mare, C., Canevari, F., Brancaglione, A., Primi, R., Klersy, C., Palo, A., Contri, E., Ronchi, V., Beretta, G., Reali, F., Parogni, P., Facchin, F., Bua, D., Rizzi, U., Bussi, D., Ruggeri, S., Oltrona, V., Luigi, Savastano, S., 2020. Out-of-hospital cardiac arrest during the Covid-19 outbreak in Italy. The New England Journal of Medicine 383, 496–498. https://doi.org/10.1056/NEJMc2010418. (pp. 197 and 306).

Balloux, F., Tan, C., Swadling, L., Richard, D., Jenner, C., Maini, M., van Dorp, L., 2022. The past, current and future epidemiological dynamic of Sars-Cov-2. Oxford Open Immunology. (p. 13).

Bamberg, S., 2006. Is a residential relocation a good opportunity to change people's travel behavior? Results from a theory-driven intervention study. Environment and Behavior 38, 820–840. (p. 186).

Bambra, C., Riordan, R., Ford, J., Matthews, F., 2020. The Covid-19 pandemic and health inequalities. Journal of Epidemiology and Community Health 74, 964–968. (p. 283).

Bandiera, L., Pavar, G., Pisetta, G., Otomo, S., Mangano, E., Seckl, J., et al., 2020. Face coverings and respiratory tract droplet dispersion. medRxiv. Available from: https://www.medrxiv.org/content/10.1101/2020.08.11.20145086v1. (p. 121).

Bandyopadhyay, S., 2022. Live free or die: the trade-offs in loosening Covid restrictions. https://www.birmingham.ac.uk/news/2022/loosening-covid-restrictions. (p. 180).

Banholzer, N., Lison, A., Vach, W., 2022. Comment on 'A literature review and meta-analysis of the effects of lockdowns on Covid-19 mortality'. Available at: https://papers.ssrn.com/sol3/papers.cfm?abstract_id=4032477. (p. 124).

Baqaee, D., Farhi, E., 2022. Supply and demand in disaggregated Keynesian economies with an application to the Covid-19 crisis. American Economic Review 112, 1397–1436. (pp. 206 and 210).

Barišić, P., Kovač, T., 2022. The effectiveness of the fiscal policy response to Covid-19 through the lens of short and long run labor market effects of Covid-19 measures. Public Sector Economics 46, 43–81. (p. 266).

Barrett, S., 2013. Economic considerations for the eradication endgame. Philosophical Transactions of the Royal Society B: Biological Sciences 368, 226–227. (p. 46).

Barrett, S., Hoel, M., 2007. Optimal disease eradication. Environment and Development Economics 12, 627–652. (p. 464).

Bauchner, H., Fontanarosa, P., Golub, R., 2020. Editorial evaluation and peer review during a pandemic: how journals maintain standards. JAMA 324, 453–454. https://doi.org/10.1001/jama.2020.11764. arXiv:https://jamanetwork.com/journals/jama/articlepdf/2767892/jama_bauchner_2020_ed_200063.pdf. (p. 327).

Baum, F., Freeman, T., Musolino, C., Abramovitz, M., De Ceukelaire, W., Flavel, J., Friel, S., Giugliani, C., Howden-Chapman, P., Huong, N.T., et al., 2021. Explaining Covid-19 performance: what factors might predict national responses? BMJ 372. (p. 58).

Bawden, D., Robinson, L., 2020. Information overload: an overview. In: Oxford Encyclopedia of Political Decision Making. Oxford University Press, Oxford, pp. 120–144. https://openaccess.city.ac.uk/id/eprint/23544/. (p. 339).

Baxter, M., King, R.G., 1993. Fiscal policy in general equilibrium. American Economic Review, 315–334. (p. 220).

Beaubien, J., 2016. How Boko Haram is keeping polio alive in Nigeria. NPR. Available from: http://www.npr.org/sections/goatsandsoda/2016/09/14/493755014/nigeria-has-to-wipe-out-polio-again. (p. 48).

Beck, S., 2011. Staging bone marrow donation as a ballot: reconfiguring the social and the political using biomedicine in Cyprus. Body & Society 17, 93–119. (p. 345).

Becker, G.S., et al., 1964. Human capital. (p. 392).

Becker, J., Starobinski, D., 2020. Now your iphone can warn you if you've been exposed to the coronavirus. The Conversation. https://theconversation.com/how-apple-and-google-will-let-your-phone-warn-you-if-youve-been-exposed-to-the-coronavirus-136597. (p. 116).

Bedford, J., Farrar, F., Ihekweazu, C., Kang, G., Koopmans, M., Nkengasong, J., 2019. A new twenty-first century science for effective epidemic response. Nature 575, 130–136. https://doi.org/10.1038/s41586-019-1717-y. (pp. 33 and 411).

Bell, D., Nicoll, A., Fukuda, K., Horby, P., Monto, A., Hayden, F., et al., 2006. Non-pharmaceutical interventions for pandemic influenza, national and community measures. Emerging Infectious Diseases 12 (1), 88–94. (p. 464).

Bellizzi, S., Nivoli, A., Lorettu, L., Farina, G., Ramses, M., Ronzoni, A.R., 2020. Violence against women in Italy during the Covid-19 pandemic. International Journal of Gynaecology and Obstetrics 150, 258–259. (p. 280).

Bengio, Y., Janda, R., Yu, Y.W., Ippolito, D., Jarvie, M., Pilat, D., Struck, B., Krastev, S., Sharma, A., 2020. The need for privacy with public digital contact tracing during the Covid-19 pandemic. The Lancet Digital Health 2, e342–e344. https://doi.org/10.1016/S2589-7500(20)30133-3. (p. 113).

Berger, L., Berger, N., Bosetti, V., Gilboa, I., Hansen, L., Jarvis, C., Marinacci, M., Smith, R., 2020. Uncertainty and decision-making during a crisis: how to make policy decisions in the Covid-19 context? The Becker Friedman Institute, Geneva. Available from: https://bfi.uchicago.edu/working-paper/uncertainty-and-decision-making-during-a-crisis-how-to-make-policy-decisions-in-the-covid-19-context/. (p. 340).

Bergstrom, C., Dean, N., 2020. What the proponents of 'natural' herd immunity don't say: try to reach it without a vaccine, and millions will die. (p. 197).

Berman, J., 2020. Anti-Vaxxers: How to Challenge a Misinformed Movement. MIT Press, Boston. (p. 142).

Betsch, C., Wieler, L., Habersaat, K., 2020. Monitoring behavioural insights related to Covid-19. The Lancet 395, 1255–1256. https://doi.org/10.1016/S0140-6736(20)30729-7. http://www.sciencedirect.com/science/article/pii/S0140673620307297. (p. 194).

Black, F.L., 1982. The role of herd immunity in control of measles. The Yale Journal of Biology and Medicine 55, 351. (p. 162).

Blanchard, O.J., Summers, L.H., 1986. Fiscal increasing returns, hysteresis, real wages and unemployment. (p. 266).

Bloom, D.E., Kuhn, M., Prettner, K., 2022. Modern infectious diseases: macroeconomic impacts and policy responses. Journal of Economic Literature 60, 85–131. (p. 465).

Bloom, J.D., Chan, Y.A., Baric, R.S., Bjorkman, P.J., Cobey, S., Deverman, B.E., Fisman, D.N., Gupta, R., Iwasaki, A., Lipsitch, M., et al., 2021. Investigate the origins of Covid-19. Science 372, 694. (p. 19).

Bloom, N., Liang, J., Roberts, J., Ying, Z., 2014. Does working from home work? Evidence from a Chinese experiment. The Quarterly Journal of Economics 130, 165–218. https://doi.org/10.1093/qje/qju032. arXiv:https://academic.oup.com/qje/article-pdf/130/1/165/30629971/qju032.pdf. (p. 19).

Bodilsen, J., Nielsen, P.B., Søgaard, M., Dalager-Pedersen, M., Speiser, L.O.Z., Yndigegn, T., Nielsen, H., Larsen, T.B., Skjøth, F., 2021. Hospital admission and mortality rates for non-Covid diseases in Denmark during Covid-19 pandemic: nationwide population based cohort study. BMJ 373. (p. 306).

Boin, A., 't Hart, P., Stern, E., Sundelius, B., 2005. The Politics of Crisis Management: Public Leadership Under Pressure. Cambridge University Press. (p. 360).

Bol, D., Giani, M., Blais, A., Loewen, P.J., 2021. The effect of Covid-19 lockdowns on political support: some good news for democracy? European Journal of Political Research 60, 497–505. (p. 179).

Bollyky, T., Patrick, S., 2020. Improving pandemic preparedness: lessons from Covid-19. The Council of Foreign Relations WHO Report. Available from: https://www.cfr.org/report/pandemic-preparedness-lessons-COVID-19/findings/. (p. 60).

Bollyky, T.J., Hulland, E.N., Barber, R.M., Collins, J.K., Kiernan, S., Moses, M., Pigott, D.M., Reiner Jr, R.C., Sorensen, R.J., Abbafati, C., et al., 2022. Pandemic preparedness and Covid-19: an exploratory analysis of infection and fatality rates, and contextual factors associated with preparedness in 177 countries, from jan 1, 2020, to sept 30, 2021. The Lancet 399, 1489–1512. (p. 281).

Bom, P.R., Ligthart, J.E., 2014. What have we learned from three decades of research on the productivity of public capital? Journal of Economic Surveys 28, 889–916. (p. 267).

Bosa, I., Castelli, A., Castelli, M., Ciani, O., Compagni, A., Galizzi, M.M., Garofano, M., Ghislandi, S., Giannoni, M., Marini, G., et al., 2022. Response to Covid-19: was Italy (un) prepared? Health Economics, Policy and Law 17, 1–13. (p. 64).

Box, G., 1976. Science and statistics. Journal of the American Statistical Association 71, 791–799. https://doi.org/10.1080/01621459.1976.10480949. https://www.tandfonline.com/doi/abs/10.1080/01621459.1976.10480949. (p. 453).

Brahmbhatt, M., Dutta, A., 2008. On Sars type economic effects during infectious disease outbreaks. (p. 47).

Braithwaite, I., Callender, T., Bullock, M., Aldridge, R.W., 2020. Automated and partly automated contact tracing: a systematic review to inform the control of Covid-19. The Lancet Digital Health 2, e607–e621. (p. 111).

Brauer, F., Castillo-Chavez, C., 2012. Mathematical Models for Communicable Diseases. SIAM. (p. 464).

Brauner, J., Mindermann, S., Sharma, M., Stephenson, A., Gavenčiak, T., Johnston, D., et al., 2020. The effectiveness of eight nonpharmaceutical interventions against Covid-19 in 41 countries. medRxiv. Available from: https://www.medrxiv.org/content/10.1101/2020.05.28.20116129v4. (p. 122).

Brennen, J., Simon, F., Howard, P., Nielsen, R., 2020. Types, sources, and claims of COVID-19 misinformation. Reuters Institute for the Study of Journalism. Available from: https://reutersinstitute.politics.ox.ac.uk/types-sources-and-claims-covid-19-misinformation. (p. 337).

Bricongne, J.C., Meunier, B., 2021. The best policies to fight pandemics: five lessons from the literature so far. Available at: https://voxeu.org/article/best-policies-fight-pandemics-five-lessons-literature-so-far. (pp. 124, 125, and 127).

Brown, N., 2018. Immunitary Life: a Biopolitics of Immunity. Springer. (p. 345).

Brown, N., Machin, L., McLeod, D., 2011. Immunitary bioeconomy: the economisation of life in the international cord blood market. Social Science & Medicine 72, 1115–1122. (p. 345).

Brown, N., Nettleton, S., 2018. Economic imaginaries of the anti-biosis: between 'economies of resistance' and the 'resistance of economies'. Palgrave Communications 4, 1–8. (p. 345).

Brzezinski, A., Kecht, V., Van Dijcke, D., Wright, A.L., 2020. Belief in science influences physical distancing in response to COVID-19 lockdown policies. Working Paper No. 2020-56. Available from: https://ssrn.com/abstract=3587990. (p. 335).

Buchanan, J.M., 1964. What should economists do? Southern Economic Journal, 213–222. (p. 192).

Buchanan, J.M., Tullock, G., 1965. The Calculus of Consent: Logical Foundations of Constitutional Democracy, vol. 100. University of Michigan Press. (p. 189).

Buckman, S.R., Glick, R., Lansing, K.J., Petrosky-Nadeau, N., Seitelman, L.M., 2020. Replicating and projecting the path of Covid-19 with a model-implied reproduction number. Infectious Disease Modelling 5, 635–651. (p. 127).

Budish, E., Kettler, H., Kominers, S.D., Osland, E., Prendergast, C., Torkelson, A.A., 2022. Distributing a billion vaccines: covax successes, challenges, and opportunities. Oxford Review of Economic Policy 38, 941–974. (p. 422).

Buechner, M.N., 1976. Frank knight on capital as the only factor of production. Journal of Economic Issues 10, 598–617. (p. 389).

Burgio, E., 2020. Covid-19: the Italian drama. Four avoidable risk factors. Wall Street Journal International. Available from: https://wsimag.com/science-and-technology/61967-covid-19-the-italian-drama. (p. 86).

Butler, C.D., Soskolne, C.L., 2013. Ecosystems, Stable and Sustainable. Springer, New York, New York, NY, pp. 649–654. Chapter 6. (p. 37).

Byrne, S.K., 2008. Healthcare avoidance: a critical review. Holistic Nursing Practice 22, 280–292. (p. 306).

Cabanes, B., Segrestin, B., Weil, B., Le Masson, P., 2014. Understanding the role of collective imaginary in the dynamics of expectations: the space industry case study. In: 21st International Product Development Management Conference. Limerick, Irlande, 2014, pp. 1–26. (p. 345).

Calisher, C., Carroll, D., Colwell, R., Corley, R., Daszak, P., et al., 2020. Statement in support of the scientists, public health professionals, and medical professionals of China combatting Covid-19. The Lancet 395, e42–e43. (p. 18).

Carfi, A., Bernabei, R., Landi, F., for the Gemelli Against COVID-19 Post-Acute Care Study Group, 2020. Persistent symptoms in patients after acute COVID-19. JAMA 324, 603–605. https://doi.org/10.1001/jama.2020.12603. arXiv:https://jamanetwork.com/journals/jama/articlepdf/2768351/jama_carf_2020_ld_200075_1596830733.59855.pdf. (p. 354).

Carlson, C.J., Albery, G.F., Merow, C., Trisos, C.H., Zipfel, C.M., Eskew, E.A., Olival, K.J., Ross, N., Bansal, S., 2022. Climate change increases cross-species viral transmission risk. Nature, 1. (p. 37).

Carlsson-Szlezak, P., Reeves, M., Swartz, P., 2020a. Understanding the economic shock of coronavirus. Harvard Business Review. https://www.nature.com/articles/s41591-020-0820-9. (pp. 7 and 205).

Carlsson-Szlezak, P., Reeves, M., Swartz, P., 2020b. What coronavirus could mean for the global economy. Harvard Business Review 3, 1–10. (p. 205).

Carpi, T., Hino, A., Iacus, S.M., Porro, G., 2021. Twitter subjective well-being indicator during Covid-19 pandemic: a cross-country comparative study. arXiv preprint. arXiv:2101.07695. (p. 302).

Carvalho, V.M., Garcia, J.R., Hansen, S., Ortiz, Á., Rodrigo, T., Rodríguez Mora, J.V., Ruiz, P., 2020. Tracking the Covid-19 crisis with high-resolution transaction data. Royal Society Open Science 8, 210218. (p. 139).

Case, A., Deaton, A., 2020. Deaths of Despair and the Future of Capitalism. Princeton University Press, Princeton, NJ. (p. 174).

Casella, B., Bolwijn, R., Moran, D., Kanemoto, K., 2019. Unctad insights: improving the analysis of global value chains: the unctad-eora database. Transnational Corporations 26, 115–142. https://doi.org/10.18356/3aad0f6a-en. https://www.un-ilibrary.org/content/journals/2076099x/26/3/5. (p. 255).

Caselli, F., Grigoli, F., Lian, W., Sandri, D., 2020. The great lockdown: dissecting the economic effects. International Monetary Fund. World Economic Outlook 65, 84. (p. 139).

Castoriadis, C., 1987. Transformación social y creación cultural. Estudios Venezolanos de, Caracas. (p. 345).

Ceballos, G., Ehrlich, P.R., Barnosky, A.D., García, A., Pringle, R.M., Palmer, T.M., 2015. Accelerated modern human–induced species losses: entering the sixth mass extinction. Science Advances 1, e1400253. (p. 387).

Cell Press, 2020. Mild Covid-19 cases can produce strong t cell response. Science Daily. Available from: http://www.sciencedaily.com/releases/2020/08/200817132331.html. (p. 378).

Cellini, N., Canale, N., Mioni, G., Costa, S., 2020. Changes in sleep pattern, sense of time and digital media use during Covid-19 lockdown in Italy. Journal of Sleep Research 29, e13074. https://doi.org/10.1111/jsr.13074. https://onlinelibrary.wiley.com/doi/abs/10.1111/jsr.13074. (p. 301).

Center for Health Security, NTI and Economist Intelligence Unit, 2019. Global health security index. Nuclear Threat Initiative, Washington, DC. Available from: http://ghsindex.org/wp-content/uploads/2020/04/2019-Global-Health-Security-Index.pdf. (p. 60).

Centers for Disease Control (CDC), et al., 1991. Measles–United States, 1990. Morbidity and Mortality Weekly Report 40, 369–372. (p. 162).

Centers for Disease Control and Prevention, 1999. Outbreak of West Nile-like viral encephalitis–New York. MMWR 48, 845–849. (p. 24).

Centers for Disease Control and Prevention, 2003. Hepatitis a outbreak associated with green onions at a restaurant–Monaca, Pennsylvania. MMWR 52, 1155–1157. (p. 23).

Centers for Disease Control and Prevention, 2015. Cdc's Ebola response in the United States and abroad. Available from: https://www.cdc.gov/budget/documents/ebola/2015-ebola-and-gh-security-activities-factsheet.pdf. (p. 47).

Cf, O., 2015. Transforming our world: the 2030 agenda for sustainable development. United Nations, New York, NY, USA. (p. 382).

Chater, N., 2020. Facing up to the uncertainties of Covid-19. Nature Human Behaviour 4, 439. https://doi.org/10.1038/s41562-020-0865-2. (pp. 194, 340, and 341).

Chetty, R., Friedman, J., Hendren, N., Stepner, M., Insights, T.T.O., 2020. The economic impacts of Covid-19: evidence from a new public database built using private sector data. NBER Working Papers. https://doi.org/10.3386/w27431. (p. 177).

Chilton, S., Nielsen, J., Wildman, J., 2020. Beyond Covid-19: how the 'dismal science' can prepare us for the future. Health Economics 29, 851–853. https://doi.org/10.1002/hec.4114. https://onlinelibrary.wiley.com/doi/abs/10.1002/hec.4114. arXiv:https://onlinelibrary.wiley.com/doi/pdf/10.1002/hec.4114. (p. 159).

Chinazzi, M., Davis, J.T., Ajelli, M., Gioannini, C., Litvinova, M., Merler, S., Pastore y Piontti, A., Mu, K., Rossi, L., Sun, K., Viboud, C., Xiong, X., Yu, H., Halloran, M.E., Longini, I.M., Vespignani, A., 2020. The effect of travel restrictions on the spread of the 2019 novel coronavirus (Covid-19) outbreak. Science 368, 395–400. https://doi.org/10.1126/science.aba9757. https://science.sciencemag.org/content/368/6489/395. arXiv:https://science.sciencemag.org/content/368/6489/395.full.pdf. (p. 122).

Chiusi, F., Fischer, S., Spielkamp, M., 2020. Automated decision-making systems in the Covid-19 pandemic: a European perspective. Automating Society Report. Special Issue of the Automating Society Report 2020. Available from: http://algorithmwatch.org/automating-society-2020-covid19. (pp. 114 and 115).

Choi, W., Shim, E., 2020. Optimal strategies for vaccination and social distancing in a game-theoretic epidemiologic model. Journal of Theoretical Biology 505, 110422. (p. 465).

Chu, D., Akl, E., Duda, S., Solo, K., Yaacoub, S., Schünemann, H., et al., 2020. Physical distancing, face masks, and eye protection to prevent person-to-person transmission of Sars-Cov-2 and Covid-19: a systematic review and meta-analysis. The Lancet 395, 1973–1987. (pp. 103 and 121).

Chudasama, Y.V., Gillies, C.L., Zaccardi, F., Coles, B., Davies, M.J., Seidu, S., Khunti, K., 2020. Impact of Covid-19 on routine care for chronic diseases: a global survey of views from healthcare professionals. Diabetes & Metabolic Syndrome: Clinical Research & Reviews 14, 965–967. (p. 309).

CISPA, 2020. Joint statement on contact tracing. CISPA. Available from: https://cispa.de/en/news-and-events/news-archive/articles/2020/joint-statement-on-contact-tracing. (pp. 116 and 120).

Claeys, G., Guetta-Jeanrenaud, L., 2022. How rate increases could impact debt ratios in the euro area's most-indebted countries. Bruegel Blog. Available from: https://www.bruegel.org/blog-post/how-rate-increases-could-impact-debt-ratios-euro-areas-most-indebted-countries. (p. 246).

Clemens, J., Veuger, S., 2020. Implications of the Covid-19 pandemic for state government tax revenues. National Tax Journal 73, 619–644. (p. 210).

Coase, R.H., 1960. The problem of social cost. The Journal of Law & Economics 3, 1–44. (p. 191).

Cobb, S., Miller, M., Wald, N., 1959. On the estimation of the incubation period in malignant disease. Journal of Chronic Diseases 9, 385–393. (p. 23).

Cohen, J., 2020. Shots of hope. (p. 135).

Coia, J., Ritchie, L., Adisesh, A., Booth, C.M., Bradley, C., Bunyan, D., Carson, G., Fry, C., Hoffman, P., Jenkins, D., et al., 2013. Guidance on the use of respiratory and facial protection equipment. The Journal of Hospital Infection 85, 170–182. (p. 121).

Coibion, O., Gorodnichenko, Y., Weber, M., 2020. The cost of the Covid-19 crisis: lockdowns, macroeconomic expectations, and consumer spending. Technical Report. National Bureau of Economic Research. (pp. 139 and 267).

Coll, P., Lindsay, A., Meng, J., Gopalakrishna, A., Raghavendra, S., Bysani, P., O'Brien, D., 2020. The prevention of infections in older adults: oral health. Journal of the American Geriatrics Society 68, 411–416. (p. 5).

Collins, A., 2018. The global risks report 2018. Technical Report. World Economic Forum, Geneva. (p. xxv).

Consortium, P., 2016. One health in action. EcoHealth Alliance. Available on: https://www.ecohealthalliance.org/wp-content/uploads/2016/10/One-Health-in-Action-Case-Study-Booklet_ENGLISH_Jan-7-2017-FINAL.pdf. (p. 382).

Costa, A.O.C., Neto, H.d.C.A., Nunes, A.P.L., de Castro, R.D., de Almeida, R.N., 2021. Covid-19: is reinfection possible? EXCLI Journal 20, 522. (p. 15).

Costa Dias, M., Joyce, R., Postel-Vinay, F., Xu, X., 2020. The challenges for labour market policy during the Covid-19 pandemic*. Fiscal Studies 41, 371–382. https://doi.org/10.1111/1475-5890.12233. https://onlinelibrary.wiley.com/doi/abs/10.1111/1475-5890.12233. arXiv:https://onlinelibrary.wiley.com/doi/pdf/10.1111. (p. 15).

Cowling, B.J., Ali, S.T., Ng, T.W., Tsang, T.K., Li, J.C., Fong, M.W., Liao, Q., Kwan, M.Y., Lee, S.L., Chiu, S.S., et al., 2020. Impact assessment of non-pharmaceutical interventions against coronavirus disease 2019 and influenza in Hong Kong: an observational study. Lancet Public Health 5, e279–e288. (p. 121).

Coyne, C.J., Duncan, T.K., Hall, A.R., 2021. The political economy of state responses to infectious disease. Southern Economic Journal 87, 1119–1137. (p. 193).

Craig, K.J.T., Rizvi, R., Willis, V.C., Kassler, W.J., Jackson, G.P., 2021. Effectiveness of contact tracing for viral disease mitigation and suppression: evidence-based review. JMIR Public Health and Surveillance 7, e32468. (p. 123).

Cunha, F., Heckman, J., 2007. The technology of skill formation. American Economic Review 97, 31–47. https://doi.org/10.1257/aer.97.2.31. https://www.aeaweb.org/articles?id=10.1257/aer.97.2.31. (pp. 276 and 396).

Cutler, D.M., 2022. The costs of long COVID. JAMA Health Forum 3, e221809. https://doi.org/10.1001/jamahealthforum.2022.1809. arXiv:https://jamanetwork.com/journals/jama-health-forum/articlepdf/2792505/cutler_2022_jf_220018_1652290314.87944.pdf. (p. 355).

Czeisler, M.É., Marynak, K., Clarke, K.E., Salah, Z., Shakya, I., Thierry, J.M., Ali, N., McMillan, H., Wiley, J.F., Weaver, M.D., et al., 2020. Delay or avoidance of medical care because of Covid-19–related concerns—United States, June 2020. Morbidity and Mortality Weekly Report 69, 1250. (p. 306).

Davis, H.E., McCorkell, L., Vogel, J.M., Topol, E.J., 2023. Long Covid: major findings, mechanisms and recommendations. Nature Reviews. Microbiology, 1–14. (p. 352).

Dawel, A., Shou, Y., Smithson, M., Cherbuin, N., Banfield, M., Calear, A.L., Farrer, L.M., Gray, D., Gulliver, A., Housen, T., et al., 2020. The effect of Covid-19 on mental health and wellbeing in a representative sample of Australian adults. Frontiers in Psychiatry 11, 579985. (pp. 174, 181, and 301).

De Figueiredo, A., Simas, C., Karafillakis, E., Paterson, P., Larson, H.J., 2020. Mapping global trends in vaccine confidence and investigating barriers to vaccine uptake: a large-scale retrospective temporal modelling study. The Lancet 396, 898–908. (p. 141).

De Filippo, O., D'Ascenzo, F., Angelini, F., Bocchino, P.P., Conrotto, F., Saglietto, A., Secco, G.G., Campo, G., Gallone, G., Verardi, R., Gaido, L., Iannaccone, M., Galvani, M., Ugo, F., Barbero, U., Infantino, V., Olivotti, L., Mennuni, M., Gili, S., Infusino, F., Vercellino, M., Zucchetti, O., Casella, G., Giammaria, M., Boccuzzi, G., Tolomeo, P., Doronzo, B., Senatore, G., Grosso Marra, W., Rognoni, A., Trabattoni, D., Franchin, L., Borin, A., Bruno, F., Galluzzo, A., Gambino, A., Nicolino, A., Truffa Giachet, A., Sardella, G., Fedele, F., Monticone, S., Montefusco, A., Omed, P., Pennone, M., Patti, G., Mancone, M., De Ferrari, G.M., 2020. Reduced rate of hospital admissions for ACS during Covid-19 outbreak in northern Italy. The New England Journal of Medicine 383, 88–89. https://doi.org/10.1056/NEJMc2009166. (p. 197).

de Garine-Wichatitsky, M., Binot, A., Ward, J., Caron, A., Perrotton, A., Ross, H., Tran Quoc, H., Valls-Fox, H., Gordon, I.J., Promburom, P., et al., 2021. 'Health in' and 'health of' social-ecological systems: a practical framework for the management of healthy and resilient agricultural and natural ecosystems. Frontiers in Public Health 8, 616328. (p. 390).

De Rosa, S., Spaccarotella, C., Basso, C., Calabro, M., Curcio, A., Filardi, P., Mancone, M., Mercuro, G., Muscoli, S., Nodari, S., et al., 2020. The ccuaig reduction of hospitalizations for myocardial infarction in Italy in the Covid-19 era. European Heart Journal 41, 2083–2088. (p. 315).

Deb, P., Furceri, D., Jimenez, D., Kothari, S., Ostry, J.D., Tawk, N., 2022. The effects of Covid-19 vaccines on economic activity. Swiss Journal of Economics and Statistics 158, 1–25. (pp. 140, 141, 148, and 149).

Deb, P., Furceri, D., Ostry, J.D., Tawk, N., 2021. The economic effects of Covid-19 containment measures. Open Economies Review, 1–32. (pp. 125 and 139).

Del Boca, D., Oggero, N., Profeta, P., Rossi, M., 2020. Women's and men's work, housework and childcare, before and during Covid-19. Review of Economics of the Household 18, 1001–1017. https://doi.org/10.1007/s11150-020-09502-1. (p. 277).

Delmastro, M., Zamariola, G., 2020. Depressive symptoms in response to Covid-19 and lockdown: a cross-sectional study on the Italian population. Scientific Reports 10, 1–10. (p. 301).

Deruelle, T., 2022. Covid-19 as a catalyst for a European health union: recent developments in health threats management. In: Vanhercke, B., Spasova, S. (Eds.), Social Policy in the European Union: State of Play 2021. Re-Emerging Social Ambitions as the EU Recovers from the Pandemic. Twenty Second Annual Report. European Trade Union Institute (ETUI), pp. 127–144. Chapter 6. (p. 239).

Deshpande, A., Velappan, N., Davis-Anderson, K., 2022. Warning signs of potential black swan outbreaks in infectious disease. Frontiers in Microbiology 150. (p. 365).

Dewatripont, M., Goldman, M., Muraille, E., Platteau, J., 2020. Rapidly identifying workers who are immune to Covid-19 and virus-free is a priority for restarting the economy. Voxeu.org. Available from: https://voxeu.org/article/rapidly-identifying-workers-who-are-immune-covid-19-and-virus-free-priority-restarting-economy. (pp. 141 and 433).

Di Pietro, M., Marattin, L., Minetti, R., 2020. Fiscal policies amid a pandemic: the response of Italy to the Covid-19 crisis. National Tax Journal 73, 927–950. (p. 221).

Diebold, F.X., et al., 2022. Real-Time Real Economic Activity: Entering and Exiting the Pandemic Recession of 2020. Technical Report. Penn Institute for Economic Research, Department of Economics, University of . . . . (p. 202).

Dimaschko, J., Shlyakhover, V., Iabluchanskyi, M., 2022. Strong correlations in a biological system as a cause of sustained epidemic waves: implications for understanding the Covid-19 pandemic. Available on: https://www.researchsquare.com/article/rs-456723/v1.pdf. (p. 14).

Dimitrijevic, A., Williams, G., Maguire, J., 2020. Global credit conditions: the k-shaped recovery. Standard & Poor Global Rating. Available from: https://www.spglobal.com/_assets/documents/ratings/research/100046623.pdf. (p. 255).

Dobson, A., Pimm, S., Hannah, L., Kaufman, L., Ahumada, J., Ando, A., Bernstein, A., Busch, J., Daszak, P., Engelmann, J., Kinnaird, M., Li, B., Loch-Temzelides, T., Lovejoy, T., Nowak, K., Roehrdanz, P., Vale, M., 2020. Ecology and economics for pandemic prevention. Science 369, 379–381. https://science.sciencemag.org/content/369/6502/379. arXiv:https://science.sciencemag.org/content/369/6502/379.full.pdf. (pp. 367, 368, and 371).

Doganoglu, T., Ozdenoren, E., 2020. Should I stay or should I go (out): the role of trust and norms in disease prevention during pandemics. Covid Economics 16, 135–160. https://www.wiwi.uni-wuerzburg.de/fileadmin/12010300/2020/covid_05052020.pdf. (pp. 361 and 362).

Domash, A., Summers, L.H., 2022. How tight are US labor markets? Technical Report. National Bureau of Economic Research. (p. 268).

Domingo, F.R., Waddell, L.A., Cheung, A.M., Cooper, C.L., Belcourt, V.J., Zuckermann, A.M., Corrin, T., Ahmad, R., Boland, L., Laprise, C., et al., 2021. Prevalence of long-term effects in individuals diagnosed with Covid-19: an updated living systematic review. MedRxiv. Available from: https://www.medrxiv.org/content/10.1101/2021.06.03.21258317v2. (p. 354).

Dunn, R.R., Davies, T.J., Harris, N.C., Gavin, M.C., 2010. Global drivers of human pathogen richness and prevalence. Proceedings of the Royal Society B: Biological Sciences 277, 2587–2595. (p. 383).

Durand, G., 1993. The implication of the imaginary and societies. Current Sociology 41, 17–32. (p. 345).

Durante, R., Guiso, L., Gulino, G., 2020. Civic capital and social distancing: evidence from Italians? Response to COVID-19. VOX CEPR Policy Portal. (pp. 361 and 362).

Ebrahim, S., Memish, Z., 2020. Covid-19? The role of mass gatherings. Travel Medicine and Infectious Disease 34, 101617. (p. 106).

ECB, 2020. Economic and monetary developments. ECB Economic Bulletin. Available from: https://www.ecb.europa.eu/pub/economic-bulletin/focus/2020/html/ecb.ebbox202006_05~d36f12a192.en.html. (pp. 263 and 264).

Eck, K., Hatz, S., 2020. State surveillance and the Covid-19 crisis. Journal of Human Rights 19, 603–612. https://doi.org/10.1080/14754835.2020.1816163. arXiv:https://doi.org/10.1080/14754835.2020.1816163. (p. 117).

EClinicalMedicine, 2020. Emerging zoonoses: a one health challenge. EClinicalMedicine 19, 100300. https://doi.org/10.1016/j.eclinm.2020.100300. http://www.sciencedirect.com/science/article/pii/S2589537020300444. (pp. 35 and 425).

Eichenbaum, M.S., Rebelo, S., Trabandt, M., 2021. The macroeconomics of epidemics. The Review of Financial Studies 34, 5149–5187. (p. 127).

El Jaouhari, M., Edjoc, R., Waddell, L., Houston, P., Atchessi, N., Striha, M., Bonti-Ankomah, S., 2021. Impact of school closures and re-openings on Covid-19 transmission. Canada Communicable Disease Report 47, 515–523. (p. 124).

El-Shabasy, R.M., Nayel, M.A., Taher, M.M., Abdelmonem, R., Shoueir, K.R., et al., 2022. Three wave changes, new variant strains, and vaccination effect against Covid-19 pandemic. International Journal of Biological Macromolecules. (p. 202).

Emanuel, E.J., Persad, G., Upshur, R., Thome, B., Parker, M., Glickman, A., Zhang, C., Boyle, C., Smith, M., Phillips, J.P., 2020. Fair allocation of scarce medical resources in the time of Covid-19. The New England Journal of Medicine 382, 2049–2055. https://doi.org/10.1056/NEJMsb2005114. arXiv:https://doi.org/10.1056/NEJMsb2005114. (p. 340).

Emmerling, J., Furceri, D., Monteiro, F.L., Loungani, M.P., Ostry, M.J.D., Pizzuto, P., Tavoni, M., 2021. Will the Economic Impact of COVID-19 Persist? Prognosis from 21st Century Pandemics. International Monetary Fund. (p. 267).

Endo, A., Centre for the Mathematical Modelling of Infectious Diseases COVID-19 Working Group, Abbott, S., et al., 2020. Estimating the overdispersion in Covid-19 transmission using outbreak sizes outside China [version 3; peer review: 2 approved]. Wellcome Open Research 5. https://doi.org/10.12688/wellcomeopenres.15842.3. (p. 463).

Eppinger, P., Felbermayr, G.J., Krebs, O., Kukharskyy, B., 2020. Covid-19 shocking global value chains. (p. 250).

Epstein, S., Lister, S., Belasco, A., Jansen, D., 2019. Fy2015 funding to counter Ebola and the Islamic State (IS). (p. 47).

Erdős, P., Rényi, A., et al., 1960. On the evolution of random graphs. Publications of the Mathematical Institute of the Hungarian Academy of Sciences 5, 17–60. (p. 467).

Ettman, C.K., Abdalla, S.M., Cohen, G.H., Sampson, L., Vivier, P.M., Galea, S., 2020. Prevalence of depression symptoms in US adults before and during the Covid-19 pandemic. JAMA Network Open 3, e2019686. (pp. 174 and 301).

European Center for Disease Prevention and Control, 2019. Health emergency preparedness for imported cases of high-consequence infectious diseases. ECDC Report. Available from: https://www.ecdc.europa.eu/en/publications-data/health-emergency-preparedness-imported-cases-high-consequence-infectious-diseases. (pp. 83 and 105).

European Center for Disease Prevention and Control, 2020a. Contact tracing: public health management of persons, including healthcare workers, having had contact with Covid-19 cases in the European Union. WHO Report, Geneva. Available from: https://www.ecdc.europa.eu/en/covid-19-contact-tracing-public-health-management. (pp. 104 and 109).

European Center for Disease Prevention and Control, 2020b. Guidance for discharge and ending isolation in the context of widespread community transmission of Covid-19. ECDC Report, Stockholm. Available from: https://www.ecdc.europa.eu/en/publications-data/covid-19-guidance-discharge-and-ending-isolation. (p. 105).

European Center for Disease Prevention and Control, 2020c. Guidelines for the implementation of non-pharmaceutical interventions against Covid-19. ECDC Report. Available from: https://www.ecdc.europa.eu/sites/default/files/documents/covid-19-guidelines-non-pharmaceutical-interventions-september-2020.pdf. (p. 103).

European Center for Disease Prevention and Control, 2020d. Resource estimation for contact tracing, quarantine and monitoring activities for Covid-19 cases in the EU/EEA. ECDC Report, Stockholm. Available from: https://www.ecdc.europa.eu/en/publications-data/resource-estimation-contacttracing-quarantine-and-monitoring-activities-covid-19. (p. 106).

European Center for Disease Prevention and Control, 2020e. Use of gloves in healthcare and non-healthcare settings in the context of the Covid 19 pandemic. ECDC Report. Available from: https://www.ecdc.europa.eu/en/publications-data/gloves-healthcare-and-non-healthcare-settings-covid-19. (pp. 104 and 105).

European Centre for Disease Prevention and Control (ECDC), 2021. COVID-19 in children and the role of school settings in transmission-second update. Technical Report. ECDC Stockholm, Sweden. Available from: https://www.ecdc.europa.eu/en/publications-data/children-and-school-settings-covid-19-transmission. Stockholm. (p. 279).

Fall, F., Fournier, J.M., 2015. Macroeconomic uncertainties, prudent debt targets and fiscal rules. Technical Report. OECD. (p. 266).

Fan, V., Jamison, D., Summers, L., 2016. The inclusive cost of pandemic influenza risk. NBER Working Papers 22137. (p. 47).

Fan, V.Y., Jamison, D.T., Summers, L.H., 2018. Pandemic risk: how large are the expected losses? Bulletin of the World Health Organization 96, 129. (p. 45).

Fana, M., Tolan, S., Torrejón, S., Brancati, C.U., Fernández-Macías, E., 2020. The COVID confinement measures and EU labour markets. Publications Office of the European Union Luxembourg. (pp. 206 and 207).

Fancourt, D., Steptoe, A., Bu, F., 2021. Trajectories of anxiety and depressive symptoms during enforced isolation due to Covid-19 in England: a longitudinal observational study. Lancet Psychiatry 8, 141–149. (p. 301).

Farboodi, M., Jarosch, G., Shimer, R., 2021. Internal and external effects of social distancing in a pandemic. Journal of Economic Theory 196, 105293. (pp. 127 and 464).

Farnsworth, K., Adenuga, A., De Groot, R., 2015. The complexity of biodiversity: a biological perspective on economic valuation. Ecological Economics 120, 350–354. (p. 391).

Farrar, J., Ahuja, A., 2021. Spike: The Virus Vs. the People-the Inside Story. Profile Books. (pp. 67 and 68).

Fatás, A., Summers, L.H., 2016. Hysteresis and fiscal policy during the global crisis. VoxEU 12, 2016. (p. 266).

Faust, C.L., Dobson, A.P., Gottdenker, N., Bloomfield, L.S., McCallum, H.I., Gillespie, T.R., Diuk-Wasser, M., Plowright, R.K., 2017. Null expectations for disease dynamics in shrinking habitat: dilution or amplification? Philosophical Transactions of the Royal Society B: Biological Sciences 372, 20160173. (p. 383).

Federation of European Heating, Ventilation and Air Conditioning Associations (REHVA), 2020. Rehva Covid-19 guidance document. Brussels. Available from: https://www.rehva.eu/fileadmin/user_upload/REHVA_COVID-19_guidance_document_ver2_20200403_1.pdf. (p. 104).

Fenichel, E.P., Horan, R.D., Hickling, G.J., 2010. Management of infectious wildlife diseases: bridging conventional and bioeconomic approaches. Ecological Applications 20, 903–914. (p. 347).

Fenichel, E.P., Kuminoff, N.V., Chowell, G., 2013. Skip the trip: air travelers' behavioral responses to pandemic influenza. PLoS ONE 8, e58249. (pp. 346 and 347).

Ferguson, N., Laydon, D., Nedjati-Gilani, G., Imai, N., Ainslie, K., Baguelin, M., Bhatia, S., Boonyasiri, A., Cucunubá, Z., Cuomo-Dannenburg, G., Dighe, A., Dorigatti, I., Fu, H., Gaythorpe, K., Green, W., Hamlet, A., 2020. Impact of non-pharmaceutical interventions (NPIS) to reduce Covid-19 mortality and healthcare demand. Imperial College COVID-19 Response Team. Available from: https://www.imperial.ac.uk/media/imperial-college/medicine/sph/ide/gida-fellowships/Imperial-College-COVID19-NPI-modelling-16-03-2020.pdf. (pp. 160 and 433).

Ferretti, L., Wymant, C., Kendall, M., Zhao, L., Nurtay, A., Abeler-Dörner, L., Parker, M., Bonsall, D., Fraser, C., 2020. Quantifying Sars-Cov-2 transmission suggests epidemic control with digital contact tracing. Science 368. https://doi.org/10.1126/science.abb6936. https://science.sciencemag.org/content/368/6491/eabb6936. arXiv:https://science.sciencemag.org/content/368/6491/eabb6936.full.pdf. (pp. 109 and 111).

Fine, P., Eames, K., Heymann, D.L., 2011. 'Herd immunity': a rough guide. Clinical Infectious Diseases 52, 911–916. (p. 162).

Fiorillo, A., Sampogna, G., Giallonardo, V., Del Vecchio, V., Luciano, M., Albert, U., Carmassi, C., Carrà, G., Cirulli, F., Dell'Osso, B., et al., 2020. Effects of the lockdown on the mental health of the general population during the Covid-19 pandemic in Italy: results from the comet collaborative network. European Psychiatry 63. (p. 301).

Fisher, B., Wit, L.A.d., Ricketts, T.H., 2021. Integrating economics into research on natural capital and human health. Review of Environmental Economics and Policy 15, 95–114. (p. 382).

Fisher, K.A., Tenforde, M.W., Feldstein, L.R., Lindsell, C.J., Shapiro, N.I., Files, D.C., Gibbs, K.W., Erickson, H.L., Prekker, M.E., Steingrub, J.S., et al., 2020. Community and close contact exposures associated with Covid-19 among symptomatic adults ≤ 18 years in 11 outpatient health care facilities—United States, July 2020. Morbidity and Mortality Weekly Report 69, 1258. (p. 124).

Fisman, D.N., Amoako, A., Tuite, A.R., 2022. Impact of population mixing between vaccinated and unvaccinated subpopulations on infectious disease dynamics: implications for Sars-Cov-2 transmission. CMAJ 194, E573–E580. (p. 347).

Fitzpatrick, T., Wilton, A., Cohen, E., Rosella, L., Guttmann, A., 2022. School reopening and Covid-19 in the community: evidence from a natural experiment in Ontario, Canada: study examines school reopening and Covid-19 in Ontario, Canada. Health Affairs 41, 864–872. (p. 280).

Fleck, L., 2012. Genesis and Development of a Scientific Fact. University of Chicago Press. (p. 348).

Flyvbjerg, B., 1998. Habermas and Foucault: thinkers for civil society? British Journal of Sociology, 210–233. (pp. 443 and 444).

Flyvbjerg, B., 2020. The law of regression to the tail: how to survive Covid-19, the climate crisis, and other disasters. Environmental Science & Policy 114, 614–618. (p. 450).

Fraser, C., Donnelly, C.A., Cauchemez, S., Hanage, W.P., Van Kerkhove, M.D., Hollingsworth, T.D., Griffin, J., Baggaley, R.F., Jenkins, H.E., Lyons, E.J., Jombart, T., Hinsley, W.R., Grassly, N.C., Balloux, F., Ghani, A.C., Ferguson, N.M., Rambaut, A., Pybus, O.G., Lopez-Gatell, H., Alpuche-Aranda, C.M., Chapela, I.B., Zavala, E.P., Guevara, D.M.E., Checchi, F., Garcia, E., Hugonnet, S., Roth, C., et al., 2009. Pandemic potential of a strain of influenza a (h1n1): early findings. Science 324, 1557–1561. https://doi.org/10.1126/science.1176062. https://science.sciencemag.org/content/324/5934/1557. arXiv:https://science.sciencemag.org/content/324/5934/1557.full.pdf. (p. 453).

Frenk, J., Bobadilla, J.L., Stern, C., Frejka, T., Lozano, R., 1991a. Elements for a theory of the health transition. Health Transition Review, 21–38. (p. 412).

Frenk, J., Frejka, T., Bobadilla, J.L., Stern, C., Lozano, R., Sepúlveda, J., José, M., 1991b. The epidemiologic transition in Latin America. Boletin de la Oficina Sanitaria Panamericana. Pan American Sanitary Bureau 111, 485–496. (p. 410).

Fuchs-Schündeln, N., Krueger, D., Ludwig, A., Popova, I., 2020. The long-term distributional and welfare effects of Covid-19 school closures. NBER Working Papers 27773. https://doi.org/10.3386/w27773. http://www.nber.org/papers/w27773. (pp. 181 and 276).

Funk, S., Salathé, M., Jansen, V.A., 2010. Modelling the influence of human behaviour on the spread of infectious diseases: a review. Journal of the Royal Society Interface 7, 1247–1256. (p. 464).

Gamhewage, G.M., 2014. Complex, confused, and challenging: communicating risk in the modern world. Journal of Communication in Healthcare 7, 252–254. (p. 87).

Gandhi, S., Huseby, T., Sonthalia, B., 2020. Managing disruption now for resilient supply chains in the future. AT Kerney. Available from: https://www.de.kearney.com/operations-performance-transformation/article/?/a/managing-disruption-now-for-resilient-supply-chains-in-the-future. (p. 254).

Gardner, B., Lally, P., Wardle, J., 2012. Making health habitual: the psychology of 'habit-formation' and general practice. British Journal of General Practice 62, 664–666. https://doi.org/10.3399/bjgp12X659466. (p. 262).

Garfin, D., Silver, R., Holman, E., 2020. The novel coronavirus (COVID-2019) outbreak: amplification of public health consequences by media exposure. Health Psychology 5, 355–357. (pp. 196 and 335).

Garrett, L., 1995. The Coming Plague: Newly Emerging Diseases in a World Out of Balance. Penguin Books, London. (p. 370).

Gates, B., 2020. The first modern pandemic. GatesNotes. Available from: https://www.gatesnotes.com/Health/Pandemic-Innovation. (p. 87).

Gebreyes, W., Dupouy-Camet, J., Newport, M., Oliveira, C., Schlesinger, L., Saif, Y., et al., 2014. The global one health paradigm: challenges and opportunities for tackling infectious diseases at the human, animal, and environment interface in low-resource settings. PLoS Neglected Tropical Diseases 8, e3257. (p. 47).

Gechert, S., 2015. What fiscal policy is most effective? A meta-regression analysis. Oxford Economic Papers 67, 553–580. (p. 219).

Gelderblom, H.R., 1996. Structure and classification of viruses. In: Medical Microbiology, 4th edition. (p. 13).

Geloso, V., Hyde, K., Murtazashvili, I., 2021. Pandemics, economic freedom, and institutional trade-offs. European Journal of Law and Economics, 1–25. (p. 182).

Gersovitz, M., Hammer, J.S., 2003. Infectious diseases, public policy, and the marriage of economics and epidemiology. The World Bank Research Observer 18, 129–157. (p. 464).

Ghebreyesus, T., 2018. Forewords. (p. xxii).

Ghebreyesus, T., 2021. Statement on the second meeting of the International Health Regulations (2005) Emergency Committee regarding the outbreak of novel coronavirus (2019-nCoV). Technical Report. WHO. (pp. 15 and 59).

Giuliano, P., Spilimbergo, A., 2013. Growing up in a recession. The Review of Economic Studies 81, 787–817. https://doi.org/10.1093/restud/rdt040. arXiv:http://academic.oup.com/restud/article-pdf/81/2/787/18390775/rdt040.pdf. (p. 186).

Global Pandemic App Watch, 2020. Covid-19 exposure notification and contact tracing apps. Available from: https://craiedl.ca/gpaw/. (p. 111).

Global Preparedness Monitoring Board, 2019a. Pandemic preparedness financing. WHO Report. Available from: https://apps.who.int/gpmb/assets/thematic_papers/tr-4.pdf. (p. 46).

Global Preparedness Monitoring Board, 2019b. A world at risk: annual report on global preparedness for health emergencies. WHO Report, Geneva. Available from: https://apps.who.int/gpmb/assets/annual_report/GPMB_annualreport_2019.pdf. (p. 45).

Global Preparedness Monitoring Board, 2019c. A World at Risk: Annual Report on Global Preparedness for Health Emergencies. Technical Report. World Health Organization and the World Bank, Geneva. (pp. 59 and 60).

Goeschl, T., Swanson, T., 2007. Designing the legacy library of genetic resources: approaches, methods and results. In: Biodiversity Economics. Cambridge University Press, Cambridge, pp. 273–292. (p. 391).

Gohdes, A., 2020. Repression technology: Internet accessibility and state violence. American Journal of Political Science 64, 488–503. (p. 117).

Gonçalves, B., 2020. Epidemic modeling 102: all Covid-19 models are wrong, but some are useful. Medium article. Available from: https://medium.com/data-for-science/epidemic-modeling-102-all-covid-19-models-are-wrong-but-some-are-useful-c81202cc6ee9. (p. 452).

Gonzalez, D., Karpman, M., Kenney, G.M., Zuckerman, S., 2021. Delayed and forgone health care for nonelderly adults during the Covid-19 pandemic. Findings from the September 11-28 Coronavirus Tracking Survey. Available from: https://www.urban.org/sites/default/files/publication/103651/delayed-and-forgone-health-care-for-nonelderly-adults-during-the-covid-19-pandemic_1.pdf. (p. 309).

Gopinath, G., 2020. The great lockdown: worst economic downturn since the great depression. IMF Blog 14. (p. 134).

Gourinchas, P., 2020. Flattening the pandemic and recession curves. In: Baldwin, R., di Mauro, B.W. (Eds.), Mitigating the COVID Economic Crisis: Act Fast and do Whatever It Takes. CEPR Press, pp. 31–40. Chapter 2. https://voxeu.org/content/mitigating-covid-economic-crisis-act-fast-and-do-whatever-it-takes. (p. 197).

Greenhalgh, T., Knight, M., A'Court, C., Buxton, M., Husain, L., 2020. Management of post-acute Covid-19 in primary care. BMJ 370. https://www.bmj.com/content/370/bmj.m3026. arXiv:https://www.bmj.com/content/370/bmj.m3026.full.pdf. (p. 354).

Greer, S.L., Rozenblum, S., Fahy, N., Brooks, E., Jarman, H., de Ruijter, A., Palm, W., Wismar, M., et al., 2022. Everything You Always Wanted to Know About European Union Health Policies but Were Afraid to Ask. World Health Organization. Regional Office for Europe. (p. 239).

Grifoni, A., Weiskopf, D., Ramirez, S., Mateus, J., Dan, J., Rydyznski, C., Rawlings, S., Sutherland, A., Premkumar, L., Jadi, R., Marrama, D., de Silva, A., Frazier, A., Carlin, A., Greenbaum, J., Peters, B., Krammer, F., Smith, D., Crotty, S., Sette, A., 2020. Targets of t cell responses to Sars-Cov-2 coronavirus in humans with Covid-19 disease and unexposed individuals. Cell 181, 1489–1501.e15. https://doi.org/10.1016/j.cell.2020.05.015. http://www.sciencedirect.com/science/article/pii/S0092867420306103. (p. 377).

Groff, D., Sun, A., Ssentongo, A.E., Ba, D.M., Parsons, N., Poudel, G.R., Lekoubou, A., Oh, J.S., Ericson, J.E., Ssentongo, P., et al., 2021. Short-term and long-term rates of postacute sequelae of Sars-Cov-2 infection: a systematic review. JAMA Network Open 4, e2128568. (p. 355).

Grossi, G., Pezone, G., Triassi, M., 2020. Salto di specie dei virus e l'approccio one health. Quotidiano Sanità. Available from: https://www.quotidianosanita.it/studi-e-analisi/articolo.php?articolo_id=84204. (pp. 8, 9, 12, 16, and 37).

Gualano, M.R., Moro, G.L., Voglino, G., Bert, F., Siliquini, R., 2021. Monitoring the impact of Covid-19 pandemic on mental health: a public health challenge? Reflection on Italian data. Social Psychiatry and Psychiatric Epidemiology 56, 165–167. (p. 301).

Guerrera, G., Picozza, M., D'Orso, S., Placido, R., Pironello, M., Verdiani, A., Termine, A., Fabrizio, C., Giannessi, F., Sambucci, M., et al., 2021. Bnt162b2 vaccination induces durable Sars-Cov-2–specific t cells with a stem cell memory phenotype. Science Immunology 6, eabl5344. (p. 379).

Guerrieri, V., Lorenzoni, G., Straub, L., Werning, I., 2022. Macroeconomic implications of Covid-19: can negative supply shocks cause demand shortages? American Economic Review 112, 1437–1474. (pp. 205 and 215).

Guiso, L., 2020. Sfiducia, protezione e populismo, le eredità difficili della crisi. Il Sole24ore. Available from: https://www.ilsole24ore.com/art/sfiducia-protezione-e-populismo-eredita-difficili-crisi-AEq33VkF. (p. 186).

Gunitsky, S., 2015. Corrupting the cyber-commons: social media as a tool of autocratic stability. Perspectives on Politics 13, 42–54. (p. 117).

Gupta, A., Madhavan, M., Sehgal, K., Nair, N., Mahajan, S., Sehrawat, T., Bikdeli, B., Ahluwalia, N., Ausiello, J., Wan, E., Freedberg, D., Kirtane, A., Parikh, S., Maurer, M., Nordvig, A., Accili, D., Bathon, J., Mohan, S., Bauer, K., Leon, M., Krumholz, H., Uriel, N., Mehra, M., Elkind, M.V., Stone, G., Schwartz, A., Ho, D., Bilezikian, J., Landry, D., 2020. Extrapulmonary manifestations of Covid-19. Nature Medicine 26, 1017–1032. https://doi.org/10.1038/s41591-020-0968-3. (p. 351).

Hafner-Fink, M., Uhan, S., 2021. Life and attitudes of Slovenians during the Covid-19 pandemic: the problem of trust. International Journal of Sociology 51, 76–85. (p. 88).

Haider, M., 2005. Global public health communication: challenges, perspectives, and strategies. Jones & Bartlett Learning. (pp. 87 and 90).

Haldane, V., De Foo, C., Abdalla, S.M., Jung, A.S., Tan, M., Wu, S., Chua, A., Verma, M., Shrestha, P., Singh, S., et al., 2021. Health systems resilience in managing the Covid-19 pandemic: lessons from 28 countries. Nature Medicine 27, 964–980. (p. 59).

Hamner, L., Dubbel, P., Capron, I., Ross, A., Jordan, A., Lee, J., et al., 2020. High Sars-Cov-2 attack rate following exposure at a choir practice - Skagit County, Washington, March 2020. Morbidity and Mortality Weekly Report 69, 606–610. (p. 104).

Han, Q., Zheng, B., Daines, L., Sheikh, A., 2022. Long-term sequelae of Covid-19: a systematic review and meta-analysis of one-year follow-up studies on post-Covid symptoms. Pathogens 11, 269. (p. 354).

Hanna, T.P., King, W.D., Thibodeau, S., Jalink, M., Paulin, G.A., Harvey-Jones, E., O'Sullivan, D.E., Booth, C.M., Sullivan, R., Aggarwal, A., 2020. Mortality due to cancer treatment delay: systematic review and meta-analysis. BMJ 371. (p. 310).

Hansen, L.P., 2014. Nobel lecture: uncertainty outside and inside economic models. Journal of Political Economy 122, 945–987. https://doi.org/10.1086/678456. arXiv:https://doi.org/10.1086/678456. (p. 340).

Harris, J.E., 2022. The repeated setbacks of HIV vaccine development laid the groundwork for Sars-Cov-2 vaccines. Health Policy and Technology, 100619. (pp. 130 and 134).

Harvey, F., Ammar, W., Endo, H., Gupta, G., Konyndyk, J., Matsoso, P., Tam, T., 2020. Interim report on WHO's response to Covid-19 January–April 2020. Independent Oversight and Advisory Committee (IOAC) for the WHO Health Emergencies Programme. Available from: https://www.who.int/publications/m/item/interim-report-on-who-s-response-to-covid--january--april-2020. (p. 60).

Hastie, C.E., Lowe, D.J., McAuley, A., Winter, A.J., Mills, N.L., Black, C., Scott, J.T., O'Donnell, C.A., Blane, D.N., Browne, S., Ibbotson, T.R., Pell, J.P., 2022. Outcomes among confirmed cases and a matched comparison group in the long-Covid in Scotland study. Nature Communications 13, 5663. https://doi.org/10.1038/s41467-022-33415-5. (p. 354).

Haug, N., Geyrhofer, L., Londei, A., Dervic, E., Desvars-Larrive, A., Loreto, V., Pinior, B., Thurner, S., Klimek, P., 2020. Ranking the effectiveness of worldwide Covid-19 government interventions. Nature Human Behaviour 4, 1303–1312. https://doi.org/10.1038/s41562-020-01009-0. (pp. 123 and 125).

Hausmann, R., Schetter, U., 2022. Horrible trade-offs in a pandemic: poverty, fiscal space, policy, and welfare. World Development 153, 105819. (p. 185).

Heckler, A.F., Mikula, B., Rosenblatt, R., 2013. Student accuracy in reading logarithmic plots: the problem and how to fix it. In: 2013 IEEE Frontiers in Education Conference (FIE), pp. 1066–1071. (p. 337).

Heckman, J.J., Pinto, R., 2022. The econometric model for causal policy analysis. Annual Review of Economics 14, 893–923. (p. 396).

Hedwall, M., 2020. The ongoing impact of Covid-19 on global supply chains. World Economic Forum. Available from: https://www.weforum.org/agenda/2020/06/ongoing-impact-covid-19-global-supply-chains/. (p. 255).

Heise, S., Karahan, F., Şahin, A., 2022. The missing inflation puzzle: the role of the wage-price pass-through. Journal of Money, Credit, and Banking 54, 7–51. (p. 258).

Herby, J., Jonung, L., Hanke, S., et al., 2022. A literature review and meta-analysis of the effects of lockdowns on Covid-19 mortality. Studies in Applied Economics. (p. 124).

Herrmann, B., Thöni, C., Gächter, S., 2008. Antisocial punishment across societies. Science 319, 1362–1367. https://doi.org/10.1126/science.1153808. https://science.sciencemag.org/content/319/5868/1362. arXiv:https://science.sciencemag.org/content/319/5868/1362.full.pdf. (p. 362).

Highlevel Panel on the Global Response to Health Crises, 2019. Protecting humanity from future health crises: Report of the High-level Panel on the Global Response to Health Crises. Technical Report. WHO. (p. 59).

Hill, R., Lakner, C., Mahler, D.G., Narayan, A., Yonzan, N., 2021. Poverty, median incomes, and inequality in 2021: a diverging recovery. Available at: https://documents1.worldbank.org/curated/en/936001635880885713/pdf/Poverty-Median-Incomes-and-Inequality-in-2021-A-Diverging-Recovery.pdf. (p. 208).

Hoffman, S., Silverberg, S., 2018. Delays in global disease outbreak responses: lessons from h1n1, Ebola, and Zika. American Journal of Public Health, 329–333. https://doi.org/10.2105/AJPH.2017.304245. (p. 340).

Holling, C.S., 1973. Resilience and stability of ecological systems. Annual Review of Ecology and Systematics, 1–23. (p. 390).

Holmdahl, I., Buckee, C., 2020. Wrong but useful: what Covid-19 epidemiologic models can and cannot tell us. The New England Journal of Medicine 383, 303–305. https://doi.org/10.1056/NEJMp2016822. arXiv:https://doi.org/10.1056/NEJMp2016822. (p. 453).

Hooghe, L., Lenz, T., Marks, G., 2019. A Theory of International Organization. Oxford University Press. (p. 416).

Hossain, A.D., Jarolimova, J., Elnaiem, A., Huang, C.X., Richterman, A., Ivers, L.C., 2022. Effectiveness of contact tracing in the control of infectious diseases: a systematic review. Lancet Public Health. (p. 123).

House of Lords, 2003. Select committee on science and technology fourth report. Science and Technology Committee Publications. Available from: https://publications.parliament.uk/pa/ld200203/ldselect/ldsctech/138/13809.htm. (p. 357).

Howell O'Neill, P., Ryan-Mosley, T., Johnson, B., 2020. A flood of coronavirus apps are tracking US. Now it's time to keep track of them. MIT Technology Review. https://www.technologyreview.com/2020/05/07/1000961/launching-mittr-covid-tracing-tracker/. (pp. 118 and 119).

Huang, Y., Yang, C., Xu, X.f., Xu, W., Liu, S.w., 2020. Structural and functional properties of Sars-Cov-2 spike protein: potential antivirus drug development for Covid-19. Acta Pharmacologica Sinica 41, 1141–1149. (p. 20).

Hubbell, S.P., 2011. The Unified Neutral Theory of Biodiversity and Biogeography (MPB-32). Princeton University Press. (p. 385).

Hupkau, C., Petrongolo, B., 2020. Work, care and gender during the Covid-19 crisis. Fiscal Studies 41, 623–651. (p. 277).

IMF, 2019. World economic outlook: global manufacturing downturn, rising trade barriers. IMF WEO. IMF Publishing, Washington, DC. Available from: https://www.imf.org/-/media/Files/Publications/WEO/2019/October/English/text.ashx. (p. 240).

IMF, 2022. World economic outlook: war sets back the global recovery. Available at: https://www.imf.org/en/Publications/WEO/Issues/2022/04/19/world-economic-outlook-april-2022. (pp. 265, 273, and 372).

Institute for Health Metrics and Evaluation (IHME), 2021. Financing global health 2020: the impact of Covid-19. (p. 373).

International Labour Office, 2021. World Employment and Social Outlook: Trends 2021. International Labour Office. (p. 267).

International Monetary Fund (IMF), 2022. Fiscal monitor: fiscal policy from pandemic to war. (pp. 184 and 265).

International Monetary Fund, 2021. Regional Economic Outlook, October 2021, Asia and Pacific. International Monetary Fund. (p. 139).

Islamaj, E., Mattoo, A., 2021. Lives versus livelihoods during the Covid-19 pandemic: how testing softens the trade-off. Covid Economics 82. (pp. 164, 178, and 180).

ISTAT, 2020a. Il valore monetario dello stock di capitale umano. ISTAT, Roma. Available from: https://www.istat.it/it/files/2014/02/Il-valore-monetario-dello-stock-di-capitale-umano.pdf. (p. 198).

ISTAT, 2020b. Speciale emergenza covid-19. ISTAT Report. Available from: https://www.istat.it/it/violenza-sulle-donne/speciale-covid-19. (p. 280).

Iwasaki, A., 2020. We still know very little about the body's immune response to the Sars Cov-2. (p. 197).

Iwasaki, A., 2021. What reinfections mean for Covid-19. Lancet. Infectious Diseases 21, 3–5. (p. 166).

Jablonski, D., 2001. Lessons from the past: evolutionary impacts of mass extinctions. Proceedings of the National Academy of Sciences 98, 5393–5398. (p. 387).

Jabr, F., 2020. How humanity unleashed a flood of new diseases. The New York Times Magazine. Available from: https://www.nytimes.com/2020/06/17/magazine/animal-disease-covid.html. (pp. 40 and 42).

Jacob, S., Lawarée, J., 2021. The adoption of contact tracing applications of Covid-19 by European governments. Policy Design and Practice 4, 44–58. (pp. 108 and 111).

Jamison, J., Bundy, D., Jamison, D., Spitz, J., Verguet, S., 2020. Comparing the impact on COVID-19 mortality of self-imposed behavior change and of government regulations: an observational analysis of 13 countries. SSRN preprint. (p. 125).

Jenkins, B., Jones, B., 2020. Reopening the world: the WHO, international institutions, and the Covid-19 response. The Brookings Institution Report. Available from: https://www.brookings.edu/blog/order-from-chaos/2020/06/16/reopening-the-world-the-who-international-institutions-and-the-covid-19-response/. (p. 418).

Jensen, N., Barry, A., Kelly, A.H., 2023. More-than-national and less-than-global: the biochemical infrastructure of vaccine manufacturing. Economy and Society 52, 9–36. (p. 423).

Jewell, N., Lewnard, J., Jewell, B., 2020. Caution warranted: using the institute for health metrics and evaluation model for predicting the course of the Covid-19 pandemic. Annals of Internal Medicine 173, 226–227. https://doi.org/10.7326/M20-1565. (p. 454).

Jiang, R.D., Liu, M.Q., Chen, Y., Shan, C., Zhou, Y.W., Shen, X.R., Li, Q., Zhang, L., Zhu, Y., Si, H.R., et al., 2020. Pathogenesis of Sars-Cov-2 in transgenic mice expressing human angiotensin-converting enzyme 2. Cell 182, 50–58. (p. 20).

Jiang, Y., Rubin, L., Peng, T., Liu, L., Xing, X., Lazarovici, P., Zheng, W., 2022. Cytokine storm in Covid-19: from viral infection to immune responses, diagnosis and therapy. International Journal of Biological Sciences 18, 459. (p. 353).

John, A., Eyles, E., Webb, R.T., Okolie, C., Schmidt, L., Arensman, E., Hawton, K., O'Connor, R.C., Kapur, N., Moran, P., et al., 2020. The impact of the Covid-19 pandemic on self-harm and suicidal behaviour: update of living systematic review. F1000Research, 9. (p. 300).

Johnson, R.C., Noguera, G., 2012. Accounting for intermediates: production sharing and trade in value added. Journal of International Economics 86, 224–236. (p. 248).

Johnson, R.C., Noguera, G., 2017. A portrait of trade in value-added over four decades. Review of Economics and Statistics 99, 896–911. (p. 383).

Jonas, O., 2014. Global health threats of the 21st century. Finance and Development 51, 16–19. (p. 47).

Jones, K., Patel, N., Levy, M., Storeygard, A., Balk, D., Gittleman, J., Daszak, P., 2008. Global trends in emerging infectious diseases. Nature 451, 990–993. (p. 7).

Jones, N., Qureshi, Z., Temple, R., Larwood, J., Greenhalgh, T., Bourouiba, L., 2020. Two metres or one: what is the evidence for physical distancing in Covid-19? BMJ 370. https://doi.org/10.1136/bmj.m3223. (p. 103).

Julkunen, H., Cichońska, A., Tiainen, M., Koskela, H., Nybo, K., Mäkelä, V., Nokso-Koivisto, J., Kristiansson, K., Perola, M., Salomaa, V., et al., 2023. Atlas of plasma NMR biomarkers for health and disease in 118,461 individuals from the UK biobank. Nature Communications 14, 604. (p. 393).

Jung, C.G., 1991. The collective unconscious. Collected Works 7, 90–113. (p. 345).

Kabir, K.A., Tanimoto, J., 2020. Evolutionary game theory modelling to represent the behavioural dynamics of economic shutdowns and shield immunity in the Covid-19 pandemic. Royal Society Open Science 7, 201095. (p. 465).

Kahn, L., 2009. Who's in Charge?: Leadership During Epidemics, Bioterror Attacks and Other Public Health Crises. Praeger Security International, Santa Barbara, CA, USA. (p. 57).

Kates, R.W., Ausubel, J.H., Berberian, M., 1985. Climate Impact Assessment. John Wiley & Sons. (p. 390).

Keating, C., 2018. The genesis of the global burden of disease study. Lancet (London, England) 391, 2316–2317. (p. 29).

Keeling, M., Hollingsworth, T., Read, J., 2020. The efficacy of contact tracing for the containment of the 2019 novel coronavirus (Covid-19). medRxiv. Available from: https://www.medrxiv.org/content/early/2020/02/17/2020.02.14.20023036. (p. 109).

Kelly, T., Machalaba, C., Karesh, W., Crook, P., Gilardi, K., Nziza, J., Uhart, M., Robles, R., 2020. Implementing one health approaches to confront emerging and re-emerging zoonotic disease threats: lessons from predict. One Health Outlook 2, 1. https://doi.org/10.1186/s42522-019-0007-9. (p. 425).

Kendall, M., Milsom, L., Abeler-Dörner, L., Wymant, C., Ferretti, L., Briers, M., Holmes, C., Bonsall, D., Abeler, J., Fraser, C., 2020. Epidemiological changes on the isle of wight after the launch of the NHS test and trace programme: a preliminary analysis. The Lancet 2, E658–E666. https://doi.org/10.1016/S2589-7500(20)30241-7. https://www.thelancet.com/journals/landig/article/PIIS2589-7500(20)30241-7/fulltext. (p. 107).

Kendzerska, T., Zhu, D.T., Gershon, A.S., Edwards, J.D., Peixoto, C., Robillard, R., Kendall, C.E., 2021. The effects of the health system response to the Covid-19 pandemic on chronic disease management: a narrative review. Risk Management and Healthcare Policy 14, 575. (p. 316).

Keogh-Brown, M., Smith, R., Edmunds, J., Beutels, P., 2010. The macroeconomic impact of pandemic influenza: estimates from models of the United Kingdom, France, Belgium and the Netherlands. The European Journal of Health Economics 11, 543–554. (p. 126).

Kessler, D., 2020. Moving towards new major trade-offs post-covid 19. WanSquare. Available from: https://www.scor.com/en/media/news-press-releases/moving-towards-new-major-trade-offs-post-covid-19. (pp. 187 and 188).

Kilpatrick, A., Randolph, S., 2012. Drivers, dynamics, and control of emerging vector-borne zoonotic diseases. Lancet (London, England) 380, 1946–1955. https://pubmed.ncbi.nlm.nih.gov/23200503. (p. 366).

King, G., Pan, J., Roberts, M., 2017. How the Chinese government fabricates social media posts for strategic distraction, not engaged argument. American Political Science Review 111, 484–501. (p. 117).

Kleinman, R., Merkel, C., 2020. Digital contact tracing for Covid-19. CMAJ 192, E653–E656. (p. 111).

Kluge, H., 2020. Behavioural insights are valuable to inform the planning of appropriate pandemic response measures. WHO Report, Geneva. Available from: http://www.euro.who.int/en/media-centre/sections/statements/2020/statementbehavioural-insights-are-valuable-to-inform-the-planning-of-appropriate-pandemic-responsemeasures. (pp. 194 and xxxi).

Knibbs, L., Morawska, L., Bell, S., Grzybowski, P., 2011. Room ventilation and the risk of airborne infection transmission in 3 health care settings within a large teaching hospital. American Journal of Infection Control 39. (p. 104).

Knight, F.H., 1944. Diminishing returns from investment. Journal of Political Economy 52, 26–47. (p. 389).

Knudsen, O.K., Scandizzo, P.L., 2005. Bringing social standards into project evaluation under dynamic uncertainty. Risk Analysis 25, 457–466. (p. 413).

Kose, M.A., Nagle, P.S.O., Ohnsorge, F., Sugawara, N., 2021. What has been the impact of Covid-19 on debt? Turning a wave into a tsunami. (pp. 243 and 244).

Koyama, M., 2021. Epidemic disease and the state: is there a tradeoff between public health and liberty? Public Choice, 1–23. (p. 190).

Kramer, A., Kramer, K.Z., 2020. The potential impact of the Covid-19 pandemic on occupational status, work from home, and occupational mobility. (p. 269).

Kresge, N., 2020. Virus may spread twice as fast as earlier thought, study say. Bloomberg. Available from: https://www.bloomberg.com/news/articles/2020-04-08/virus-may-spread-twice-as-fast-as-earlier-thought-study-says. (p. 196).

Kruining, N.v., et al., 2021. The interplay of the sociotechnical imaginaries and policies of personalized healthcare in the Netherlands. (p. 349).

Kucharski, A.J., Klepac, P., Conlan, A.J., Kissler, S.M., Tang, M.L., Fry, H., Gog, J.R., Edmunds, W.J., Emery, J.C., Medley, G., et al., 2020. Effectiveness of isolation, testing, contact tracing, and physical distancing on reducing transmission of Sars-Cov-2 in different settings: a mathematical modelling study. Lancet. Infectious Diseases 20, 1151–1160. (p. 122).

Kudlay, D., Svistunov, A., 2022. Covid-19 vaccines: an overview of different platforms. Bioengineering 9, 72. (p. 134).

Kumar, V., Alshazly, H., Idris, S.A., Bourouis, S., 2021. Evaluating the impact of Covid-19 on society, environment, economy, and education. Sustainability 13, 13642. (p. 266).

Kunitz, S.J., 1990. Public policy and mortality among indigenous populations of northern America and Australasia. Population and Development Review, 647–672. (p. 410).

Kuo, L., 2020. China confirms human-to-human transmission of coronavirus. The Guardian. Available from: https://www.theguardian.com/world/2020/jan/20/coronavirus-spreads-to-beijing-as-china-confirms-new-cases. (p. 69).

Kurowski, C., Evans, D.B., Tandon, A., Eozenou, P.H.V., Schmidt, M., Irwin, A., Cain, J.S., Pambudi, E.S., Postolovska, I., 2021. From Double Shock to Double Recovery-Implications and Options for Health Financing in the Time of COVID-19. World Bank Group. (p. 374).

Kydland, F.E., Prescott, E.C., 1982. Time to build and aggregate fluctuations. Econometrica, 1345–1370. (p. 220).

Lacey, E., Massad, J., Utz, R., 2021. A review of fiscal policy responses to Covid-19. Available at: https://openknowledge.worldbank.org/bitstream/handle/10986/35904/A-Review-of-Fiscal-Policy-Responses-to-COVID-19.pdf?sequence=1&isAllowed=y. (p. 236).

Lakha, F., King, A., Swinkels, K., Lee, A., 2022. Are schools drivers of Covid-19 infections—an analysis of outbreaks in Colorado, USA in 2020. Journal of Public Health 44, e26–e35. (p. 126).

Lancet, 2020. Covid-19: a stress test for trust in science. The Lancet 396, 799. https://doi.org/10.1016/S0140-6736(20)31954-1. http://www.sciencedirect.com/science/article/pii/S0140673620319541. (p. 331).

Lane, J., 2006. Mass vaccination and surveillance/containment in the eradication of smallpox. In: Mass Vaccination: Global Aspects—Progress and Obstacles, pp. 17–29. (p. 162).

Lane, S., MacDonald, N., Marti, M., Dumolard, L., 2018. Vaccine hesitancy around the globe: analysis of three years of WHO/UNICEF joint reporting form data-2015–2017. Vaccine 36,

3861–3867. https://doi.org/10.1016/j.vaccine.2018.03.063. http://www.sciencedirect.com/science/article/pii/S0264410X18304195. (p. 142).

Lania, A., Sandri, M., Cellini, M., Mirani, M., Lavezzi, E., Mazziotti, G., 2020. Thyrotoxicosis in patients with Covid-19: the Thyrcov study. European Journal of Endocrinology 183, 381–387. https://eje.bioscientifica.com/view/journals/eje/aop/eje-20-0335/eje-20-0335.xml. (p. 351).

Larson, H., Jarrett, C., Eckersberger, E., Smith, D., Paterson, P., 2014. Understanding vaccine hesitancy around vaccines and vaccination from a global perspective: a systematic review of published literature, 2007–2012. Vaccine 32, 2150–2159. https://doi.org/10.1016/j.vaccine.2014.01.081. http://www.sciencedirect.com/science/article/pii/S0264410X14001443. (p. 142).

Lattouf, A., 2020. Domestic violence spikes during coronavirus as families trapped at home. Network 10 Daily. (p. 280).

Lazarevic, N., Barnett, A.G., Sly, P.D., Knibbs, L.D., 2019. Statistical methodology in studies of prenatal exposure to mixtures of endocrine-disrupting chemicals: a review of existing approaches and new alternatives. Environmental Health Perspectives 127, 026001. (p. 393).

Lazzerini, M., Barbi, E., Apicella, A., Marchetti, F., Cardinale, F., Trobia, G., 2020. Delayed access or provision of care in Italy resulting from fear of Covid-19. The Lancet Child & Adolescent Health 4, e10–e11. (p. 306).

Lazzerini, M., Putoto, G., 2020. Covid-19 in Italy: momentous decisions and many uncertainties. The Lancet Global Health 8, e641–e642. https://doi.org/10.1016/S2214-109X(20)30110-8. https://pubmed.ncbi.nlm.nih.gov/32199072. (p. 340).

Le Bert, N., Tan, A., Kunasegaran, K., Tham, C., Hafezi, M., Chia, A., Chng, M., Lin, M., Tan, N., Linster, M., Chia, W., Chen, M., Wang, L., Ooi, E., Kalimuddin, S., Tambyah, P., Low, J., Tan, Y., Bertoletti, A., 2020. Sars-Cov-2-specific t cell immunity in cases of Covid-19 and Sars, and uninfected controls. Nature 584, 457–462. https://doi.org/10.1038/s41586-020-2550-z. (pp. 350, 377, and 378).

Lee, A., Challen, K., Gardois, P., Mackway-Jones, K., Carley, S., Phillips, W., Booth, A., Walter, D., Goodacre, S., 2011. Emergency planning in health: scoping study of the international literature, local information resources and key stakeholders. NIHR Service Delivery and Organisation programme. Available from: https://www.researchgate.net/publication/256484404_Emergency_planning_in_health_Scoping_study_of_the_international_literature_local_information_resources_and_key_stakeholders. (p. 56).

Lee, L., Ostroff, S., McGee, H., Jonson, D., Downes, F., Cameron, D., et al., 1991. A outbreak of shigellosis at an outdoor music festival. American Journal of Epidemiology 133, 608–615. (p. 24).

Leitner, S., 2021. Number of e-Card consultations: analysis of eCard consultations during the pandemic/during the lockdown in 2020. Business Intelligence im Gesundheitswesen. (p. 311).

Lettieri, E., Masella, C., Radaelli, G., 2009. Disaster management: findings from a systematic review. Disaster Prevention and Management: An International Journal 18, 117–136. (p. 56).

Li, Q., Guan, X., Wu, P., Wang, X., Zhou, L., Tong, Y., 2020a. Early transmission dynamics in Wuhan, China, of novel coronavirus? Infected pneumonia. The New England Journal of Medicine 382, 1199–1207. (p. 105).

Li, Y., Campbell, H., Kulkarni, D., Harpur, A., Nundy, M., Wang, X., Nair, H., 2021. The temporal association of introducing and lifting non-pharmaceutical interventions with the time-varying reproduction number (R) of Sars-Cov-2: a modelling study across 131 countries. Lancet. Infectious Diseases 21, 193–202. (pp. 122 and 125).

Li, Y., Qian, H., Hang, J., Chen, X., Hong, L., Liang, P., et al., 2020b. Evidence for probable aerosol transmission of Sars-Cov- 2 in a poorly ventilated restaurant. medRxiv. Available from: https://www.medrxiv.org/content/10.1101/2020.04.16.20067728v1. (p. 104).

Link, B.G., Phelan, J., 1995. Social conditions as fundamental causes of disease. Journal of Health and Social Behavior, 80–94. (p. 283).

Liu, X., Xu, X., Li, G., Xu, X., Sun, Y., Wang, F., Shi, X., Li, X., Xie, G., Zhang, L., 2020. Differential impact of non-pharmaceutical public health interventions on COVID-19 epidemics in the United States. Research Square. (p. 125).

Lloyd-Smith, J., Schreiber, S., Kopp, P., et al., 2005. Superspreading and the effect of individual variation on disease emergence. Nature 438, 355–359. (p. 463).

Loayza, N., 2020. Costs and Trade-Offs in the Fight Against the Covid-19 Pandemic: A Developing Country Perspective. World Bank Research and Policy Briefs. https://openknowledge.worldbank.org/handle/10986/33764. (pp. 184 and 185).

Long, Q., Tang, X., Shi, Q., Li, Q., Deng, H., Yuan, J., Hu, J., Xu, W., Zhang, Y., Lv, F., Su, K., Zhang, F., Gong, J., Wu, B., Liu, X., Li, J., Qiu, J., Chen, J., Huang, A., 2020. Clinical and immunological assessment of asymptomatic Sars-Cov-2 infections. Nature Medicine 26, 1200–1204. https://doi.org/10.1038/s41591-020-0965-6. (p. 377).

Lovelock, J., 1995. New statements on the Gaia theory. Microbiologia (Madrid, Spain) 11, 295–304. (p. 389).

Lovelock, J.E., 1972. Gaia as seen through the atmosphere. In: Atmosphere Environment, pp. 579–580. (p. 34).

Lu, J., Gu, J., Li, K., Xu, C., Su, W., Lai, Z., et al., 2020. Covid-19 outbreak associated with air conditioning in restaurant, Guangzhou, China, 2020. Emerging Infectious Diseases 26. (p. 104).

Lucifora, C., 2020. La 'fase 2' inizia dalla formazione dei lavoratori. lavoce.info. Available from: https://www.lavoce.info/archives/65679/la-fase-2-inizia-dalla-formazione-dei-lavoratori/. (p. 436).

Lund, S., Manyika, J., Woetzel, J., Barriball, E., Krishnan, M., Alicke, K., Birshan, M., George, K., Smit, S., Swan, D., Hutzler, K., 2020. Risk, resilience, and rebalancing in global value chains. McKinsey Global Institute. Available from: https://www.mckinsey.com/business-functions/operations/our-insights/risk-resilience-and-rebalancing-in-global-value-chains. (pp. 250 and 253).

MacIntyre, C., Chughtai, A., 2020. A rapid systematic review of the efficacy of face masks and respirators against coronaviruses and other respiratory transmissible viruses for the community, healthcare workers and sick patients. International Journal of Nursing Studies 108, 103629. (p. 121).

Mackenbach, J.P., 2022. Omran's 'epidemiologic transition' 50 years on. International Journal of Epidemiology 51, 1054–1057. (p. 412).

Maffesoli, M., 1996. Ordinary Knowledge: an Introduction to Interpretative Sociology. Polity Press. (p. 345).

Mahler, D.G., Yonzan, N., Lakner, C., Aguilar, R.A.C., Wu, H., 2021. Updated estimates of the impact of Covid-19 on global poverty: turning the corner on the pandemic in 2021? World Bank Blogs 24. (pp. 209 and 210).

Mahmoodi, F., Blutinger, E., Echazú, L., Nocetti, D., 2021. Covid-19 and the health care supply chain: impacts and lessons learned. CSCMP's Supply Chain Quarterly. February 17. Available online at: https://www.supplychainquarterly.com/articles/4417-covid-19-and-the-health-care-supply-chain-impacts-and-lessons-learned. (p. 257).

Mann, G., 2015. Poverty in the midst of plenty: unemployment, liquidity, and Keynes's scarcity theory of capital. Critical Historical Studies 2, 45–83. (p. 82).

Marantz Henig, R., 1993. Dancing Matrix: How Science Confronts Emerging Viruses. Vintage Publisher, New York. (p. 370).

Marazzi, F., Piano Mortari, A., Belotti, F., Carrà, G., Cattuto, C., Kopinska, J., Paolotti, D., Atella, V., 2022. Psychotropic drug purchases during the Covid-19 pandemic in Italy and their relationship with mobility restrictions. Scientific Reports 12, 19336. (pp. 174, 181, and 303).

Maringe, C., Spicer, J., Morris, M., Purushotham, A., Nolte, E., Sullivan, R., Rachet, B., Aggarwal, A., 2020. The impact of the Covid-19 pandemic on cancer deaths due to delays in diagnosis in England, UK: a national, population-based, modelling study. Lancet Oncology 21, 1023–1034. (pp. 306 and 310).

Martinez-Martin, N., Wieten, S., Magnus, D., Cho, M., 2020. Digital contact tracing, privacy, and public health. Hastings Center Report 50, 43–46. (p. 113).

Masterman, C., Viscusi, W., 2018. The income elasticity of global values of a statistical life: stated preference evidence. Journal of Benefit-Cost Analysis 9, 407–434. (p. 198).

Mateus, J., Grifoni, A., Tarke, A., Sidney, J., Ramirez, S., Dan, J., Burger, Z., Rawlings, S., Smith, D., Phillips, E., Mallal, S., Lammers, M., Rubiro, P., Quiambao, L., Sutherland, A., Yu, E., da Silva Antunes, R., Greenbaum, J., Frazier, A., Markmann, A., Premkumar, L., de Silva, A., Peters, B., Crotty, S., Sette, A., Weiskopf, D., 2020. Selective and cross-reactive Sars-Cov-2 t cell epitopes in unexposed humans. Science 370, 89–94. https://science.sciencemag.org/content/early/2020/08/04/science.abd3871. arXiv:https://science.sciencemag.org/content/early/2020/08/04/science.abd3871.full.pdf. (p. 377).

Mathews, B., Fragos, T., Bollyky, T., Patrick, S., 2020. Improving pandemic preparedness. Lessons from Covid-19. The Brookings Institution Report. Available from: https://www.brookings.edu/blog/order-from-chaos/2020/06/16/reopening-the-world-the-who-international-institutions-and-the-covid-19-response/. (pp. 90, 419, and 420).

Mazza, C., Ricci, E., Biondi, S., Colasanti, M., Ferracuti, S., Napoli, C., Roma, P., 2020. A nationwide survey of psychological distress among Italian people during the Covid-19 pandemic: immediate psychological responses and associated factors. International Journal of Environmental Research and Public Health 17. https://doi.org/10.3390/ijerph17093165. https://www.mdpi.com/1660-4601/17/9/3165. (p. 301).

Mbembe, A., Shread, C., 2021. The universal right to breathe. Critical Inquiry 47, S58–S62. (p. 424).

McAdams, D., 2020. Economic epidemiology in the wake of Covid-19. Economics 82120, 1–40. (pp. 465 and 467).

McCannon, B.C., Hall, J.C., 2021. Stay-at-home orders were issued earlier in economically unfree states. Southern Economic Journal 87, 1138–1151. (p. 182).

McGowan, V.J., Bambra, C., 2022. Covid-19 mortality and deprivation: pandemic, syndemic, and endemic health inequalities. Lancet Public Health 7, e966–e975. (p. 316).

McKibbin, W., 2004. Economic modeling of Sars: the g-cubed approach. (p. 47).

McNabb, S., Jajosky, R.A., Hall-Baker, P.A., Adams, D.A., Sharp, P., Anderson, W.J., Javier, A.J., Jones, G.J., Nitschke, D.A., Worshams, C.A., et al., 2007. Summary of notifiable diseases—United States, 2005. Morbidity and Mortality Weekly Report 54, 1–92. (p. 162).

Menachery, V.D., Yount, B.L., Debbink, K., Agnihothram, S., Gralinski, L.E., Plante, J.A., Graham, R.L., Scobey, T., Ge, X.Y., Donaldson, E.F., et al., 2015. A Sars-like cluster of circulating bat coronaviruses shows potential for human emergence. Nature Medicine 21, 1508–1513. (p. 20).

Menge, D., MacPherson, A., Bytnerowicz, T., Quebbeman, A., Schwartz, N., Taylor, B., Wolf, A., 2018. Logarithmic scales in ecological data presentation may cause misinterpretation. Nature Ecology & Evolution 2, 1393–1402. https://doi.org/10.1038/s41559-018-0610-7. (p. 337).

Milgram, S., 1967. The small world problem. Psychology Today 2, 60–67. (p. 467).

Miller, G.W., Jones, D.P., 2014. The nature of nurture: refining the definition of the exposome. Toxicological Sciences 137, 1–2. (p. 392).

Moawad, A.M., El Desouky, E.D., Salem, M.R., Elhawary, A.S., Hussein, S.M., Hassan, F.M., 2021. Violence and sociodemographic related factors among a sample of Egyptian women during the Covid-19 pandemic. Egyptian Journal of Forensic Sciences 11, 1–9. (p. 280).

Mogharab, V., Ostovar, M., Ruszkowski, J., Hussain, S.Z.M., Shrestha, R., Yaqoob, U., Aryanpoor, P., Nikkhoo, A.M., Heidari, P., Jahromi, A.R., et al., 2022. Global burden of the Covid-19 associated patient-related delay in emergency healthcare: a panel of systematic review and meta-analyses. Globalization and Health 18, 1–18. (p. 306).

Montella, R., 2020. The Covid-19 pandemic: multilateralism and parliaments. Open Democracy. Available from: https://www.opendemocracy.net/en/can-europe-make-it/covid-19-pandemic-multilateralism-and-parliaments/. (p. 416).

Monzani, D., Vergani, L., Pizzoli, S.F.M., Marton, G., Pravettoni, G., et al., 2021. Emotional tone, analytical thinking, and somatosensory processes of a sample of Italian tweets during the first phases of the Covid-19 pandemic: observational study. Journal of Medical Internet Research 23, e29820. (pp. 302 and 303).

Mora, C., Tittensor, D.P., Adl, S., Simpson, A.G., Worm, B., 2011. How many species are there on Earth and in the ocean? PLoS Biology 9, e1001127. (p. 384).

Morens, D., Daszak, P., Taubenberger, J., 2020. Escaping pandora's box - another novel coronavirus. The New England Journal of Medicine 382, 1293–1295. (p. 18).

Morens, D., Fauci, A., 2012. Emerging infectious diseases in 2012: 20 years after the institute of medicine report. MBio 3. (p. 412).

Morrison, S., 2020. The Apple-Google contact tracing tool gets a beta release and a new risk level feature. Voxeu.org. Available from: https://www.vox.com/recode/2020/4/24/21234420/apple-google-contact-tracing-exposure-notification-update. (p. 116).

Moss, P., 2022. The t cell immune response against Sars-Cov-2. Nature Immunology 23, 186–193. (p. 378).

Mougang, Y.K., Di Zazzo, L., Minieri, M., Capuano, R., Catini, A., Legramante, J.M., Paolesse, R., Bernardini, S., Di Natale, C., 2021. Sensor array and gas chromatographic detection of the blood serum volatolomic signature of Covid-19. iScience 24, 102851. (p. 396).

Muller, I., Cannavaro, D., Dazzi, D., Covelli, D., Mantovani, G., Muscatello, A., Ferrante, E., Orsi, E., Resi, V., Longari, V., Cuzzocrea, M., Bandera, A., Lazzaroni, E., Dolci, A., Ceriotti, F., Re, T., Gori, A., Arosio, M., Salvi, M., 2020. Sars-Cov-2-related atypical thyroiditis. The Lancet Diabetes & Endocrinology 8, 739–741. (p. 351).

Mulligan, K., Van Nuys, K., Peneva, D., Ryan, M., Joyce, G., 2020. The value of treatment for COVID-19. White Papers Coronavirus (COVID-19). USC Schaeffer Center. Available from: https://healthpolicy.usc.edu/research/the-value-of-treatment-for-covid-19/. (p. 357).

Mummert, A., Weiss, H., Long, L.P., Amigó, J.M., Wan, X.F., 2013. A perspective on multiple waves of influenza pandemics. PLoS ONE 8, e60343. (p. 347).

Murray, E.J., 2020. Epidemiology's time of need: Covid-19 calls for epidemic-related economics. The Journal of Economic Perspectives 34, 105–120. (p. 350).

Murray, K.A., Daszak, P., 2013. Human ecology in pathogenic landscapes: two hypotheses on how land use change drives viral emergence. Current Opinion in Virology 3, 79–83. (p. 383).

Nabbe, M., Brand, H., 2021. The European health union: European Union's concern about health for all. Concepts, definition, and scenarios. In: Healthcare. MDPI, p. 1741. (p. 320).

Nabben, K., Poblet, M., Barca, J., 2020. What is known from a network? Digital contact tracing, privacy, and pandemics in the digital age. Mercatus Working Paper. Mercatus Center at George Mason University. Available from: https://papers.ssrn.com/sol3/papers.cfm?abstract_id=3719288. (p. 117).

Nadarajah, R., Wu, J., Hurdus, B., Asma, S., Bhatt, D.L., Biondi-Zoccai, G., Mehta, L.S., Ram, C.V.S., Ribeiro, A.L.P., Van Spall, H.G., et al., 2022. The collateral damage of Covid-19 to cardiovascular services: a meta-analysis. European Heart Journal. (p. 308).

Nan, X., Thompson, T., 2021. Introduction to the special issue on 'public health communication in an age of Covid-19'. Health Communication 36, 1–5. (p. 87).

National Collaborating Centre for Methods and Tools, 2021. Living Rapid Review Update 17: What Is the Specific Role of Daycares and Schools in Covid-19 Transmission? McMaster University, Hamilton (ON). Available from: https://www.nccmt.ca/covid-19/covid-19-rapid-evidence-service/19. (p. 279).

National Intelligence Council, 2018. Global trends 2025: a transformed world. (p. xxiv).

Ndugga, N., Pham, O., Hill, L., Artiga, S., Alam, R., Parker, N., 2021. Latest data on Covid-19 vaccinations race/ethnicity. Kais Family Found. (p. 281).

Ndwandwe, D., Wiysonge, C.S., 2021. Covid-19 vaccines. Current Opinion in Immunology 71, 111–116. (p. 424).

Newhagen, J., Bucy, E., 2020. Overcoming resistance to Covid-19 vaccine adoption: how affective dispositions shape views of science and medicine. The Harvard Kennedy School (HKS) Misinformation Review. https://doi.org/10.37016/mr-2020-44. https://misinforeview.hks.harvard.edu/article/overcoming-resistance-to-covid-19-vaccine-adoption-how-affective-dispositions-shape-views-of-science-and-medicine/. (p. 142).

Newman, M.E., Watts, D.J., Strogatz, S.H., 2002. Random graph models of social networks. Proceedings of the National Academy of Sciences 99, 2566–2572. (p. 467).

NHS-UK, 2022. Delivery plan for tackling the Covid-19 backlog of elective care. Available at: https://www.england.nhs.uk/coronavirus/wp-content/uploads/sites/52/2022/02/C1466-delivery-plan-for-tackling-the-covid-19-backlog-of-elective-care.pdf. (p. 314).

NICE, 2021. Bed occupancy. Available from: https://www.nice.org.uk/guidance/ng94/evidence/39.bed-occupancy-pdf-172397464704. (p. 291).

Noar, S.M., Austin, L., 2020. (Mis) communicating about Covid-19: insights from health and crisis communication. Health Communication 35, 1735–1739. (p. 87).

Nozick, R., 1974. Anarchy, State, and Utopia, vol. 5038. Basic Books, New York. (p. 259).

Nussbaumer-Streit, B., Mayr, V., Dobrescu, A., Chapman, A., Persad, E., Klerings, I., et al., 2020. Quarantine alone or in combination with other public health measures to control Covid-19: a rapid review. Cochrane Database of Systematic Reviews 4. (pp. 105 and 122).

O Murchu, E., Byrne, P., Carty, P.G., De Gascun, C., Keogan, M., O'Neill, M., Harrington, P., Ryan, M., 2022. Quantifying the risk of Sars-Cov-2 reinfection over time. Reviews in Medical Virology 32, e2260. (p. 379).

OECD, 2020a. Health at a Glance 2021: OECD Indicators. OECD Publishing, Paris. (p. 310).

OECD, 2020b. Tax and fiscal policy in response to the coronavirus crisis: strengthening confidence and resilience. Available from: https://read.oecd-ilibrary.org/view/?ref=128_128575-o6raktc0aa&title=Tax-and-Fiscal-Policy-in-Response-to-the-Coronavirus-Crisis. (pp. 337, 339, 434, and 435).

OECD, 2021a. OECD Health Statistics. OECD Publishing, Paris. https://www.oecd.org/els/health-systems/health-data.htm. (pp. 135, 144, 314, and 315).

OECD, 2021b. OECD Science, Technology and Innovation Outlook 2021: Times of Crisis and Opportunity. OECD Publishing, Paris. Available from: https://www.oecd-ilibrary.org/docserver/75f79015-en.pdf?expires=1654875677&id=id&accname=ocid45116923&checksum=B704F6136F1FC85CAE7E5E602BEDA0A2. (p. 128).

OECD, 2021c. Tackling the Mental Health Impact of the COVID-19 Crisis: an Integrated, Whole-of-Society Response. Technical Report. Organisation for Economic Co-Operation and Development. (p. 300).

OECD, 2022a. Household savings. https://www.oecd-ilibrary.org/content/data/cfc6f499-en. (p. 266).

OECD, 2022b. OECD health statistics 2022. Available at: https://stats.oecd.org/Index.aspx?ThemeTreeId=9. (pp. 290, 293, 304, and 305).

OECD Policy Responses to Coronavirus (COVID-19), 2021. Rising from the Covid 19 crisis: policy responses in the long-term care sector. OECD. Available from: https://read.oecd-ilibrary.org/view/?ref=1122_1122652-oyri4k81cp&title=Rising-from-the-COVID-19-crisis-policy-responses-in-the-long-term-care-sector. (pp. 91, 92, and 359).

OECD/European Observatory on Health Systems and Policies, 2021a. France: country health profile 2021, state of health in the EU. (p. 310).

OECD/European Observatory on Health Systems and Policies, 2021b. Italy: country health profile 2021, state of health in the EU. (p. 310).

Okell, L., Verity, R., Watson, O., Mishra, S., Walker, P., Whittaker, C., Katzourakis, A., Donnelly, C., Riley, S., Ghani, A., Gandy, A., Flaxman, S., Ferguson, N., Bhatt, S., 2020. Have deaths from Covid-19 in Europe plateaued due to herd immunity? The Lancet 395, e110–e111. (p. 127).

Omram, A.R., 1971. The epidemiologic transition: a theory of the epidemiology of population change. Millbank Memorial Fund Quarterly 49, 509–538. (pp. 28 and 410).

Omran, A.R., 1998. The epidemiologic transition theory revisited thirty years later. World Health Statistics Quarterly 53, 99–119. (p. 410).

Organization, W.H., et al., 1986. Ottawa charter for health promotion, 1986. Technical Report. World Health Organization. Regional Office for Europe. (p. 382).

Ostfeld, R.S., Keesing, F., 2017. Is biodiversity bad for your health? Ecosphere 8, e01676. (p. 383).

Oswald, M., Grace, J., 2021. The Covid-19 contact tracing app in England and 'experimental proportionality'. Public Law 1, 27–37. (p. 111).

Our World in Data, 2017. Deaths from terrorism, 2017. Our World in Data grapher. Available from: http://ourworldindata.org/grapher/fatalities-from-terrorism. (p. 428).

Oxford, J., Sefton, A., Jackson, R., Innes, W., Daniels, R., Johnson, N., 2002. World war I may have allowed the emergence of 'Spanish' influenza. Lancet. Infectious Diseases 2, 111–114. (p. 36).

Ozerskaya, S., 2008. OECD best practice guidelines for biological resource centers. OECD, 2007, 115 p. (p. 393).

Paganetto, L., Scandizzo, P.L., 2020. Is globalization sustainable? In: Capitalism, Global Change and Sustainable Development. Springer, pp. 261–270. (p. 260).

Painter, E.M., Ussery, E.N., Patel, A., Hughes, M.M., Zell, E.R., Moulia, D.L., Scharf, L.G., Lynch, M., Ritchey, M.D., Toblin, R.L., et al., 2021. Demographic characteristics of persons vaccinated during the first month of the Covid-19 vaccination program—United States, December 14, 2020–January 14, 2021. Morbidity and Mortality Weekly Report 70, 174. (p. 281).

Palayew, A., Norgaard, O., Safreed-Harmon, K., Andersen, T., Rasmussen, L., Lazarus, J., 2020. Pandemic publishing poses a new Covid-19 challenge. Nature Human Behaviour 4, 666–669. https://doi.org/10.1038/s41562-020-0911-0. (p. 327).

Pan, W., Tyrovolas, S., Vazquez, I., Raj, R., Fernandez, D., Zaitchik, B., Lantos, P., Woods, C., 2020. COVID-19: effectiveness of non-pharmaceutical interventions in the United States before phased removal of social distancing protections varies by region. medRxiv preprint. (p. 125).

Parri, N., Lenge, M., Buonsenso, D., 2020. Children with Covid-19 in pediatric emergency departments in Italy. The New England Journal of Medicine 383, 187–190. (p. 124).

Parvez, M., Parveen, S., 2017. Evolution and emergence of pathogenic viruses: past, present, and future. Intervirology 60, 1–7. (p. 8).

Pearce, W., 2020. Trouble in the trough: how uncertainties were downplayed in the UK's science advice on Covid-19. Humanities & Social Sciences Communications 7, 1–6. (p. 350).

Perrin, P., McCabe, O., Everly, G., Links, J., 2009. Preparing for an influenza pandemic: mental health considerations. Prehospital and Disaster Medicine 24, 223–230. https://doi.org/10.1017/S1049023X00006853. (p. xxiv).

Perrings, C., 2014. Our Uncommon Heritage: Biodiversity Change, Ecosystem Services, and Human Well-Being. Cambridge University Press. (p. 464).

Perrings, C., Castillo-Chavez, C., Chowell, G., Daszak, P., Fenichel, E.P., Finnoff, D., Horan, R.D., Kilpatrick, A.M., Kinzig, A.P., Kuminoff, N.V., et al., 2014. Merging economics and epidemiology to improve the prediction and management of infectious disease. EcoHealth 11, 464–475. (p. 346).

Perrings, C., Espinoza, B., 2021. Mobility restrictions and the control of Covid-19. Ecology, Economy and Society-the INSEE Journal 4, 31–43. (p. 346).

Perrow, C., 2011. The Next Catastrophe: Reducing Our Vulnerabilities to Natural, Industrial, and Terrorist Disasters. Princeton University Press, Princeton, NJ. (pp. 84 and xxvii).

Perscheid, C., Benzler, J., Hermann, C., Janke, M., Moyer, D., Laedtke, T., Adeoye, O., Denecke, K., Kirchner, G., Beermann, S., Schwarz, N., Tom-Aba, D., Krause, G., 2018. Ebola outbreak containment: real-time task and resource coordination with sormas. Frontiers in ICT 5, 7. https://doi.org/10.3389/fict.2018.00007. https://www.frontiersin.org/article/10.3389/fict.2018.00007. (p. 108).

Petersen, E., Wilson, M., Touch, S., McCloskey, B., Mwaba, P., Bates, M., Dar, O., Mattes, F., Kidd, M., Ippolito, G., Azhar, E., Zumla, A., 2016. Rapid spread of Zika virus in the Americas - implications for public health preparedness for mass gatherings at the 2016 Brazil Olympic games. International Journal of Infectious Diseases 44, 11–15. https://doi.org/10.1016/j.ijid.2016.02.001. http://www.sciencedirect.com/science/article/pii/S1201971216000217. (p. 47).

Phadke, V.K., Bednarczyk, R.A., Salmon, D.A., Omer, S.B., 2016. Association between vaccine refusal and vaccine-preventable diseases in the United States: a review of measles and pertussis. JAMA 315, 1149–1158. (p. 162).

Philipson, T., 2000. Economic epidemiology and infectious diseases. Handbook of Health Economics 1, 1761–1799. (pp. 347, 348, and 464).

Pieh, C., O' Rourke, T., Budimir, S., Probst, T., 2020. Relationship quality and mental health during Covid-19 lockdown. PLoS ONE 15, e0238906. (pp. 174 and 301).

Pierce, M., Hope, H., Ford, T., Hatch, S., Hotopf, M., John, A., Kontopantelis, E., Webb, R., Wessely, S., McManus, S., Abel, K., 2020. Mental health before and during the Covid-19 pandemic: a longitudinal probability sample survey of the UK population. Lancet Psychiatry 7, 883–892. https://doi.org/10.1016/S2215-0366(20)30308-4. (p. 181).

Pigou, A.C., 1920. Some problems of foreign exchange. The Economic Journal 30, 460–472. (p. 191).

Pike, J., Bogich, T., Elwood, S., Finnoff, D.C., Daszak, P., 2014. Economic optimization of a global strategy to address the pandemic threat. Proceedings of the National Academy of Sciences 111, 18519–18523. (pp. 47, 414, and 415).

Pike, J., Shogren, J., Aadland, D., Viscusi, W., Finnoff, D., Skiba, A., Daszak, P., 2020. Catastrophic risk: waking up to the reality of a pandemic? EcoHealth 17, 217–221. (p. 428).

Plohl, N., Musil, B., 2020. Modeling compliance with Covid-19 prevention guidelines: the critical role of trust in science. Psychology, Health & Medicine 0, 1–12. https://doi.org/10.1080/13548506.2020.1772988. arXiv:https://doi.org/10.1080/13548506.2020.1772988. pMID: 32479113. (p. 335).

Plowright, R., Parrish, C., McCallum, H., et al., 2017. Pathways to zoonotic spillover. Nature Review, Microbiology 15, 502–510. (pp. 9 and 10).

Preston, R., 1995. The Hot Zone: The Terrifying True Story of the Origins of the Ebola Virus. Bantam Doubleday Dell Publishing Group, New York. (p. 370).

Primessnig, U., Pieske, B.M., Sherif, M., 2021. Increased mortality and worse cardiac outcome of acute myocardial infarction during the early Covid-19 pandemic. ESC Heart Failure 8, 333–343. (p. 315).

Proietti, C., Santini, M., Probst, P., Annunziato, A., De Groeve, T., Fonio, C., 2021. Uplifting of COVID-19 containment measures in Europe. Publications Office of the European Union. (p. 125).

Puaschunder, J., Gelter, M., 2022. The law, economics, and governance of generation Covid-19 long-haul. Indiana Health Law Review 19, 47. (p. 224).

Puntmann, V.O., Carerj, M.L., Wieters, I., Fahim, M., Arendt, C., Hoffmann, J., Shchendrygina, A., Escher, F., Vasa-Nicotera, M., Zeiher, A.M., et al., 2020. Outcomes of cardiovascular magnetic resonance imaging in patients recently recovered from coronavirus disease 2019 (Covid-19). JAMA Cardiology 5, 1265–1273. (p. 354).

Putnam, R., 1993. Making Democracy Work. Princeton University Press, Princeton, NJ. (p. 362).

Puyeo, T., 2020. Coronavirus: the hammer and the dance. What the next 18 months can look like, if leaders buy us time. Available from: https://tomaspueyo.medium.com/coronavirus-the-hammer-and-the-dance-be9337092b56. (p. 196).

Qi, H., Liu, B., Wang, X., Zhang, L., 2022. The humoral response and antibodies against Sars-Cov-2 infection. Nature Immunology 23, 1008–1020. (p. 350).

Quammen, D., 2012. Spillover: Animal Infections and the Next Human Pandemic. W.W. Norton & Co, New York. (p. 370).

Radin, E., Eleftheriades, C., 2021. Financing Pandemic Preparedness and Response. Technical Report. The Independent Panel for Pandemic Preparedness and Response. (p. 59).

Ragab, D., Salah Eldin, H., Taeimah, M., Khattab, R., Salem, R., 2020. The Covid-19 cytokine storm; what we know so far. Frontiers in Immunology 11, 1446. https://www.frontiersin.org/article/10.3389/fimmu.2020.01446. (p. 353).

Rajpal, V.R., Sharma, S., Kumar, A., Chand, S., Joshi, L., Chandra, A., Babbar, S., Goel, S., Raina, S.N., Shiran, B., 2022. 'Is omicron mild'? Testing this narrative with the mutational landscape of its three lineages and response to existing vaccines and therapeutic antibodies. Journal of Medical Virology. (p. 350).

Ratliff, E., 2020. We can protect the economy from pandemics. Why didn't we? Wired. (p. 369).

Ravaghi, H., Guisset, A.L., Elfeky, S., Nasir, N., Khani, S., Ahmadnezhad, E., Abdi, Z., 2023. A scoping review of community health needs and assets assessment: concepts, rationale, tools and uses. BMC Health Services Research 23, 1–20. (p. 428).

Reisman, D.A., 1998. Adam Smith on market and state. JITE. Journal of Institutional and Theoretical Economics (Zeitschrift für die gesamte Staatswissenschaft), 357–383. (p. 189).

Reluga, T.C., 2010. Game theory of social distancing in response to an epidemic. PLoS Computational Biology 6, e1000793. (p. 465).

Riera, R., Bagattini, Â.M., Pacheco, R.L., Pachito, D.V., Roitberg, F., Ilbawi, A., 2021. Delays and disruptions in cancer health care due to Covid-19 pandemic: systematic review. JCO Global Oncology 7, 311–323. (pp. 309 and 310).

Rochwerg, B., Parke, R., Murthy, S., Fernando, S.M., Leigh, J.P., Marshall, J., Adhikari, N.K., Fiest, K., Fowler, R., Lamontagne, F., et al., 2020. Misinformation during the coronavirus disease 2019 outbreak: how knowledge emerges from noise. Critical Care Explorations 2. (p. 328).

Rockström, J., Steffen, W., Noone, K., Persson, Å., Chapin, F.S., Lambin, E.F., Lenton, T.M., Scheffer, M., Folke, C., Schellnhuber, H.J., et al., 2009. A safe operating space for humanity. Nature 461, 472–475. (p. 38).

Rogers, E.M., 2004. A prospective and retrospective look at the diffusion model. Journal of Health Communication 9, 13–19. (p. 283).

Rogers, E.M., Agarwala-Rogers, R., 1976. Communication in organizations. (p. 284).

Romanelli, C., Cooper, D., Campbell-Lendrum, D., Maiero, M., Karesh, W.B., Hunter, D., Golden, C.D., 2015. Connecting global priorities: biodiversity and human health: a state of knowledge review. (p. 383).

Romano, A., Sotis, C., Dominioni, G., Guidi, S., 2020. The scale of Covid-19 graphs affects understanding, attitudes, and policy preferences. Health Economics 29, 1482–1494. (pp. 336 and 337).

Romano, G., Schivardi, F., 2020. Decreto liquidità, l'importante è fare in fretta. lavoce.info. Available from: https://www.lavoce.info/archives/65428/decreto-liquidita-tutto-dipende-dalla-velocita-di-attuazione/. (p. 435).

Roser, M., Ritchie, H., Spooner, F., 2021. Burden of disease. Our World in Data. https://ourworldindata.org/burden-of-disease. (p. 31).

Ross, A.A., 2006. Coming in from the cold: constructivism and emotions. European Journal of International Relations 12, 197–222. (p. 83).

Ross, R., 1916. An application of the theory of probabilities to the study of a priori pathometry — part I. Proceedings of the Royal Society of London. Series A, Containing Papers of a Mathematical and Physical Character 92, 204–230. (p. 453).

Rossi, R., Socci, V., Talevi, D., Mensi, S., Niolu, C., Pacitti, F., Di Marco, A., Rossi, A., Siracusano, A., Di Lorenzo, G., 2020. Covid-19 pandemic and lockdown measures impact on mental health among the general population in Italy. Frontiers in Psychiatry, 790. (p. 301).

Rothe, C., Schunk, M., Sothmann, P., Bretzel, G., Froeschl, G., Wallrauch, C., et al., 2020. Transmission of 2019-ncov infection from an asymptomatic contact in Germany. The New England Journal of Medicine 382, 970–971. (p. 104).

Rozmus, E., 2023. Determining the functional significance of kcnh2 variants of uncertain significance identified in a large patient biobank. Biophysical Journal 122, 242a–243a. (p. 393).

Russell, B., 2020. Apple and Google pledge to shut down coronavirus tracker when pandemic ends. The Verge. Available from: https://www.theverge.com/2020/4/24/21234457/apple-google-coronavirus-contact-tracing-tracker-exposure-notification-shut-down. (p. 116).

Ryan, M.P., 1992. Gender and public access: womens politics in nineteenth century America. In: Calhoun, Craig (Ed.), Habermas and the Public Sphere. (p. 443).

Ryan, W., Evers, E., 2020. Logarithmic axis graphs distort lay judgment. PsyArXiv. Working Paper. Available from: https://ssrn.com/abstract=3605872. (p. 337).

Rydland, H.T., Friedman, J., Stringhini, S., Link, B.G., Eikemo, T.A., 2022. The radically unequal distribution of Covid-19 vaccinations: a predictable yet avoidable symptom of the fundamental causes of inequality. Humanities & Social Sciences Communications 9, 1–6. (p. 283).

Sabat, I., Neumann-Böhme, S., Varghese, N.E., Barros, P.P., Brouwer, W., van Exel, J., Schreyögg, J., Stargardt, T., 2020. United but divided: policy responses and people's perceptions in the EU during the Covid-19 outbreak. Health Policy 124, 909–918. (p. 179).

Sah, R.K., Stiglitz, J.E., 1984. The architecture of economic systems: hierarchies and polyarchies. Technical Report. National Bureau of Economic Research. (pp. 55 and 76).

Sah, R.K., Stiglitz, J.E., 1985. Human fallibility and economic organization. American Economic Review 75, 292–297. (p. 56).

Saldana, J., 2009. The Coding Manual for Qualitative Researchers. SAGE Publications, London, UK. (p. 56).

Sanders, J., Monogue, M., Jodlowski, T., Cutrell, J., 2020. Pharmacologic treatments for coronavirus disease 2019 (COVID-19): a review. JAMA 323, 1824–1836. https://doi.org/10.1001/jama.2020.6019. arXiv:https://jamanetwork.com/journals/jama/articlepdf/2764727/jama_sanders_2020_rv_200005.pdf. (p. 358).

Santi, L., Golinelli, D., Tampieri, A., Farina, G., Greco, M., Rosa, S., Beleffi, M., Biavati, B., Campinoti, F., Guerrini, S., et al., 2021. Non-Covid-19 patients in times of pandemic: emergency department visits, hospitalizations and cause-specific mortality in northern Italy. PLoS ONE 16, e0248995. (p. 306).

Santomauro, D.F., Herrera, A.M.M., Shadid, J., Zheng, P., Ashbaugh, C., Pigott, D.M., Abbafati, C., Adolph, C., Amlag, J.O., Aravkin, A.Y., et al., 2021. Global prevalence and burden of depressive and anxiety disorders in 204 countries and territories in 2020 due to the Covid-19 pandemic. The Lancet 398, 1700–1712. (p. 300).

Santos, B., Silva, I., Ribeiro-Dantas, M., Alves, G., Endo, P., Lima, L., 2020. Covid-19: a scholarly production dataset report for research analysis. Data in Brief 32, 106178. http://www.sciencedirect.com/science/article/pii/S2352340920310726. (pp. 196, 328, and 329).

Santosa, A., Wall, S., Fottrell, E., Högberg, U., Byass, P., 2014. The development and experience of epidemiological transition theory over four decades: a systematic review. Global Health Action 7, 23574. (p. 410).

Saunders, R., Buckman, J.E., Fonagy, P., Fancourt, D., 2021. Understanding different trajectories of mental health across the general population during the Covid-19 pandemic. Psychological Medicine, 1–9. (p. 301).

Scandizzo, P.L., 2009. Science and technology in world agriculture: the world development report as an example of narratives on achievements. Science and Technology in World Agriculture, 1000–1013. (p. 345).

Scandizzo, P.L., 2018. The search for sustainable and inclusive development. In: The Search for Sustainable and Inclusive Development, pp. 89–123. (pp. 260 and 261).

Scandizzo, P.L., 2020. Coronavirus, salvare vite umane o l'economia? Formiche. Available from: https://formiche.net/2020/03/misure-economia-italia-coronavirus/. (pp. 197 and 198).

Scandizzo, P.L., Knudsen, O.K., 1980. The evaluation of the benefits of basic need policies. American Journal of Agricultural Economics 62, 46–57. (p. 413).

Scandizzo, P.L., Knudsen, O.K., 2022. The New Normalcy under a Pandemic: a Real Options Approach. Mimeo. (pp. 221, 222, and 269).

Scandizzo, P.L., Pierleoni, M.R., et al., 2020. Short and long-run effects of public investment: theoretical premises and empirical evidence. Theoretical Economics Letters 10, 834. (pp. 220 and 221).

Scandizzo, P.L., Savastano, S., 2010. The adoption and diffusion of gm crops in United States: a real option approach. AgBioForum 13, 142–157. (p. 284).

Scarpetta, S., Pearson, M., Colombo, F., Lopert, R., Dedet, G., Wenzel, M., 2021. Access to Covid-19 vaccines: global approaches in a global crisis. (p. 281).

Schellekens, P., Sourrouille, D.M., 2020. Covid-19 mortality in rich and poor countries: a tale of two pandemics? World Bank Policy Research Working Paper. (pp. 32 and 33).

Schellekens, P., Wadhwa, D., 2021. Relative severity of Covid-19 mortality: a new indicator on the world bank's data platform. Data Blog. (p. 281).

Schivardi, F., 2020. Come evitare il contagio finanziario alle imprese. lavoce.info. Available from: https://www.lavoce.info/archives/65428/decreto-liquidita-tutto-dipende-dalla-velocita-di-attuazione/. (p. 435).

Schmidt, K.A., Ostfeld, R.S., 2001. Biodiversity and the dilution effect in disease ecology. Ecology 82, 609–619. (p. 383).

Scholthof, K.B.G., 2007. The disease triangle: pathogens, the environment and society. Nature Reviews. Microbiology 5, 152–156. (p. 365).

Scientists, U.B., 2020. Open letter to the UK government regarding Covid-19. Available from: https://sites.google.com/view/covidopenletter/home. (p. 194).

Scquizzato, T., Landoni, G., Paoli, A., Lembo, R., Fominskiy, E., Kuzovlev, A., Likhvantsev, V., Zangrillo, A., 2020. Effects of Covid-19 pandemic on out-of-hospital cardiac arrests: a systematic review. Resuscitation 157, 241–247. (p. 316).

Seeger, M.W., Pechta, L.E., Price, S.M., Lubell, K.M., Rose, D.A., Sapru, S., Chansky, M.C., Smith, B.J., 2018. A conceptual model for evaluating emergency risk communication in public health. Health Security 16, 193–203. (p. 90).

Segreto, R., Deigin, Y., McCairn, K., Sousa, A., Sirotkin, D., Sirotkin, K., Couey, J.J., Jones, A., Zhang, D., 2021. Should we discount the laboratory origin of Covid-19? Environmental Chemistry Letters 19, 2743–2757. (pp. 19 and 20).

Sekine, T., Perez-Potti, A., Rivera-Ballesteros, O., Strolin, K., Gorin, J., Olsson, A., Llewellyn-Lacey, S., Kamal, H., Bogdanovic, G., Muschiol, S., Wullimann, D., Kammann, T., Emgord, J., Parrot, T., Folkesson, E., et al., 2020. Robust t cell immunity in convalescent individuals with asymptomatic or mild Covid-19. Cell 183, 158–168. https://doi.org/10.1016/j.cell.2020.08.017. http://www.sciencedirect.com/science/article/pii/S0092867420310084. (p. 378).

Seow, J., Graham, C., Merrick, B., Acors, S., Steel, K., Hemmings, O., O'Bryne, A., Kouphou, N., Pickering, S., Galao, R., Betancor, G., Wilson, H., Signell, A., Winstone, H., Kerridge, C., Temperton, N., Snell, L., Bisnauthsing, K., Moore, A., Green, A., Martinez, L., Stokes, B., Honey, J., Izquierdo-Barras, A., Arbane, G., Patel, A., OConnell, L., O Hara, G., MacMahon, E., Douthwaite, S., Nebbia, G., Batra, R., Martinez-Nunez, R., Edgeworth, J., Neil, S., Malim, M., Doores, K., 2020. Longitudinal observation and decline of neutralizing antibody responses in the three months following Sars-Cov-2 infection in humans. Nature Microbiology 5, 1598–1607. (p. 377).

Seric, A., Görg, H., Mösle, S., Windisch, M., 2020. Managing Covid-19: how the pandemic disrupts global value chains. UNIDO's Department of Policy Research and Statistics. UNIDO. Available from: https://iap.unido.org/articles/managing-covid-19-how-pandemic-disrupts-global-value-chains. (p. 252).

Serviço Nacional de Saúde, 2021. Consultas médicas nos cuidados de saúde primários. Available from: https://transparencia.sns.gov.pt/explore/dataset/evolucao-das-consultas-medicas-nos-csp/export/?sort=tempo. (p. 309).

Sharma, R., Malaviya, P., 2023. Ecosystem services and climate action from a circular bioeconomy perspective. Renewable & Sustainable Energy Reviews 175, 113164. (p. 399).

Shet, A., Carr, K., Danovaro-Holliday, M.C., Sodha, S.V., Prosperi, C., Wunderlich, J., Wonodi, C., Reynolds, H.W., Mirza, I., Gacic-Dobo, M., et al., 2022. Impact of the Sars-Cov-2 pandemic on routine immunisation services: evidence of disruption and recovery from 170 countries and territories. The Lancet Global Health 10, e186–e194. (p. 309).

Shimoyama, T., Matsuda, K., Kamatani, Y., Nagata, Y., Yamaguchi, H., Kimura, K., 2023. Abstract tp202: sex-specific differences in risk profiles for cancer among 19702 Japanese patients with ischemic stroke: the Biobank Japan project. Stroke 54, ATP202. (p. 393).

Simon, D., Tascilar, K., Krönke, G., Kleyer, A., Zaiss, M.M., Heppt, F., Meder, C., Atreya, R., Klenske, E., Dietrich, P., Abdullah, A., Kliem, T., Corte, G., Morf, H., Leppkes, M., Kremer, A., Ramming, A., Pachowsky, M., Schuch, F., Ronneberger, M., Kleinert, S., Maier, C., Hueber, A., Manger, K., Manger, B., Berking, C., Tenbusch, M., Überla, K., Sticherling, M., Neurath, M., Schett, G., 2020. Patients with immune-mediated inflammatory diseases receiving cytokine inhibitors have low prevalence of Sars-Cov-2 seroconversion. Nature Communications 11, 3774. https://doi.org/10.1038/s41467-020-17703-6. (p. 353).

Singh, S., Bartos, M., Abdalla, S., Legido-Quigley, H., Nordström, A., Sirleaf, E.J., Clark, H., 2021. Resetting international systems for pandemic preparedness and response. BMJ 375. (p. 58).

Sinha, P., Matthay, M., Calfee, C., 2020. Is a 'cytokine storm' relevant to COVID-19? JAMA Internal Medicine 180, 1152–1154. arXiv:https://jamanetwork.com/journals/jamainternalmedicine/articlepdf/2767939/jamainternal_sinha_2020_ed_200008.pdf. (p. 353).

Sirleaf, E.J., Clark, H., 2021. Report of the independent panel for pandemic preparedness and response: making Covid-19 the last pandemic. The Lancet 398, 101–103. (p. 59).

Sironi, V.A., Inglese, S., Lavazza, A., 2022. The 'one health' approach in the face of Covid-19: how radical should it be? Philosophy, Ethics, and Humanities in Medicine 17, 1–10. (p. 390).

Smil, V., et al., 2008. Global Catastrophes and Trends: The Next 50 Years. Mit Press. (p. xxiv).

Smith, A., 2010. The Wealth of Nations: An Inquiry into the Nature and Causes of the Wealth of Nations. Harriman House Limited. (p. 189).

Smith, D.R., 2019. Herd immunity. Veterinary Clinics: Food Animal Practice 35, 593–604. (p. 162).

Smith, G.J., Vijaykrishna, D., Bahl, J., Lycett, S.J., Worobey, M., Pybus, O.G., Ma, S.K., Cheung, C.L., Raghwani, J., Bhatt, S., et al., 2009. Origins and evolutionary genomics of the 2009 swine-origin h1n1 influenza a epidemic. Nature 459, 1122–1125. (p. 368).

Smith, K.F., Goldberg, M., Rosenthal, S., Carlson, L., Chen, J., Chen, C., Ramachandran, S., 2014. Global rise in human infectious disease outbreaks. Journal of the Royal Society Interface 11, 20140950. https://doi.org/10.1098/rsif.2014.0950. (pp. 381 and 411).

Solow, A., Polasky, S., Broadus, J., 1993. On the measurement of biological diversity. Journal of Environmental Economics and Management 24, 60–68. (p. 391).

Sønderskov, K.M., Dinesen, P.T., Santini, Z.I., Østergaard, S.D., 2020. The depressive state of Denmark during the Covid-19 pandemic. Acta Neuropsychiatrica 32, 226–228. (pp. 174 and 301).

Stafoggia, M., Breitner, S., Hampel, R., Basagaña, X., 2017. Statistical approaches to address multi-pollutant mixtures and multiple exposures: the state of the science. Current Environmental Health Reports 4, 481–490. (p. 393).

Stall, N.M., Johnstone, J., McGeer, A.J., Dhuper, M., Dunning, J., Sinha, S.K., 2020. Finding the right balance: an evidence-informed guidance document to support the re-opening of Canadian nursing homes to family caregivers and visitors during the coronavirus disease 2019 pandemic. Journal of the American Medical Directors Association 21, 1365–1370. (p. 91).

Stingone, J.A., Buck Louis, G.M., Nakayama, S.F., Vermeulen, R.C., Kwok, R.K., Cui, Y., Balshaw, D.M., Teitelbaum, S.L., 2017. Toward greater implementation of the exposome research paradigm within environmental epidemiology. Annual Review of Public Health 38, 315–327. (p. 392).

Suárez-Álvarez, A., López-Menéndez, A.J., 2022. Is Covid-19 vaccine inequality undermining the recovery from the Covid-19 pandemic? Journal of Global Health 12. (p. 184).

Suleman, M., Sonthalia, S., Webb, C., Tinson, A., Kane, M., Bunbury, S., Finch, D., Bibby, J., 2021. Unequal pandemic, fairer recovery. The Health Foundation. (pp. 279 and 314).

Sundararajan, K., Bi, P., Milazzo, A., Poole, A., Reddi, B., Mahmood, M.A., 2022. Preparedness and response to Covid-19 in a quaternary intensive care unit in Australia: perspectives and insights from frontline critical care clinicians. BMJ Open 12, e051982. (p. 428).

Szreter, S., 1993. The idea of demographic transition and the study of fertility change: a critical intellectual history. Population and Development Review, 659–701. (p. 28).

Taleb, N.N., 2007. The Black Swan: The Impact of the Highly Improbable, vol. 2. Random House. (p. 365).

Tang, F., Quan, Y., Xin, Z.T., Wrammert, J., Ma, M.J., Lv, H., Wang, T.B., Yang, H., Richardus, J.H., Liu, W., et al., 2011. Lack of peripheral memory b cell responses in recovered patients with severe acute respiratory syndrome: a six-year follow-up study. The Journal of Immunology 186, 7264–7268. (p. 378).

Task force COVID-19 del Dipartimento Malattie Infettive e Servizio di Informatica, 2022. Epidemia Covid-19. Aggiornamento nazionale: 9 febbraio 2022. Available at: https://www.epicentro.iss.it/coronavirus/bollettino/Bollettino-sorveglianza-integrata-COVID-19_9-febbraio-2022.pdf. (pp. 144, 145, and 146).

Taylor, P.A., 2009. Critical theory 2.0 and im/materiality: the bug in the machinic flows. Interactions: Studies in Communication & Culture 1, 93–110. (p. 345).

Taylor-Clark, K.A., Viswanath, K., Blendon, R.J., 2010. Communication inequalities during public health disasters: Katrina's wake. Health Communication 25, 221–229. (p. 87).

Teichman, D., Underhill, K., 2020. Infected by bias: behavioral science and the legal response to Covid-19. American Journal of Law and Medicine. Forthcoming. Available from: https://ssrn.com/abstract=3691822. (pp. 193 and 195).

Tenforde, M., Kim, S., Lindsell, C., Billig Rose, E., et al., 2020. Symptom duration and risk factors for delayed return to usual health among outpatients with Covid-19 in a multistate health care systems network. CDC - Morbidity and Mortality Weekly Report (MMWR), Washington, DC. Available from: https://www.cdc.gov/mmwr/volumes/69/wr/mm6930e1.htm. (p. 354).

Thampy, L.M., Niveditha, A.C., Sadasivan, S., Deepthy, P., 2023. The role of oxidative stress and cytokine storm in the pathogenesis, oral manifestation, progression and adverse effects related to Sars-Cov2 infection-a review. Oral and Maxillofacial Pathology Journal 14, 66–69. (p. 353).

The Kaiser Family Foundation, 2020. The u.s. government and the world health organization. Kaiser Family Foundation. Available from: http://kff.org/coronavirus-covid-19/fact-sheet/the-u-s-government-and-the-world-health-organization. (p. 419).

The Medicines Utilisation Monitoring Centre, 2021. National Report on Medicines use in Italy. Technical Report. Italian Medicines Agency, Rome. (p. 302).

Thèves, C., Crubézy, E., Biagini, P., 2016. History of smallpox and its spread in human populations. Microbiology Spectrum 4, 4. (p. 162).

Topley, W.W.C., Wilson, G.S., 1923. The spread of bacterial infection. The problem of herd-immunity. Epidemiology and Infection 21, 243–249. (p. 162).

Tounkara, K., Nwankpa, N., 2017. Rinderpest experience. Revue Scientifique Et Technique 36, 569–578. (p. 162).

Toxvaerd, F., 2020. Externalities: why do we need coordinated public action in the pandemic? Economic Observatory. Available from: https://www.coronavirusandtheeconomy.com/question/externalities-why-do-we-need-coordinated-public-action-pandemic. (p. 192).

Tse, L.V., Hamilton, A.M., Friling, T., Whittaker, G.R., 2014. A novel activation mechanism of avian influenza virus h9n2 by furin. Journal of Virology 88, 1673–1683. (p. 20).

Tversky, A., Kahneman, D., 1983. Extensional versus intuitive reasoning: the conjunction fallacy in probability judgment. Psychological Review 90, 293. (p. 370).

UN Women, 2021. Covid-19 and ending violence against women and girls. United Nations Entity for Gender Equality and the Empowerment of Women (UN Women). Available at: https://www.unwomen.org/en/digital-library/publications/2020/04/issue-brief-covid-19-and-ending-violence-against-women-and-girls. (p. 278).

UNCTAD, 2020. World investment report 2020: international production beyond the pandemic. UN Report. Available from: https://unctad.org/webflyer/world-investment-report-2020. (p. 255).

UNICEF, 2014. Ebola virus disease: personal protective equipment and other Ebola-related supplies. United Nations International Children's Emergency Fund, Washington, DC. Available from:

https://reliefweb.int/sites/reliefweb.int/files/resources/UNICEF_Ebola_SuppliesInformationNote_1Sept2014.pdf. (p. 48).

UNICEF, et al., 2021. Preventing a lost decade. urgent action to reverse the devastating impact of Covid-19 on children and young people. United Nations Children's Fund. Available at: https://www.unicef.org/reports/unicef-75-preventing-a-lost-decade. (p. 279).

UNWTO, 2020. Tourism and Covid-19. https://webunwto.s3.eu-west-1.amazonaws.com/s3fs-public/2020-04/COVID19_NewDS_.pdf. (p. 184).

Uribe, P., Basu, P., Linfelow, M., 2021. Preparing for the next pandemic: what will it take? Technical Report. The World Bank. (p. 58).

Ursell, L.K., Metcalf, J.L., Parfrey, L.W., Knight, R., 2012. Defining the human microbiome. Nutrition Reviews 70, S38–S44. (p. 4).

Vaithianathan, R., Ryan, M., Anchugina, N., Selvey, L., Dare, T., Brown, A., 2020. Digital contact tracing for Covid-19: a primer for policymakers. Centre for Social Data Analytics: Auckland University of Technology & the University of Queensland. Available from: https://apo.org.au/node/306445. (p. 113).

Valabhji, J., Barron, E., Gorton, T., Bakhai, C., Kar, P., Young, B., Khunti, K., Holman, N., Sattar, N., Wareham, N.J., 2022. Associations between reductions in routine care delivery and non-Covid-19-related mortality in people with diabetes in England during the Covid-19 pandemic: a population-based parallel cohort study. The Lancet Diabetes & Endocrinology 10, 561–570. (p. 309).

van Boven, M., Weissing, F.J., 2004. The evolutionary economics of immunity. The American Naturalist 163, 277–294. (p. 346).

Van Doren, P., 2020. When and how we should 'trust the science'. Pandemics and Policy. The Cato Institute. Available from: https://www.cato.org/publications/pandemics-policy/when-how-we-should-trust-science. (p. 332).

van Dorp, L., Mislav, A., Richard, D., Shaw, L.P., Ford, C.E., Ormond, L., Owen, C.J., Pang, J., Tan, C., Boshier, F., Torres Ortiz, A., Balloux, F., 2020. Emergence of genomic diversity and recurrent mutations in Sars-Cov-2. Infection. Genetics and Evolution 83, 104351. https://doi.org/10.1016/j.meegid.2020.104351. http://www.sciencedirect.com/science/article/pii/S1567134820301829. (p. 67).

Van Lente, H., 1993. Promising technology. In: The Dynamics of Expectations in Technological Developments. Enschede. (p. 345).

Vandersmissen, A., Welburn, S., 2014. Current initiatives in one health: consolidating the one health global network. Revue Scientifique Et Technique 33, 421–432. https://doi.org/10.20506/rst.33.2.2297. (p. 426).

Verick, S., Schmidt-Klau, D., Lee, S., 2022. Is this time really different? How the impact of the Covid-19 crisis on labour markets contrasts with that of the global financial crisis of 2008–09. International Labour Review 161, 125–148. (p. 267).

Verplanken, B., Roy, D., 2016. Empowering interventions to promote sustainable lifestyles: testing the habit discontinuity hypothesis in a field experiment. Journal of Environmental Psychology 45, 127–134. https://doi.org/10.1016/j.jenvp.2015.11.008. http://www.sciencedirect.com/science/article/pii/S0272494415300487. (p. 186).

Verplanken, B., Walker, I., Davis, A., Jurasek, M., 2008. Context change and travel mode choice: combining the habit discontinuity and self-activation hypotheses. Journal of Environmental Psychology 28, 121–127. (p. 186).

Vinceti, M., Filippini, T., Rothman, K.J., Ferrari, F., Goffi, A., Maffeis, G., Orsini, N., 2020. Lockdown timing and efficacy in controlling Covid-19 using mobile phone tracking. EClinicalMedicine 25, 100457. (p. 127).

Vineis, P., Robinson, O., Chadeau-Hyam, M., Dehghan, A., Mudway, I., Dagnino, S., 2020. What is new in the exposome? Environment International 143, 105887. (p. 395).

Viner, R., Russell, S., Croker, H., Packer, J., Ward, J., Stansfield, C., et al., 2020. School closure and management practices during coronavirus outbreaks including Covid-19: a rapid systematic review. The Lancet Child & Adolescent Health 4, 397–404. (p. 124).

Waldby, C., Mitchell, R., 2006. Tissue Economies: Blood, Organs, and Cell Lines in Late Capitalism. Duke University Press. (p. 345).

Walker, I., Thomas, G., Verplanken, B., 2015. Old habits die hard: travel habit formation and decay during an office relocation. Environment and Behavior 47, 1089–1106. (p. 186).

Wanqing, Z., 2020. Domestic violence cases surge during Covid-19 epidemic. Sixth Tone 2, 846–848. (p. 280).

Watts, D., 2020. Covidsafe, Australia's digital contact tracing app: the legal issues. Australia's Digital Contact Tracing App: the Legal Issues (May 2, 2020). https://ssrn.com/abstract=3591622. (pp. 109 and 111).

Wei, Y., Wei, L., Liu, Y., Huang, L., Shen, S., Zhang, R., et al., 2020. A systematic review and meta-analysis reveals long and dispersive incubation period of Covid-19. medRxiv. Available from: https://www.medrxiv.org/content/10.1101/2020.06.20.20134387v1. (p. 105).

Weiss, H.H., 2013. The sir model and the foundations of public health. Materials Matematics 3, 1–17. (p. 456).

Weisz, G., Olszynko-Gryn, J., 2010. The theory of epidemiologic transition: the origins of a citation classic. Journal of the History of Medicine and Allied Sciences 65, 287–326. (p. 29).

Weitzman, M.L., 1995. Diversity functions. In: Perrings, C., Mäler, K.G., Folk, C., Holling, C., Jansson, B.O. (Eds.), Biodiversity Loss: Economic and Ecological Issues. Cambridge University Press, pp. 21–43. Chapter 1. (p. 391).

Wenxiao, T., Houlin, T., Fangfang, C., Yinong, W., Tingling, X., Kaiju, L., et al., 2020. Epidemic update and risk assessment of 2019 novel coronavirus - China, January 28, 2020. China CDC Weekly. Available from: https://www.fmprc.gov.cn/ce/cgmu/ger/zt/Coronavirus/P020200209690499160210.pdf. (p. 105).

Whitaker, M., Elliott, J., Chadeau-Hyam, M., Riley, S., Darzi, A., Cooke, G., Ward, H., Elliott, P., 2022. Persistent Covid-19 symptoms in a community study of 606,434 people in England. Nature Communications 13, 1–10. (p. 355).

White, D., Chang, H., Benach, J., Bosler, E., Meldrum, S., Means, R., et al., 1991. Geographic spread and temporal increase of the lyme disease. Epidemic. JAMA 266, 1230–1236. (p. 24).

Whitman, E., 2016. How Zika virus–carrying Aedes aegypti mosquitos were eradicated, and then returned. International Business Times. Available from: http://www.ibtimes.com/how-zika-virus-carrying-aedes-aegyptimosquitoes-were-eradicated-then-returned-2309666. (p. 47).

WHO, 2017. What is 'one health'? WHO Report. Available from: https://www.who.int/news-room/questions-and-answers/item/one-health. (p. 382).

WHO, 2018a. Managing epidemics: key facts about major deadly diseases. WHO Report. Available from: https://apps.who.int/iris/handle/10665/272442. (pp. 7, 21, 25, 26, 80, and xxii).

WHO, 2018b. World health statistics 2018: monitoring health for the SDGS. WHO Report, Geneva. Available from: https://www.who.int/gho/publications/world_health_statistics/2018/en/. (p. 425).

WHO, 2019. Non-pharmaceutical public health measures for mitigating the risk and impact of epidemic and pandemic influenza. WHO Report, Geneva. Available from: https://apps.who.int/iris/bitstream/handle/10665/329438/9789241516839-eng.pdf?ua=1. (p. 105).

WHO, 2020a. Contact tracing in the context of Covid-19: interim guidance. WHO Report, Geneva. Available from: https://apps.who.int/iris/handle/10665/332049. (pp. 104, 107, and 116).

WHO, 2020b. Managing the COVID-19 infodemic: promoting healthy behaviours and mitigating the harm from misinformation and disinformation. Joint statement by WHO, UN, UNICEF, UNDP, UNESCO, UNAIDS, ITU, UN Global Pulse, and IFRC. Available from: https://www.who.int/news/item/23-09-2020-managing-the-covid-19-infodemic-promoting-healthy-behaviours-and-mitigating-the-harm-from-misinformation-and-disinformation. (p. 337).

WHO, 2020c. Pulse survey on continuity of essential health services during the COVID-19 pandemic: interim report, 27 August 2020. Technical Report. World Health Organization. (p. 300).

WHO, 2020d. Surveillance strategies for Covid-19 human infection: interim guidance. WHO Report, Geneva. Available from: https://apps.who.int/iris/handle/10665/332051. (pp. 104 and 107).

WHO, 2020e. Who guidelines on hand hygiene in health care: a summary. WHO Report, Geneva. Available from: https://www.who.int/gpsc/5may/tools/who_guidelines-handhygiene_summary.pdf. (p. 104).

WHO, 2021. Covid-19 and NCDS. Available at: https://www.who.int/publications/m/item/rapid-assessment-of-service-delivery-for-ncds-during-the-covid-19-pandemic. (p. 309).

WHO, 2022a. Mental Health and COVID-19: early evidence of the pandemic's impact. Scientific brief - 2 March 2022. Technical Report. World Health Organization. (pp. 300 and 301).

WHO, 2022b. Third round of the global pulse survey on continuity of essential health services during the COVID-19 pandemic: November–December 2021: interim report, 7 February 2022. Technical Report. World Health Organization. (pp. 300 and 308).

WHO, et al., 2005. The World health report: 2005: make every mother and child count. World Health Organization. (p. xxv).

WHO, et al., 2021. Strategy to achieve global Covid-19 vaccination by mid-2022. (p. 373).

WHO, et al., 2022. Third round of the global pulse survey on continuity of essential health services during the COVID-19 pandemic. Interim report. Technical Report. World Health Organization, Geneva. (pp. 306, 307, 316, and 317).

Wicke, P., Bolognesi, M.M., 2020. Framing Covid-19: how we conceptualize and discuss the pandemic on Twitter. PLoS ONE 15, e0240010. (p. 349).

Wild, C.P., 2005. Complementing the genome with an 'exposome': the outstanding challenge of environmental exposure measurement in molecular epidemiology. Cancer Epidemiology Biomarkers & Prevention 14, 1847–1850. (pp. 392 and 393).

Willis Towers Watson, 2021. Actions to restore stability survey. Available from: https://www.willistowerswatson.com/en-US/Insights/2020/07/ac. (p. 150).

Wolfe, N.D., Dunavan, C.P., Diamond, J., 2007. Origins of major human infectious diseases. Nature 447, 279–283. (p. 8).

Woodford, M., 2011. Simple analytics of the government expenditure multiplier. American Economic Journal: Macroeconomics 3, 1–35. (pp. 220 and 222).

World Bank, 2010. People, pathogens and our planet: volume 1—towards a one health approach for controlling zoonotic diseases. WB Report, Washington, DC. Available from: http://documents1.worldbank.org/curated/en/214701468338937565/pdf/508330ESW0whit1410B01PUBLIC1PPP1Web.pdf. (p. 47).

World Bank, 2015. Ebola: most African countries avoid major economic loss but impact on Guinea, Liberia, Sierra Leone remains crippling. Available from: http://www.worldbank.org/en/news/press-release/2015/01/20/ebola-most-african-countries-avoid-major-economic-loss-butimpact-on-guinea-liberia-sierra-leone-remains-crippling. (p. 47).

World Bank, 2016a. 2014–2015 West Africa Ebola crisis: impact update. (p. 47).

World Bank, 2016b. The short-term economic costs of Zika in Latin America and the Caribbean. WB Report, Washington, DC. Available from: http://pubdocs.worldbank.org/en/410321455758564708/The-short-termeconomic-costs-of-Zika-in-LCR-final-doc-autores-feb-18.pdf. (p. 47).

World Bank, 2020a. Implications of Covid-19 for commodities. April 2020 commodity markets outlook. https://openknowledge.worldbank.org/bitstream/handle/10986/33624/CMO-April-2020.pdf. (p. 184).

World Bank, 2020b. World bank predicts sharpest decline of remittances in recent history. https://www.worldbank.org/en/news/press-release/2020/04/22/world-bank-predicts-sharpest-decline-of-remittances-in-recent-history. (p. 184).

World Bank, 2020c. World development report 2020: trading for development in the age of global value chains. The World Bank. (p. 247).

World Bank, 2020d. Global economic prospects. WB Report, Washington, DC. Available from: https://www.worldbank.org/en/publication/global-economic-prospects. (p. 255).

World Wide Fund Italy, 2020. The loss of nature and the rise of pandemics. protecting human and planetary health. WWF, Roma. Available from: https://wwfint.awsassets.panda.org/downloads/the_loss_of_nature_and_rise_of_pandemics___protecting_human_and_planetary_health.pdf. (p. 367).

Wu, J., MacCann, A., Katz, J., Peltier, E., Singhu, K., 2020. The pandemic's hidden toll: half a million deaths. The New York Times. Available from: https://www.nytimes.com/interactive/2020/04/21/world/coronavirus-missing-deaths.html. (p. 196).

Wu, L.P., Wang, N.C., Chang, Y.H., Tian, X.Y., Na, D.Y., Zhang, L.Y., Zheng, L., Lan, T., Wang, L.F., Liang, G.D., 2007. Duration of antibody responses after severe acute respiratory syndrome. Emerging Infectious Diseases 13, 1562. (p. 378).

Xiao, H., Dai, X., Wagenaar, B.H., Liu, F., Augusto, O., Guo, Y., Unger, J.M., 2021. The impact of the Covid-19 pandemic on health services utilization in China: time-series analyses for 2016–2020. The Lancet Regional Health-Western Pacific 9, 100122. (p. 306).

Xie, T., Liu, W., Anderson, B.D., Liu, X., Gray, G.C., 2017. A system dynamics approach to understanding the one health concept. PLoS ONE 12, e0184430. (p. 382).

Yared, P., 2019. Rising government debt: causes and solutions for a decades-old trend. The Journal of Economic Perspectives 33, 115–140. https://doi.org/10.1257/jep.33.2.115. https://www.aeaweb.org/articles?id=10.1257/jep.33.2.115. (p. 235).

Yazdanpanah, F., Hamblin, M.R., Rezaei, N., 2020. The immune system and Covid-19: friend or foe? Life Sciences 256, 117900. (p. 349).

Yonzan, N., Lakner, C., Mahler, D.G., 2021. Is Covid-19 increasing global inequality? World Bank Blogs 24. (p. 209).

Zeng, L.P., Gao, Y.T., Ge, X.Y., Zhang, Q., Peng, C., Yang, X.L., Tan, B., Chen, J., Chmura, A.A., Daszak, P., et al., 2016. Bat severe acute respiratory syndrome-like coronavirus wiv1 encodes an extra accessory protein, ORFX, involved in modulation of the host immune response. Journal of Virology 90, 6573–6582. (p. 20).

Zhang, L., Hua, N., Sun, S., 2008. Wildlife trade, consumption and conservation awareness in southwest China. Biodiversity and Conservation 17, 1493–1516. (p. 40).

Zhang, S., Diao, M., Yu, W., Pei, L., Lin, Z., Chen, D., 2020. Estimation of the reproductive number of novel coronavirus (Covid-19) and the probable outbreak size on the Diamond Princess cruise ship: a data-driven analysis. International Journal of Infectious Diseases 93, 201–204. (p. 15).

Zhongming, Z., Linong, L., Xiaona, Y., Wangqiang, Z., Wei, L., et al., 2021. Strengthening Economic Resilience Following the COVID-19 Crisis. Technical Report. OECD, Paris. (pp. 206, 207, and 208).

Ziauddeen, N., Woods-Townsend, K., Saxena, S., Gilbert, R., Alwan, N.A., 2020. Schools and Covid-19: reopening pandora's box? Public Health in Practice 1, 100039. (p. 279).

Zinsstag, J., Kaiser-Grolimund, A., Heitz-Tokpa, K., Sreedharan, R., Lubroth, J., Caya, F., Stone, M., Brown, H., Bonfoh, B., Dobell, E., et al., 2023. Advancing one human–animal–environment health for global health security: what does the evidence say? The Lancet. (p. 425).

Zucchi, K., 2018. What countries spend on antiterrorism. Investopedia. Available from: http://investopedia.com/articles/investing/061215/what-countries-spend-antiterrorism.asp. (p. 428).

List of acronyms

Acronym	Definition
AGENAS	Agenzia nazionale per i servizi sanitari regionali
ACE2	Angiotensin Converting Enzime 2
ACT	Automated Contact Tracing
AD	Anxiety Disorder
ADM	Automatic Decision Making
AE	Advanced Economy
AIDS	Acquired Immunodeficiency Syndrome
AIFA	Italian Drug Agency (Agenzia Italiana del Farmaco)
AMC	Anticipated Market Contract
ANS	Autonomic Nervous System
API	Application Programming Interface
ARDS	Acute Respiratory Distress Syndrome
ATC	Anatomic, Therapeutic, Clinical Classification
ATC-A	Access to COVID-19 Tools Accelerator
ATP	Adenosine Triphosphate
ATS	Automatic Tracing System
BCG	Boston Consulting Group
BLE	Bluetooth Low Energy
CCDC	China Center for Disease Control
CD	Communicable Disease
CDC	Center for Disease Control
CEDS	Community Emission Data System
CEPI	Coalition for Epidemic Preparedness Innovations
CFR	Case Fatality Rate
CIA	Central Intelligence Agency
CNR	Italian National Research Council
CONSIP	Concessionaria Servizi Informativi Pubblici
COP27	UN Conference on climate changes
COVID-19	Novel Coronavirus Disease
CPD	Civil Protection Department
CTAs	Contact Tracing Apps
CTT	COVID Tracing Tracker
DAH	Development Assistance for Health
DALY	Disability Adjusted Life Years
DNA	Desossiribonucleic Acid

continued on next page

Acronym	Definition
DoI	Diffusion of Innovation
DPCM	Decreto del Presidente del Consiglio dei Ministri
DRC	Democratic Republic of Congo
DSGE	Dynamic Stochastic General Equilibrium models
ECB	European Central Bank
ECDC	European Center for Disease Control
ECERI	European Cancer and Environmental Research Institute
ED	Emergency Department
EE	Economic Epidemiology
EEA	European Economic Area
EHU	European Health Union
EID	Emerging Infectious Diseases
EMA	European Medicine Agency
EMDE	Emerging and Developing Economies (EMDE)
EMMIE	Emerging Market and Middle Income Economy
ESS	European Social Survey
EU	European Union
FAO	Food and Agricultural Organization
FCT	Fundamental Cause Theory
FDA	Food and Drug Administration
FDI	Foreign Direct Investment
GBD	Great Barrington Declaration
GDB	Global Disease Burden
GDP	Gross Domestic Product
GDPR	General Data Protection Regulation
GEC	Great Financial Crisis
GHSA	Global Health Security Agenda
GHSI	Global Health Security Index
GPs	General Practitionners
GPS	Global Positioning System
GVC	Global Value Chains
HA	Hemagglutinin
HANK	Heterogeneous Agent New Keynesian
HERA	Health Emergency Preparedness and Response Authority
HFMD	Hand, Foot, and Mouth Disease
HIT	Heard Immunity Threshold
HIV	Human Immunodeficiency Virus
ICU	Intensive Care Unit
IHR	International Health Regulation
IMF	International Monetary Fund
IMHE	Institute for Health Metrics and Evaluation
IMST	Incidence Management Support Team
IPAC	Infection Prevention And Control
IRR	Internal Rate of Return

continued on next page

Acronym	Definition
ISS	Italian National Health Institute (Istituto Superiore di Sanità)
ISTAT	Istituto Nazionale di Statistica
JAMA	Journal of American Medical Association
JEE	Joint External Evaluation
JSM	John Snow Memorandum
LDC	Less Developed Country
LMIC	Low and Middle Income Countries
LSE	London School of Economics
LTC	Long Term Care
LTCF	Long Term Care Facility
MDD	Major Depressive Disorder
ME/CFS	Myalgic Encephalomyelitis/Chronic Fatigue Syndrome
MERS	Middle-East Respiratory Syndrome
MFF	Multi-annual Financial Framework
mRNA	messenger Ribonucleic Acid
MRS	Marginal Rate of Substitution
NA	Neurominidase
NAPHS	National Action Plan for Health Security
NCD	Non Communicable Disease
NEJM	New England Journal of Medicine
NGEU	Next Generation EU
NGO	Non Governative Organization
NIC	National Intelligence Council
NICE	National Institute for Clinical Excellence
NK	New Keynesian
NK-DSGE	New Keynesian Dynamic Stochastic General Equilibrium models
NPI	Non-Pharmaceutical Intervention
NPP	National Prevention Plan
NPV	Net Present Value
OECD	Organisation for Economic Co-operation and Development
OH	One Health
OIE	Office International des Epizooties (or World Organisation for Animal Health)
OSMED	Italian Observatory for Drug Consumption
OT	Omics Technologies
PEPP	Pandemic Emergency Purchase Programme
PEPP-PT	Pan-European Privacy Preserving Proximity Tracing
PHEIC	Public Health Emergency of International Concern
PPE	Personal Protective Equipment
PPR	Pandemic Preparedness and Response
PSPP	Public Sector Purchase Programme
QALY	Quality Adjusted Life Years
RBC	Real Business Cycle
RBC-DSGE	Real Business Cycle Dynamic Stochastic General Equilibrium
RBD	Receptor Binding Domain
RCU	Regional Crisis Unit

continued on next page

Acronym	Definition
RFF	Recovery and Resilience Framework
RNA	Ribonucleic Acid
SAGE	Scientific Advisory Group for Emergencies
SARS	Severe Acute Respiratory Syndrome
SDG	Sustainable Development Goals
SEIR	Susceptible, Exposed, Infected, Recovered
SEIRS	Susceptible, Exposed, Infected, Recovered, Susceptible
SGP	Stability and Growth Pact
SIMG	Società Italiana di Medicina Generale
SIR	Susceptible, Infected, Recovered
SME	Small and Medium Enterprise
SNS	Strategic National Stockpile
TFEU	Treaty on the Functioning of EU
TFP	Total Factor Productivity
TMT	Tecnology, Media e Telecomunications
TRS	Technical Rate of Substitution
UCLA	University of California at Los Angeles
UCR	Unità di Crisi Regionale (Regional Crisis Unit)
UHC	Universal Health Coverage
UN	United Nation
UNCTAD	United Nations Conference for Trade And Development
UNDESA	United Nations Department of Social Affairs
USC	University of Southern California
WTO	World Trade Organization
WVS	World Value Survey
YLL	Years of Life Lost

Alphabetical index

Author index

Printed in the United States
by Baker & Taylor Publisher Services